D1544628

INSECT
BIOLOGY

Contributing Authors

J. Wayne Brewer
MONTANA STATE
UNIVERSITY

John L. Capinera
COLORADO STATE
UNIVERSITY

Rex G. Cates
THE UNIVERSITY OF
NEW MEXICO

George C. Eickwort
CORNELL UNIVERSITY

George M. Happ
THE UNIVERSITY
OF VERMONT

INSECT
BIOLOGY

A Textbook of Entomology

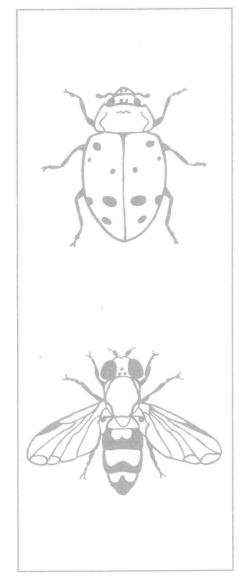

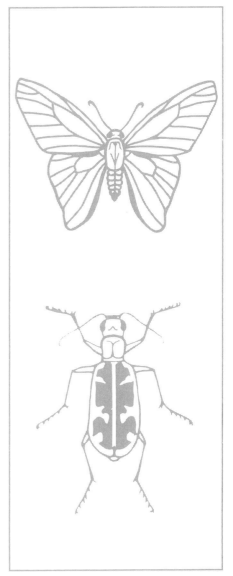

Howard E. Evans
COLORADO STATE
UNIVERSITY

▲▼ ADDISON - WESLEY PUBLISHING COMPANY

READING, MASSACHUSETTS   MENLO PARK, CALIFORNIA   LONDON   AMSTERDAM   DON MILLS, ONTARIO   SYDNEY

Library of Congress Cataloging in Publication Data

Evans, Howard Ensign.
    Insect biology.

    Includes index.
    1. Entomology. I. Brewer, J. Wayne (Jesse Wayne),
1940–   . II. Title.
QL463.E93   1984   595.7   83-15481
ISBN 0-201-11981-1

ABCDEFGHIJ-HA-8987654

Foreword

Entomology is that branch of biology that is concerned with insects. As abundantly successful organisms, insects play major roles in all terrestrial and freshwater ecosystems. In spite of their small size, they also have profound effects on our economy and even on our politics—witness the furor in 1981–1982 over the Mediterranean fruit fly in California and the periodic public outcry over the poisoning of songbirds, waterfowl, and fish as a result of the misuse of pesticides. Some of the most exciting recent research in ecology, behavior, genetics, and endocrinology has involved insects—research often conducted by persons who regard themselves as biologists rather than entomologists.

This text is directed toward the study of living insects as integral members of the biosphere and as objects of much fascinating research. It is designed for students who have had a sound course in basic biology and who seek further demonstration of biological principles, as well as for students who wish to examine entomology as a possible career. In so vast a subject, one is bound to call on specialists in various subdisciplines, and this we have done. Users of this text may wish to expand on certain topics and to skim others, depending on their backgrounds and interests. References at the end of each chapter should provide an introduction to some of the key literature.

Contents

PART THREE
Behavior 172

PART FOUR
The Relationships of Plants and Insects 214

PART FIVE
The Relationships of Insects with Animals   272

PART SIX
Insects and Their Environments   322

PART SEVEN
The Natural and Artificial Regulation of Insect Populations 364

INSECT

BIOLOGY

Why study insects? Surely these small-brained and sometimes repulsive creatures are the ultimate in insignificance, no more than minor and occasional nuisances in our technological world. Is an ant worthy of our consideration, a beetle, a cockroach, a caterpillar? Entomologists think so, but of course they are biased: Insects are their living. They would claim that in a world now far more densely populated than ever before, when we are faced with a struggle to provide food and energy for all, insects cannot be tolerated as major competitors and must be cherished for the help they can sometimes provide us. They would claim, too, that insects are far from insignificant, making up in numbers and in mindless efficiency what they lack in size and ingenuity.

Insects are the most abundant of all living things in terms of numbers of kinds—at least a million species, making up several quite dissimilar groups. And in terms of numbers of individuals, they are the most abundant of terrestrial animals. According to one estimate, insects exist at an average of about 400 pounds per acre in the United

The Structure and Diversity of Insects

P A R T

States, as compared to the weight of humanity, said to be about 14 pounds per acre. Insects feed on our crops and forest and shade trees, and on products in storage; they attack us directly and plague our domestic animals; they transmit diseases of both plants and animals. Yet we could scarcely live a good life without them: They provide a major source of food for birds, fish, and many mammals; they pollinate many plants; they assist in degrading litter and wastes; they assist in controlling populations of harmful insects; they continue to teach us much about the underlying processes of life (where would the science of genetics be without *Drosophila?*).

How does one account for the great success of insects as inhabitants of our planet? Their small size is clearly one of their greatest assets, for the world is full of small places in which to live. Insects live inside seeds, between the upper and lower surfaces of leaves, under bark, inside other insects, and in a myriad of other habitats larger animals could not invade. They are masters of rapid reproduction and development. A pioneer American entomologist, C. V. Riley, once calcu-

O N E

lated that a pregnant female aphid might, in the course of a season, leave 10^{108} descendants, assuming all survived to reproduce. Some insects are remarkably resistant to environmental extremes: Certain fly larvae live in hot springs, others in lakes with a high salt content; the petroleum fly lives in pools of crude oil. They often adapt rapidly to changing conditions, as evidenced by the many species that have evolved strains resistant to certain insecticides. Above all, insects profit by being winged. Their ability to fly permits them to disperse to the farthest reaches of the earth—and makes it likely that if one plants cabbages, the aphids and cabbageworms will find them!

In Chapter 1 we shall suggest several other features that may help explain the great success of insects as a group. We shall take a close look at their structure—their flight, feeding, and reproductive mechanisms; their unique skeletal structure; their ways of packing so much into such small bodies. Even though some of the terms used are adapted from human anatomy—words such as pharynx, femur, abdomen—the body plan of insects is vastly different from ours.

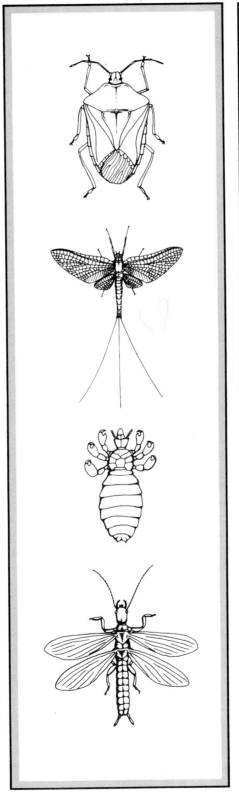

In Chapter 2 we shall survey the major groups into which the insects have evolved. It is not easy to deal with a group that assumes so many forms and lifestyles! The unexpected is forever violating our complacency and our smug hypotheses. But that is the challenge and the fascination of entomology, the reason why so many people are committed to insects professionally or as amateurs. The collecting of butterflies has diverted many persons, among them Winston Churchill and Vladimir Nabokov. It is said that Count Dejean, an aide-de-camp of Napoleon, required his troops to collect beetles for him, even in battle. Charles Darwin, an avid beetle collector in his youth, tells in his *Autobiography* how he found two rare beetles under some bark. He seized one in each hand, then saw a third "which I could not bear to lose, so that I popped the one which I held in my right hand into my mouth." We do not expect that most of you who read this text will develop quite that much enthusiasm for insects. We do hope that you will gain a greater appreciation of their diversity and their importance in our world.

Insect Structure

Figure 1–1
The lubber grasshopper, *Romalea microptera.* (Photograph by Edward S. Ross, California Academy of Sciences.)

The roach that eludes us in the kitchen, the wasp that bothers us on a picnic, and the butterfly that enchants us in the garden have faced problems in their evolution similar to those of mammals such as ourselves. We must all support ourselves and be able to move; we must be able to bring oxygen to our body cells and remove carbon dioxide without drying out; we must be able to locate, ingest, and digest food and defecate that which we can't digest; we must be able to circulate nutrients to our cells and remove excretory products; we must be able to perceive our environment and coordinate our responses; and we must be able to reproduce. Insects consequently have anatomical systems that are analogous in function to those of mammals, but neither homologous in evolution nor particularly similar in structure.

In this chapter we shall concentrate on the skeletomuscular system of insects. We shall be describing external anatomy, that which we see without dissection and that we principally use in classifying insects, as well as some of the muscles that are involved in major movements of the body. We shall first describe the basic structure of the lubber grasshopper (*Romalea microptera*) (Fig. 1–1), an insect commonly used in laboratories of introductory entomology courses. A brief account follows of some of the important modifications found in other insect orders, although we cannot hope to do justice to the anatomical diversity that occurs among the million or so species of insects. Finally, we shall very briefly describe the anatomy of the internal organs, whose functions will be considered in greater detail in Part II. The basic structure of the internal organs is illustrated by the cockroach *Periplaneta americana,* also commonly used in introductory laboratories.

The Skeletomuscular System

The Integument.   Insects and other arthropods are supported by **exoskeletons,** in which the skeleton is on the outside of the body and the muscles attach inside, the opposite of the vertebrate body plan. An exoskeleton superficially resembles a medieval suit of armor, and both a suit of armor and a dry insect on a pin retain their shapes indefinitely. The exoskeleton is actually a noncellular covering, the **cuticle,** that is secreted by the single outer cell layer of the body, the **epidermis.** Together these form the **integument.** The chemical structure of the cuticle is covered in Chapter 3.

As a suit of armor must be flexible at the joints for a knight to move, so the hard, rigid plates of exoskeleton, called **sclerites,** are joined by soft, flexible **membranes.** The exact shapes of these sclerites and how they articulate with each other are functionally important to an insect, just as the exact shapes and articulations of bones are crucial to us. Their configurations consequently are relatively stable within an insect species and vary slightly but consistently among related species, so the integument is of prime importance in taxonomy. The sclerites are frequently marked by pits that de-

"*I told you not to turn over that rock.*"

(Drawing by H. Martin; copyright 1970, *The New Yorker Magazine*, Inc.)

note rodlike invaginations called **apodemes** and by indented lines called **sutures** that mark internal strengthening ridges.

The muscles attach inside the integument at specific points (often the apodemes and internal ridges), so the sclerites act as levers operated by sets of antagonistic muscles. An insect's skeletomuscular system is more energetically efficient and permits more precise movements than does a worm's hydrostatic skeleton. The vertebrate endoskeleton shares these beneficial characteristics; so why don't insects have their skeletons inside, as we do?

For terrestrial animals, endoskeletons are more efficient at larger body sizes and exoskeletons at smaller body sizes. Indeed, there is little size overlap between insects and other terrestrial arthropods on one hand and mammals, birds, and reptiles on the other. From an engineer's viewpoint, an insect is a set of hollow tubes. For the same weight, a tube is more resistant to buckling than is a solid rod. However, as an animal increases in size, mass (weight) increases by the power of 3, while surface area increases by the power of 2, so the exoskeleton would have to be prohibitively thick to support effectively a mammal-sized arthropod. Moreover, the skeletons of larger animals are more subject to stress injury from impact, and an endoskeleton is cushioned by the surrounding soft tissues.

One keynote to success for insects, then, is their small size. We shall see this theme repeated in other organ systems. The insect respiratory and circulatory systems, for example, are both efficient and "cheap" in terms of energy and materials but would be totally inadequate for a large animal. Small size also gives insects the appearance of remarkable strength, because the power of a muscle is proportional to the area of its cross section, while the mass it moves is proportional to volume. Consequently a flea can high-jump 8 inches (20 cm), which corresponds to a leap of 800 feet (245 meters) for a human! The comic-strip hero Spiderman is said to have superstrength because the bite of a radioactive spider gave him a spider's proportionate strength. However, a man-sized spider would have less strength than a human because of the greater weight of the exoskeleton.

A less conspicuous element of an insect's support system is the **hydrostatic skeleton.** Relaxed membranous areas can be extended when muscles compress the rest of the blood-filled body, because the blood itself maintains a constant volume. The hydrostatic skeleton is important in maintaining the shape of soft-bodied larvae like caterpillars and maggots and is also important in expanding the appendages after an insect molts its cuticle.

The insect integument fulfills the functions of both the skin and the skeleton of a vertebrate. As a skeleton, it provides support for the animal and attachment points for the muscles. As a skin, it protects the body from mechanical injury and acts as a barrier against pathogens, against chemicals (such as pesticides), and against water loss. Moreover, invaginations of the integument form the oxygen-conducting system, most of the glands, and even the ends of the gut.

Most colors that characterize insects, whether the green of a katydid, the orange of a monarch's wings, or the beautiful iridescent blue of a morpho butterfly, are produced in the integument, either as pigments or as structural features of the cuticle itself that selectively scatter, reflect, or (rarely) diffract light of given wave lengths. The stimuli to which an insect reacts—whether touch, sound, taste, smell, or electromagnetic radiation such as light—must pass through the integument to be received by the nervous system. All sense organs are modifications of the integument itself, in the form of **sensilla,** each adapted to perceive a certain set of stimuli (Fig. 1–2). The most abundant sensilla are those hairs or **setae** on the body that contain nerve endings. The types of sensory organs are considered in Chapter 6.

Sclerites cannot be stretched, so an insect inside its cuticle has little room for growth. In order to grow, an insect must **molt** a small cuticle in exchange for a larger one. The insect cannot walk around naked, of course, so the new, larger cuticle must be produced by the epidermis underneath the old cuticle. The new cuticle need not be exactly the same form as the old, so the insect has the potential to change its form, or undergo **metamorphosis,** as it grows. Molting and metamorphosis are considered in depth in Chapter 4, because they are key processes in understanding the evolutionary success of insects.

Segmentation and Tagmata.   The insect embryo is divided along its long axis into **segments.** Each segment is a center of development, so components of most organ systems are repeated in each segment. We say that the components of each system are serially homologous among the segments. **Serial homology** can be most clearly ob-

Figure 1–2
Scanning electron micrograph of a portion of a wasp's antenna. The external "armor" of insects is by no means a barrier to sensing the outside world. In this micrograph many modifications of the cuticle for sensory perception are visible: a pit, three plate organs, and setae (further details in Chapter 6). (Steven Alm, New York State Agricultural Experiment Station at Highlands.)

served in an earthworm. If we open up a worm, we see that ganglia of the nervous system, excretory organs, and blood vessels, for example, are represented in each segment. In the postembryonic insect, the divisions between the segments are often obscured, and components of the different systems are lost in some segments and well developed or modified for different functions in others. For instance, the embryos of insects typically develop a pair of limb buds in each segment, but as the embryo continues to develop, some of these buds regress while others enlarge in different ways. The insect that hatches from the egg has only three pairs of walking limbs, but it also has three sets of mouthparts, each of which developed from a pair of limb buds and is serially homologous with each pair of legs, although it is different in form and function from the legs.

As an embryo develops, the segments of the body group into clusters called **tagmata.** Each tagma contains a specific set of segments and is specialized for functions different from those of the other tagmata. From front to rear, an insect's tagmata are called the **head, thorax,** and **abdomen.**

Before we consider these tagmata, we should define some directional terms. Set your insect with its feet on the ground (Fig. 1–1). **Ventral** refers to its lower surface, and **dorsal** refers to its upper surface, so a dorsal sclerite is on top of (or on the back of) the insect. **Lateral** refers to the side, and **medial** (or **median**) refers to the center, usually to the longitudinal midline of the insect. **Anterior** indicates toward the head end, and **posterior** the reverse, toward the abdomen. On an appendage like a leg, **distal** or **apical** indicates toward the tip, and **proximal** or **basal** the reverse, toward the body.

The Head.　　The head of an insect serves the same functions as do heads of most bilaterally symmetric animals: It contains the mouth and its associated food-handling organs; the principal sensory organs; and the brain, a major center of nervous integration and of memory. The cuticle is fused into a strong capsule that protects the brain and provides stable origins for the strong muscles that operate the mouthparts. The regions of the head capsule are named in Fig. 1–3. The sutures mostly indicate strengthening internal ridges and do not mark the boundaries of segments, with one possible exception, the **postoccipital** suture. This suture marks an internal ridge that surrounds and braces the hole in the back of the head capsule through which the internal organs pass to go to the thorax. The neck, or **cervix,** of an insect is largely membranous except that often there are small sclerites on each side connecting the head capsule to the thorax.

Internally, most head capsules are braced by an anterior and a posterior pair of apodemes that usually meet centrally to form the **tentorium.** In some insects with exceptionally strong capsules, such as beetles, most sutures and the tentorium are lacking.

A head-on view of an insect reveals two important sets of sensory organs. The conspicuous visual receptors are the two **compound eyes.** Under high magnification, each can be seen to be composed of many hexagonal facets, which are the surfaces of the functional units, the **ommatidia** (Fig. 6–14). The internal structure of an ommatidium and the way in which an insect sees the world are considered in Chapter 6. In most adults and many immature insects, a less conspicuous set of visual organs, the **ocelli,** are also present. There may be as many as three ocelli, which do not perceive images but are sensitive to small changes in light intensity. The **antennae** are the pair of "feelers" in the front of the face, but they are better called "smellers" because they are major olfactory organs (Fig. 6–10). When viewed under a scanning electron microscope, a remarkable array of sensilla is apparent on the antenna (Fig. 1–2). Their inter-

Figure 1–3
Head of *Romalea microptera*. (a) Anterior view; (b) lateral view; (c) posterior view; (d) dorsoposterior view, with head capsule cut away to expose tentorium.

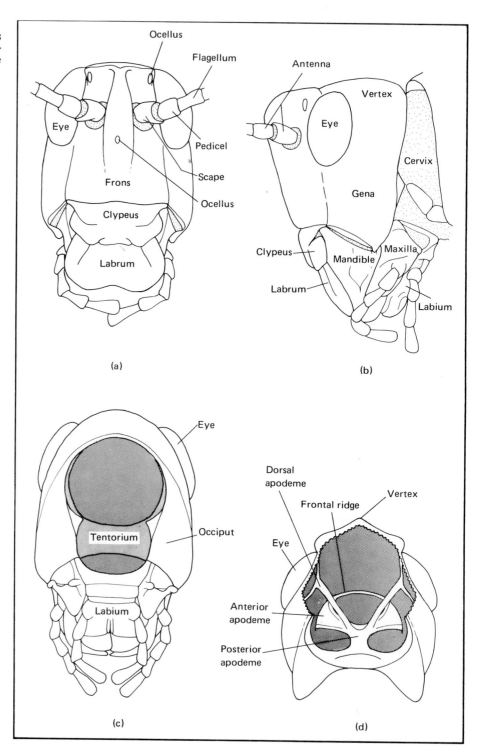

(a)

(b)

(c)

(d)

nal structure and function are also covered in Chapter 6. Each antenna consists of three segments, a basal **scape,** a small **pedicel,** and a typically elongate **flagellum** that is frequently subdivided into **flagellomeres.**

The mouthparts are usually directed downward (hypognathous) in those insects that stand over their food, like grasshoppers. Those insects that chase and grasp their prey with their jaws (Fig. 12–1) and the burrowers that use their jaws to chew their way have mouthparts that are directed anteriorly, with the ventral region of the head capsule consequently sclerotized (prognathous). The mouthparts consist of three pairs of appendages that are serially homologous with the legs. We shall use a generalized insect, the grasshopper, in the discussion that follows (Fig. 1–4).

The most anterior pair of mouthparts is the **mandibles.** These are a pair of strongly sclerotized, unsegmented jaws that operate transversely, the opposite of ours, to bite or crush food. Each is hinged at two points on the head capsule, with a strong internal ridge on the capsule between the hinge points that continues across the front of the head. The frontal ridge is marked externally by a suture that separates the **clypeus** from the rest of the head. The large muscles that close the mandibles insert on each mandible inside the hinge line and fan out to originate far up in the head capsule. The smaller muscles that open the mandibles insert outside the hinge line.

The jointed **maxillae** join the head capsule behind the mandibles and also work transversely. They function as manipulatory organs, helping to guide the food toward the mouth. The **labium** is actually a fused second pair of maxillae that forms the "underlip" to the food-intake or **buccal cavity.** Both the maxillae and the labium are equipped with segmented **palpi** that serve as sensory organs to distinguish food. Anteriorly, the buccal cavity is closed by the flaplike **labrum;** and laterally, by the mandibles and maxillae. When food is swallowed by a grasshopper, the bulging "tongue," or **hypopharynx,** on the anterior surface of the labium helps push food into the true mouth located at the base of the mandibles.

From a basic set of biting mouthparts like that of the grasshopper illustrated in Fig. 1–4, insects have evolved a remarkable variety of feeding organs to match their varied diets (Fig. 1–5). Indeed, no other group of animals has evolved such diversity in one organ system. This adaptive radiation, coupled with modifications of the gut and digestive physiology, is one important reason why in insects there are so many more species than occur in any other group of animals.

Many biting insects exhibit only slight modifications over the grasshopper plan. Predators typically have long, sharply pointed mandibles with which they capture prey. Larval ant lions and aphid lions have a grooved mandible, with the maxilla fitting over this groove to form a closed canal. These larvae impale their prey with the mandibles and suck out their insides through the grooves. A dragonfly larva has an enlarged, jointed labium that is normally held retracted under its head (Fig. 1–5a). When a potential prey swims nearby, the labium is shot forward with incredible speed, the victim grasped with the modified palpi, and the labium retracted to bring the prey into contact with the biting mandibles. We usually think that fish eat insects, but large larvae may turn the table and capture small fish with the underwater analogue of a frog's tongue!

Biting mouthparts are not efficient in acquiring liquid food. Major groups of insects have a beak, or **proboscis,** that functions to suck up liquids, coupled with a strongly muscular pump that draws the liquid up into the buccal cavity or anterior gut. The nectar of flowers is an especially valued liquid food (see Chapter 11). In bees the maxillae and labium are fused basally into a single organ, and the distal sections fit together to form a functional tube when sucking nectar (Fig. 1–5b). The joint between

Figure 1–4
Mouthparts of *Romalea microptera.* (a) Mandibles,
anterior-lateral views; (b) right maxilla, posterior
view; (c) labium and hypopharynx, lateral view;
(d) labium, posterior view.

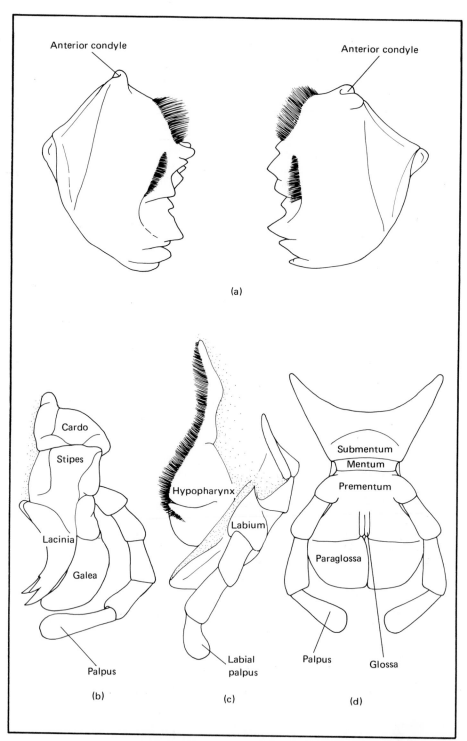

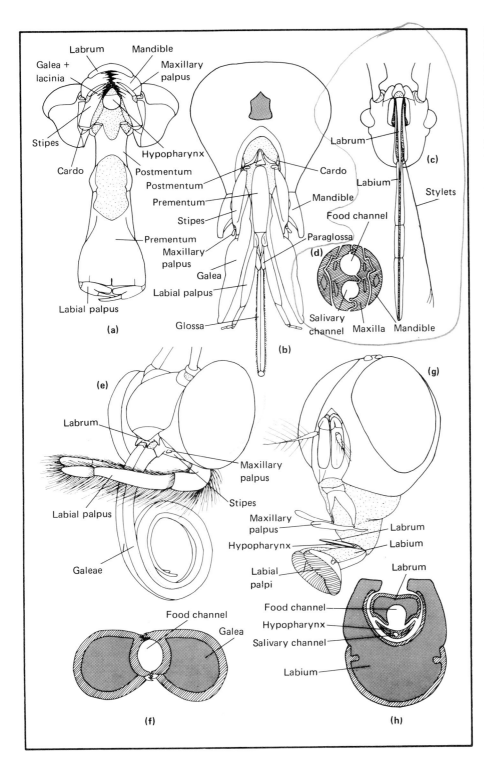

(a)

(b)

(c)

(d)

(e)

(f)

(g)

(h)

Figure 1-5
Mouthpart modifications in various insects. (a) Dragonfly larva, ventral view with labium extended; (b) honey bee, posterior view; (c) squash bug, ventral view with stylets lifted from labial sheath; (d) transverse section of stylets of squash bug (modified from Tower, 1914); (e) monarch butterfly, anterior-lateral view; (f) transverse section of proboscis of butterfly; (g) house fly, anterior-lateral view with labium lifted to expose hypopharynx; (h) transverse section of distal portion of proboscis of house fly. (f modified from Eastham and Eassa, 1955; h modified from Matheson, 1951.)

the basal and distal sections allows a bee to fold its proboscis neatly under its head when not in use. The mandible retains the basic biting form and is used for digging or handling wax.

Nectar-sucking mouthparts of a very different form have evolved in the butterflies and most moths. A long, coiled proboscis is formed only from the interlocking **galeae,** the apical lobes of the maxillae (Fig. 1–5e,f). Most other mouthpart components are absent, except for the labial palpi.

The proboscis of a house fly (Fig. 1–5g) is modified for lapping food that has been predigested by salivary enzymes "spit" onto it by the fly. The saliva is ejected via a channel inside the slender, sclerotized hypopharynx (Fig. 1–5h). Most of the proboscis is formed from the labium, with the labrum forming a flap that covers the hypopharynx. The two lobes at the end of the proboscis are modified labial palpi and are lined by grooves that collect the liquid and direct it into the proboscis. Mandibles are absent, and the maxillae are represented only by palpi. Like the honey bee proboscis, the house fly proboscis is jointed and can be retracted when not in use.

Sap, blood, and cell contents of living plants and animals are all potential liquid nutrients, shielded from a hungry sucking insect by skin or **epidermis.** Major groups of insects have highly modified mouthparts that form sets of **stylets,** extraordinarily thin, sclerotized lances that can pierce the protective armor of an animal or a plant. The stylets contain or collectively form channels that conduct salivary enzymes into the wound and siphon nutrients up to the mouth. Those notorious pests and disease vectors, the mosquitoes, belong to this guild of blood suckers. Their method of feeding is described in Chapter 13. It takes a careful eye to detect the homologies between house fly (Fig. 1–5g) and mosquito (Fig. 13–2) proboscises.

The success of the true bugs and their relatives (aphids, leafhoppers, cicadas, scale insects) is in large part due to their remarkable beaks (Fig. 1–5c). The conspicuous part of the beak is the labium, which forms an anteriorly grooved sheath for the stylets. The four stylets are the paired mandibles and maxillae, grooved to fit together and enclosing an upper food channel and lower salivary channel (Fig. 1–5d). When a bug feeds, the labial sheath folds back and the maxillary stylets move alternately to penetrate the host. Saliva is injected via the salivary channel, and the digested insides of the host are sucked up through the food channel. The hosts of true bugs include plants, arthropods, and vertebrates.

With a few exceptions, a piercing-sucking insect cannot fold up its proboscis as can a house fly or bee, because of the sclerotized stylets. For an insect that does much walking, as do most true bugs, this poses the same awkward problem that a mounted knight faces—what do you do with your lance when you aren't piercing someone? The knight's solution, give it to a squire, is obviously not open to a bug. Instead, a bug directs the entire beak backward under its body, so that it passes between the leg bases.

In their immature stages, insects that will undergo considerable metamorphosis may have strikingly modified heads. Compound eyes and ocelli are absent, but rudimentary images can be formed by caterpillars and some other larvae by means of clusters of simple eyes (**stemmata**). A well-formed head capsule is lacking in many larvae that live inside their food; the fly maggot is an extreme case in which the only sclerotized structures are a mouth hook and the internal skeletal apparatus with which it articulates.

The number of segments that comprise an insect's head is hotly debated among morphologists. All agree that the mandibles, maxillae, and labium represent the appendages of three segments, but the evidence for segments anterior to these is less convincing. Embryonic limb buds and a pair of nerve ganglia that fuse with the brain support the presence of an additional segment with no externally recognizable components after hatching. The antennae do not resemble legs, but they could represent appendages of still another anterior segment. A few morphologists even argue for a preantennal segment whose embryonic limb buds form the labrum or the compound eyes.

The Thorax.   The central tagma of an insect's body, the thorax, is specialized for locomotion. In adult and most immature insects, the thorax bears three pairs of jointed appendages: the **legs,** each pair arising from one of the three segments that comprise this tagma: the **prothorax,** the **mesothorax,** and the **metathorax** (Fig. 1–6). To support the legs and their muscles, each thoracic segment forms a strong box, with its lateral walls sclerotized to form **pleura.** An internal ridge, marked externally by the **pleural suture,** runs upward on the pleuron from its articulation with the leg. This ridge invaginates to form an apodeme that meets a second apodeme invaginating from the ventral sclerite, the **sternum.** The arch thus formed provides a strong internal brace for the leg (Fig. 1–6b). In many advanced insects, the legs are attached close together ventrally; the Y-shaped sternal apodeme invaginates medially between them; and there is no separate sternum, the pleura appearing to meet at the ventral midline.

The prothorax may be either movably or tightly connected to the remainder of the thorax, with its dorsal sclerite, the **pronotum,** often forming a shield that covers most of the segment, as in grasshoppers (Fig. 1–6a). In most adult insects, the meso- and metathorax (together called the **pterothorax**) bear **wings;** these segments are usually tightly fused to each other to provide support for the wings.

Each leg (Fig. 1–6c) is composed of six segments, or **podites.** A podite is a hollow, sclerotized tube that articulates with other podites or the body at membranous **joints** and is moved by muscles that insert at its base and originate in a more basal podite (thus, an **intrinsic** muscle) or within the thorax itself (an **extrinsic** muscle). The basal podite is the short **coxa** that articulates with the end of the pleural suture and sometimes also with the sternum. Its large extrinsic muscles permit the leg to be brought forward and backward along the body axis. The second podite, the **trochanter,** is also small and also powered by extrinsic muscles. Its two articulations with the coxa permit only an up-and-down movement that raises and lowers the entire leg in relation to the body. The two longest podites, the **femur** and **tibia,** follow. The femur is nearly fused with the trochanter and has little or no independent movement, but it is the largest podite because it houses the origins of the intrinsic muscles of the tibia (Fig. 1–6c). The tibial-femoral joint, also with two articulations, forms the "knee" or "elbow" of an insect's leg. The **tarsus** is subdivided into **tarsomeres** to form a flexible foot, and the last podite, the **pretarsus,** is represented only by a pair of **claws** and sometimes a pad, the **arolium,** between them. The intrinsic muscles that move the tarsus and pretarsus originate in the tibia and sometimes in the femur.

The insect "foot" is a remarkable tool. The tarsus spreads out when weight is placed on it, and its ventral surface, like the arolium, is often sticky because of a secretion or surface sculpture. In addition, the pretarsal muscle pulls the claws downward to grasp a rough surface. A fly therefore has the remarkable ability to climb vertical surfaces and can even walk on ceilings. This is of course due to a fly's being fly sized, where adhesive forces (based on surface area) are relatively great; a human-sized fly could not climb walls.

From our unstable perch on two legs, we have to admire an insect's six-legged stance (Fig. 1–1). The six-legged mode of locomotion has evidently evolved indepen-

Figure 1–6
Thorax of *Romalea microptera*. (a) Lateral view, with pleural sclerites underlying pronotum and wings shown with dashed lines; (b) internal view of metathorax to show apodemes bracing leg and wing; (c) hind leg, with femur cut to expose muscles that move tibia.

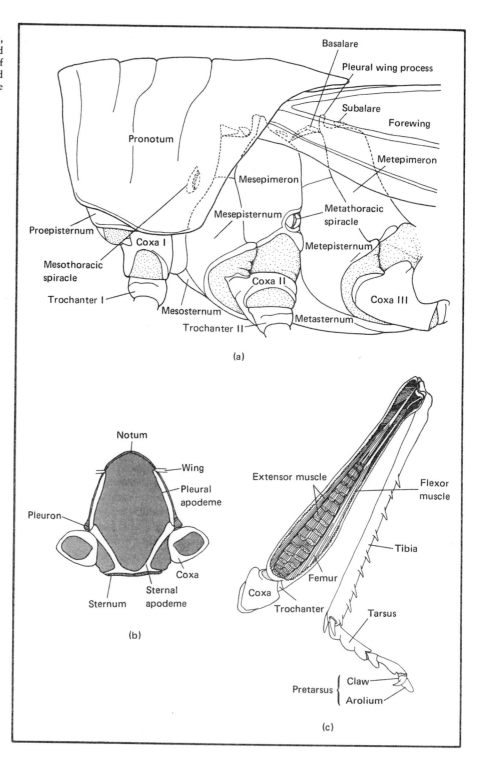

dently several times among the arthropods (see Chapter 2 and the Appendix). As analyzed by the British arthropodologist S. M. Manton, having three pairs of legs near the head provides a stable support for feeding and permits efficient movement and diverse gaits. The insect "hangs down" from its femoral-tibial joints so its center of gravity is low and within the legs.

The legs are of different lengths, so the middle legs step outside the forelegs and hind legs. The hind legs are longest; they and the middle legs push, while the forelegs pull the body forward. G. Hughes of the University of Bristol notes two rules that insects follow in walking or running: Each leg alternates in stepping with its opposite on the same segment, and no leg is raised until the leg behind it is in a supporting position, so there is always a triangle of support on the ground. The time required to bring a leg forward remains constant, but the time necessary to thrust the leg against the substrate decreases as the speed of the insect (correlated with the frequency of stepping) increases. A slow-moving roach has a recovery:power–stroke time ratio of about 1:5, and the legs are lifted one at a time. A fast-moving roach has a recovery:power–stroke time ratio of about 1:1, and the legs are lifted in sets of alternating triangles, each consisting of the foreleg and hind leg on one side and the middle leg on the other. The ability of roaches to accelerate and change direction rapidly gives us humans the impression that they are speedy animals, but the record speed for an American cockroach is slightly less than three miles per hour.

The diversity of life habits among the insects is reflected in the diversity of their legs. In particular, the forelegs are least necessary for locomotion and often serve additional functions, like prey grasping in mantids, digging in mole crickets, and even assisting in copulation.

Jumping is an escape mechanism evolved by such insects as grasshoppers, fleas, leafhoppers, and flea beetles. Insects usually jump by simultaneously extending their hind legs, although which joint provides the extension differs among insects. In grasshoppers it is the femoral-tibial joint, and the femur is consequently greatly enlarged (Fig. 1–6c). A grasshopper can "high-jump" 45 cm while it is "long-jumping" 90 cm, or 10 and 20 times its body length, respectively. This is equivalent to a 6-foot human high-jumping 60 feet (higher than a five-story building) and broad-jumping 120 feet from a standing start! To do this, the grasshopper reaches a take-off velocity of 340 cm/second, exerting a force of over 800 grams on each tibia, which translates into a thrust of 22 grams against the ground. In comparison to its body weight (3 grams), this is about four times the thrust possible for a human leg. The acceleration is too rapid to be produced by the direct contraction of muscles. The force developed by the huge tibial extensor muscle is stored as elastic energy, the leg being held in a flexed position by a cuticular lock. At the appropriate moment, this lock is released, allowing the tibia to extend rapidly. The leg thus operates like a catapult.

In aquatic beetles and bugs, the hind legs, and sometimes the middle legs, provide the oars for swimming (Fig. 7–23). These legs are moved synchronously (except in some water-scavenger beetles). Because the swimming insect is completely surrounded by water, the recovery stroke of the leg counteracts the thrust stroke. The swimming legs are flattened and fringed with hinged hairs. These surfaces are maximally exposed during the rapid backstroke, then the leg is twisted so it presents as little surface area to the water as possible during the slower recovery stroke. Fast-swimming water beetles and bugs are streamlined to the same quality as a good race car, worthy companions of fish in our lakes and streams.

Finally, those larvae that live in their food often lack legs altogether, and the thorax is soft, so the segments resemble those of the abdomen.

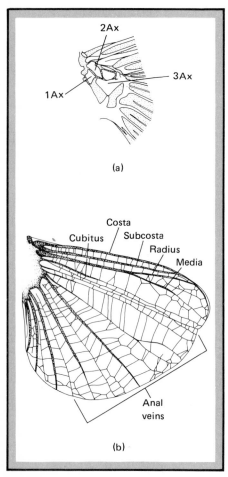

Figure 1–7
Hind wing of *Romalea microptera.* (a) Axillary
sclerites of flattened wing; (b) veins.

The Wings and Mechanics of Flight. The ability of adult insects to fly is unique among invertebrate animals. Insects share the air only with birds, bats, and at one time pterodactyls, but unlike these creatures, their wings are not modified legs. In this way insects resemble angels, at least according to presumably authoritative drawings. Wings arise only from the mesothorax and metathorax, as outgrowths of the integument at the borders of the dorsal sclerites (**nota**) and the pleura. They are functional only in the adult in all of today's insects except for the mayflies, which have two winged stages.

A wing consists of a network of sclerotized braces (**veins**) that enclose regions of thin, transparent cuticle called **cells.** The arrangement of veins is quite constant within a species but differs among groups of species, so characters of venation are important in taxonomy. Before the turn of the century, John Henry Comstock and J. G. Needham, two pioneers of American entomology at Cornell University, developed a system of nomenclature of the principal wing veins that can be applied throughout the insects. This system has become the standard for most orders and is used, with slight modification, to name the veins of a grasshopper wing in Fig. 1–7(b).

In several groups of insects, the front wings play a major role in protecting the more delicate hind wings and a minimal role in flight. In the grasshopper, as well as in cockroaches and some other insects, the front wings are somewhat parchmentlike and are called **tegmina** (singular, tegmen). The very hard front wings of most beetles are called **elytra** (singular, elytron). When only the basal part of the front wings is thickened, as it is in many true bugs (Hemiptera), they are spoken of as **hemelytra.**

The wing, as an extension of the body, is a living structure. The hollow veins contain blood, nerves, and tubes for air. The arrangement of the veins is crucial for efficient flight. The strongest veins occur toward the leading edge of the fore wing, while the trailing margin of the hind wing is flexible and may even be expanded fanlike as in the grasshopper. At the base of the wing, nearly hidden from view, is a set of small **axillary sclerites** (Fig. 1–7a) that connect to the bases of the longitudinal veins and articulate with the thorax. Incredibly, there are no muscles that extend into the wing itself. The wing is actually a first-class lever, with an axillary sclerite that sits on the fulcrum formed by the **pleural wing process** at the top of the ridge indicated by the pleural suture.

The **dorsoventral muscles** that power the upstroke insert dorsally on the thoracic notum and originate ventrally on the sternum or in the coxae or trochanters (Fig. 1–8). The contraction of these muscles moves the edge of the notum downward. Because the notal margin articulates with the axillary sclerites, this movement also brings these sclerites down below the pleural wing process. The wing is like a see-saw, with one end much longer than the other, so the downward movement of the short end creates a much greater upward movement of the long end.

In dragonflies and cockroaches, the downstroke is powered by sets of muscles that insert on two small sclerites, the **basalare** and **subalare,** in the pleuron that in turn articulate with the axillary sclerites outside the pleural wing process (Fig. 1–8a and b). The basalare and subalare muscles originate on the pleuron or the sternum or in the coxae. These muscles are antagonists of the dorsoventral muscles—their contraction pulls the long end of the wing lever downward. This is spoken of as a **direct flight mechanism,** since the downstroke is produced by muscles attaching close to the wing bases. Dragonflies are often said to have a "primitive" flight mechanism, and it is indeed an ancient one, preceding the dinosaurs, but it is far from inefficient. More than any other animal, a dragonfly is a creature of the air, capturing and eating its prey, mating, and even laying eggs without landing (Fig. 7–20). Each of the dragonfly's four wings can be moved independently, providing the insect with unrivaled control and mobility.

In other insects, most of the power for the downstroke does not come from the basalare and subalare muscles. These insects possess an **indirect flight mechanism,** in which downstroke power is provided indirectly to the wings through **dorsal longitudinal muscles** that change the shape of the thorax. These muscles insert on large apodemes that extend inward from the anterior and posterior ends of the mesothoracic and metathoracic nota (Fig. 1–8c). When the dorsal longitudinal muscles contract, the apodemes are brought closer to each other, and this stress is conducted by sutures to the edges of the nota, which in turn are raised in relation to the pleural wing processes. These muscles are also antagonists of the dorsoventral muscles.

The flight system depends on the elastic nature of the thorax. Most of the power generated by the dorsoventral muscles (86% in a grasshopper) during the recovery upstroke is actually stored in elastic energy, to be released during the power downstroke.

Elastic energy is also stored to produce a ''**click mechanism**'' that helps drive the wings of flies, grasshoppers, beetles, and undoubtedly many other insects. The operation of this mechanism can best be visualized in a schematic diagram of an insect's thorax (Fig. 1–9). When the wings are brought toward the midpoint of either the up- or the downstroke, the pleural wing processes are necessarily pushed outward. This movement is resisted by the elasticity of the pleura and the contraction of muscles connecting the pleural and sternal apodemes. When the wing passes this midpoint, these elastic structures release their energy as the pleural wing processes move inward. The muscles that power the strokes are therefore putting much of their energy into stretching elastic structures and need to contract for only half the wing beat.

You can demonstrate both the click mechanism and the action of the indirect flight muscles with a little carbon tetrachloride. Capture a live fly and place it in a closed jar with a paper towel moistened with carbon tetrachloride. This chemical contracts normal muscles but does not affect the indirect flight muscles. Now remove the dead fly (don't breathe the carbon tetrachloride!). Notice that the wings are in either an up or a down position, not in between. If the wings are up, take a pair of forceps and squeeze the front and back of the thorax, duplicating the effect of the dorsal longitudinal muscles. If you're accurate, the wings will pop through the downstroke. Now squeeze the dorsum and venter of the thorax together, duplicating the effect of the dorsoventral muscles. The wings should move through the upstroke.

Two pairs of wings both beating the air could result in inefficient flight, because the first pair would disturb the air for the second pair. Some insects, especially dragonflies, have solved the problem by beating the wings out of phase. Most other efficient fliers have solved it by using only one pair of wings to power flight. The segment that bears these wings is much better developed than the other thoracic segments. The mesothoracic wings are the power wings in flies, bees and wasps, butterflies and moths, and true bugs and their relatives. The hind wings are often linked to the fore wings in flight, essentially creating a single functional wing. Most moths have long bristles on the hind wing that hook into the base of the fore wing, while bees and wasps have a row of hooks on the leading edge of the hind wing that engage the posterior edge of the fore wing. In flies, the hind wings are reduced to small knobs, the **halteres** (Fig. 1–10), that vibrate in opposite phase to the fore wings and act as gyroscopes to main-

Figure 1–8
Flight muscles. (a) Dragonfly, longitudinal section; (b) dragonfly, transverse section, looking anteriorly at mesothorax; (c) grasshopper, longitudinal section.   (a and b modified from Hatch, 1966; c modified from Snodgrass, 1929.)

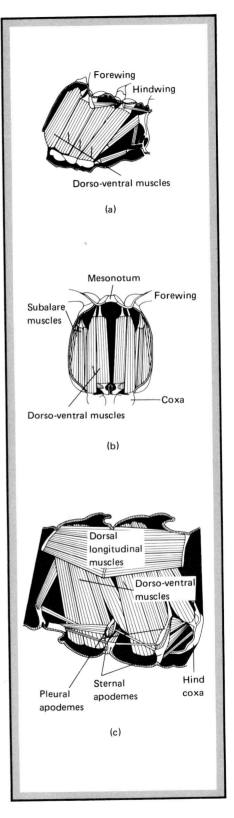

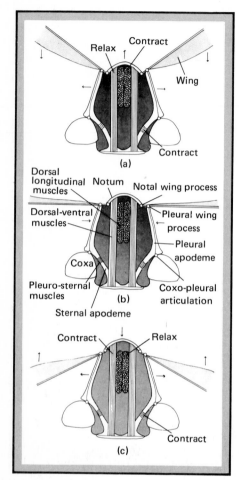

Figure 1–9
Diagrammatic transverse section of an insect thorax, illustrating the click mechanism. (a) Wings in up position, with arrows indicating movement at beginning of downstroke; (b) wings in unstable midposition; (c) wings in down position, with arrows indicating movement at beginning of upstroke.

tain flight stability. In beetles, the hind wings provide most of the power, the fore wings being hardened elytra that are simply held at an angle to the body.

In insects other than dragonflies and mayflies, a small muscle that inserts on an axillary sclerite pulls the wing back over the abdomen when the insect isn't flying. This eliminates the hindrance that outstretched wings would cause when the insect is walking and also allows the mesothoracic wings to protect the metathoracic wings and abdomen when not in use. The fore wings are often sclerotized and protectively colored for this function, as in beetles, grasshoppers, true bugs, and earwigs.

An insect does not fly through a simple up-and-down flapping of its wings. An insect's wing is more like a helicopter rotor than an airplane wing. It must simultaneously provide lift, thrust, and control. The wingbeat frequency of a given insect is dependent on the size of the insect and its wings and on other properties of its "flight motor," and it remains relatively constant, no matter whether the insect is hovering or flying fast. Control is provided by changing the angle or amplitude of the strokes of one or both wings.

Werner Nachtigall of the University of Saarlandes, Germany, has analyzed the flight of the black blow fly *Phormia regina* by using ultrahigh speed photography (8000 pictures per second) to "stop" the wings, which beat about 200 times per second. He describes his research in a remarkable book, *Insects in Flight*. Figure 1–11 diagrams the complicated path the wing tip takes relative to the body as it cuts the air. In the downstroke, the wing moves obliquely forward while it twists sharply, so the angle of attack is minimal in the middle of the downstroke, when the wing is at maximum speed. As it completes the downstroke, it decelerates and begins to rotate in the opposite direction. At the bottom of the stroke, the wing pauses for a few ten-thousandths of a second while the twisting continues, so the upper surface during the downstroke is now facing downward. The wing is jerked upward and obliquely backward, then the upstroke continues at an angle of 45°, in a path behind that of the downstroke. At the top of the stroke the wing is twisted back again to begin the downstroke. This results in maximum life and thrust in the power downstroke, with minimum drag in the recovery upstroke.

In all but six orders of insects, the frequency at which wings beat is limited by the frequency at which nerves can fire, usually less than 30 times per second. These flight muscles are **neurogenic,** each contraction being stimulated by a separate nerve discharge, as is true in all other flying animals, including hummingbirds. These insects are relatively large and have relatively long wings in comparison with the six orders containing the beetles, flies, true bugs, thrips, bark lice, and bees and wasps that have

Figure 1–10
The haltere (modified hind wing) of a fly. Point of attachment to the thorax is at the left; adjacent to it are two sets of campaniform sensilla in rows. (From H. E. Evans, 1968, *Life on a Little-Known Planet*, Dutton.)

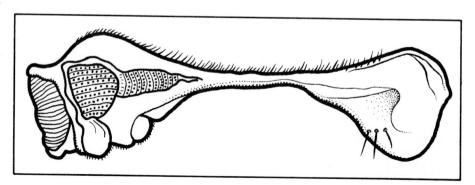

myogenic flight muscles. In myogenic muscles, each contraction is initiated by stretching, not by a nerve impulse. Since the contractions are not synchronized with the nerve impulses, the muscles are also said to be **asynchronous.** The rate of contraction depends on the stress put on the muscles by the flight system itself. The nerves act like the throttle control on an automobile: They adjust the power that may be drawn from the flight motor but do not control the individual cycles of motion. Asynchronous flight muscles can contract at rates well over 100 times per second, producing a correspondingly high wingbeat frequency. We have seen that a blow fly beats its wings over 200 times per second, while a mosquito wing beat is over 500 times per second, and the midge *Forcipomyia* has been recorded at 1047 beats per second! The brief, powerful contraction of an asynchronous flight muscle is necessarily short—only 1% or 2% of the muscle's length during a fly's wing stroke.

Like all insect muscles, asynchronous muscles are **striated;** they appear banded when viewed under a light microscope, as do our skeletal muscles but not our "smooth" visceral muscles. Each muscle cell, or **fiber,** consists of contractile units called **fibrils** that are arranged parallel in the cell, separated by cytoplasm that contains mitochondria and nuclei. Different insect muscles vary in the arrangement of the fibrils, mitochondria, and nuclei. Asynchronous fibers are very large (up to 1.8 mm in diameter); the fibrils are few, cylindrical, and also very large (up to 5 μm in diameter); and the cytoplasm is almost completely filled by large mitochondria, which fill about 30% of the volume of the fiber (Fig. 1–12). In both vertebrates and insects, the fibrils contain protein filaments of actin and myosin, precisely arranged to produce the striated appearance. The sliding of these filaments produces muscle contraction.

The force exerted by an asynchronous muscle is not unusual, but its remarkable rate of contraction is the result of a correspondingly high power output (energy generated per unit time). Indeed, a honey bee's 3.84 horsepower per kilogram is some thirty times greater than our own leg muscle's power output and over twice that of the much vaunted hummingbird, and is roughly equivalent to that of an old-fashioned airplane piston engine! Not surprisingly, asynchronous muscles have the highest metabolic rate of any known animal tissue—in a honey bee 1300–2200 kilocalories per kilogram per hour, a 100-fold increase over the resting level.

How fast can an insect fly? An old report, widely repeated in textbooks, estimated that a bot fly attained 400 yards per second. This translates into about 800 miles per hour, greater than the speed of sound! Irving Langmuir, in debunking this estimate in 1938, noted that a fly would have to take in one and a half times its own weight in food *per second* to sustain flight, and if it hit a human (as bot flies sometimes do), it would have the impact of a pistol bullet. Langmuir calculated a possible flight speed of 25 miles per hour, still remarkably fast for an insect. A large dragonfly can attain a cruising speed of 18 mph and a honey bee about 14 mph, certainly not Grand Prix velocities. But insects are small, and a more legitimate measure of speed might be how many body lengths an animal attains per second. A human may make 5 lengths per second, roughly equivalent to a Volkswagen at highway speed, while a jet fighter at three times the speed of sound attains 100 lengths per second. Compare this to a fly, which makes 250–300 body lengths per second!

Insects exhibit a diversity of flight characteristics, correlated with the diversity in wing types. We can readily appreciate the diversity of bird flight when we consider that the wing of a soaring albatross is 100 times longer than that of a buzzing hummingbird, but the wing of a large dragonfly is 1000 times longer than that of a tiny wasp. Many insects with large wings, like butterflies, dragonflies, and locusts, are capable of gliding and soaring on outstretched wings, while short-winged insects with asyn-

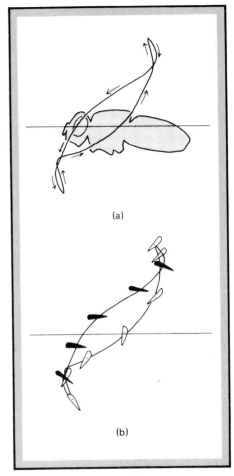

Figure 1–11
Path of the wing tip of a blow fly in flight, projected onto a flat surface. (a) Movement of the wing tip relative to the fly's body. Notice that the downstroke path is in front of the upstroke path. (b) Plane of the wing through the same path. The leading edge of the wing is shown thicker than the trailing edge. The upstroke is represented by hollow symbols, and the downstroke is represented by solid symbols. The body axis of the fly is represented by a horizontal line in both figures. (Modified from W. Nachtigall, 1974.)

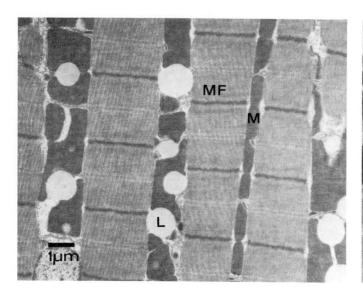

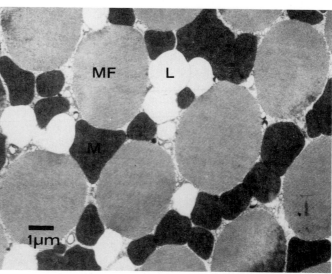

Figure 1–12
Electron micrograph of asynchronous muscle, showing arrangement of microfibrils (MF), mitochondria (M), and lipid droplets (L). Left, longitudinal section; right, transverse section. (From Ashhurst and Cullen, 1977, in R. T. Tregear, ed. *Insect Flight Muscle*, Elsevier Biomedical Press, Amsterdam.)

chronous flight muscles must keep their wings beating. Hovering is an ability of many, if not most, insects, and is especially well developed in dragonflies, hawk moths, and hover flies. Torkel Weis-Fogh, a Danish native, and Charles Ellington have studied hovering at Cambridge University and described different ways in which the wings are moved to generate lift. Perhaps the most interesting is the "clap-fling" mechanism first described in the tiny wasp *Encarsia formosa*. The wings are clapped together over the dorsum, and then the anterior margins are flung open while the posterior margins retain contact, like flinging open a book, thereby creating vortices of air that provide lift as the wings move downward.

Finally, consider the flight of the smallest insects, those less than 0.1 mm long. Such pygmy insects have thin wings lined with long hairs, shaped much like the swimming legs of an aquatic insect. Indeed, these insects use their wings like oars, because air for tiny creatures has a relative viscosity like that of water for larger animals. The exact way in which such pygmies "row" is unclear, because unlike in a rowboat the "oars" cannot be lifted from the medium on the recovery stroke.

The Abdomen.   The terminal tagma in an insect's body, the abdomen, is specialized for "visceral" functions—the storing and processing of nutrients, the circulation of blood, the pumping of oxygen, the development of eggs and their deposition, and the production of sperm and copulation. In contrast to the other tagmata, the abdomen is flexible and capable of expansion and contraction (Figs. 5–22, 7–18).

Eleven abdominal segments are apparent in an insect embryo. The first segment joins the abdomen to the thorax and may be quite reduced. Segments 2 through 7 are quite similar to each other; indeed, it is here that the segmental nature of an insect is most apparent. In most adult insects each of these segments bears a dorsal sclerite, the **tergum,** and a ventral sclerite, the **sternum** (Fig. 1–13a). The lateral area is largely or wholly membranous, allowing the segment to expand and contract dorsoventrally. The terga and sterna of adjoining segments are connected by membranes, allowing anterior–posterior expansion, contraction, and twisting that are controlled by muscles

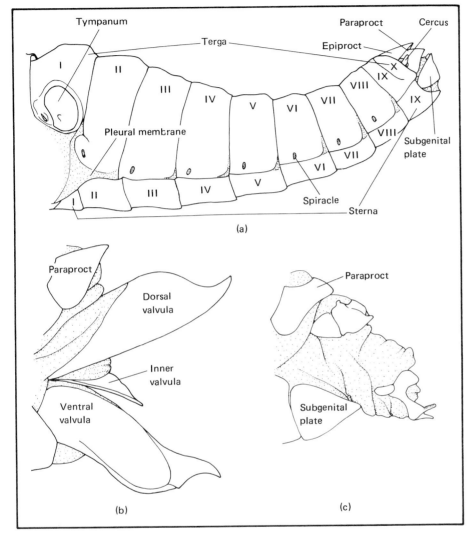

Figure 1–13
Abdomen of *Romalea microptera.* (a) Lateral view of male; (b) ovipositor of female, lateral view; (c) external male genitalia (everted), lateral view.

inserting on apodemes near the anterior edge of each sclerite. In all adult insects except bristletails (see Chapter 2), these segments lack appendages. Consequently muscles do not occupy much of the space inside the abdomen, which instead is filled with the gut, gonads, fat body, and other visceral organs.

The genital ducts open ventrally on segment 8 or 9 of the abdomen, and these segments usually bear external organs that assist in reproduction. Some female insects, like mayflies and stoneflies, lack any special apparatus to assist in egg deposition. However, most females have an **ovipositor** to manipulate eggs and often to place them in a protected location. In the "lower" insects like bristletails, grasshoppers, dragonflies, and cicadas, as well as in bees, wasps, and their relatives, the ovipositor is formed from appendages of segments 8 and 9 that are probably serially homologous with the thoracic legs. Two sets of **valves,** which form the shaft of the ovipositor, connect to basal sclerites called **valvifers,** which in turn articulate with the abdominal terga (Figs. 1–13b, 1–14a). The valves are held together by an interlocking mechanism that permits them to slide back and forth on one another (Fig. 1–14b), and their inner surfaces contain ridges or spines that help move the egg toward the tip. Most insects with an appendicular ovipositor make a hole in a firm substrate and place the egg at its bottom. For instance, grasshoppers dig a hole in the soil, cicadas lay eggs inside twigs, and many parasitic wasps puncture their hosts' cuticle. In bees, ants, and predatory and social wasps, the ovipositor has been modified to form a **sting** that injects venom, the egg exiting from the base of the shaft instead of from the tip (Figs. 1–14a, 8–13). Incidentally, male bees and wasps therefore can't sting, which enables a person to impress onlookers by fearlessly capturing male wasps with bare hands. Also, mosquitoes, black flies, deer flies, and their relatives bite and don't sting; the itch results from saliva that they inject while they suck up blood with the proboscis.

Most "higher" insects—those with complete metamorphosis such as flies, beetles, and moths and butterflies—lack an appendicular ovipositor. Instead, the terminal abdominal segments are narrowed and form a telescoping tube that is used to place

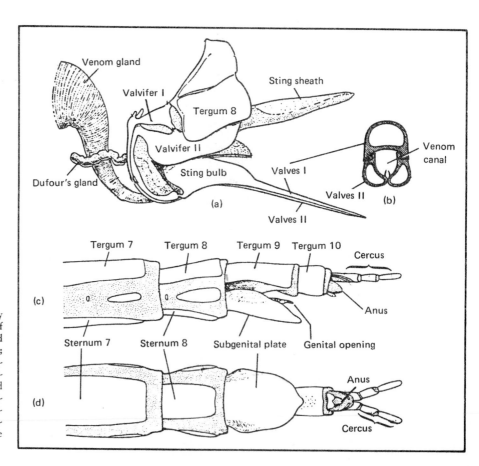

Figure 1–14
Ovipositors. (a) Appendicular ovipositor of honey bee, modified as a sting; (b) transverse section of shaft of sting; (c) lateral view of ovipositor formed from the end of the abdomen of a scorpionfly; (d) same, ventral view.   (a and b from R. E. Snodgrass, 1935, *Principles of Insect Morphology.* By permission of McGraw-Hill Book Company. c and d from G. Mickoleit, 1975, "Die genital und postgenital segmente der Mecoptera-Weibchen," in *Zeitschrift für Morphologie der Tiere,* vol. 80, by permission of the publisher, Springer-Verlag, and the author.)

eggs (Fig. 1–14c,d). In moths and butterflies, this probe is accompanied by a pair of lobes that manipulate eggs.

All male insects except bristletails and dragonflies **copulate** with females; they transfer sperm directly from their reproductive tract into the reproductive tract of the female. In most insects, the **penis** or **aedeagus,** which opens behind segment 9, is accompanied by elaborate **external genitalia,** appendages of that segment that hold the female during copulation (Figs. 1–13c, 1–15). The structure of these genitalia differs widely among the orders of insects, and whether or not they are serially homologous with the thoracic legs has been hotly debated by morphologists. External

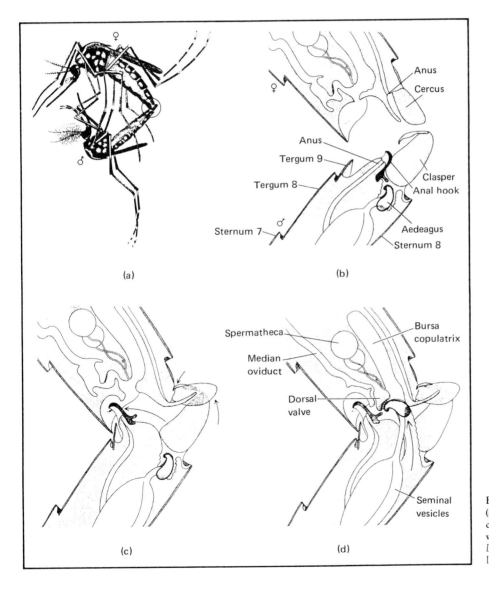

Figure 1–15
(a) Copulation of mosquitoes. (b–c) Stages in clasping of female and transfer of sperm (shown with arrows in d). (From J. C. Jones, *The Sexual Life of a Mosquito*, © 1968 by Scientific American, Inc. All rights reserved.)

genitalia also differ among closely related species, and thus they are valuable to taxonomists. The function of these differences is not clear. At one time entomologists imagined a "lock-and-key" arrangement in which females had a specific "lock" that only the correct male mechanical "key" (his genitalia) could operate. For most insects, the importance of correct sensory cues, rather than mechanical fit, seems to be a better hypothesis.

We cannot begin to cover the variety of external genitalia here. In many insects, the abdominal terga and sterna are also modified to help in copulation. The basic mating position in insects is male above female, with the male's abdomen twisted around to make genital contact (Fig. 14–2). End-to-end copulation is made possible through asymmetric genitalia, as in cockroaches, or by rotating the genitalia or the whole end of the abdomen 180°, as in true bugs, sawflies, and mosquitoes (Fig. 1–15). However, house flies rotate the end of the abdomen 360°! Finally, the dragonflies have a unique system of transferring sperm. Before mating, the male bends his abdomen to place sperm with his penis into accessory genitalia on his second and third abdominal segments. When dragonflies mate, the female picks up the sperm from the accessory genitalia with the end of her abdomen while the male holds her thorax or head with the tip of his abdomen.

Abdominal segments 10 and 11, the last of the insect's body, are quite small and often fused or not externally visible. In many insects segment 11 bears a pair of appendages, the **cerci.** These are usually sensory organs, functioning like rear-guard detectors, although in some insects they are modified to aid in copulation or function as defensive or prey-catching organs, like the pincers of earwigs.

Of course, abdomens vary considerably among the different orders of insects. Bristletails (Fig. 4–2, left column) bear appendages anterior to segment 8 in the form of movable **styli** that act to elevate the abdomen as it slides over the ground. These segments also bear **vesicles,** membranous sacs that can be everted by hydrostatic pressure and are used to absorb water. In true bugs and beetles, the hard fore wings protect the abdominal dorsum, which in turn is membranous. In bees, ants, and wasps, the first abdominal segment is closely fused to the thorax, and there is a constriction between the first and second segments so the apparent abdomen actually begins with segment 2.

The abdomens of immature insects often differ considerably from those of their adults. Those of aquatic immatures often bear external gills (Fig. 2–4). The abdomens of soft-bodied larvae of advanced insects have few or no sclerites and depend on a hydrostatic skeleton for support and movement. Many muscles are necessary to maintain and move the hydrostatic skeleton. The soft cuticle allows expansion as the larva feeds and also allows most of the old cuticle to be reclaimed at a molt (see Chapter 4). For these larvae, the abdomen is as important as the thorax for locomotion, and the abdominal segments often bear unjointed appendages called **prolegs** to grasp the substrate (Fig. 14–12). On your next encounter with a caterpillar, take the time to watch how it moves as it attempts to escape your grasp. All in all, the skeletomuscular system is adapted for very different lifestyles in the larva and in the adult.

The Internal Systems

In ancient times, it was supposed that insects had no well-formed internal organs; rather, they were believed to be filled with fluid and amorphous soft tissue. With the invention of the microscope, Renaissance scientists such as Marcello Malpighi (1628–1694) and Jan Swammerdam (1637–1685) found that this was far from the case.

Using simple instruments, Malpighi and Swammerdam dissected even the smallest of insects, such as lice, and discovered that indeed their internal organs were as complex and intricate as those of much larger animals. Now that we have much more elaborate instrumentation, including electron microscopes, there seems no end to the discoveries that can be made concerning the internal structure and function of insects. In this section we shall review the major organ systems only briefly, reserving a more detailed discussion of function for Chapters 3–6.

The Tracheal System. All terrestrial animals face a problem in their efforts to survive in a dry environment. They must be able to supply oxygen to their cells without allowing those cells to dry out. The insect solution to this problem depends on a waterproof cuticle that allows air to enter at only a few places, the **spiracles** (Fig. 1–16c). There are at most ten segmental pairs of spiracles, on the mesothorax, metathorax, and first eight abdominal segments. The spiracles mark points of invagination of the epidermis during development. These invaginations form cuticle-lined, air-conducting tubes called **tracheae.** The tracheae form longitudinal **tracheal trunks** that interconnect among segments. The tracheae repeatedly branch as they extend to all parts of the body (Fig. 3–6). They end very close to the cells in very fine **tracheoles,** which are the principal sites of gas exchange. The total extent of the tracheal system is truly amazing: A silkworm has 1,500,000 tracheoles!

The Circulatory System. We humans have a "closed" circulatory system, in which blood is transported to the cells via tiny capillaries, so the blood cells (**hemocytes**) (Fig. 3–10) never leave the circulatory system. Although an insect has a functional equivalent of the heart, the **dorsal blood vessel** (Fig. 1–16b), the blood leaves this vessel to percolate through the body cavity and reach the cells without benefit of capillaries. We say that an insect has an "open" circulatory system, and the body cavity is called a **hemocoel,** because it contains the blood. Moreover, there is no separate lymphatic system, and the blood also has the functions of vertebrate lymph, so it is often called **hemolymph.**

The dorsal blood vessel is a simple musculated tube. The portion in the abdomen, the **heart,** contains segmentally arranged valves, called **ostia,** which aspirate hemolymph from the abdomen. The thoracic portion, the **aorta,** is a valveless tube that simply conducts the hemolymph to the head, where it is released near the brain. From there the hemolymph percolates rearward through the hemocoel until it is again picked up by the heart.

The heart is connected to a **dorsal diaphragm,** whose movements help pull the blood into the hemocoel surrounding the ostia (Fig. 3–9). In addition, a **ventral diaphragm** is often present, whose movements help circulate hemolymph past the ventral nerve cord. Small pumps help circulate the hemolymph in legs, wings, and antennae. The best developed of these are the **thoracic accessory pulsatile organs,** which pull hemolymph out of the posterior longitudinal wing veins. The resulting flow of hemolymph is far from spectacular.

The Alimentary Canal. The **gut,** or alimentary canal, of an insect (Fig. 1–16a) can be visualized as a tube that runs through the animal from mouth to anus. The food that is taken into this tube is still outside the body itself, and part of it will remain so as it passes through the tube to exit as feces. The function of the gut is to digest the nutrients in this food and to absorb these nutrients across the gut walls into the hemocoel. The insect gut consists of embryonic integumental invaginations from the mouth and

Figure 1–16
Internal organs of the thorax and abdomen of the
cockroach *Periplaneta americana.* (a) Dorsal view
with gut partially removed to show central nervous
system; (b) ventral view of dorsal blood vessel and
dorsal diaphragm (roaches and other orthopteroid
insects are unusual in having the heart extend into
the thorax and in having lateral vessels); (c) ventral
view of principal dorsal and lateral tracheae;
(d) female reproductive system, dorsal view;
(e) male reproductive system, dorsal view.

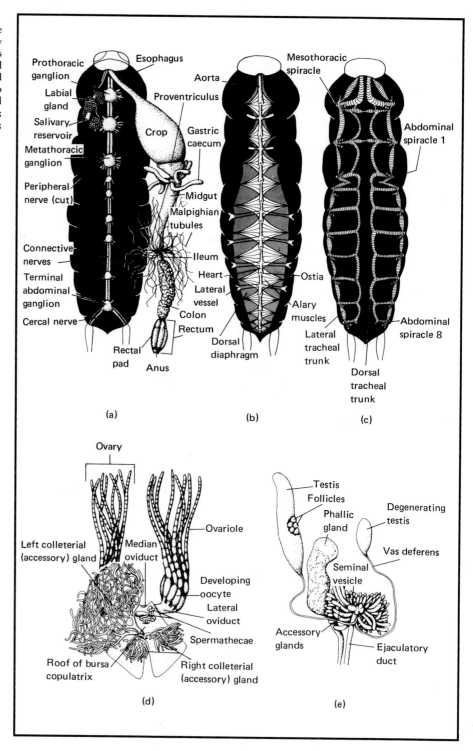

anus that develop to form the **foregut** and **hindgut,** respectively, and a connecting nonintegumental section called the **midgut.** The foregut and hindgut are lined by cuticle, and all sections are coated by muscles that move the food through the gut.

Let us follow a chunk of food as it passes through the gut of a cockroach. As it is manipulated by the mouthparts, it is covered by saliva secreted by the **labial glands.** The food is ingested through the mouth, sucked up by the strongly muscular **pharynx,** and passed by the tubular **esophagus** through the thorax and into the abdomen. There the foregut is expanded into a **crop,** a collapsible sac that holds the food chunk until it can be passed into the midgut. The foregut ends in the **proventriculus,** which in the roach is strongly musculated with sclerotized teeth that grind the chunk into smaller bits.

The midgut is the principal center of digestion and absorption. Its cellular lining both secretes enzymes and absorbs the digested nutrients and passes them into the hemocoel. Anteriorly, the midgut is produced into fingerlike **gastric caeca** that are especially important in absorption.

Undigested particulates leave the midgut and enter the hindgut, where they are once again surrounded by cuticle. The first portion is called the **ileum.** In the cockroach, the anterior hindgut is further subdivided into a **colon.** This leads to the expanded **rectum,** which contains six regions of thick cells called **rectal pads** (Fig. 3–18). The rectal pads are the chief centers of water and ion absorption in the gut. Once through the rectum, the undigested wastes become part of the feces and leave the gut through the anus.

Alimentary canals vary greatly among insects and are correlated with special diets and mouthparts. Insects with sucking mouthparts have a powerful pharynx, and liquid-feeding insects typically have a very large crop, often connected to the rest of the foregut by a separate duct. The midgut may have specialized regions, especially in the true bugs and their relatives, culminating in the **filter chamber** of the leafhoppers that allows most water to bypass the midgut (Fig. 1–17). In termites and scarab beetles the ileum contains symbiotic microbes and is the principal center of digestion.

The Excretory System and the Fat Body.

Excretion is the removal of the waste products of cellular metabolism so an animal is not poisoned by its own machinery. We passed the principal excretory organs when we went from the midgut to the hindgut (Fig. 3–16). These are the **malpighian tubules,** named for the seventeenth-century anatomist who discovered them. There may be from 2 to 250 malpighian tubules in an insect, although the basic number is probably 6. They absorb waste products from the hemolymph and deposit them in the hindgut, from which they are passed out of the insect.

When we dissect a roach, we first encounter white, seemingly amorphous **fat body** surrounding the gut and body wall. Our first impulse is to remove the fat body to get to the more interesting organs it conceals, but we should not treat it so cavalierly. These more or less well-defined masses of large cells are not only important for the storage of fats, carbohydrates, and proteins, but are also the principal centers of intermediary metabolism. The fat body is thus functionally equivalent to our liver and is a principal target of a physiologist's study.

The Reproductive Systems.

The organs that produce the eggs and sperm are rather similar in their basic structure. Both consist of a pair of **gonads,** which are subdivided into tubes in which the **gametes** are produced. Each gonad is connected by a

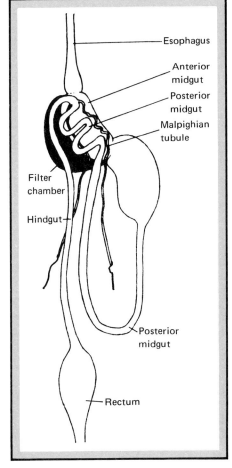

Figure 1–17
Filter chamber of a spittle bug (Homoptera). (Adapted from R. F. Chapman, 1971, *The Insects: Structure and Function,* by permission of the publisher, Hodder & Stoughton, Ltd.; after Snodgrass, 1935.)

tube to a central cuticle-lined duct that joins the gamete-delivering apparatus near the end of the abdomen.

Let us look at the male system first (Figs. 1–16e, 4–10). The gonads, or **testes,** are each subdivided into a set of sperm tubes, or **follicles,** in which the formation of sperm takes place. The follicles of each testis are enclosed in a sheath, which sometimes encloses both testes, forming a single organ. The follicles join the **vas deferens,** which is often expanded into a **seminal vesicle** that stores the mature sperm. The vasa deferentia join a common **ejaculatory duct** that ends in the external genitalia. **Accessory glands** send secretions that accompany the sperm during insemination and often form a packet called a **spermatophore** that encloses the sperm.

In the female, the gonads, or **ovaries,** are subdivided into egg tubes, or **ovarioles,** varying in number from 1, in aphids, to 2000 in those immense African termite queens that churn out an egg every two to three seconds (Fig. 1–16d). Unlike the male system, the ovarioles are rarely enclosed in a sheath. The ovarioles join the **lateral oviduct,** which in turn meets a **median oviduct** that often exits in an ovipositor. Before oviposition, mature eggs may be stored in the bases of the ovarioles, lateral oviducts, or median oviduct. A portion of the median oviduct may be expanded to form a **bursa copulatrix,** a pouch that receives the male intromittent organ, the **aedeagus,** during copulation, and in butterflies and most moths the bursa has a separate opening to the outside. Sperm deposited in the bursa make their way to a storage organ called a **spermatheca,** where they are kept until an egg is sent down the median oviduct. **Accessory glands** typically join the median or lateral oviducts. Their secretions accompany the egg and often make a "glue" that attaches it to a substrate or forms a protective coat over it.

The Nervous and Endocrine Systems.   We shall end our discussion of insect anatomy with the communication systems that coordinate the actions and growth of the animal and enable it to respond to the outside world. The nervous system consists of both a **central nervous system** from which **peripheral nerves** branch and extend to the sensory organs, muscles, and glands, and a **visceral nervous system** that develops from the fore- and hindguts in the embryo. The endocrine system that produces the hormones that control development is included in or closely connected to the nervous system, so it is best considered here.

We vertebrates have our central nervous system located dorsally, enclosed in a backbone; but in an insect the central nervous system is mostly located ventrally. It consists of a dorsal **brain** in the head that is connected by a pair of nerves around the foregut to a series of **ventral nerve cord ganglia** (Fig. 1–16a). In the embryo, each body segment gives rise to a pair of ganglia which then fuse to form a single ganglion. The ganglia are connected between segments by pairs of **connective nerves.** In all insects, the three pairs of ganglia from the mouthpart segments fuse to form a **subesophageal ganglion,** and the ganglia from the last four abdominal segments also fuse, so that the maximum number of discrete ganglia in the ventral nerve cord is twelve (one in the head, three in the thorax, and eight in the abdomen) (Fig. 6–2). The ganglia are further fused among the segments in most insects, especially short compact ones; the epitome of fusion occurs in house flies, in which all thoracic and abdominal ganglia form one mass.

The brain of an insect (Fig. 6–25) consists of three fused sets of ganglia. The largest portion of the brain is the **protocerebrum,** the principal memory center in the nervous system and the ganglion that sends nerves to the eyes. The **deutocerebrum,** just behind it, sends nerves to the antennae. The **tritocerebrum** is the small pair of ganglia that connect to the ventral nerve cord.

The visceral nervous system consists of three parts. One develops from the hindgut and innervates that and the reproductive system. A second part consists of a set of small nerve fibers that run between the connectives of the ventral nerve cord. The third part is the **stomodeal nervous system,** a set of small ganglia lying on the surface of the foregut (Fig. 6–25). It is connected to the protocerebrum and also to endocrine organs called the **corpora allata** (Fig. 4–5). These and the closely associated **prothoracic gland** produce hormones important in development, as we shall discuss in Chapter 4.

Summary

Insects, like other arthropods, are supported by an exoskeleton, in which the secreted outer covering of the body, the cuticle, is composed of hard plates connected by flexible membranes. Muscles attach to the inside of the exoskeleton, the opposite of our skeleton. An exoskeleton is more efficient in a small animal, one of many structural reasons why insects remain small and their size range barely overlaps that of vertebrates.

An insect is divided along its long axis into segments. Each segment acts as a center of development for most organ systems, so organs tend to be repeated, with modification, along the body. The segments are clustered into three functional tagmata, the head, thorax, and abdomen.

The head, composed of about five tightly fused segments, houses the brain, the principal sensory organs (eyes and antennae), and the mouth. Three pairs of appendages, the mandibles, maxillae, and labium, assist in moving food into the mouth. These mouthparts are modified into a remarkable variety of organs for the diets of different insects, culminating in various kinds of beaks for sucking liquid food.

The thorax, composed of three segments, is the locomotory center of the body. Three pairs of jointed legs provide a stable, efficient mode of terrestrial locomotion. A major key to the success of insects is the development of wings on the second and third thoracic segments. The wings of most insects are powered by indirect flight muscles that move the wings indirectly by changing the shape of the thorax. In most strongly flying insects, only one pair of wings provides flight power, and the other pair of wings is linked to it in flight or is reduced. Fast-flying, short-winged insects like flies, bees, beetles, and true bugs have asynchronous flight muscles that do not require neural stimulation for each contraction, so the wingbeat frequency surpasses 100 beats per second.

The abdomen, composed of 11 segments, is a flexible tagma that houses the visceral organs. The genital ducts open on segment 8 or 9. Most females have an ovipositor to assist in depositing eggs, composed either of two pairs of valves that puncture a substrate or of the telescoping end of the abdomen itself. Most males have external genitalia that hold the female during copulation and place sperm into her genital tract.

Insects do not (with very rare exceptions) transport oxygen in their blood, but instead use a tracheal system of air-filled cuticular tubes. The circulatory system is open, in that blood (or hemolymph) flows freely through the body cavity (hemocoel) instead of being enclosed in capillaries. A dorsal blood vessel is the most important pumping organ. The gut consists of two cuticle-lined invaginations, the foregut and the hindgut, joined by a midgut. The midgut is the principal center of digestion and absorption in most insects, while the rectum of the hindgut is the principal center of water and ion absorption. The malpighian tubules are excretory organs that draw nitrogenous wastes from the hemolymph and deposit them into the hindgut. Reproductive organs, located in the abdomen, are basically similar in the two sexes. A pair of gonads is composed of tubes in which the gametes are produced. Females can store sperm for a long time after copulation, and fertilization is internal. The central nervous system consists of a ventral set of segmental ganglia connected to a dorsal brain. It is closely associated with the endocrine system that controls development.

Selected Readings

Barrington, E. J. W. 1979. *Invertebrate Structure and Function.* Second edition. New York: Wiley. 765 pp.

Chapman, R. F. 1971. *The Insects: Structure and Function.* Second edition. New York: American Elsevier. 819 pp.

Dalton, S. 1975. *Borne on the Wind.* New York: Reader's Digest Press. 160 pp.

Matsuda, R. 1965. *Morphology and Evolution of the Insect Head.* Memoir of the American Entomological Institute, no. 4. 334 pp.

_____. 1970. *Morphology and Evolution of the Insect Thorax.* Memoir of the Entomological Society of Canada, no. 76. 431 pp.

_____. 1976. *Morphology and Evolution of the Insect Abdomen.* Elmsford, N.Y.: Pergamon Press. 534 pp.

Nachtigall, W. 1974. *Insects in Flight.* New York: McGraw-Hill. 153 pp.

Richards, O. W., and R. G. Davies. 1977. *Imms' General Textbook of Entomology.* Tenth edition. Vol. 1: *Structure, Physiology, and Development.* London: Chapman & Hall. 418 pp.

Rockstein, M., ed. 1973–1974. *The Physiology of Insecta.* Second edition. Six volumes (especially vol. 3). New York: Academic Press.

Scudder, G. G. E. 1971. "Comparative morphology of insect genitalia." *Annual Review of Entomology*, vol. 16, pp. 379–406.

Snodgrass, R. E. 1935. *Principles of Insect Morphology.* New York: McGraw-Hill. 667 pp.

Insect Diversity

The basic body plan of insects has proved a most successful one. Not long after life first appeared on land, in the mid-Paleozoic era, somewhat more than 300 million years ago, insectlike creatures began to inhabit the earth. By the Upper Carboniferous period, when most of our major coal deposits were formed, insects were already abundant, diverse, and fully winged. Wings have enabled insects to disperse widely and to find new sources of food; along with their diverse mouthparts and other unique attributes, they have been able to exploit the earth more fully than any other type of organism. Today at least half of the kinds of organisms existing on earth are insects (Fig. 2–1).

The diversity of form among insects boggles the mind. There is not much resemblance among a louse, a butterfly, and a beetle. Even the life stages of single species may appear utterly dissimilar (look ahead to Fig. 2–5). A beginning entomologist is likely to vacillate between bewilderment and fascination with this vast array of forms.

Insect Classification

The discovery of patterns among all this diversity constitutes the science of **taxonomy,** and the patterns that are discovered provide the framework of a **classification.** To be meaningful and useful, a classification must be as **natural** as possible; that is, each group must (so far as we can judge) have been derived from a common ancestor. The study of evolutionary lineages is called **phylogeny,** and it is phylogeny that provides the basis for natural classifications.

The importance of having a natural classification is obvious. Truly related groups of organisms have many attributes in common. A classification is essentially a filing system for information about organisms; under each group we file data on structure, function, life histories, and so forth. Thus it is also an information retrieval system. If we place an insect in the order Hymenoptera, for example, we know it has mouthparts, wings, and an ovipositor of a certain basic pattern; that it undergoes complete metamorphosis; and so forth. If we decide it is a bee, we can be sure it fed on pollen and nectar as a larva, that the adult has branching body hairs, and so forth. Identification of the species enables us to determine from published accounts precisely what role it plays in nature. Unfortunately there are so many species of insects that accurate identification is often a task for specialists on particular groups. And all too often we discover that, in fact, relatively little is known about the species we are concerned with. But if something is known about other species in that group, we can sometimes extrapolate information cautiously from one to the other.

The higher groupings of insects, shown in Fig. 2–6, are, in descending order: **class, subclass, infraclass, series, superorder,** and **order.** Each order may be divided into **suborders** and **superfamilies,** as necessary, and in any case into **families, subfamilies,**

Figure 2–1
Relative size of the major groups of organisms on earth, exclusive of bacteria, fungi, and algae. Figures are in thousands. (T. R. E. Southwood, 1978, from "The components of diversity," *Symposia of the Royal Entomological Society of London*, no. 9, pp. 19–40, used with permission of Blackwell Scientific Publications, Ltd.)

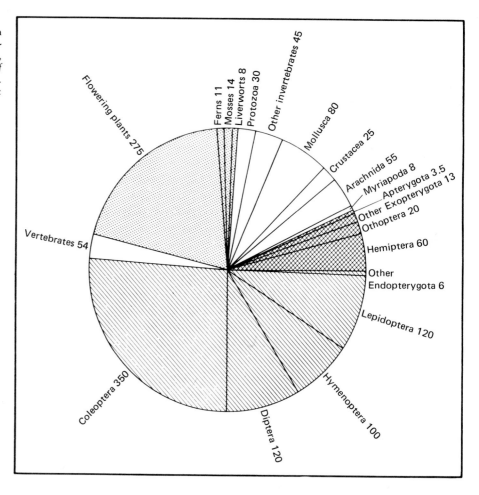

genera, and **species**. In practice, it is not usually necessary to use all these categories. As an example, the grasshopper used as a model in Chapter 1 and shown in Fig. 1–1 might be classified as shown in the box on page 33 (the more commonly used categories are indicated with an asterisk).

Each step as we descend the hierarchy tells us something more precise about the insect. For example, all members of the order Orthoptera have chewing mouthparts; small and widely separated coxae; and slender, thickened front wings. All members of the family Acrididae have short antennae; a short, horny ovipositor; and an auditory tympanum on each side of the first abdominal segment. The information we hang on our classificatory framework is by no means merely academic: It may be of great practical value. For example, the structure and behavior of most Acrididae render them susceptible to similar control measures.

Our system of naming species is one inherited from the great Swedish naturalist Linnaeus, who first employed it consistently in 1758. It consists of the combined generic and specific names and is termed **binominal nomenclature** (literally, "naming with two names"). The scientific names of species are standard the world over, and the

```
*Phylum Arthropoda
  *Class Insecta
     Subclass Pterygota
        Infraclass Neoptera
           Series Exopterygota
              Superorder Orthopterodea
                 *Order Orthoptera
                    Suborder Caelifera
                       *Family Acrididae
                          Subfamily Romaleinae
                             *Genus: Romalea
                                *Species: microptera
                                   Scientific name: Romalea micruptera
                                      Common name: Eastern lubber grasshopper
```

published information on any one species, in whatever language, can be looked for under the same Latinized scientific name. On the other hand, common names differ from country to country, or even in different parts of the same country, and the same common name is sometimes applied to different insects. In the United States the common names of many important insect species have been standardized by the Entomological Society of America, which from time to time publishes a list of such names. In this text we have used the established common names where these exist; otherwise, the scientific names.

It will be noted from Fig. 2–6 and Table 2–1 that the names of most orders end in -ptera (Greek for "wing"), but there are some exceptions. Names of families always end in -idae. Names of superfamilies have also been standardized as -oidea; subfamilies, as -inae. We will have little need to use these last two categories in the text, but there are a few cases in which several related families are difficult to separate, so we have left them grouped into superfamilies.

It is important that students of insects be familiar with the major groups, especially the orders. Sometimes the names are easier to remember if you are aware of their derivation (Table 2–1). In the larger orders, knowledge of the major families is also important. Identification of an insect as a member of the order Coleoptera may be fairly meaningless—there are over 100 families and 200,000 known species of beetles! Identification of insects can be greatly aided by the use of **keys.** Many keys to the orders and families of insects are available, and several are listed at the end of this chapter. Our synopsis of major groups of insects is in the form of a series of tables—first for orders (Tables 2–3 and 2–4), then for the major suborders of certain orders (Table 2–5), and finally for the major families of several of the larger orders (Tables 2–6 to 2–13). Each family is cross-referenced to discussions of members of that family in the text. Several small orders of insects that are rarely encountered in nature are omitted from the tables but are discussed briefly in the following section.

Unfortunately there are many uncertainties concerning the natural classification of the insects and indeed of the phylum to which the insects belong, the Arthropoda. There are several reasons for this, apart from the magnitude of the task of making critical studies of each group in so large a complex of organisms. For one thing, insects have been evolving for several million years, and the forms we see about us are the end

---

Table 2–1.   Derivation of insect ordinal names

Subclass Apterygota: Greek *a* (without) + *pteron* (wing)
        Order Archeognatha: Green *archaios* (primitive) + *gnathos* (jaw)
        Order Thysanura: Greek *thysanos* (bristle) + *oura* (tail)
Subclass Pterygota: Greek *pteron* (wing)
    Infraclass Paleoptera: Greek *palaios* (ancient) + *pteron* (wing)
        Order Ephemeroptera: Greek *ephemeros* (short-lived) + *pteron* (wing)
        Order Odonata: Greek *odon* (tooth) (referring to the mandibles)
    Infraclass Neoptera: Greek *neos* (new) + *pteron* (wing)
     Series Exopterygota: Greek *exo* (outside) + *pteron* (wing)
        Order Plecoptera: Greek *plecos* (plaited) + *pteron* (wing)
        Order Embioptera: Greek *embios* (lively) + *pteron* (wing)
        Order Dictyoptera: Greek *dictyon* (net) + *pteron* (wing)
        Order Grylloblattodea: Latin *gryllus* (cricket) + *blatta* (cockroach)
        Order Orthoptera: Greek *orthos* (straight) + *pteron* (wing)
        Order Dermaptera: Greek *derma* (skin) + *pteron* (wing)
        Order Zoraptera: Greek *zoros* (pure) + *a* (without) + *pteron* (wing)
        Order Psocoptera: Latin *psocus* (book louse) + Greek *pteron* (wing)
        Order Hemiptera: Greek *hemi* (half) + *pteron* (wing)
        Order Thysanoptera: Greek *thysanos* (fringe) + *pteron* (wing)
        Order Phthiraptera: Greek *phtheir* (louse) + *a* (without) + *pteron* (wing)
     Series Endopterygota: Greek *endo* (inside) + *pteron* (wing)
        Order Megaloptera: Greek *megalo* (large) + *pteron* (wing)
        Order Neuroptera: Greek *neuron* (nerve) + *pteron* (wing)
        Order Coleoptera: Greek *coleos* (sheath) + *pteron* (wing)
        Order Mecoptera: Greek *mecos* (length) + *pteron* (wing)
        Order Trichoptera: Greek *trichos* (hair) + *pteron* (wing)
        Order Lepidoptera: Greek *lepido* (scale) + *pteron* (wing)
        Order Hymenoptera: Greek *hymen* (membrane) + *pteron* (wing)
        Order Diptera: Greek *di* (two) + *pteron* (wing)
        Order Siphonaptera: Greek *siphon* (tube) + *a* (without) + *pteron* (wing)

---

products of patterns of evolution that can often be only dimly perceived by the best comparative methods. For another, the chitinous bodies of arthropods do not fossilize nearly as well as the calcareous shells of mollusks or the bones of vertebrates. Thus not only is the fossil record very incomplete, but the existing fossils often consist of tantalizing fragments. Niels Kristensen, of the University of Copenhagen, Denmark, concluded a recent review of insect phylogeny with the comment that "it is obvious that we are very far from having a satisfactory knowledge of the phylogeny of the [major groups of] insects." Kristensen's outline makes no pretense of being definitive, but it summarizes many of the most recent thoughts on the subject. It is, with minor modifications, followed here.

The Ancestry and Evolution of Insects

Insects form a class of the phylum **Arthropoda**. Arthropods have an even more ancient lineage, trilobites and crustacealike organisms being abundant in the oceans as long as 500 million years ago. Although the trilobites are now extinct, Crustacea are still very much with us, chiefly in the oceans and in freshwater lakes and streams. The earlier occupancy of the seas by great numbers of crustaceans may explain, in part, why insects have not occupied the oceans to any appreciable extent.

Insects are by no means the only arthropods occurring on land. Spiders, ticks, mites, scorpions, centipedes, and millipedes are all arthropods. These and several other groups all share the basic arthropod design: a chitinous exoskeleton, jointed legs, ventral nerve cord, and other common features. Spiders, mites, ticks, and scorpions belong to a major group (subclass) in which antennae are lacking and the major mouthparts consist of **chelicerae,** which are not believed to be homologous to the mandibles of insects. These basically four-legged arthropods belong to an evolutionary line that diverged from the insect lineage shortly after the arthropods first appeared on land, in the mid-Paleozoic era. Although they differ in important ways from insects, entomologists are often asked to identify them and to control them. We shall defer discussion of them to the Appendix, where references are given to some of the important literature on these groups.

Centipedes, millipedes, and two other smaller groups, the symphylids and pauropods, are elongate, many-legged creatures, often called **myriapods** (which literally means "innumerable feet"). We shall also discuss these groups further in the Appendix, but since they are generally believed to belong to the same lineage as the insects, we must say a few words about them here. Myriapods have antennae, mandibles, and maxillae, as well as a tracheal system not unlike that of insects. However, they have only two tagmata, head and body, and the body has a relatively large number of segments, most of them bearing legs. Most myriapods are, however, born with only a few body segments and a few pairs of legs; as they grow and molt, additional segments and legs are added, a condition termed **anamorphosis.** When myriapods move, they typically do so with lateral twists of the body and with numerous legs touching the ground at one time, producing a wavelike motion that passes down the length of the body.

Early in the evolution of the myriapod lineage, certain groups appeared in which segments and legs were not added at molts; that is, they were not anamorphic but **epimorphic.** Legs were retained on the three segments behind the head, and the remainder of the body (which at first bore reduced leglike appendages) included only 11 segments. The leg-bearing segments tended to become larger and more rigid, providing the leg musculature with space and firm points of attachment. Relieved of its locomotor function, the remainder of the body became specialized for reproduction and for containing the major parts of the internal organs. Thus these arthropods had three tagmata: head, thorax, and abdomen. They were relatively straight-bodied organisms that moved very differently from myriapods, the thoracic legs functioning in tripod fashion. That is, at one step the fore- and hind legs of one side and the middle legs of the other side touched the ground; on the next step, the same legs of the opposite side (Chapter 1, p. 15).

At one time all six-legged arthropods were considered to have had a common origin and were grouped with the insects, but this is now questioned by most authorities. Springtails (Collembola), for example, retain a form of embryonic development different from insects and more like myriapods, and they have only six abdominal segments, some of which bear locomotory appendages. Another group, the proturans, like myriapods, are anamorphic; and unlike either myriapods or insects, lack antennae. Finally, diplurans, like springtails, have the segments of the antennal flagellum individually musculated. All three groups are closely associated with the soil and are blind, or nearly so, and generally pale in color and weakly sclerotized; all three have unsegmented tarsi and much reduced malpighian tubules or none at all. Their mouthparts are retracted into pouches in the head, with only the tips protruding during feeding. For this reason they are often called **Entognatha** (meaning internal mouthparts), though whether these three groups are truly related or represent diverse lineages that have survived by living obscure lives in the soil is a matter of dispute. That these are truly ancient groups is shown by the fact that fossil springtails have been found in Devonian rocks, dating from over 350 million years ago, well before the appearance of true insects. We shall defer further discussion of these six-legged "near insects," also, to the Appendix.

The Class Insecta. It is generally agreed that the remaining six-legged arthropods share a common ancestry and constitute a well-defined class, the Insecta. An insect may be defined as an arthropod having the following basic features:

1. External mouthparts that include mandibles, maxillae, and labium;
2. Antennae with two musculated segments, scape and flagellum, the flagellum subdivided into several to many flagellomeres;
3. Compound eyes and ocelli present;
4. Thorax bearing three pairs of legs, the tarsus of each leg divided into two to five tarsomeres;
5. Abdomen with 11 segments (at least in the embryo), with the gonopore on segment 8 or 9 ventrally and with cerci on segment 11;
6. Malpighian tubules well developed;
7. Embryonic development by superficial cleavage (as described in Chapter 4, p. 93);
8. Epimorphic, that is, born with the full complement of body segments.

If we survey the insects as a whole, we find exceptions to many of these statements. For example, some insects are blind, and many immature insects as well as some adults lack ocelli; tarsi are undivided in many immatures; and indeed thoracic legs may be lacking altogether. However, there is reason to believe that these and other modifications are secondarily acquired; that is, they represent evolutionary developments of a basic insect plan well represented by grasshoppers and cockroaches, which we discussed in Chapter 1.

It will be noted that the presence of wings is not listed as a basic insectan feature. Insects did acquire wings early in their evolution, and some, such as lice and fleas, lost them secondarily. However, two groups exist in the contemporary fauna that are believed to represent a stage in evolution before the acquisition of wings. Although they are not often seen and are of little importance in our economy, they are worth considering as prototypes of a body plan that later became very successful.

Primitively Wingless Insects. The two groups of which we speak are together called bristletails, since in both cases the body terminates in three slender, bristlelike "tails" (the cerci and a similar, slender prolongation of the last abdominal segment). Bristletails were at one time placed in a single order, but most authorities now place them in two orders, the **Archeognatha** (sometimes called Microcoryphia) and the **Thysanura** (sometimes called Zygentoma). In neither group is there evidence of wings or evidence that they are derived from ancestors that had wings. Thus they are grouped in a separate subclass from the winged insects, called the **Apterygota** (from the Greek *a*, "without," *pteron*, "wing"). The major features of the two subclasses of insects and the two orders of Apterygota are summarized in Table 2–2. It should be

Table 2–2.   Major features of the two subclasses of insects and the two orders of Apterygota

Subclass Apterygota. Wings absent; thorax with three more or less equal, independent segments; abdomen with limb rudiments ventrally. Sperm transfer is external to body. Molting continues after sexual maturity.

    Order Archeognatha. One mandibular condyle; compound eyes large, contiguous, ocelli present; coxae bear styli (short, articulated appendages); spiracles present on abdominal segments 2–8; cerci shorter than terminal filament. Body cylindrical or laterally compressed.

    Order Thysanura. Two mandibular condyles; compound eyes small or absent, ocelli often absent; coxae without styli; spiracles present on abdominal segments 1–8; cerci about as long as terminal filament. Body flattened.

Subclass Pterygota. Wings present in adult, or lost secondarily; thoracic segments of adult closely consolidated, the second two segments forming a boxlike pterothorax with well-formed pleural sclerites; abdomen of adult without ventral limb rudiments; mandibles with two condyles unless modified for piercing. Insemination is internal, by copulation. Molting ceases when sexual maturity is reached.

noted that in both cases leg rudiments are retained on the abdominal segments and that sperm transfer is external, as it is in other arthropods such as centipedes, symphilids, and springtails (see Fig. 4–12).

Members of the order Archeognatha are sometimes called machilids or jumping bristletails, since one of the common genera is *Machilis* and they do, in fact, jump readily (Fig. 4–2, left). These very archaic insects are usually found on rocks in damp places, where they feed on algae, lichens, and other plant material. One species is often abundant on rocks lashed by ocean surf. The order Thysanura is best known from several species that occur in houses and libraries, one called the silverfish because of its dense, silvery scales (Fig. 2–2), another the firebrat because of its tendency to thrive in warm places such as furnace rooms and bakehouses. At times, these insects do considerable damage to books, as they feed on paper and on the bindings.

The Origin of Wings. The concentration of walking appendages in the midbody region, the thorax, was undoubtedly an important prerequisite for the later development of wings, for this required the evolution of a boxlike, well-musculated thorax. Winged insects, and those that have lost their wings secondarily, are grouped in the subclass **Pterygota.** The origin of wings is a matter of some dispute. Insect wings, as we saw in Chapter 1, are not modified appendages but are entirely new structures arising as outgrowths of dorsal parts of the integument of the mesothorax and metathorax. Jarmila Kukalova-Peck, of Carleton University, Ottawa, Canada, has hypothesized that the ancestral pterygote insects were aquatic and evolved movable, musculated gill plates on most body segments, comparable to those on the abdomen of immature mayflies today (Fig. 2–4c). Living in swampy forests, they may have found it advantageous to climb out of the water to feed on vegetation or escape enemies, and the gill plates could have helped to break falls and to glide to other pools with a flapping motion. Those gill plates toward the center of balance eventually became enlarged to serve as wings. This is sometimes called the "flying fish" theory of wing origin.

Figure 2–2
A silverfish (order Thysanura). (Photograph by Edward S. Ross, California Academy of Sciences.)

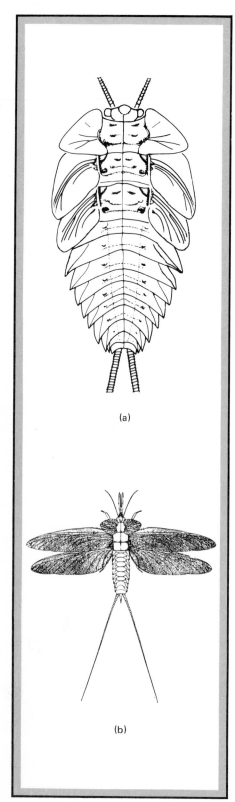

(a)

(b)

In contrast, proponents of the "flying squirrel" theory believe that the ancestral pterygote insects were terrestrial and arboreal and developed lateral flanges of the thorax, which at first served for gliding from tree to tree or to the ground; later they developed the hinges and musculature necessary for true flight. It is noteworthy that some of the bristletails are able to jump and have lateral extensions of the sides of the thorax. Another generalized and ancient insect group, the cockroaches, includes many arboreal species having lateral thoracic flanges in the immature stages and on the prothorax as adults.

Still other theories have been propounded. Many early fossil insects have conspicuous color patterns on the wings, suggesting that the wings may have served in sexual display. Matthew M. Douglas, of Boston University, has suggested that the lateral lobes that were precursors of wings may have served a role in thermoregulation (Fig. 2–3-a). That is, they may have served as plates to absorb heat, which was then transferred to the leg muscles, enabling the insect to run to seek food or escape from enemies at lower temperatures than might otherwise be possible—an ancient type of "solar collector." Using freshly killed alfalfa butterflies, Dr. Douglas removed the wings and attached wing fragments of appropriate size to the sides of the thorax. Small thermocouples placed inside the thorax showed that, under radiant heat, the thoracic temperature was 55% higher than in similar specimens without such lobes.

Of course, these theories are not mutually exclusive. Thoracic lobes serving originally for thermoregulation might also have served in sexual display and for gliding about in vegetation; then, at a later stage, they may have developed sufficient size and an adequate hinge mechanism and musculature to permit true flight. Even today insects do sometimes bask in the sun with their wings spread, and some do use wing patterns as mating signals. Whatever the final answer, there is no question that the acquisition of wings was a major event in insect evolution. For 100 million years the insects were the only winged animals, and today they remain the only winged invertebrates and by far the most abundant of winged animals.

Primitive Winged Insects. The earliest winged insects are believed to have held the wings somewhat stiffly from the sides of the body, rather like modern dragonflies. This is suggested by the fact that many fossils from Paleozoic rocks in which insects first made their appearance are preserved in a flattened position, with the wings outstretched. Some of these fossils are indeed much like dragonflies, and some are quite large, with wingspans up to nearly 30 inches. These insects are placed in the order **Protodonata,** and were probably ancestral to the true dragonflies. Paleozoic rocks include not only Protodonata and several other groups that are similar to insects living today, but also several orders of insects that became extinct at the end of the Paleozoic era and are believed to be evolutionary "dead ends," giving rise to no other groups. The best known of these is the order **Palaeodictyoptera** (Fig. 2–3). Members of this order and two other extinct orders had long, piercing mouthparts, which they may

Figure 2–3

(a) An immature terrestrial Paleozoic insect (reconstructed), showing broad, articulated lobes on the thorax. (b) An adult Paleozoic insect possessing sucking mouthparts, broad prothoracic lobes, and a network of delicate venules on the wings. Both belong to the extinct order Palaeodictyoptera. (J. Kukalova-Peck, 1978, from "Origin and evolution of insect wings and their relation to metamorphosis, as documented by the fossil record," *Journal of Morphology*, vol. 156, pp. 53–97, used with permission of Alan R. Liss, Inc.)

have used for feeding on the sap of trees that lived in the forests that now provide fossil fuel for modern society.

Very early in geologic time (in these same, Paleozoic strata) insects evolved the capacity to hold the wings vertically above the back, in the manner of modern Ephemeroptera (mayflies) and some Odonata (damselflies). This undoubtedly involved a more fluttering type of flight as well as the capacity to settle on small perches more readily. These two orders of insects have a great many veins and cross-veins in the wings and little tendency toward a strengthening of the veins toward the anterior wing margins. They also lack the ability to fold the wings close to the body; they must be held extended, either laterally or vertically. The two orders Odonata and Ephemeroptera have several similarities (for example, short, bristlelike antennae and aquatic larvae) but also many differences, including a very different manner of flight. Nevertheless they are believed to represent a similar stage of evolution, when insects had acquired the power of flight but retained wings with a complex venation and no mechanism permitting them to be folded close to the body. These two orders, along with the Palaeodictyoptera and other extinct orders, are therefore grouped in an infraclass called the **Paleoptera** (meaning "primitive wings").

The Neoptera: Exopterygota.

One of the remarkable features of most modern insects is that although they are winged as adults, they are nevertheless able to become inconspicuous and to enter small spaces—in the soil, in borings and crevices, and so forth—by folding the wings flat or rooflike against their abdomen. Insects with this capacity are said to be neopterous (infraclass **Neoptera,** meaning "new wings"). This advance also occurred very early in geologic time; indeed both Neoptera and Paleoptera appeared at the same time in the Upper Carboniferous period (Fig. 2–7). Wing folding involved the development of a more complex hinge mechanism, including a third axillary sclerite (Fig. 1–7b) and associated musculature as well as an appropriate folding or reduction of the hind wings so that they would fit beneath the front wings. The front wings then (in some cases) became thickened, serving to protect the hind wings when folded, as in beetles, grasshoppers, and true bugs. Many Neoptera exhibit a reduction in the number of wing veins and a tendency for those toward the anterior margin to be crowded and strengthened, permitting a stronger forward thrust. The development of wing flexion—the capacity to draw the wings flat against the body—was undoubtedly a major step in insect evolution, the third of the steps we have been discussing (Fig. 2–6).

The earliest neopterous insects to evolve underwent a gradual, external development of their wing pads through the immature stages, as is also true of paleopterous insects (Fig. 2–4; Fig. 4–2, center). Neoptera of this type belong to a major series called **Exopterygota** (exo meaning "exterior to the body"). All the neopterous insects of the Upper Carboniferous period were Exopterygota, including some extinct orders as well as cockroaches and grasshoppers (Fig. 2–7). These two orders (Dictyoptera and Orthoptera) have biting mouthparts basically like those of the grasshopper (Fig. 1–4) and form the nucleus of a superorder, the **Orthopterodea** (Fig. 2–6). This is a diverse group of insects that has proved difficult to classify in a satisfactory way. Some entomologists argue that walking sticks, for example, should be placed in a separate order from the

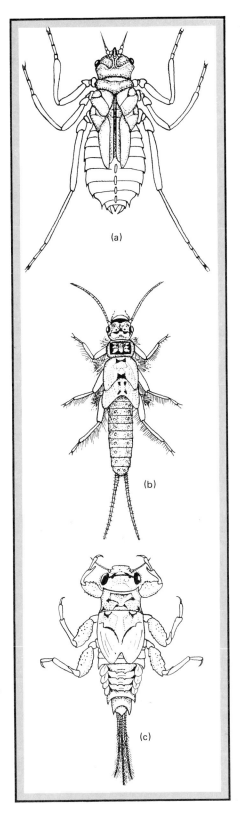

(a)

(b)

(c)

Figure 2–4
Immatures of three orders of Exopterygota inhabiting fresh water. (a) A dragonfly (Odonata); (b) a stonefly (Plecoptera); (c) a mayfly (Ephemeroptera).

grasshoppers and that mantids should be placed in a separate order from the cockroaches. We feel that the resemblances are strong enough to justify considering the walking sticks and mantids to belong to suborders of their respective orders (Table 2–5). The termites (Isoptera) provide a particularly controversial problem. That they evolved from cockroaches is well established, but some feel that they have evolved so many differences that they deserve to be placed in an order of their own. Problems such as this may never be solved to everyone's satisfaction; there will always be "lumpers" and "splitters," as well as those who weigh convenience and tradition more heavily than phylogenetic relationship.

Although the major groups of Orthopterodea are included in Tables 2–3 and 2–5, we have deliberately omitted two orders of rarely encountered insects, both of which were described fairly recently (1913 and 1914). The order **Grylloblattodea** includes a very few species that occur at high altitudes in western North America and eastern Asia, often closely associated with glacial ice. Grylloblattids have some features in common with crickets (Gryllidae) and some in common with cockroaches (Blattidae); thus the order is appropriately named. It is believed that the group is an ancient one, perhaps a "missing link" between the orders Orthoptera and Dictyoptera. Possibly grylloblattids have survived in this unusual habitat as a result of lack of competition and important natural enemies.

The order **Zoraptera** includes minute insects associated with wood or with termite nests. They occur in many parts of the world and are occasionally encountered in the southern United States. They bear some resemblance to termites and like termites shed their wings along a basal fracture (although many species lack wings altogether); they are gregarious but not social like termites. The mouthparts and tarsal segmentation suggest the bark lice (Psocoptera), and it may be that these obscure insects represent another "missing link," in this case between the Psocoptera and the more typical Orthopterodea.

A second major group of Exopterygota centers around the order Hemiptera and is referred to as the superorder Hemipterodea. The mouthparts of this group are diverse but for the most part consist of a sucking proboscis; the group also differs from Orthopterodea in having shorter antennae, simpler wing venation, fewer malpighian tubules, and other features. Again, there is some disagreement as to the best classification. Some prefer to regard the Homoptera as an order distinct from the Hemiptera. We prefer to regard it as a suborder (Table 2–5), since we feel that the differences between the two groups are not of a magnitude that justifies the status of an order. Somewhat the same comments apply to the lice (Phthiraptera), which are sometimes split into separate orders for the biting and sucking lice.

The Endopterygota. The fourth and final major step in insect evolution provided insects with further opportunities to exploit their environments. They developed the capacity to retain their wing pads internally, as **imaginal discs** (Chapter 4, p. 95); hence they are called **Endopterygota** (*endo* meaning "internal"). Since wing development is suppressed in the immature stages, the wings must develop rapidly prior to emergence of the adult, winged form. Thus an additional, sedentary stage, the **pupa,** is interposed between larva and adult. Endopterygote insects are said to have **complete metamorphosis.**

While the more generalized members of this group (the order Megaloptera, for example) show few differences between larva and adult other than the presence or absence of wings, the evolution of complete metamorphosis set in motion a remarkable twofold pattern of evolution. Presence of a pupal stage permitted the suppression not only of wings but even of legs, eyes, mouthparts, glands, and many other structures, requiring extensive reorganization during the pupal stage. Larvae were able to develop special structures of their own, such as gills, prolegs on the abdomen, specialized mouthparts, glands, and so forth, later to be replaced by adult structures that had been held in embryonic state. Larvae became increasingly specialized for exploiting diverse sources of food, while adults became specialists in dispersal and reproduction. Although the pupa is relatively defenseless, insects evolved a variety of mechanisms for its protection: concealed pupal cells, cryptic form and color, or cocoons. Insect metamorphosis is discussed more fully in Chapter 4.

The degree of metamorphosis among the more advanced endopterygotes is almost beyond belief. Maggots of house flies lack most of the external structures we think of as belonging to insects: There are no legs, no sclerites, no color pattern, not even a discernible head aside from the mouth hooks at the anterior end (Fig. 2–5). Yet, after a short pupal stage inside the hardened last larval integument (or "puparium"), a totally dissimilar insect is produced: sclerotized, bristly, with elaborate mouthparts, an efficient flight mechanism, and a remarkable reproductive capacity. Under ideal conditions, this entire life cycle can be completed in no more than a week!

The success of each of the four major steps in insect evolution is well shown by the numbers of orders and the number of species resulting from each step (Figs. 2–1 and 2–6). The first step, the development of the insect body ground plan, resulted in the eventual evolution of some 30 orders of insects. Some of these became extinct, some developed into relatively small groups, while others became enormously successful and rich in species. The second step, the development of wings, was of course a most important one. Of the 24 orders of living insects, 22 are pterygotes, and these include over 99% of the species. Those with the capacity to flex the wings over the abdomen (Neoptera; step 3) comprise 20 of the 22 winged orders, only the dragonflies (Odonata) and mayflies (Ephemeroptera) retaining the paleopterous condition in the modern fauna. Those with complete metamorphosis (Endopterygota; step 4) belong to only nine orders, but these include the largest groups, such as the beetles, flies, moths, and wasps. Probably at least 80% of living insect species have complete metamorphosis (Fig. 2–1). It is interesting to speculate what might have happened had insects failed to develop wings or to develop the means of flexing the wings or of retaining the developing wings internally in the immature stages. Probably they would constitute at most a few paragraphs in treatises on invertebrate zoology.

Evolution and Classification of the Endopterygota.

The Endopterygota appeared in the fossil record somewhat later than other major groups, at the beginning of the Permian period, about 270 million years ago. The Megaloptera (dobsonflies and snakeflies) are believed to be the most generalized, having somewhat active pupae and undergoing a less marked metamorphosis than most Endopterygota. The Coleoptera (beetles) are believed to have evolved from Megaloptera, as evidenced chiefly by a marked similarity of the larvae of more generalized beetles to those of dobsonflies. The Neuroptera also have many features in common with Megaloptera in the adult stage, although the larvae have evolved quite different mouthparts.

The six remaining orders share certain features, such as the presence of silk glands opening on the labium of the larvae, which are used to form the cocoon (although

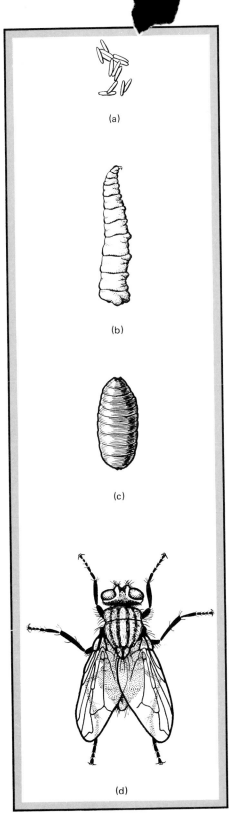

(a)

(b)

(c)

(d)

Figure 2–5
Stages in the development of an endopterygote insect, the house fly. (a) Egg; (b) larva (or maggot);
(c) puparium; and (d) adult. (H. E. Evans, 1968, from *Life on a Little-Known Planet*, Dutton.)

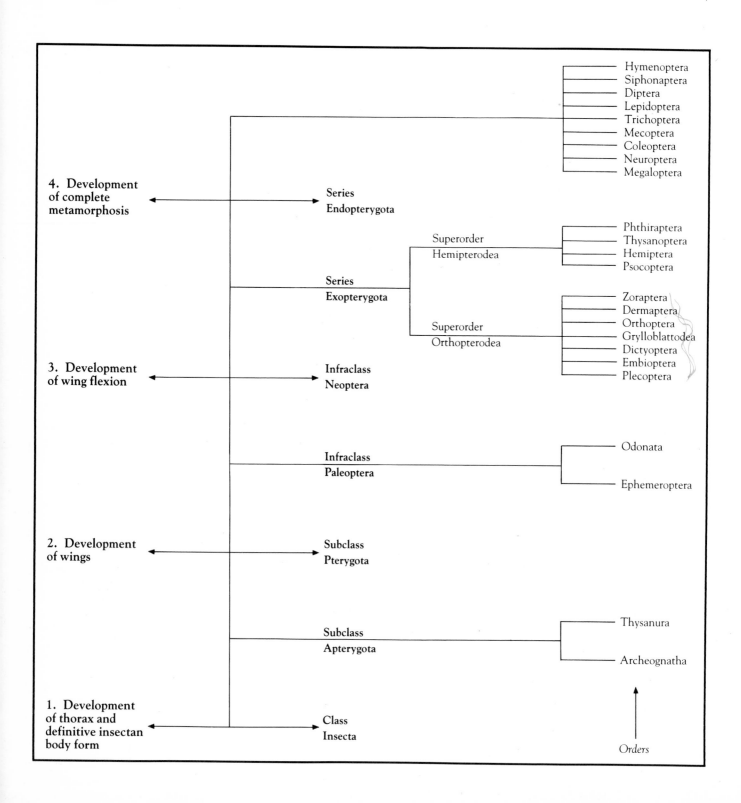

these have been lost secondarily in some groups). The Mecoptera (scorpionflies) have a long fossil record and are perhaps the most generalized of these orders. There are no major differences in wing venation between Mecoptera and Diptera (true flies), and fossil wings are often difficult to place in one or the other. A close resemblance between the larvae of Siphonaptera (fleas) and those of Diptera suggests that fleas evolved from a flylike ancestor. Trichoptera (caddisflies) and Lepidoptera (butterflies and moths) have similar wing venation and other common features; they are also believed to have evolved from a Mecopteralike ancestor. This leaves only the Hymenoptera, which are something of an enigma, as they retain an ovipositor quite comparable to that of Orthopterodea and have a unique wing venation.

Indeed there are many puzzles concerning the evolution of the higher insects. Interested students are referred to Kristensen's recent review paper and the books by H. B. Boudreaux and W. Hennig (listed at the end of this chapter). They will find many points of disagreement! One of the most intractable problems concerns a small group of insects that are parasites of other insects but have larviform females that do not leave the host. The males have wings of unusual appearance, giving the group the name "twisted winged insects." Students of Coleoptera accept these insects as very specialized beetles (as we do), but there is no universal agreement on this point, and many entomologists place them in a separate order, the Strepsiptera. The life history of one of these unusual insects is outlined in Fig. 13–10.

Further Comments on the Fossil Record and Insect Evolution. It is a surprising fact that what we think of as "higher insects" actually had their origin a very long time ago (Fig. 2–7). Beetles, caddisflies, and several other groups of Endopterygota were present in the Permian period, roughly 250 million years ago. Indeed, if we could somehow return to the Permian to collect insects, it would be an exciting experience—no less than 23 orders were present! Two other major orders, the Diptera and Hymenoptera, appeared in the next geologic period, the Triassic, when dinosaurs were beginning to roam the earth. But one major event had not yet occurred: the evolution of the flowering plants, which took place primarily in the Cretaceous period, beginning about 130 million years ago. This event set the stage for the appearance of the Lepidoptera and the bees, which underwent an intimate coevolution with the flowering plants (Chapter 11). The proliferation of the birds and mammals in the Cenozoic similarly set the stage for such groups as the lice, fleas, and biting flies, which live at the expense of these groups (Chapter 13).

The first social insects also made their appearance in the Cretaceous. Several kinds of ants are known from this period, and some are clearly workers, suggesting that ants may have been evolving social behavior earlier than this. A termite wing has also been found in rocks in Labrador dating from the Cretaceous. Fossil bees related to the honey bee and wasps apparently related to the yellow jackets have been found in mid-Cenozoic deposits, but both groups may have become social earlier than this. We can at least say that insects lived in complex societies millions of years before humans.

Many fossil insects are known from the most recent geologic era, the Cenozoic. Baltic amber, dating from about 35 million years ago, is especially rich in well-preserved insect specimens. Amber consists of hardened lumps of fossil plant resin, in

Figure 2–6
Diagrammatic representation of the four major steps in insect evolution and the way they are reflected in insect classification.

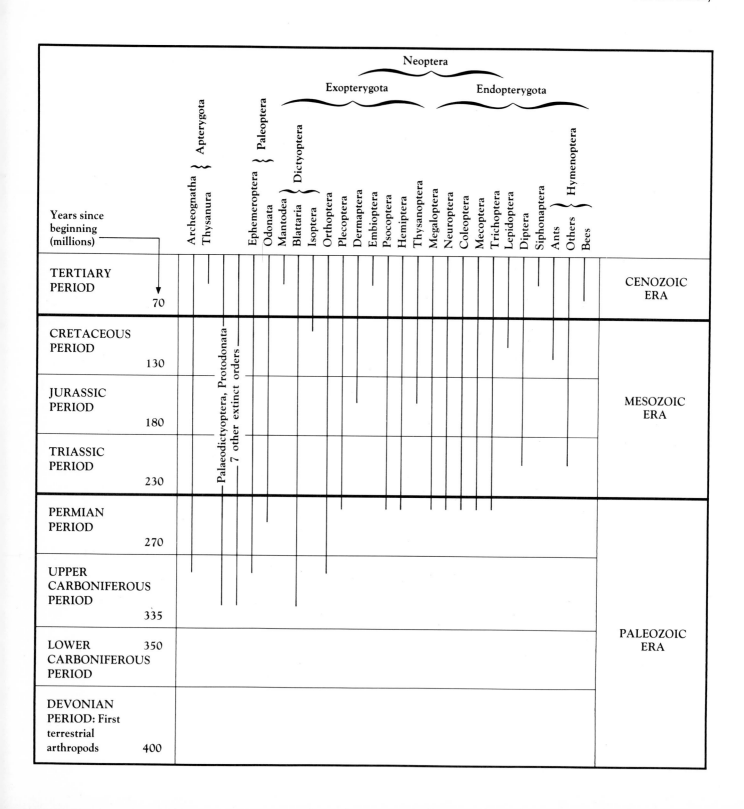

which insects have in some cases become entrapped (Fig. 2–8). Insects found in Baltic amber have a distinctly modern appearance; in fact many of them belong to existing genera, although not to existing species. Most evolution during the Cenozoic has been at the level of the family, genus, and species. There is reason to believe that some groups of insects are at present undergoing rather rapid evolution. Some groups of flies, for example, are very difficult to classify, suggesting that they have diversified very recently. Studies of the genetics of *Drosophila*, as well as observations on the development of resistance to insecticides, confirm the fact that insects, despite their ancient lineage, are capable of rapid evolution under appropriate conditions. (See also Chapter 15, p. 332.)

Summary

The diversity of insects is so great that their classification presents many problems. To be meaningful and useful, a classification must be based on the pattern of evolution so far as this can be deduced from available evidence. A sound classification is a filing system for information about organisms and also an information retrieval system.

Insects are classified in a series of groupings of decreasing scope, beginning with the highest level, the phylum, and proceeding through the order and family to the level of the species. Each step as we descend the hierarchy tells us something more precise about the insect. Species names consist of the combined generic and specific names and are standard the world over. Common names have been established for most North American insects of importance to humans. The names of most insect orders end in -*ptera*. It is important that students of insects be familiar with the orders and with the major families of the larger orders. The characteristics of the orders, major suborders, and many families are summarized in Tables 2–3 to 2–13.

Insects belong to the phylum Arthropoda, a group that also includes spiders, mites, centipedes, and a variety of other organisms. At least two and perhaps as many as four groups of many-legged arthropods became six-legged, with three tagmata: head, thorax, and abdomen. True insects have the following features in common (although these may be modified or lost in some cases): (1) external mouthparts with mandibles, maxillae, and labium; (2) antennae with two musculated segments; (3) compound eyes and ocelli; (4) thorax with three pairs of legs, tarsi divided into two to five tarsomeres; (5) abdomen with 11 segments; (6) malpighian tubules well developed; (7) superficial cleavage of embryo; and (8) epimorphic development.

Two groups of insects are believed to represent a stage in evolution before the development of wings. These are the orders Archeognatha and Thysanura, together called bristletails. These two orders comprise the subclass Apterygota, in contrast to the winged insects, subclass Pterygota (some Pterygota, such as lice and fleas, have secondarily lost their wings). There are several theories as to the origin of wings. They may have originally been gills, those on the thorax becoming enlarged to serve as wings. Or they may have been lateral outgrowths of the thorax that were originally used in gliding. In the early stages of their evolution, wings may also have been used for thermoregulation or in sexual display.

The most primitive winged insects, including mayflies, dragonflies, and several extinct groups, could not fold the wings back over their bodies. These constitute an

Apterygota.

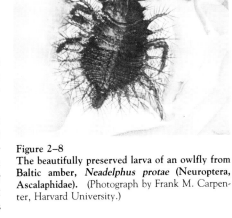

Figure 2–8
The beautifully preserved larva of an owlfly from Baltic amber, *Neadelphus protae* (Neuroptera, Ascalaphidae). (Photograph by Frank M. Carpenter, Harvard University.)

(Text continues on p. 51)

Figure 2–7
The appearance of the insect orders in the fossil record.

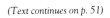

Table 2–3. The major orders of exopterygote insects

Order	Common Name	Example	Front Wings	Hind Wings
Ephemeroptera	Mayflies		Triangular, membranous; many veins and cross-veins	Smaller, rounded (may be absent)
Odonata	Dragonflies, damselflies		Long, slender, membranous; many veins and cross-veins	Similar to front wings
Plecoptera	Stoneflies		Slender, membranous, with numerous veins	Usually wider than front wings; vannus present
Embioptera	Webspinners		Slender, membranous but often smoky; few veins present	Very similar to front wings
Dictyoptera	Cockroaches, mantids, termites		Elongate, often thickened; usually with many veins	Wider than front wings and with vannus (except in most termites)
Orthoptera	Grasshoppers, crickets, katydids, walking sticks		Long and slender, thickened, with many veins (may be absent)	Wider than front wings, membranous, with vannus (may be absent)
Dermaptera	Earwigs		Very short, padlike, leathery (may be absent)	Large, membranous, folding fanlike under front wings (may be absent)
Psocoptera	Barklice, booklice		Membranous, with few veins (may be absent)	Similar to but somewhat smaller than front wings (may be absent)
Hemiptera	True bugs, cicadas, leafhoppers, aphids, etc.		Membranous or thickened, with few veins (may be absent)	Membranous, shorter but often somewhat wider than front wings (may be absent)
Thysanoptera	Thrips		Very slender, with a wide fringe of hairs, few veins or none (may be absent)	Same as front wings
Phthiraptera	Lice		None	None

Antennae	Mouthparts	Caudal Appendages	Other Features
Very short, bristlelike	Biting type, but nonfunctional in adult	2 slender, many-segmented cerci and often a median filament	Larvae are aquatic and have lateral abdominal gills (Fig. 2–4c)
Very short, bristlelike	Biting, with sharp teeth on mandibles and maxillae	2 or more short, unsegmented appendages	Larvae are aquatic, without lateral gills but with a prehensile labium (Figs. 1–5a, 2–4a)
Long, many-segmented, thread-like	Biting, but weakly developed in adult	Short-to-long, many-segmented paired cerci; no ovipositor	Larvae are aquatic, with or without feathery gills on head, thorax, or base of abdomen (Fig. 2–4b)
Moderately long, many-segmented, threadlike	Biting; not unlike Orthoptera	A pair of short, 2-segmented cerci; no ovipositor	Front tarsi swollen, with silk glands; gregarious insects, living in silken galleries
Short to very long, threadlike, many-segmented	Biting; not unlike Orthoptera	A pair of short-to-moderately-long cerci; no external ovipositor	Coxae large, close together; see the 3 suborders, Table 2–5
Moderately to very long, threadlike, many-segmented	Biting type, as described for grasshopper, Chapter 1 (Fig. 1–4)	Short, unsegmented cerci; an external ovipositor usually present in female	Coxae small, widely spaced; see the 3 suborders, Table 2–5
Moderately long, threadlike, many-segmented	Biting; not unlike Orthoptera	A pair of unsegmented, forcepslike cerci; no ovipositor	
Long, many-segmented, thread-like	Biting, with a chisellike development of the maxillae	No cerci; ovipositor of female small or absent	Most species less than 6 mm in length
Short to long, several to many segments, threadlike	Piercing-sucking, without palpi, labium forming a sheath for stylets (Fig. 1–5c,d)	No cerci; ovipositor distinct in female	See the 2 suborders, Table 2–5
Short, with 6 to 10 segments	Piercing-sucking, asymmetrical, palpi present	No cerci; female with ovipositor or end of abdomen tubelike	Most species less than 5 mm in length
Short, with 3 to 5 segments	Either biting or sucking; see Table 2–5	No cerci; female with or without a small ovipositor	External parasites of warm-blooded animals; see Table 2–5 for suborders

Table 2–4. The major orders of endopterygote insects

Order	Common Name	Example	Front Wings	Hind Wings
Megaloptera	Dobsonflies, snakeflies		Elongate, membranous, with numerous veins and cross-veins	Similar to or somewhat wider than front wings
Neuroptera	Lacewings, ant lions		Similar to Megaloptera but often with much branching near margin	Similar to front wings but sometimes slightly smaller
Coleoptera	Beetles		Hardened, protective "elytra," which meet in a straight line on back	Membranous, fold complexly beneath front wings (may be absent)
Mecoptera	Scorpionflies		Membranous, slender (especially basally), with numerous cross-veins	Similar to front wings (both pairs may be absent or reduced and modified)
Trichoptera	Caddisflies		Elongate, with few cross-veins, covered with hairs	Similar to front wings but broader and slightly shorter
Lepidoptera	Moths, butterflies		Slender to rather broad, clothed with scales; relatively few cross-veins	Similar to front wings but usually shorter, broader, and more rounded; often attached to front wings
Diptera	Flies, gnats, midges		Membranous, with relatively few veins and cross-veins (may be absent)	Absent as functional wings; forming small, knobbed "halteres"
Siphonaptera	Fleas		Absent	Absent
Hymenoptera	Sawflies, wasps, ants, bees		Membranous, with relatively few veins but usually several cross-veins (may be absent)	Smaller than front wings and capable of attachment to them by a series of hooklets

Antennae	Mouthparts	Larval Name	Example of Larva	Larval Features	Notes
Long, many-segmented, threadlike	Biting; prognathous	Hellgramites (in part)		Biting mouthparts, prognathous; may have lateral gills	Many are aquatic
Many-segmented; short to rather long; may be clubbed	Biting; hypognathous	Ant lions, aphid lions		Sucking tube between the elongate mandibles and maxillae	Most are terrestrial
Usually have 11 segments; may be clubbed, serrate, lamellate	Biting	Grubs		With or without true legs; without ventral prolegs; head complete	Major families, Tables 2–8 and 2–9
Long, many-segmented, threadlike	Biting, located at end of a beak			Caterpillarlike; ventral prolegs without hooklets; 20 or more ommatidia in lateral eyes	See text, p. 194 (Fig. 7–18)
Long, many-segmented, threadlike	Basically of biting type, with large palpi, but mandibles reduced or absent	Caddisworms		Aquatic, with a pair of hooks at tail end; most build cases	See text, p. 325
Long, many-segmented; may be feathery or clubbed	Coiled sucking tube formed from maxillae; mandibles reduced or absent	Caterpillars		Ventral abdominal prolegs bearing hooklets (some exceptions)	Major families, Table 2–10
Long, many-segmented, or if short with a terminal bristle	Sucking, sometimes piercing, of variable structure	Maggots, wigglers, grubs		No thoracic legs; prolegs variable or absent; head often incomplete	Major families, Tables 2–12 and 2–13
Short, stout, 3 segments	Piercing-sucking; palpi present			Hairy; head complete but no legs or ventral prolegs	Adults are parasites of warm-blooded animals
Several to many segments; usually threadlike	Chewing, but maxillae and labium often elongated for sucking also			Head complete; legs and ventral prolegs present or absent, never with hooklets	Major families, Table 2–11

Table 2–5. Major suborders of some orders of Exopterygota

Order	Suborder	Example	Distinguishing Features	Notes
Dictyoptera	Blattaria (Cockroaches)		Flattened, with a shieldlike pronotum; 2 ocelli; legs fitted for running; cerci with several segments	See text, pages 142, 182 (Figs. 4–15, 6–2)
	Mantodea (Mantids)		Elongate; front legs modified for grasping prey; 3 ocelli; cerci with several segments	See text, pages 147, 177 (Fig. 14–16)
	Isoptera (Termites)		Front and hind wings similar, with a basal fracture line; 2 ocelli or none; cerci very short	Social, with sterile workers and soldiers; see text, pages 207–208
Orthoptera	Ensifera (Crickets, long-horned grasshoppers)		Jumping hind legs; very long antennae; tympanic organs on front tibiae; long ovipositor	See text, pages 175, 197 (Figs. 7–2, 7–22)
	Caelifera (Short-horned grasshoppers)		Jumping hind legs; shorter antennae; tympanic organs on abdomen; short, horny ovipositor	See text, pages 357–359 (Figs. 1–1, 14–1)
	Phasmida (Walking sticks)		Legs slender, fitted for walking; no sound production or tympanic organs; ovipositor small, mostly concealed	See text, pages 134, 319 (Fig. 14–2, 5–20)
Hemiptera	Heteroptera (True bugs)		Wings folding flat on abdomen, fore wings thickened basally and membranous apically; beak arises toward front of head	Major families, Table 2–6
	Homoptera (Leafhoppers, cicadas, aphids, etc.)		Wings usually sloping over sides of body, fore wings of uniform thickness; base of beak close to front coxae	Major families, Table 2–7
Phthiraptera	Mallophaga (Biting lice)		Biting mouthparts; tarsi with 1 or 2 segments, terminating in 1 or 2 claws	See text, pages 296–297
	Anoplura (Sucking lice)		Piercing-sucking mouthparts that can be withdrawn into head when not in use; tarsi with 1 segment and 1 large claw	See text, pages 297–299 (Figs. 13–8, 13–9)

50

infraclass called Paleoptera. All other winged insects possess a wing-flexing mechanism that enables them to hold the wings flat or rooflike against their body. These belong to the infraclass Neoptera. The Neoptera in turn fall into two groups depending on whether the wing pads develop externally during the immature stages (Exopterygota) or are retained internally as imaginal discs (Endopterygota). In the latter group, a pupal stage is interposed between larva and adult, and these insects are said to have complete metamorphosis.

The success of each of the four major steps in insect evolution is well shown by the numbers of orders and the number of species resulting from each step. (1) The development of the insect body ground plan resulted in the eventual evolution of some 30 orders of insects. (2) The development of wings resulted in the evolution of 22 orders of living insects, which include over 99% of the species. (3) The development of wing flexion resulted in the evolution of 20 orders, which include more than 95% of living species. (4) Insects with complete metamorphosis include only 9 orders, but these contain at least 80% of living species.

There are many problems and uncertainties as to the best classification of the insects. The fossil record is not always helpful in resolving these problems, as it is very incomplete, and the origin of most major groups is buried deep in the past, more than 230 million years ago. The appearance of the flowering plants late in the Mesozoic, however, set the stage for the evolution of many groups that are now associated with these plants, and the radiation of the birds and mammals in the Cenozoic made possible the development of such groups as lice, fleas, and biting flies that now live at the expense of these groups.

Selected Readings

Classification and Evolution

Boudreaux, H. B. 1979. *Arthropod Phylogeny with Special Reference to Insects.* New York: Wiley. 320 pp.

Carpenter, F. M. 1976. "Geological history and evolution of the insects." *Proceedings of the 15th International Congress of Entomology,* Washington, D.C., pp. 63–70.

Hennig, W. 1981. *Insect Phylogeny.* Translated and edited by A. C. Pont. New York: Wiley. 514 pp.

Kristensen, N. P. 1981. "Phylogeny of insect orders." *Annual Review of Entomology,* vol. 26, pp. 135–57.

Books Useful for Insect Identification

Arnett, R. H., Jr.; N. M. Downie; and H. E. Jaques. 1980. *How to Know the Beetles.* Second edition. Dubuque, Iowa: Brown. 416 pp.

Bland, R. G., and H. E. Jaques. 1978. *How to Know the Insects.* Third edition. Dubuque, Iowa: Brown. 409 pp.

Borror, D. J.; D. M. DeLong; and C. A. Triplehorn. 1981. *An Introduction to the Study of Insects.* Fifth edition. Philadelphia: Saunders. 827 pp.

Borror, D. J., and R. E. White. 1970. *A Field Guide to the Insects of America North of Mexico.* Boston: Houghton Mifflin. 404 pp.

Brues, C. T.; A. L. Melander; and F. M. Carpenter. 1964. "Classification of insects." *Bulletin of the Museum of Comparative Zoology, Harvard University,* vol. 108. 907 pp.

Chu, H. F. 1949. *How to Know the Immature Insects.* Dubuqe, Iowa: Brown. 234 pp.

Daly, H. V.; J. T. Doyen; and P. R. Ehrlich. 1978. *An Introduction to Insect Biology and Diversity.* New York: McGraw-Hill. 664 pp.

Ehrlich, P. R., and A. H. Ehrlich. 1961. *How to Know the Butterflies.* Dubuque, Iowa: Brown. 262 pp.

(Text continued on p. 67)

Table 2-6. Some important families of Hemiptera, suborder Heteroptera

Family	Common Name	Example	No. Segments in Beak	No. of Ocelli	Antennae
Belostomatidae	Giant water bugs		3	0	4 segments; inconspicuous
Notonectidae	Backswimmers		3 or 4	0	4 segments; inconspicuous
Gerridae	Water striders		4	2	4 segments; moderately long
Cimicidae	Bedbugs		3	0	4 segments; moderately long
Reduviidae	Assassin bugs		3	0 or 2	4 segments; thread-like
Coreidae	Squash bug and relatives		4	2	4 segments; moderately long; inserted high on head
Lygaeidae	Chinch bug and relatives		4	2	4 segments; moderately long; inserted low on head
Miridae	Plant bugs		4	0	4 segments; moderately long, slender
Pentatomidae	Stink bugs		4	2	5 segments; moderately long

Special Features	Feeding Behavior	Notes
Aquatic; front legs fitted for grasping prey; middle and hind legs fitted for swimming; tarsi with claws	Predaceous	See text, page 198 (Fig. 7–23)
Aquatic; middle and front legs rather short, with claws; hind legs long, oarlike, without claws	Predaceous	Swim upside-down
Swim on surface of water; legs very long; claws inserted before end of tarsi	Predaceous or feed on dead insects	
Flattened; wings absent or represented by stubs	Blood feeders on birds and mammals	See text, page 293
Terrestrial; front legs used for grasping prey; wings usually present, but may be short or absent	Predaceous or blood feeders on warm-blooded animals	See text, pages 293–294 (Fig. 4–2b)
Terrestrial; with scent glands; membranous part of front wings with numerous branching veins	Phytophagous	
Terrestrial; with scent glands; membranous part of front wings with 4 or 5 simple veins	Mostly phytophagous but a few are predaceous	
Terrestrial; small and often brightly colored; membranous part of front wings with 2 cells closed off by a vein	Mostly phytophagous but a few are predaceous	
Terrestrial, with a large scutellum behind pronotum, reaching middle of abdomen or beyond; scent glands present	Phytophagous or predaceous	(Fig. 12–10)

Table 2–7. Some important groups of Hemiptera, suborder Homoptera

Family or Superfamily	Common Name	Example	Antennae	Wings
Cicadidae	Cicadas		Very small, with a terminal bristle, arising between eyes	Membranous, usually clear and translucent
Cicadellidae	Leafhoppers		As in Cicadidae	Front wings somewhat thickened and with a color pattern
Membracidae	Treehoppers		As in Cicadidae	Membranous, transparent, often partly covered by pronotum
Cercopidae	Spittlebugs		As in Cicadidae	Front wings somewhat thickened and colored
Fulgoroidea	Planthoppers		Small, with a terminal bristle, arising below eyes	Membranous, transparent, or front wings thickened and colored
Psyllidae	Jumping plantlice		Long, threadlike, 9–11 segments	Front wings slightly thicker than hind wings, without a stigma
Aleyrodidae	Whiteflies		Threadlike, usually with 7 segments	Opaque, whitish, with very few veins, no stigma
Aphidoidea	Aphids, plantlice		Threadlike, with 3–6 segments	Transparent, with a stigma on front wings (dark spot near tip); wings may be absent
Coccoidea	Scale insects		Threadlike, with up to 13 segments, but in females usually very short or absent	Males with 1 pair of membranous wings, or wingless; females always wingless

Number of Tarsal Segments	Special Features	Notes
3	3 ocelli; sound-producing organs beneath abdomen of male; larvae live in soil and have front legs fitted for digging	
3	2 ocelli; hind tibiae armed with rows of spines	See text, pages 233–234
3	Pronotum very large, prolonged backward over abdomen	
3	Hind tibiae armed with 1 or 2 stout spines and an apical fringe of spines; larvae of many species live in a mass of froth produced from abdomen	
3	Head sometimes with a prolongation in front	
2	Hind legs fitted for jumping	
2	Very small, under 2 mm; larvae immobile, scalelike	
2	Often have cornicles (paired, projecting tubules) near end of abdomen	See text, pages 351, 353 (Figs. 5–15, 16–2)
0 or 1	Larvae as well as adult females sedentary, located beneath a waxy scale that they secrete	See text, pages 218, 276

Table 2–8. Some important families of Coleoptera (see also Table 2–9)

Family	Common Name	Example	Antennae	Special Features
Carabidae	Ground beetles		Threadlike	Robust, long-legged, with large mandibles; usually black or brown with little or no color pattern
Cicindelidae	Tiger beetles		Threadlike	Long-legged, very active predators; often with an attractive color pattern
Gyrinidae	Whirligig beetles		Short and thick	Eyes divided into upper and lower sections; legs fitted for swimming on surface film
Hydrophilidae	Water scavenger beetles		Clubbed, shorter than maxillary palpi	Legs flattened and fringed, fitted for swimming
Staphylinidae	Rove beetles		Threadlike or somewhat clubbed	Elytra very short, exposing most of abdomen
Silphidae	Carrion beetles		Clubbed	Elytra often somewhat short, exposing tip of abdomen; may have color pattern
Lampyridae	Fireflies		Threadlike or pectinate	Somewhat soft-bodied; often with light-producing organs beneath abdomen
Elateridae	Click beetles		Serrate	Pro- and mesosterna forming a clicking mechanism
Buprestidae	Metallic wood borers		Serrate	Hard-bodied; head sunken into prothorax
Dermestidae	Dermestids; skin beetles		Clubbed	Oval or elongate, often clothed with short hairs or scales; may have color pattern

Tarsal Formula	Example of Larva	Legs of Larva	Other Larval Features	Notes
5-5-5		Long, with 5 segments, 2 claws; often with an anal proleg	Prognathous; with movable cerci	Usually predaceous; see text, pages 133, 135 (Figs. 5–21, 12–1)
5-5-5		With 5 segments, 2 claws	Live in vertical burrows in soil, using head as stopper; large hooks on 5th abdominal segment	Predaceous
5-5-5		Long, with 5 segments, 2 claws	Lateral, abdominal gill filaments, terminal abdominal hooks	Aquatic; predaceous
5-5-5		Long, with 4 segments, 1 claw	May have lateral or terminal abdominal gills	Mostly aquatic; predators or scavengers
5-5-5		Long, with 4 segments, 1 claw; often with an anal proleg	Prognathous; with movable cerci	Terrestrial; predators or scavengers (Fig. 5–25)
5-5-5		With 4 segments, 1 claw	With sclerotized dorsal plates, movable cerci	Carrion feeders and predators (Fig. 7–24)
5-5-5		With 4 segments, 1 claw	With sclerotized dorsal plates, no cerci; may have light organs	Predaceous; see text, page 316 (Fig. 14–10)
5-5-5		Short, with 4 segments, 1 claw	Typical wireworms, slender, sclerotized; cerci fixed or absent	Most feed on roots of plants; a few predaceous or scavengers
5-5-5		Absent	Pale, unsclerotized; head sunken into thorax; prothorax wider than rest of body	Borers
5-5-5		With 4 segments, 1 claw	Covered densely with hairs; fixed cerci present or absent	Scavengers, chiefly on animal products

Family	Common Name	Example	Antennae	Special Features
Scarabaeidae	Scarab beetles June beetles		With a lamellate club	Robust, small to quite large
Lyctidae	Powderpost beetles		Clubbed	Small, body somewhat parallel-sided
Tenebrionidae	Darkling beetles		Threadlike or clubbed	Rather hard-bodied, dark in color; head fitting into thorax, without a "neck"
Meloidae	Blister beetles		Threadlike	Relatively soft-bodied, variously colored; head constricted behind, forming a "neck"
Coccinellidae	Lady beetles		Clubbed	Small, convex, hemispherical; often brightly colored, spotted
Chrysomelidae	Leaf beetles		Short, simple, or slightly clubbed	Small, often with a color pattern
Cerambycidae	Longhorned beetles		Slender or serrate, long, often longer than body	Medium to large in size, often with a color pattern
Bruchidae	Seed beetles		Short, may be serrate or slightly clubbed	Elytra not reaching tip of abdomen, which is exposed above
Curculionidae	Weevils Snout beetles		Elbowed and clubbed in most species	Mouthparts small, at the end of a short-to-long beak
Scolytidae	Bark beetles		Short, clubbed	Small, cylindrical; elytra sometimes hollowed and toothed posteriorly

Table 2–9. Some important families of Coleoptera (see also Table 2–8)

Tarsal Formula	Example of Larva	Legs of Larva	Other Larval Features	Notes
5-5-5		With 4 segments, 1 claw	C-shaped, pale, with a dark head	Phytophagous or dung-feeders; see text, pages 198, 304 (Fig. 13–15)
5-5-5 but basal segment very short		Small, with 3 segments and reduced claws	C-shaped, pale in color	Borers
5-5-4		With 4 segments, 1 claw	Elongate, often wireworm-like; with or without fixed cerci	Phytophagous or scavengers; see text, pages 133, 315 (Fig. 14–9)
5-5-4		Reduced to stubs except in the active first instar	Undergo hypermetamorphosis; later instars sluggish parasitoids	Adults are phytophagous, larvae parasitoids of Orthoptera, Hymenoptera (Fig. 5–18b)
3-3-3		With 4 segments, 1 claw	Active, soft-bodied but often dark-colored larvae	Mostly predaceous but one group phytophagous; see text pages 275–277 (Fig. 12–4)
5-5-5 but 4th segment very small		Short, with 4 segments, 1 claw	Soft-bodied, sluggish; mandibles often broad, with 4–5 teeth	Phytophagous; see text, pages 223, 239 (Fig. 9–1)
Like preceding, seem to be 4-4-4		Extremely short or absent	Elongate, straight-bodied; head somewhat sunken into thorax	Borers
Similar to preceding		Minute and without claws; or absent	Somewhat C-shaped	Larvae live in seeds
Similar to preceding		Absent	Soft-bodied except head; may be straight or C-shaped	Phytophagous but diverse in feeding behavior; see text, pages 282–283 (Fig. 12–13)
Similar to preceding		Absent	Similar to preceding; occur mainly in galleries under bark	Borers; see text, pages 123–125

Table 2–10. Some important groups of Lepidoptera

Family or Superfamily	Common Name	Wings	Characteristics of Wings	Other Adult Features
Tineoidea	Tineoids		Wings slender, with a long fringe of hairs; hind wings often no wider than front wings	Slender-bodied, delicate, mostly quite small
Tortricoidea	Tortricoids Bell-wings		Wings moderately wide, hind wings wider than front wings; fringe very short	Slender-bodied; delicate, small; palpi small or absent
Pyraloidea	Pyraloids Grass moths Snout moths		As above; first 2 veins of hind wings stalked to near wingtip	Slender and delicate; palpi often large and protuberant
Noctuidae	Noctuids Owlet moths Cutworm moths		Only 2 vannal veins in hind wing; first 2 veins of hind wing diverge near base of wing	Body robust; wings usually with dingy colors
Arctiidae	Tiger moths		As above, but first 2 veins of hind wing fused above discal cell	Body robust; often with a pattern of bright colors
Lasiocampidae	Tent caterpillar moths		Hind wing with a humeral lobe but no frenulum; humeral lobe with 2 short veins	Rather small, heavy-bodied moths with pectinate antennae; tongue reduced or absent
Saturniidae	Giant silk moths		As above, but humeral lobe with only 1 vein; frenulum absent or vestigial	Large, heavy-bodied moths, with pectinate antennae; tongue reduced or absent
Sphingidae	Hawkmoths		Front wings rather slender and pointed; frenulum present	Tongue well developed, long; antennae somewhat thickened, may have hook at end
Geometridae	Geometers		First vein of hind wing has a sharp angle at base; frenulum present	Body rather slender; abdomen with a tympanal organ at base
Papilionoidea	Butterflies Skippers		Hind wing with humeral lobe but no frenulum	Diurnal, often brightly colored; antennae slender, clubbed, may have terminal hook

Larval Integument	Example of Larva	Hooklets on Prolegs	Notes
Essentially "naked", i.e., only about 8 small setae each side of each abdominal segment; pale in color		Variable, but never in a single transverse series; may be absent	Larvae are concealed feeders, often leaf miners, case bearers, leaf tiers, internal feeders
As above; 3 setae in front of prothoracic spiracle; 2 setae below abdominal spiracles close together			Larvae are internal feeders; may be leaf rollers, living inside fruit, etc.
As above but with only 2 setae in front of each prothoracic spiracle		Variable, but not as below	Mostly concealed feeders, often leaf rollers or webbers, root feeders, or feed on stored grain
As above; may have color pattern; may have additional, "secondary" setae			Larvae often called "cutworms," mainly leaf feeders or feed on fruit or roots; see text, pages 149–150
Densely covered with clumps of long hairs, mostly of about equal length		Single series as above, but central hooklets may be longer than others	Larvae are "woolly bears," generally external leaf feeders
Irregularly covered with hairs, some much longer than others		Similar to Saturniidae	Larvae are external feeders on trees and shrubs, often spin large webs
With projecting knobs or processes or with tufts of urticating hairs, usually also with many short setae			Quite large; larvae feed on grasses, trees, shrubs; adult males have large bushy antennae; see text, pages 104–105
Covered with minute setae and with a large terminal "horn," sometimes replaced by a disc or spot		Similar to Saturniidae	Larvae are "hornworms," external leaf feeders; adults swift fliers, diurnal or nocturnal; see text, pages 96, 282
Without knobs, warts, or horns, but often with numerous short setae on underside; often sticklike		Similar to Saturniidae	Only 2 pairs of prolegs, and move by looping
With an abundance of short setae or large, spinose processes			External leaf feeders; pupa is a chrysalid, no true cocoon; adults are diurnal; see text, pages 189, 193, 314

Table 2–11. Some important groups of Hymenoptera

Family or Superfamily	Common Name	Example	Antennae	Ovipositor Female
Tenthredinoidea	Sawflies Leafwasps		Threadlike, pectinate, or clubbed, 3–32 segments	Flattened, sawlike
Siricoidea	Horntails Woodwasps		Threadlike, many-segmented	Strong, cylindrical, adapted for boring
Cephidae	Stem sawflies		Threadlike, many-segmented	Short, cylindrical, retractile
Ichneumonoidea	Ichneumons Braconids		Threadlike, many-segmented (more than 16 segments)	Short or long, needlelike, issuing from before end of abdomen
Chalcidoidea	Chalcid wasps		Elbowed beyond a long basal segment, 6–13 segments in all	Short, needlelike, issuing from before end of abdomen
Cynipoidea	Cynipids Gall wasps		Threadlike, not elbowed, 12–16 segments	Short, retractile, issuing from before end of abdomen
Formicidae	Ants		Usually elbowed beyond basal segment, 4–13 segments	Short, needlelike, modified for stinging (may be absent)
Scolioidea	Scolioid wasps		Threadlike, 12 segments in female, 13 in male	Short, needlelike, modified for stinging, issues from end of abdomen
Vespoidea	Potter wasps Social wasps Hornets Yellow jackets		Threadlike, 12 segments in female, 13 in male	As above
Sphecidae	Digger wasps		As above	As above
Apoidea	Bees		As above	As above

Pronotum Reaches Tegulae	Other Adult Features	Example of Larva	Notes
+	Abdomen broadly joined to thorax; wings with rather numerous veins and cross-veins		Larvae are phytophagous, have thoracic legs and often have abdominal prolegs; see text, pages 219, 331
+	Abdomen broadly joined to thorax; last abdominal segment has a "horn"		Larvae are borers, have very short thoracic legs and no prolegs, but have a terminal "horn"
+	Abdomen compressed but rather broadly joined to thorax		Larvae are stem borers, have vestigial thoracic legs and no prolegs, slender and with paired terminal appendages
+	Wings with several veins and prominent cross-veins, also with a stigma on front wing (dark spot on front margin); "wasp waist" developed		Larvae grublike, legless parasitoids, with sclerotized framework surrounding mostly fleshy mouthparts; see text, pages 283, 326 (Fig. 12–9)
−	Wings usually each with a single vein, front wing with a stigma at end of vein, slightly below wing margin; "wasp waist" present; very small		Larvae grublike parasitoids (rarely phytophagous); no framework surrounding the mostly fleshy mouthparts; see text, pages 284–285 (Fig. 12–11)
+	Wings with simple venation, no stigma but a closed cell toward outer wing margin (front wing); abdomen strongly compressed, flealike; very small		Larvae grublike, parasitoids or living in galls; as above, but mouthparts somewhat more distinct; see text, page 220 (Fig. 9–5)
+	With a double or triple deep incision between apparent thorax and apparent abdomen, thus with 1 or 2 "nodes" here; very often wingless		Live in nests socially and have a caste system; larvae hairy and have antennal orbits high on head; see text, pages 126, 206 (Fig. 8–10)
+	Rather heavy-bodied but with short "wasp waist"; front wings have stigma; wings sometimes absent		Larvae grublike parasitoids, have strong 3- or 4-toothed mandibles, occur mostly in soil
+	Wings tend to fold longitudinally when at rest; "wasp waist" present; some species are social		Larvae grublike, legless, occur in nests made by parent
−	"Wasp waist" well developed, sometimes rather long; females often have "rake" on front legs for digging		As above; larvae often have paired, projecting spinnerets on labium; see text, pages 186, 327 (Figs. 7–11, 7–21)
−	Rather hairy, some of hairs branched or feathery; females collect pollen in masses on hind legs or beneath abdomen		Larvae grublike, legless, occur in nests made by parent; mandibles specialized for pollen feeding; see text, pages 203, 258 (Figs. 8–1, 11–7)

Table 2–12. Some important families of Diptera (see also Table 2–13)

Family	Common Name	Example	Antennae	Ocelli	Mouthparts
Tipulidae	Crane flies		Long, slender, 6–40 segments	0	Nonpiercing, large palpi, on prolongation of head
Culicidae	Mosquitoes		Slender, hairy, plumose in males, 14–15 segments	0	Piercing-sucking, stylets quite long; palpi long
Psychodidae	Moth flies Sand flies		Slender, hairy, 12–16 segments	0	Piercing-sucking or short and nonfunctional
Simuliidae	Black flies		Quite short, with 11 segments	0	Short, piercing beak
Ceratopogonidae	Biting midges Punkies No-see-ems		Slender, somewhat hairy, 13–15 segments	0	Short, piercing beak
Cecidomyidae	Gall midges		Slender, beadlike, with whorls of hair; 10–36 segments	0 (rarely present)	Short, nonpiercing, usually nonfunctional
Tabanidae	Horse flies Deer flies		Short to moderately long, variable but without a terminal bristle	0 or 3	Piercing, usually quite short
Bombyliidae	Bee flies		Rather short, 3 segments, with or without a terminal bristle	3	Slender proboscis for sucking nectar
Asilidae	Robber flies		As above	3	Stout, horny proboscis
Syrphidae	Flower flies		Short, with 3 segments, last segment with a bristle	3	Short, sucking proboscis

Other Adult Features	Example of Larva	Head of Larva	Spiracles of Larva	Notes
Legs very long and slender; thorax with a dorsal, V-shaped suture		Deeply sunken into thorax and incomplete behind; mandibles work side to side	1 pair, terminal, may be replaced by gills	Larvae often aquatic
Wings bear scales along veins and margin		Complete and free; opposable mandibles and often with mouth brushes	1 pair, terminal, may be at end of breathing tube	Larvae and pupae aquatic; see text, pages 288–293 (Figs. 6–7, 13–2)
Very small; legs, wings, and body clothed with hairs		Complete and free, with opposable mandibles	Usually 2 pairs, 1 on prothorax, 1 terminal, both small	Aquatic or in damp places; see text, page 291
Very small, somewhat humpbacked		Complete and free, with opposable mandibles and mouth brushes	Minute and non-functional	Larvae aquatic, have thoracic proleg and terminal sucker; see text, pages 288, 291
Very small; may have patterned wings		Complete and free, with opposable mandibles	Absent or very minute	Larvae aquatic or in damp places, often wormlike; see text, pages 287–288
Small, delicate; wings with very few veins		Small, pale in color, not easily distinguished	9 pairs or fewer, very small	Larvae often form galls, often have a dark bar beneath thorax; see text, pages 219–220 (Fig. 9–4)
May be rather large; tarsi terminate in 3 pads		Small, retractile; mandibles hooklike, work in a vertical plane	1 pair, terminal, inside a vertical slit	Larvae aquatic or in wet places; see text, pages 290–291
Stout, hairy, rather bee-like		Small, partly retracted; mandibles hooklike, work in a vertical plane	2 pairs, 1 on prothorax, 1 on next-to-last segment	Larvae predaceous or parasitoids, somewhat C-shaped
May be large; slender; vertex hollowed between the prominent eyes		Small, partly retracted, incomplete behind; mandibles work in a vertical plane	As above	Larvae usually in soil, believed to be predators; see text, page 275 (Fig. 12–3)
Often beelike or wasplike; wings have a "false vein" (crease) near middle		Largely retracted and reduced except for mouth hooks	2 pairs, 1 on prothorax, 1 pair contiguous spiracles on last segment	Larvae diverse in appearance and habitat; see text, page 267

Table 2–13. Some important families of "higher flies"

Family	Common Name	Example	Notes
Tephritidae	Fruit flies		Wings usually prominently banded; ovipositor somewhat horny; squamae at wing bases quite small; see text, pages 190, 332
Drosophilidae	Vinegar flies		Wings clear; squamae small; antennal bristle usually feathery; larvae have terminal spiracles at ends of short tubes; see text, pages 103, 180
Anthomyidae	Anthomyids Root maggots		Wings clear; squamae somewhat enlarged; underside of scutellum hairy; no bristles on hypopleura; see text, page 221
Muscidae	House flies Stable flies		As above but no bristles under scutellum; life stages, Fig. 2–5; see text, pages 41, 305
Calliphoridae	Blow flies		Usually blue or green in color; squamae large; hypopleura has stiff bristles; see text, pages 151, 268
Sarcophagidae	Flesh flies		Usually gray and somewhat striped; squamae large; hypopleura has stiff bristles
Tachinidae	Tachina flies		Variously colored; squamae large; a large swelling beneath scutellum (= postscutellum); abdomen often very bristly; parasitoids; see text, pages 280–281 (Fig. 12–10)
Oestridae	Bot flies Warble flies		Hairy, somewhat beelike; mouth opening small, mouthparts reduced; larvae robust, as figured; internal parasites
Hippoboscidae	Louse flies		Wings present or absent; flattened, with widely separated coxae; head somewhat sunken into thorax; external parasites (Fig. 13–6)

All flies listed here have short, usually 3-segmented antennae with a terminal bristle and with rare exceptions have 3 ocelli. Larvae are "maggots," with no externally visible head except for mouth hooks, and generally with 2 pairs of spiracles, 1 pair on prothorax, and 1 larger pair terminal.

Helfer, J. R. 1963. *How to Know the Grasshoppers, Cockroaches, and Their Allies*. Dubuque, Iowa: Brown. 353 pp.

Imms, A. D. 1977. *General Textbook of Entomology*. Tenth edition. Revised by O. W. Richards and R. G. Davies. Two volumes. London: Methuen.

Merritt, R. W., and K. W. Cummins, eds. 1978. *An Introduction to the Aquatic Insects of North America*. Dubuque, Iowa: Kendall Hunt. 441 pp.

Pennack, R. W. 1978. *Fresh-water Invertebrates of the United States*. New York: Wiley. 803 pp.

Slater, J. A., and R. M. Baranowski. 1978. *How to Know the True Bugs (Hemiptera—Heteroptera)*. Dubuque, Iowa: Brown. 256 pp.

Usinger, R. L., ed. 1956. *The Aquatic Insects of California*. Berkeley: University of California Press. 508 pp.

Some Other Good Books on Insects

Eisner, T., and E. O. Wilson, eds. 1977. *The Insects: Readings from Scientific American*. San Francisco: Freeman. 334 pp.

Evans, H. E. 1968. *Life on a Little-Known Planet*. New York: Dutton. 318 pp. (Reprinted in paperback by Dell, 1970, and by Dutton, 1978.)

Farb, P., and the editors of *Life*. 1962. *The Insects*. Life Nature Library. New York: Time. 191 pp.

Lanham, U. N. 1964. *The Insects*. New York: Columbia University Press. 292 pp.

Wigglesworth, V. B. 1964. *The Life of Insects*. Cleveland: World. 360 pp.

All multicellular animals must have organ systems that provide the necessities of life: energy, waste disposal, and reproduction. Among the Insecta, the organ systems that serve these needs have certain distinctive features, such as the following:

1. The cuticular exoskeleton, including its invaginations, is an integral component of many organ systems. Relatively small size is dictated by the mechanical properties of that exoskeleton.

2. The insect body plan tends to segregate physiological functions. The head is a center for ingestion and for sensory perception and integration; the thorax is devoted to locomotion; and the abdomen acts as an expandable caboose for metabolism, reproduction, and storage.

3. The organ systems that provide for respiration, circulation, and excretion are, in contrast to those of vertebrates, relatively independent of one another.

4. Insects (and other arthropods) are composed of relatively few cells. If we were to compare a 5-gram cockroach with a 5-gram mouse, we

Function and Development

P A R T

T W O

would find that the mouse has a million times more cells. Insects are cell conservatives, while in comparison vertebrates are cell spendthrifts. Mitotic divisions are much more common in vertebrates than in insects, and for analogous functions vertebrate organs contain many more cells of smaller size than do insect organs.

Our understanding of the physiology of insects is far from complete, and there are many variations on each generalized organ design. An extended discussion of that variation is beyond the scope of this part. We will be selective and consider a few examples in some detail, partly because of space limitations and also because most insect physiologists have concentrated their studies on a few well-known species in order to ask about the general mechanisms of organ function.

In Chapter 3 we shall consider major physiological systems; in Chapter 4, the endocrine system and metamorphosis; in Chapter 5, the chemical signaling systems; and in Chapter 6, the nervous system.

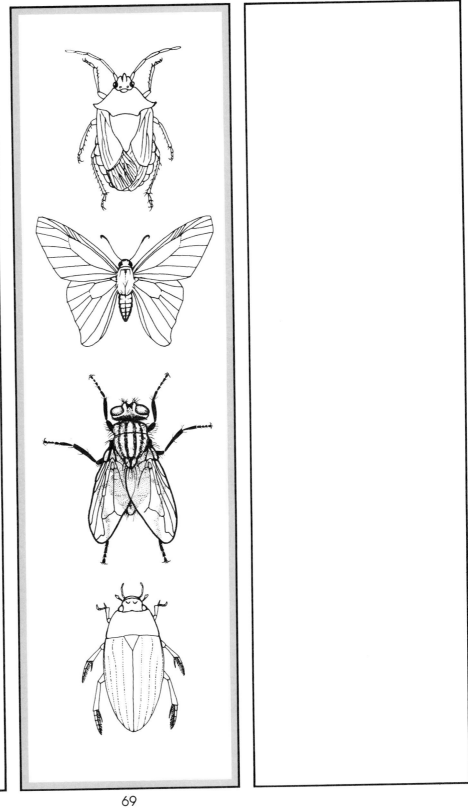

Major Life Systems

L ike animals in other phyla, arthropods move, respire, digest, excrete wastes, and conserve water. To understand the insect solution to these physiological challenges, we must begin with a dominant arthropod feature: the cuticle. The cuticle represents the interface of the insect with its environment; thus we must understand its structure and chemistry if we wish to understand how insects survive and flourish in their environment. The cuticle may be extremely hard (as in some beetles) or soft and flexible (as in caterpillars). Nevertheless its basic properties are similar throughout the insects.

The Cuticle

Let us compare the integument of a mammal with that of an insect. Mammalian skin is a mosaic of cellular remnants, whereas insect cuticle is like a tapestry—a continuous sheet of extracellular fibers. The outer portions of mammalian skin are flattened husks of dead cells. The shrunken, thickened cell membranes enclose coagulated keratins —special proteins that are insoluble because they are strongly cross-linked by sulfur bridges (Fig. 3–1a). In the innermost layers of the mammalian epidermis, there is continual cell division. The daughter cells migrate upward as they accumulate keratin; they become tightly stuck to one another and in synchrony commit cellular suicide. The dead layers move toward the surface, where eventually the cell husks are flaked off.

The surface of an insect is noncellular. The integument consists of several layers of extracellular cuticle and an underlying sheet of epidermal cells. The epidermal sheet is but one cell thick, and cell division in this monolayer is not common except for periods of growth or wound healing. Cells of the epidermal sheet export proteins, lipids, carbohydrates, and water, which are assembled to form cuticle in the space above those cells. And the insect's "skin" is also its skeleton.

Insect cuticles are diverse in their permeabilities and in their mechanical properties. Cuticle may be transparent or opaque; it may be rigid or elastic; it may be quite impermeable or may have special channels for movement of molecules through its layers.

Cuticle is a laminate, like plywood. It is divisible into two major zones: the epicuticle and the procuticle. The outer zone, the **epicuticle,** is rich in lipid and protein (Fig. 3–2). The **procuticle,** which lies beneath the epicuticle, is thicker and contains mostly carbohydrate and protein. The hard, dark outer portion of the procuticle is called the **exocuticle,** while the inner **endocuticle** is softer and lighter in color.

The sclerites, described in Chapter 1, are rigid plates with a thick layer of exocuticle. The thin, flexible intersegmental membranes may have both epicuticle and endocuticle but little or no exocuticle. Beneath the endocuticle is the tissue respon-

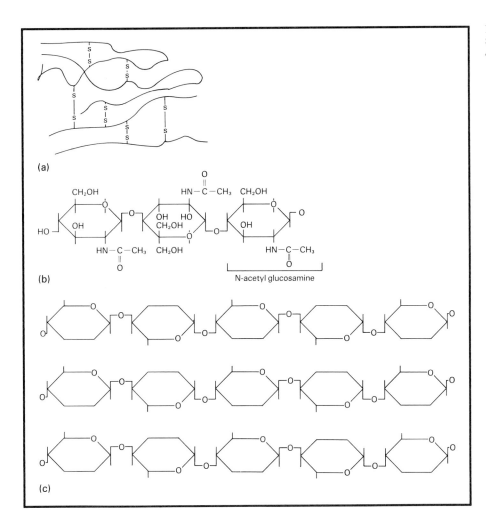

Figure 3–1
Structures of keratin (a), chitin (b), and the parallel
chains of chitin in a microfibril (c).

sible for its production, the sheet of epidermal cells. Often the epidermal cells appear
to abut directly against the inner surface of the endocuticle, but during times of molting a narrow zone of **subcuticular space** opens between the epidermal cells and the
endocuticle.

Epicuticle.   The epicuticle may be smooth or it may be sculpted into elaborate patterns. Two structural sublayers are found in all epicuticles. The **cuticulin** (Fig. 3–3)
layer is very thin (10 nm),* and it has not been possible to obtain a pure fraction of it
for detailed chemical analysis. We know only that cuticulin is tough, insoluble, and
nonelastic and that it contains cross-linked lipid and protein molecules. The thicker
(100 nm) **inner protein epicuticle** (Fig. 3–3) contains some lipid but mostly protein

---

*A nanometer (nm) is defined as one-millionth of a millimeter, or 1×10^{-9} meters.

Figure 3–2
A general diagram of the structure of the insect integument, including the cuticle, the epidermis, and the basement membrane. (After R. H. Hackman, 1971, "The integument of Arthropoda," in *Chemical Zoology*, M. Florkin and B. T. Scheer, eds., New York: Academic Press. pp. 1–62.)

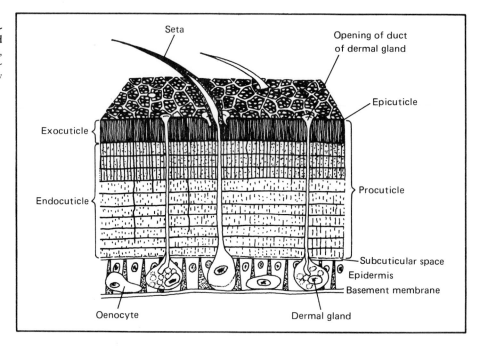

and provides mechanical support for the cuticulin. Both layers of epicuticle are thought to lack chitin, the carbohydrate that is characteristic of the procuticle.

Dense lipoprotein **epicuticular filaments** often run up from the epidermal cells, through pore canals in the procuticle, to fuse with the inner edge of the cuticulin (Fig. 3–3). These filaments and the pore canals are conduits for transport of materials from the epidermal cells to the procuticle and the epicuticle. In some cases they may transport wax to the outer surface of the animal. The amounts of wax produced are sometimes prodigious. Chemical analyses of waxes, such as beeswax from honeycombs or shellac secreted by lac insects, reveal that they are mixtures of hydrocarbons, esters, fatty acids, and other lipoid materials. These waxes are probably special cases of the surface grease that is common on the surface of insect cuticle. The greases are often soft and fluid, and they allow oil-soluble molecules, such as contact insecticides, to spread rapidly from one small spot of application and thus to coat the entire insect.

As long as the layers of epicuticle are intact, insects lose relatively little water across their body surfaces. But when the epicuticle has been abraded or dissolved away, an insect is likely to be desiccated rapidly if it is in a dry environment.

Procuticle. Procuticle contains **chitin,** proteins, and small amounts of lipid. Chitin is a colorless, insoluble structural polysaccharide, each molecule of which consists of a long chain of sugar subunits (Fig. 3–1b). The individual subunits, known as N-acetyl-glucosamines, are attached to one another by β-glycosidic linkages. A similar β-linkage is found in other structural polysaccharides with long, straight chains such as cellulose. In contrast to structural polysaccharides, food storage polysaccharides (like starch and glycogen) have α-glycosidic linkages, and the chains are much coiled to make the overall molecule more compact.

Chitin chains are clumped in microfibrils (Fig. 3–1c), and at any one level within the cuticle, the microfibrils are aligned parallel to one another. The parallel microfibrils are only loosely bound side to side, and thus they can easily be pried apart by stress *across* their long axis, but they are very resistant to stretch *along* their long axis. In each patch of insect cuticle, the microfibril orientation compensates for stress forces. In apodemes (sites of muscle attachment) the stress is in but one direction, and all the microfibrils throughout the thickness of the apodeme are aligned in the same direction. This increases the effectiveness of apodemes as rigid anchors against which muscles exert their force. In the abdominal terga of a cockroach, the stresses may come from many directions, and the orientation of microfibrils once again resists the stress. In these terga the microfibrils are arranged in layers, applied one after another to the inner surface of the endocuticle. Their long axes are parallel to the surface of the cuticle, but they shift slightly with each new layer deposited. Over many layers the ordered orientation rotates around a circle. In electron micrographs, procuticle appears to consist of alternating bands, called laminae, each 0.5–2.0 μm* in thickness, which reflect this complete rotation (Fig. 3–3). At high magnifications fine, dense lines can be seen to radiate out, like parabolic brush strokes, into intervening, lightly staining zones. The parabolic pattern is a striking artifact resulting from the shifting orientations of microfibrils, which appear darkest when viewed from the side and lightest when viewed end on. Since the axes of the microfibrils continue to rotate in successive laminae, the resistance to stretch is high in all directions within the plane of the cuticle.

---

*A micrometer (μm) is one-thousandth of a millimeter, or 1×10^{-6} meters.

Figure 3–3
Transmission electron micrograph of the larval silk moth, showing the cuticulin, dense inner epicuticle, epicuticular filaments, and several lamellae of the procuticle (a), and detail of the pattern of microfibrils in the procuticle (b). (B. K. Filshie, Commonwealth Scientific and Industrial Research Organization, Canberra, Australia.)

(a)   (b)

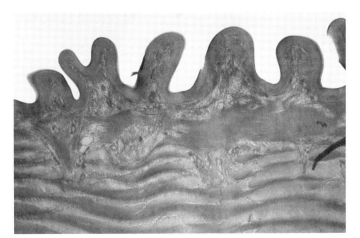
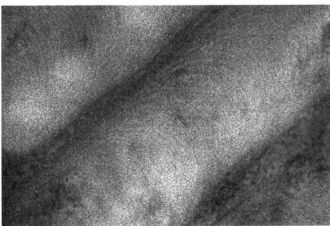

Although cuticle has only one major kind of structural polysaccharide, it has many kinds of proteins. No one knows how many, and study of these proteins has been very frustrating. The protein chains run between, around, and across the chitin microfibrils. Some proteins are loosely bound to chitin or to other proteins by salt or hydrogen bonds, while almost half the proteins in many cuticles are held in place by strong covalent bonds. Biochemists cannot extract these proteins without breaking many covalent bonds, and the proteins in the extraction mixture are heterogeneous in size and in enzymatic activity. In most cases it is not possible to be sure which of the protein components are normal cuticular constituents and which are their degradation products. But one result is consistent: Sulfur-containing amino acids are very rare, so we know that most of these proteins are very unlike vertebrate keratins.

Few proteins from cuticle have been obtained in a form pure enough that they can be rigorously studied. There is one interesting exception, **resilin,** an unusual natural rubber. In the late 1950s, the late Torkel Weis-Fogh, of Cambridge University, set out to analyze the mechanical properties of the "flight motor" in the thorax of *Schistocerca gregaria*, the desert locust. In the locust, as in other insects with indirect flight muscles, the thorax is deformed as the wings flap up and down. Weis-Fogh used a simple preparation: He cut off the locust's head and abdomen and mounted the thorax on a pedestal so that the wings stuck out to the side (Fig. 3–4). Weis-Fogh raised the wings and

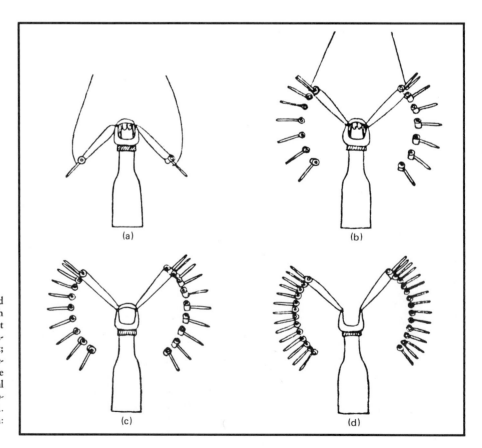

Figure 3–4
The elastic recoil of the fore wings in the isolated thorax of a desert locust drawn from multiple-flash photographs. Wider spacing between subsequent exposures indicates faster movement. (a) Unstrained, intact thorax; (b) recoil in intact thorax; (c) recoil after removal of the wing muscles; (d) recoil of the rubberlike wing hinges. The wings were mutilated and provided with an artificial mass. (After T. Weis-Fogh, 1961, "Power in flapping flight," in *The Cell and the Organism*, J. A. Ramsay and V. B. Wigglesworth, eds. London: Cambridge University Press, pp. 283–300.)

then released them; they promptly snapped back down to the equilibrium position. At first he assumed that the energy responsible for the elastic recoil was stored by deformation of the thoracic box itself. However, his meticulous analysis revealed that only about 70% of the energy was stored in the thoracic box, and that the remaining 30% was stored in the hinge of cuticle connecting the wing to the thorax. The hinge lacks chitin and is almost pure protein. Weis-Fogh and his co-workers isolated the natural rubber from the hinge and named it resilin.

When a force pulls on cuticle that contains resilin, the cuticle stretches; when the force abates, the cuticle retracts to its original shape and size. Weis-Fogh then asked himself: What kind of a molecular arrangement could account for such efficient elastic recoil? His solution was quite straightforward. Resilin forms a kinky network. Resilin, like most polypeptide chains, twists about in space and has a preferred three-dimensional conformation of lowest energy. At their configurations of lowest energy when the cuticle is not stretched, resilin chains are kinked and twisted. When stress is applied, the kinks are pulled out, and the stretched resilin chains store the energy, much as does a stretched rubber band. The energy could be lost if the strained protein molecules could slowly relax and thus reacquire their kinks by flowing past one another. But resilin does not relax in that manner. Like the rubber tires of the 1940s, resilin cuticle is "vulcanized" by cross-linking molecules (Fig. 3–5). The cross-linking takes place as the cuticle is laid down. The protein chains cannot slip past one another because they are periodically linked side to side in a three-dimensional network. Resilin releases 90% of the energy invested to stretch it; in this sense, it is the most efficient rubber found in nature.

Resilin is not unique to the locust wing hinge, but it is never found in large amounts. It is restricted to small patches, such as those in leg hinges of crayfish and flight muscle tendons of flies and dragonflies. The jumping of fleas is enhanced by a pocket of resilin at the base of each hind leg. When a flea is about to jump, the hind legs are "cocked" by pulling them up so that they rest against a ridge on the body. This compresses the resilin. When the leg is suddenly extended, it snaps from the catch and at the same time permits the resilin to expand, providing the extra bounce.

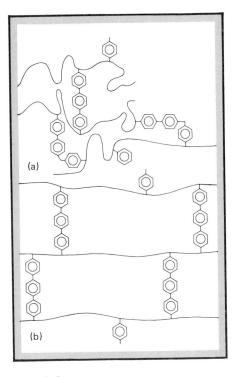

Figure 3–5
A diagram of a portion of a relaxed (a) and a stretched (b) portion of the resilin chains in cuticle.

Mechanisms of Hardening and Darkening.   If you have watched a moth emerge from its pupal cocoon and expand its wings, you will remember that at first its cuticle was pale and soft. Within an hour or two, the cuticle became much darker and harder as it was stabilized. This process of stabilization is called tanning, or **sclerotization.** The tanning reactions take place in relatively complex, structured mixtures of proteins, chitin, and lipids. The cuticle is stabilized because proteins become tightly cross-linked to one another. The cross-linking agents are called **quinones.**

Quinones are very strong oxidizing agents that readily form covalent bonds with any electron-rich group, such as the amino groups of proteins. By a multistep process, a quinone can cross-link two protein chains. Free quinone molecules oxidize aromatic amino acids in cuticular proteins, which then form cross-links, such that a rigid three-dimensional network of structural proteins is formed within minutes. The resulting cuticle is hard and inelastic. Quinones also react with each other and with aromatic amino acids to form the black pigment melanin. Thus quinone production leads to both darkening and hardening.

The tanning quinones are so reactive as to be unstable. They are generated just before they are used by enzymes (diphenol oxidases that are part of the structural proteins of the outer procuticle. The substrates for these enzymes are diphenols that come from the blood, pass through the epidermal cells, and then diffuse up through the cuticle to reach the oxidative enzymes. Once produced, the tanning quinones diffuse away into the surrounding cuticle and react with proteins and with each other. The hardness and darkness of the cuticle will vary with the amounts of quinones generated and delivered to the proteins.

Tanned cuticular protein is called **sclerotin.** In fact, many different kinds of proteins are tanned when cuticle is sclerotized, so that sclerotin is a mixture of proteins. Sclerotized cuticle is irreversibly hardened. The insect cannot retrieve the protein or chitin from it. Thus the rigid sclerotized blocks of cuticle must be shed to permit growth.

The inner portion of the procuticle, the endocuticle, is untanned. Since proteins and chitin of the endocuticle are not tightly cross-linked, endocuticular layers can be resorbed before exocuticle is shed or when the insect needs the energy reserves stored in that chitin and protein.

The Epidermis and the Basement Membrane.

The insect epidermis (Fig. 3–2) is a sheet of cells that communicate with one another to allow metabolites to surge back and forth from one cell to the next and thus ensure coordination of the cells within a neighborhood. The epidermis forms a continuous layer that seals a large compartment, the hemocoel, from a small one, the subcuticular space. When other cells breach this sheet as they run to the surface (gland ducts or nerve processes), the epidermal cells form tight collars around the intruding cell to maintain the integrity of the two compartments.

The epidermal cells are polarized for export of materials to the outside. The microvilli project from this apical surface toward the cuticle, and sometimes narrow outgrowths of the cells penetrate far into the procuticle within special **pore canals** (Fig. 3–2). At the basal surfaces (toward the hemocoel), the plasma membrane is often infolded for absorption of materials from the hemolymph. Between the epidermal cells and the hemolymph is the **basement membrane**—a noncellular sheath that is formed mostly by secretion from hemocytes (blood cells).

Respiration

Insects obtain energy from foodstuffs by oxidizing the sugars, lipids, and amino acids to produce adenosine triphosphate (ATP). In aerobic animals, respiration uses oxygen as a final dumping site for the low-energy electrons from foodstuffs and gives off carbon dioxide as a waste product. Transport of oxygen and carbon dioxide is required for an active lifestyle.

Basic physical laws govern the movements of oxygen and carbon dioxide between tissues and the atmosphere. We need to review these processes in order to understand insect respiration. Net transfer of each gas occurs only in one direction, down its particular pressure or concentration gradient. **Gas exchange** between two compartments, the animal and its environment, depends on the area of the respiratory surface. **Gas transport** within an animal can take place by diffusion through the tissue fluids, but simple diffusion is efficient only for short distances; the maximum possible thickness for an animal that depends on diffusion alone is less than 1 mm. Larger animals have special adaptations to facilitate both gas exchange and gas transport. Insects take advantage of the fact that diffusion of oxygen through air is some 3 million times faster than through water or through tissue fluids, and thus efficient oxygen transport can be achieved by running a gas-filled tube from the surface to each metabolically active tissue. The tracheal system is this system of conducting tubes.

In a fresh dissection of an insect, the **tracheal system** is a mass of silvery white tubes and their finer branches that fill the inside of the animal (Fig. 3–6). The major trunks arise from spiracles that open to the outside at the margin of abdominal and thoracic segments (Figs. 1–6, 1–13). Gases diffuse in and out of the spiracles and into the major trunks and air sacs. Each trachea has a relatively simple epicuticle composed of only cuticulin and inner epicuticle. Tracheae have no wax layer, no pore canals, and no lipid filaments. The endocuticle has a regular set of thickened ridges termed **taenidia,** which run around the ducts something like the corrugations in a highway conduit (Fig. 3–7). The corrugations of the tracheal cuticle make the tracheal trunks and air-filled sacs resistant to collapse. In addition, the chitin microfibrils are oriented for mechanical strength. They are tangentially disposed in the ridges to resist lateral compression and longitudinally oriented in the valleys to resist overextension.

The site of gas exchange between the tracheal system and the tissue fluids is the **tracheole**—the finest tubular element. Tracheae run in from the spiracles, continue to divide, and taper until they are only about 1.5–2.0 μm in diameter. Then each fine tracheal branch connects with a small radiation of even finer tubes, the tracheoles (Fig. 3–7), which run deep into muscles, glands, and other tissues. Each tracheolar radiation is enclosed within a single cell, the tracheolar end cell, and this cell ensheaths the tubular processes as they penetrate tissues.

The ends of the tracheoles usually contain fluid. In times of increased metabolic demands, the fluid within the tracheoles is somehow withdrawn so that the air-filled

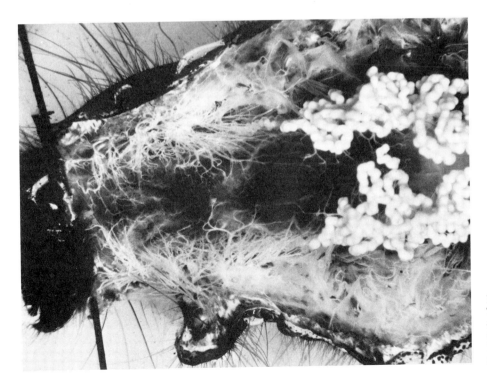

Figure 3–6
The tracheal system in the abdomen of a tent caterpillar. Note the many fine tracheal branches that convey oxygen to the organs. (G. M. Happ, 1973, from *Elements of Biology*, by C. K. Levy, Addison-Wesley.)

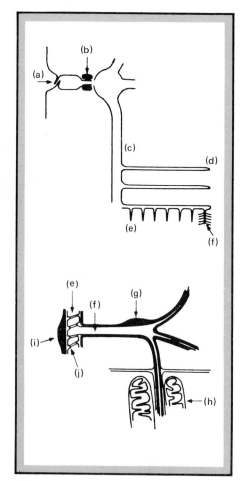

Figure 3–7
Diagram of the tracheal system of an insect. (a) Spiracle; (b) spiracular valve; (c) primary trachea; (d) secondary trachea; (e) tertiary trachea; (f) tracheole; (g) tracheolar end cell; (h) mitochondrion; (i) tracheal epithelium; (j) taenidium. (With permission from E. Bursell, 1970, *An Introduction to Insect Physiology.* Copyright: Academic Press, Inc. [London] Ltd.)

columns come close to the tissue. This withdrawal can be seen in a very simple, dramatic experiment, such as that reported by V. B. Wigglesworth, of Cambridge University, England, in 1931. Wigglesworth examined living mealworm larvae under transmitted light in a microscope and was able to focus carefully on the tracheal network over the surface of the gut. Tracheoles are visible as black threads when they contain air and are invisible when they contain fluid. When the mealworms were asphyxiated by being immersed in water, the system of black threads rapidly enlarged as fluid was withdrawn from the tracheoles, and the air-filled spaces approached the metabolically active tissues. To function in this way, tracheolar cuticle must be permeable to oxygen and carbon dioxide (to allow gas exchange) and to aqueous fluids (to allow metabolic withdrawal).

The tracheal system provides many sites for water loss, but desiccation is forestalled by a sphincter associated with each spiracle. In many insects the passive elastic properties of a cuticular valve hold the sphincter closed. When the spiracle is closed, little water is lost, but over time the oxygen is depleted and carbon dioxide accumulates within the closed tracheal system. The spiracles are then opened briefly by muscle activity to allow gas exchange. The flea, a small insect whose respiration was studied by V. B. Wigglesworth, illustrates the phenomenon well. The flea has two thoracic spiracles and eight abdominal ones, as do most insects. At rest, most of the spiracles are closed except for the first and last abdominal pairs, which are rhythmically opened and closed every 5–10 seconds. When there are high metabolic demands, all spiracles open and close rhythmically. For a small insect diffusion alone will allow enough oxygen to reach the tissues and will permit the elimination of carbon dioxide. But for larger insects more active ventilation is necessary.

Mechanical pumping is commonly used to increase air flow in large insects, such as locusts, cockroaches, and dragonflies. In a resting insect the pumping depends not on the thorax but on the abdomen, and air flow is from front to back. During inspiration (breathing in), the thoracic spiracles are opened and the abdominal ones are closed while the abdomen gradually increases in volume; during expiration (breathing out), the thoracic spiracles are closed while the abdominal ones are open and the abdomen is compressed to force the air out.

It costs a lot of energy to fly. The oxygen consumption during flight is 10- to 100-fold over resting levels, and many insects can change from a flying to a nonflying state and back again very quickly. There must be very efficient respiratory adaptations. In locusts Weis-Fogh has shown that the thorax is alternatively compressed and expanded as the flight muscles work. This change in thoracic volume pumps air in and out of the tracheal trunks and air sacs; thus the actions of the flight muscles are responsible for providing oxygen for those same muscles. The flight muscles are coated with air sacs and riddled with tracheoles so that no mitochondrion is more than 5 μm from a tracheole. The oxygen is delivered so fast that flight muscles can achieve the highest metabolic rates known for any animal tissue.

Respiration in Aquatic Insects.

Aquatic insects must also obtain oxygen. They face additional problems since oxygenated water contains only one-thirtieth as much oxygen as does air at the same pressure. When most aquatic insects are submerged, their tracheal systems are still filled with air that has oxygen and nitrogen at the same gas pressure as in the surrounding water. The spiracles exclude water because of a wreath of oily, water-repellent hairs. Carbon dioxide is easily eliminated because it readily dissolves into the surrounding water. Oxygen supply is the limiting factor. Oxygen is renewed by periodic visits to the surface for a breath of new air, or it is extracted from the water by physical or tracheal gills.

Physical gills are bubbles, or pockets of air, that adhere to the animal. The bubble is exposed to the surrounding water and is continuous with the tracheal air space. Some insects, such as water boatmen and backswimmers, can remain underwater for a rather long time, long after the oxygen originally in the bubble has been used up. This is possible because the bubble contains not pure oxygen but air, which is a mixture of nitrogen and oxygen. The total pressure in the bubble is the sum of the partial pressures of oxygen and nitrogen. When the oxygen is withdrawn from the bubble to meet metabolic needs, its partial pressure falls so that more oxygen diffuses in from the surrounding air-saturated water. A little nitrogen diffuses out as the oxygen enters. The oxygen is repeatedly replenished, and simultaneously some nitrogen is lost, so that the bubble shrinks slightly. Eventually the bubble is so small that the insect must come to the surface to renew it. In 1915 R. Ege demonstrated that an insect can remain submerged longer when the bubble is air and the water is air-saturated than when the bubble is pure oxygen and the water is oxygen-saturated. A backswimmer lives only 35 minutes in oxygen-saturated water but as long as 6 hours in air-saturated water. Why is air better than pure oxygen? Because in pure oxygen the bubble is consumed without replenishment, whereas the presence of nitrogen in an air bubble allows it to function as a physical gill. To increase the rate of exchange, aquatic insects at rest actively move water across the surface of the bubble.

A **respiratory plastron** (Fig. 3–8) can allow an insect to remain submerged indefinitely. Like bubble respiration, plastron respiration depends on diffusion of oxygen from the surrounding water into an air store. The air is a thin film held within the plastron, a framework of stiff water-repellent hairs or a cuticular meshwork. The volume of air space is fixed because the hairs are stiff and do not collapse. Since the surface area of the plastron is large, gas exchange between the air-filled plastron and the water is rapid. When oxygen is used by the insect, a small drop in gas pressure (essentially a slight vacuum) occurs within the plastron, and more oxygen enters. Nitrogen plays no part in this process. Respiratory plastrons are found in some beetles and bugs that live in environments that are alternately dry and flooded. Plastrons are very widespread on insect eggs, many of which are immersed in water whenever it rains.

Other aquatic insects have a **closed tracheal** system, in which the air-filled tracheal system is sealed away from the watery environment. Gas exchange takes place across the general body surface or at special heavily tracheated flaps or filaments of the body wall, which are called **tracheal gills.** These tracheal gills may be lateral or ventral expansions of the thorax or abdomen. In dragonfly larvae, the rectum is filled with flaplike tracheal gills. Oxygen-rich water is pumped in, and once exchange has taken place, the oxygen-depleted water is pumped out. A dragonfly larva "breathes" through its anus by a self-administered enema.

A very few insects have hemoglobin in their blood plasma, serving as a respiratory pigment. Best known of these are the bloodworms (larvae of midges; Diptera, Chironomidae), which often live in water that is deficient in oxygen. Insect hemoglobin contains two heme groups, half as many as vertebrate hemoglobin, and has a molecular weight about half that of vertebrates. Bloodworms are able to live for many days in the near absence of oxygen, apparently because of the affinity of their hemoglobin for the limited oxygen available. When oxygen pressure is restored, they return to respiration via the tracheal system.

Figure 3–8
(a) The plastron of *Dicranomyia,* a crane fly larva, consists of a narrow space roofed by a thin sheet of porous cuticle and floored by heavier cuticle. The roof is supported by a series of vertical cuticular struts. (b) Egg plastron of a blow fly, showing a dorsal view of the egg (left) and the plastron network of the median area (right). (a: With permission from H. E. Hinton, 1968, *Advances in Insect Physiology,* vol. 5. Copyright: Academic Press Inc. [London] Ltd. b: Reprinted with permission from *Journal of Insect Physiology,* vol. 4, H. E. Hinton, "Plastron respiration in the eggs of blowflies." Copyright 1960, Pergamon Press, Ltd.)

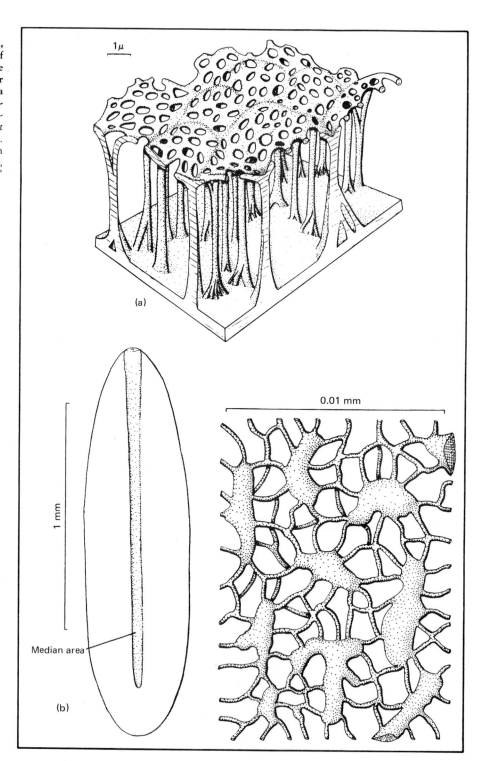

Circulatory System and Fat Body

Aside from bloodworms and a very few other insects, the blood of insects serves no function in gas exchange. It is a watery fluid, often somewhat yellowish or greenish in color, containing small molecules and proteins in addition to blood cells or **hemocytes.** To emphasize its differences from the blood of vertebrates, it is often called **hemolymph.** It is pumped and jostled about by the heart, or **dorsal vessel;** by accessory pulsatile organs; and by movements of the other organs. Unlike the familiar vertebrate system in which blood flows in an orderly fashion from the heart to periphery in the arterial system and back again in the venous system, the insect circulatory system is an open system, involving irregular spaces within which blood moves about. Blood movement depends on the dorsal vessel, suspended in the top of the hemocoelic cavity, which pumps blood forward to the head. The blood eventually returns by percolating backward, around, and through spaces in the viscera. The blood serves as a vehicle for transport of nutrients, metabolic products, and hormones.

The dorsal vessel is a muscular tube that is divisible into a posterior **heart,** which contains valves or ostia, and an anterior **aorta,** a closed tube that traverses the thorax (Fig. 1–16b). The heart is suspended from the dorsal body wall by connective tissue filaments (Fig. 3–9). Beneath the heart lies the **dorsal diaphragm,** a porous membrane attached to the lateral body wall that loosely defines a space around the heart—the **pericardial sinus.** Within the dorsal diaphragm are muscle fibers known as **alary muscles.** The muscles pull the diaphragm downward, forcing the heart to dilate and blood to flow upward into the ostia. Then a steady wave of contraction forces the blood forward through the aorta and into the head, where it is discharged.

The isolated heart (or even a fragment of it) continues to beat rhythmically; thus we know that the source of the regular heartbeat lies in the properties of the muscle itself. But the rate of heartbeat can be modified by nerves that run to the alary muscles and to the heart itself.

About 10% of the volume of the hemolymph is occupied by blood cells. Many of these cells, or hemocytes, are circulating about in the hemocoel, but many others adhere to the surfaces of tissue. Blood cells are very diverse (Fig. 3–10), but common features include fingerlike projections from the cell membranes, an absence of any surrounding basement membrane, and an abundance of enclosed vesicles and globules.

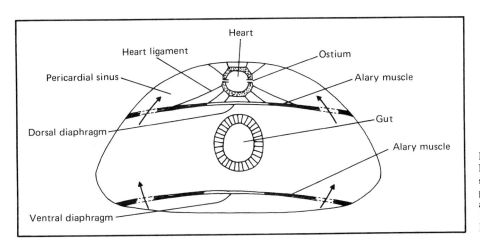

Figure 3–9
Diagram of the cross section of an insect abdomen, showing the relation of the heart, dorsal diaphragm, and ventral diaphragm to the body wall and body cavity. (Modified from J. A. Ramsay, 1952, *Physiological Approach to the Lower Animals.* London: Cambridge University Press. 150 pp.)

Figure 3–10
Pupal and adult hemocytes of a blow fly. Several cell types are shown: (a) small, apparently undifferentiated cell, in contact with surface, 72-hour pupa; (b) small, similar cell free floating, 72-hour pupa; (c) cell from late pupa, intermediate between pupal and adult types; (d) lipid containing cell, late pupa; (e) similar cell from newly emerged adult; (f) adult nongranular hemocyte; (g) same cell after it had come to rest on a surface; (h) granular hemocyte from newly emerged adult; (i) vacuolated hemocyte from newly emerged adult. (A. C. Crossley, 1964, from *Journal of Experimental Zoology*, vol. 157, with permission of Alan R. Liss, Inc.)

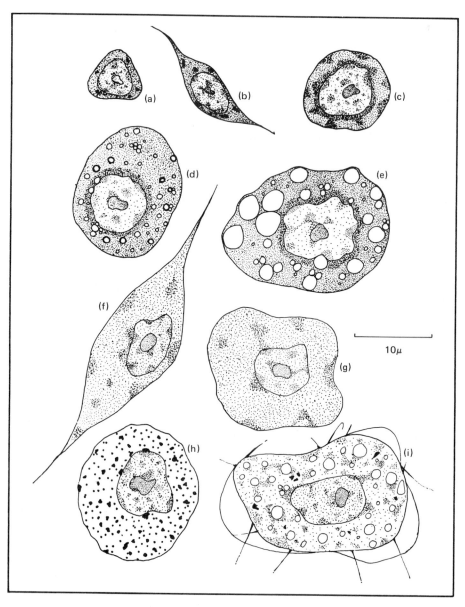

After insects are wounded, the blood clots and the clot blackens. Clotting begins when blood cells collect and exude a network of sticky threads. A variety of materials precipitate and clump about the network, and the mass darkens, owing in part to the formation of quinones that cross-link and tan the protein in the clot. As in the cuticle, quinones are formed by oxidation of phenols, and the phenol-oxidizing enzymes are common in some hemocytes. Other blood cells attach to bacteria or fungi and attempt to consume them by phagocytosis. They also, in some cases, encapsulate internal parasites (Fig. 12–14). Hemocytes are also responsible for cleaning up the residue as tissues and cells die during progressive development.

The noncellular part of hemolymph is called the **plasma.** The blood plasma contains inorganic ions (sodium and choride predominate), amino acids, monosaccharides, glycerides, metabolic wastes, hormones, proteins, lipoproteins, and many others. These plasma components are maintained at fairly constant concentrations, but they are repeatedly being exchanged for corresponding molecules in fixed cells and tissues.

The Fat Body.   In spite of the lethargy that its name might imply, the **fat body** is a highly dynamic tissue. Its cells are scattered in patches of varying sizes and rinsed constantly by the circulating hemolymph. The cells store glycogen, fats, and proteins; regulate blood sugar; metabolize sugars, lipids, and proteins; and synthesize the major blood proteins. During periods of feeding and growth, fat body cells specialize in synthesis of proteins and their secretion into the hemolymph. Some of the accumulated proteins are reabsorbed and stored during nonfeeding stages. **Vitellogenins,** female-specific proteins that are taken up by maturing oocytes, are some of those synthesized by the body fat. **Lipoprotein I,** a protein of locusts that shuttles diglycerides from one tissue to another, is another. Over the course of growth and maturation, the cells of the fat body grow, divide, and change their structure and their characteristic patterns of protein synthesis. Many of these progressive changes are regulated by the developmental hormones that will be discussed in Chapter 4. But there are also metabolic hormones, which affect short-term behavior of the fat body.

Hormones from the corpora cardiaca (Fig. 4–5) increase the amount of chemical fuel for sustained activity. Desert locusts fly for many hours. Within a few minutes after they take to the air, their fat bodies begin to release diglycerides into the hemolymph, and their flight muscle metabolism switches toward more oxidation of diglyceride lipids and less use of carbohydrates. These effects on fat body and flight muscle are brought about by the **adipokinetic hormone,** a peptide hormone released from the corpora cardiaca. Other lipid-mobilizing substances with similar properties have been reported from five orders of insects. **Trehalagon,** a hormone stored in the corpora cardiaca of cockroaches, raises the circulating level of trehalose, the major blood sugar. Recent evidence suggests that hormones very like vertebrate insulin and vertebrate glucagon are present in insects. These metabolic hormones deserve much more study than they have received up to the present.

Thermoregulation

Animals in land environments thrive within the fairly narrow temperature zone between 0° and 40°C. Body temperature has a profound effect on physiological function. Most insects that survive temperatures below 0°C do so in torpor, and those few species that are active move sluggishly. For insects at rest, body temperature approximates that of the surroundings, and metabolic rates increase markedly with rising environmental temperatures. Certain activities, such as flying or the singing of katydids, require very high metabolic rates to sustain the behavior, and the temperature of the muscles responsible determines whether or not the insects can fly or can sing. For very small insects, such as midges and fruit flies, wingbeat frequency and flight speed are limited by the environmental temperature. Although the flight muscles of these tiny insects produce heat, their mass is so small relative to the body surface that heat is lost almost as fast as it is produced. Larger insects generate heat faster than they lose it, and thus large flying insects are generally warmer than their surroundings. In many species, the act of flying helps to provide the temperature necessary to continue flying.

Insects achieve thoracic temperatures sufficient for takeoff by both behavioral and physiological means. The physiological mechanisms for heat gain all depend on the heat output from a metabolic furnace, the flight muscles. Within the last few years, the mechanisms of heat production and regulation have been investigated by a number of laboratories, but particularly those of Bernd Heinrich at the University of Vermont and George Bartholomew at the University of California, Los Angeles. When ambient temperature is low, large beetles, locusts, bees, wasps, butterflies, moths, and flies go through a warm-up phase. During warm-up, the insects shiver. They rapidly contract their flight muscles to generate enough heat to raise the thoracic temperature and thus to raise the rate of muscle contraction so that flight is possible. In some species the wings flutter and quiver somewhat during the process, but in most insects the flight muscles are "disengaged" from the wings so that the flight motor runs fast without driving its normal mechanical load.

Heinrich and Bartholomew began their investigations with *Manduca sexta*, a nocturnal hawk moth that weighs 1–2 grams and supports that weight in flight at wingbeat frequencies of 25–30 beats per second. In order to fly, the moth must have a thoracic temperature of about 38°C, but it can fly well at an environmental temperature of 15°C after warm-up. The scaly moths are well insulated, and their metabolic rate remains high and constant in flight. We might expect their thoracic temperature would rise as the environment warms up. Yet when Heinrich captured moths that were flying at ambient temperatures between 15° and 30°C and quickly stabbed them in the thorax with a thermocouple, the thoracic temperature was always about 40°C. Metabolic rates and thoracic temperatures were constant over the wide range of environmental temperatures. The moths are effectively "warm-blooded," increasing their body temperatures by the heat of metabolism, and even more interesting, dumping excess heat so that body temperatures do not get too high.

Moths control their thoracic temperatures by exploiting the fact that the insect body has three major regions. When thoracic temperatures rise beyond a set-point, moths dump heat into the abdomen. During preflight the thoracic temperature rises rapidly, while the abdominal temperature remains near that of the environment. During flight at low temperatures, the abdomen remains much cooler (up to 20°C cooler) than the heavily insulated thorax. But at high ambient temperatures, the temperature of the lightly insulated abdomen rises, and heat is lost into the surroundings (by radiation) and into the air that passes through the large abdominal air sacs. Heat is transferred from thorax to abdomen by movement of warm blood. The heart pumps cool blood forward from the abdomen; the blood in the dorsal vessel picks up heat as it passes through the thoracic muscle mass; and finally the blood moves backward to the abdomen. Its movement backward is accelerated by the ventral diaphragm (Fig. 3–11), a membrane with intrinsic muscle fibers similar to the dorsal diaphragm that we have already discussed. The rate of heartbeat regulates the rate of heat transfer. Heartbeats rise sharply as thoracic temperature rises above 40°C.

Bumble bees use a similar control strategy. By balancing shivering and blood flow, they regulate their body temperatures when flying and when walking on flowers. Bumble bees even warm their young. Within the nest, incubating queen bumble bees shiver and transfer the heat to the abdomen in order to maintain high temperatures in the brood clump (Fig. 3–12). Honey bees also use hot blood to transfer heat. But the

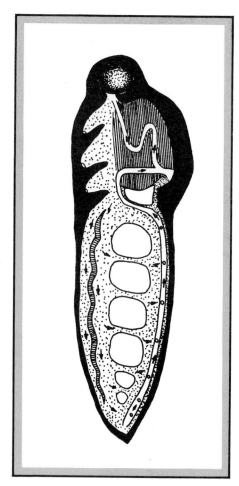

Figure 3–11
Circulatory system of a sphinx moth. The dorsal heart pumps blood forward from the abdomen and through the thoracic musculature. The blood flows backward into the abdomen with the help of the ventral diaphragm. If the thorax becomes too warm, more warm blood is pumped back to the abdomen. Heat is lost to the environment from the abdomen. (B. Heinrich, 1971, from *Journal of Experimental Biology*, vol. 54, with permission of The Company of Biologists, Ltd.)

honey bee dumps excess heat *forward* into the head, where it is lost by evaporative cooling due to the vaporization of regurgitated crop contents. As Heinrich describes it, "honey bees keep a cool head."

Digestion and Absorption

Almost every terrestrial or freshwater animal and plant is food for some insect. Food is processed in the gut, a linear sequence of special-purpose segments (Fig. 1–16a and 3–13). Food enters the **buccal cavity** and is sucked backward by the muscular **pharynx.** The blood-sucking bug *Rhodnius* ingests many times its original weight within a feeding time of 15 minutes. To accomplish this feat, the pharynx generates a suction pressure of 3–6 atmospheres so that the blood rushes through the stylet food canal (which is only 10 μm in diameter) at a velocity of 4 meters/second. The blood and salivary

Figure 3–12
Bumble bee queen incubating her initial brood clump. Her dorsal surface is heavily insulated, but the ventral side of the abdomen, near the brood, is lightly insulated. Heat is produced in the thorax and flows into the brood by way of the abdomen. The brood temperature was maintained between 24°C and 34°C, while the environmental temperatures ranged from 3°C to 33°C. (Photograph by B. Heinrich, University of Vermont.)

Figure 3–13
Diagram of the gut and of solute and fluid movement in the midgut. Sodium is actively transported from the blood into the posterior region of the midgut, and water passively follows. The water flows forward, counter to the movement of the food particles (open circles). In the anterior portion of the midgut and in the gastric caeca, sodium is actively reabsorbed into the blood. Water and solutes passively follow the sodium. (Modified from M. J. Berridge, 1970, "A structural analysis of intestinal absorption," in *Insect Ultrastructure*, A. C. Neville, ed., *Symposia of the Royal Entomological Society of London*, no. 5, pp. 135–151. By permission of Blackwell Scientific Publications Ltd. and the author.)

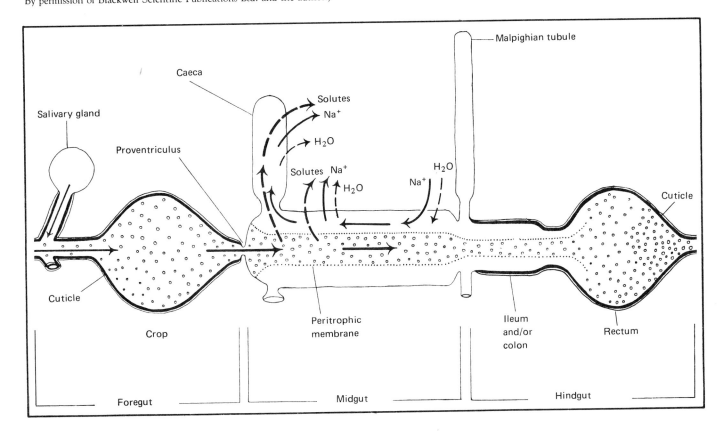

Figure 3–14
Isolated peritrophic membrane of a mayfly (*Cloeon*
***dipterum*) as seen in the electron microscope, mag-**
nified 68,000 times. Each strand of the meshwork
consists of several fine fibrils. (Werner Peters,
University of Dusseldorf, Germany.)

secretions then pass backward through the narrow **esophagus** to the crop. The crop is a
storage site. In *Rhodnius*, the huge blood meal collects there and then is slowly passed
backward into the midgut. In some ants, food can be stored in the crop for years. Between the crop and the midgut is a valve called the **proventriculus.** The proventriculus may also act to break up the crop contents, as in cockroaches (Fig. 1–16a), where it
is armed with grinding and crushing teeth.

The **midgut** is the largest surface of insect tissue that is exposed to the "outside
world" and that is not covered by cuticle. Nor is it lubricated by thick mucus secretions, as are the analogous guts of molluscs and vertebrates. However, in most insects
the solid gut contents are entrapped within a delicate tubular sheath—the **peritrophic
membrane.** This membrane is a gelatinous, cylindrical fishnet, formed of protein, chitin, and other polysaccharides (Fig. 3–14). It is believed that the peritrophic membrane protects the midgut cells from abrasion, and it may also act as an ultrafilter,
allowing digestive enzymes to enter and digestion products to exit. The peritrophic
membrane is produced from special cells at the anterior end of the midgut near the
proventriculus or from the cells of the midgut epithelium as a whole. Once produced,
the membrane moves backward with the bolus of food and is eventually eliminated in
the feces.

Absorption of the products of digestion takes place mostly in the anterior portion
of the midgut and the large outpocketings called **gastric caeca** (Fig. 3–13). It is believed that sodium is actively pumped from the midgut cells into the blood, and that
water rapidly follows. Thus the lumen of the gut is water depleted, and solutes are concentrated there, which aids absorption. At the posterior end of the midgut, water flows
in from the hemocoel so that there is a slow stream of water moving forward, counter
to the movement of food. This stream tends to sweep the products of digestion forward
from the posterior midgut to the absorptive segment at the anterior end and into the
gastric caeca.

In some insects salivary secretions allow digestion to occur outside the animal or
in the crop, but the bulk of the digestive enzymes are produced by the midgut in most
species. Even such tough materials as wood and wool are digested by some insects. In
certain cases the enzymes may be provided by symbiotic microorganisms. The best-known examples are cellulose digestion in primitive cockroaches and termites, which
is performed by protozoa that live in a special section of the hindgut.

Food debris from the midgut mixes with the secretions of the malpighian tubules
and passes backward into the hindgut. In the hindgut there is some food absorption,
but for the most part the hindgut is concerned with excretion and the regulation of
osmotic pressure.

Excretion

Concentrations of body water and of salts must be maintained fairly constant. On
land, insects are repeatedly depleted of water and threatened with desiccation. In fresh
water, there is constant danger of flooding and of outward movement of salts. In addition, toxic metabolic products, most of which contain nitrogen, must be eliminated.
The three major forms in which nitrogen is excreted are ammonia, urea, and uric acid.
Ammonia is highly toxic; uric acid is low in both solubility and toxicity; and urea is
intermediate. For terrestrial animals that lay eggs, such as birds, reptiles, and insects,
the preferred product is often uric acid. The primary insect organs that eliminate nitrogenous wastes and maintain ion balance are the **malpighian tubules** and the **rectum.**

NH_3

Ammonia

$C=O$ with NH_2 groups

Urea

Uric acid structure

Uric acid

The Malpighian Tubules.   The malpighian tubules lie in the hemocoel and are attached to the gut at the junction between the midgut and hindgut (Fig. 3–13). Each tubule is like a hollow finger that opens into the gut. Malpighian tubules absorb solutes and water from the hemolymph. As the fluid passes through the tubules and the hindgut, much of the water and solutes are retrieved back into the hemolymph. The residue is eliminated from the anus.

In vertebrates, fluids are driven into the glomerulus of the kidney by hydrostatic pressure of the blood. In the open circulatory system of insects, such sustained filtration pressure is not achieved. Rather, malphigian tubules are secretory: They actively absorb ions (principally potassium) and allow water and other solutes to enter passively.

Malpighian tubules vary in number: coccids have only a single pair, honey bees about 100, and some Orthoptera as many as 200. The elongate shape provides a large surface area for exchange between the lumen of the tubule and the hemolymph. Each tubule wall is a single layer of large cells (Fig. 3–15). Only a few (two to eight) are needed to surround the lumen. At their basal edges (toward the blood), the cell membranes of the cells are much infolded, and at their apical margin (toward the lumen) the cells bear numerous microvilli. Tracheation is usually extensive, and mitochondria are associated with both the apical and the basal membranous folds (Fig. 3–15).

In the 1950s J. A. Ramsay, of Cambridge University, performed a very careful series of investigations of malpighian tubule function. Ramsay used a fine micropipette to sample the fluid from several insect species that differed in diets and habitats. He then determined the concentrations of sodium and potassium in the fluid from the malpighian tubules and in the blood. In all the species the concentration of potassium was higher in the secretion of the malpighian tubules than in the blood. How does potassium accumulate against the concentration gradient? Could the potassium, which is a positive ion, have entered merely because the inside of the tubule was electrically negative? Ramsay measured the potential difference across the wall of the tubules in seven of the species, and he found that the inside of the tubule was electrically positive relative to the blood in most cases. Therefore potassium flows into the tubule against both the concentration gradient and the electrical gradient. Further experiments by Ramsay showed that potassium movement is correlated with the rate of secretion.

Recent studies of malpighian tubules in a variety of insects largely support these conclusions. In *Rhodnius,* a sucking bug (Hemiptera) that feeds on sodium-rich vertebrate blood, there is an active transport of sodium as well. But *Rhodnius* is of even more interest because it produces such copious amounts of urine. *Rhodnius* feeds infrequently. When the bug can have undisturbed access to a rabbit, it may take a blood meal of 10 or more times its own prefeeding weight. During feeding, which takes about 15 minutes, a hormone is released that causes the abdominal cuticle to stretch so that the crop can expand to hold the meal. Immediately after feeding, the bug begins to void urine from its anus. Simon Maddrell at Cambridge University showed that this diuresis (urine production) is under hormonal control.

Rhodnius has but four malpighian tubules, which open into an ampulla next to the rectum (Fig. 3–16). All four tubules and part of the rectum can be removed from the bug and immersed in a droplet of hemolymph or culture medium that is covered with mineral oil. The rectal cuticle is hydrophobic and spreads itself open on the oil–water boundary. As secretions form, they ooze out into the mineral oil. It is easy to measure the diameter of the droplet of the secretion and by successive measurements to calculate the rate of secretion. If the malpighian tubules and rectum are isolated a few minutes after a blood meal, the rate of secretion is high. Thereafter secretory rate falls off

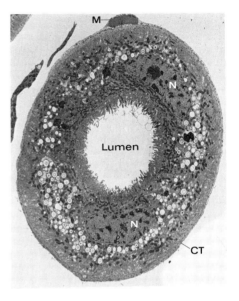

Figure 3–15
Cross section through the malpighian tubule of a cockroach. Nuclei (N) of two secretory cells are visible in this section. Note the basal infoldings toward the hemolymph and the microvilli toward the lumen. Muscle (M) is embedded in the connective tissue sheath (CT) around the outside. (B. J. Wall, J. L. Oschman, and B. A. Schmidt, 1975, from *Journal of Morphology,* vol. 146, with permission of Alan R. Liss, Inc.)

Figure 3–16
(a) The excretory system of *Rhodnius*, showing only one complete malpighian tubule. (b) The junction between the distal and proximal section of a tubule, seen from the surface and in longitudinal section. (R. H. Stobbart and J. Shaw, 1964, after V. B. Wigglesworth, from *The Physiology of Insecta*, edited by M. Rockstein; used with permission of Academic Press, Inc., and the author.)

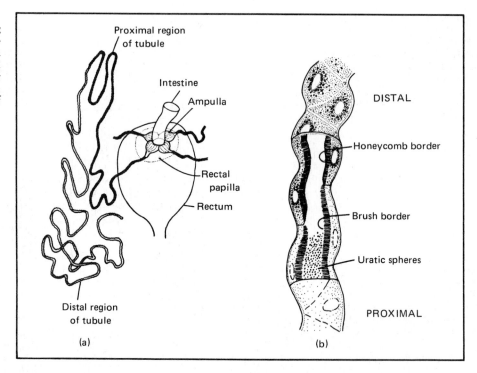

rapidly, until secretion is undetectable 150 minutes later. If hemolymph is then added from a freshly fed insect, the initial high level of activity is restored, only to decline again (Fig. 3–17).

Maddrell concluded that a blood-borne diuretic hormone is present during post-feeding diuresis and that it is rapidly degraded in the isolated preparation. Within 1 minute after the start of feeding, stretch receptors in the abdominal wall are stimulated, and they cause release of the hormone. By meticulous dissection and addition of each portion to the isolated malpighian tubules, he demonstrated that the hormone is produced by a few cells in the posterior part of the metathoracic ganglion and is passed into the blood from abdominal nerves. It increases the secretory rate of the malpighian tubules by 1000 times. Similar diuretic hormones have been reported from other blood-feeding species, such as the tsetse fly.

Once water, ions, and organic solutes have been sequestered within the malpighian tubules, the fluid begins to flow toward the rectum, and the process of reabsorption can begin. The early stages of reabsorption may occur in the proximal malpighian tubules or in the anterior hindgut. In *Rhodnius*, where the malpighian tubules open almost directly into the rectum, the tubules are divided into two distinct segments. The distal segment secretes the excretory fluid, whereas the proximal one is thought to be involved in retrieving ions and desirable metabolites back into the hemocoel. In many species where there is a distinct ileum or colon between the malpighian tubules and the rectum (Fig. 1–16a), reabsorption probably takes place in the anterior hindgut. But these segments of the gut have been little studied. It is in the rectum that the recovery of water and solutes is best understood.

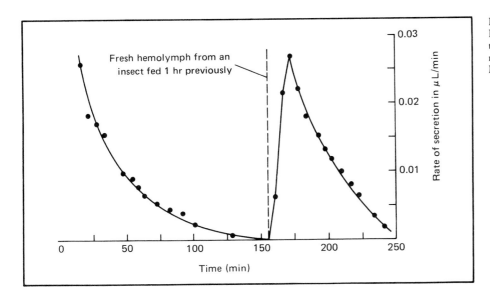

Figure 3–17
Production of secretion by the isolated malpighian tubules of *Rhodnius*. The insect finished its blood meal at zero minutes. (Modified from S. H. P. Maddrell, 1963.)

The Rectum.   Proceeding from the outside in, the heavily tracheated wall of the rectum consists of muscle, an underlying sinus, the **rectal epithelium,** and cuticle. There are typically six **rectal pads.** Each is a longitudinal infolding of cuticle containing thickened patches of epithelium and many tracheal branches (Fig. 3–18).

The functioning of the rectum is well illustrated by the studies of J. E. Phillips, of the University of British Columbia, on the locust *Schistocerca*. In this species almost all the fluid output of the malpighian tubules is reabsorbed in the rectum, and fluid is lost from the anus only when feces are eliminated. Phillips immobilized each locust and used human hair to ligate the rectum at its anterior end. Thus the rectum was isolated from the rest of the gut but bathed in hemolymph, with its tracheation intact. Phillips inserted a micropipette through the anus and glued it in place so that he could give the locust an enema, add fluid of known composition, and take samples of rectal contents. In an early experiment he introduced a pure sugar solution that was slightly hypertonic to the blood. The rectal contents became increasingly more concentrated, and after 18 hours almost all fluid was gone. Only a small amount of pasty material could be seen in the folds between the rectal pads.

Hemolymph composition remains constant over a wide range of diets and hydration stress. The primary site for the regulation of hemolymph water content is probably the rectum. Solutes, such as chloride, sodium, potassium, phosphate, amino acids, acetate, and probably calcium and magnesium, are actively absorbed from rectal fluids. Scattered evidence suggests that some of these rectal processes are under hormonal control. Hormones that inhibit or accelerate rectal diuresis have been found in extracts of the neuroendocrine centers of several species. In locusts Phillips has partially purified a peptide hormone that stimulates chloride transport. In many insects there are neurosecretory endings of unknown function within the rectal pads. The chemical nature and mechanism of action of these homeostatic hormones are unknown.

As the fluids flow from distal malpighian tubules to the rectum, they undergo acidification. Acidification may be due either to transport of hydrogen ions into the lumen or to removal of bicarbonate ions from the lumen and has a very important con-

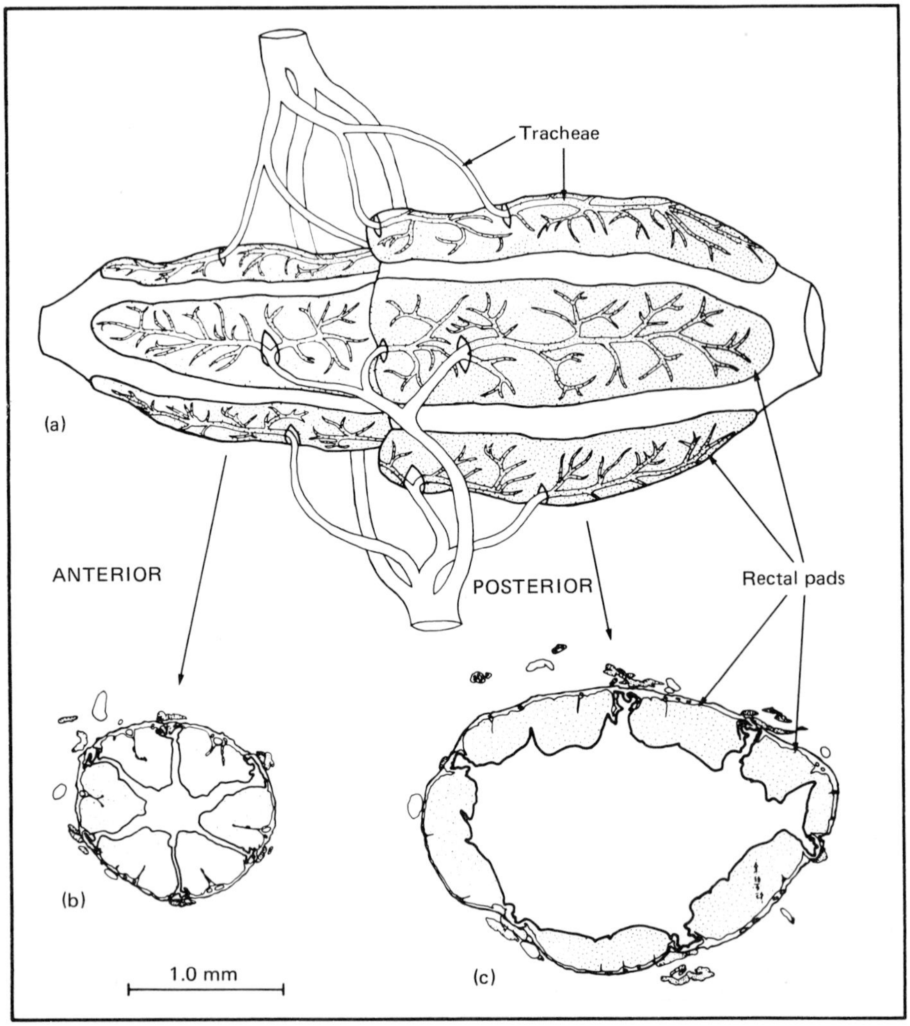

Figure 3–18
The rectum of the American cockroach. (a) The entire rectum; (b) a section through the anterior region; (c) a section of the posterior region. The rectal pads are stippled. Two main tracheal trunks send numerous branches over and into the rectal pads. (Oschman and Wall, 1969, from *Journal of Morphology,* vol. 127, with permission of Alan R. Liss, Inc.)

sequence for nitrogen excretion: It causes a change in the solubility of uric acid. Uric acid is more soluble at high pH's than at low pH's; thus it precipitates as the pH declines during passage through the hindgut. Excretion of both nitrogenous wastes and salts requires very little accompanying water.

Insects regulate their water and ion balance with other organs as well, among them the salivary glands, the midgut, and special structures like anal papillae of mosquito larvae. Each of these organs has elaborate specializations to permit active transport of solutes, and we expect that many of these processes are under hormonal control.

Among the most dramatic adaptations to survive extreme water stress is the ability of certain fly larvae to tolerate desiccation. These larvae, found in northern Nigeria and Uganda, live in temporary pools that form only in rainy seasons. When the pools dry up, the larvae become torpid and then lose up to 90% or more of their body water.

The late Howard Hinton, of Bristol University, England, heated some dry larvae to 60°C and cooled others to −270°C (liquid helium) and then rehydrated them at room temperature. The majority survived. A small percentage survived heating to the temperature of boiling water! Dried larvae were stored for years at low humidities, and upon rehydration, some metamorphosed to normal adults.

The primary factor that allowed the Insecta to thrive in terrestrial environments is their capacity to avoid desiccation. It is hardly surprising that the general mechanisms are efficient, and among the over 1 million species, there are a host of special devices that we cannot cover here. Their function and regulation deserve much more study.

Summary

Respiration, digestion, excretion, water conservation, and other physiological processes involve special modifications of the insect cuticle. Cuticle is composed of several layers, all of which contain protein. The epicuticle also has large amounts of lipids. The epicuticle consists of the outer cuticulin and the inner protein epicuticle. Beneath the epicuticle is the procuticle, which contains chitin (a large structural polysaccharide) in addition to protein and some lipids. The outer portion of the procuticle is quinone tanned and is called the exocuticle. Proteins in the exocuticle are cross-linked to form rigid sclerites. The untanned inner layers of the procuticle comprise the endocuticle.

Oxygen for respiration enters via a system of air-filled cuticular invaginations that form the tracheal system. Each tracheal trunk contains taenidia that prevent its collapse. The finest tracheae connect with tracheoles that allow the oxygen to diffuse through a watery medium to reach the tissues. Carbon dioxide is lost by the same route. The tracheal system opens to the atmosphere through spiracles. In small insects, gases enter and leave purely by diffusion, but in larger insects, the tracheae are mechanically pumped so their gases surge in and out. Aquatic insects use physical gills, respiratory plastrons, or tracheal gills to extract oxygen from their watery environment.

Insect blood (hemolymph) is pumped forward by the dorsal vessel and moves backward from the head through the spaces between the tissues. The hemolymph contains a variety of blood cells in addition to small metabolites and proteins. The fat body cells are closely associated with the hemocytes. The fat body stores protein, sugar, and lipid and proteins that are used by other tissues (vitellogenins) or that transport small molecules through the hemolymph (lipoprotein I). Hormones from the corpora cardiaca control the metabolism of the fat body.

Large insects use the heat produced by their flight muscles to increase muscle efficiency and thus to allow flight in cool environments. At low ambient temperatures muscles are warmed by shivering until they achieve a sufficient temperature to support flight. As excess heat is produced in the thorax, it is carried by warm blood into the head (where honey bees lose heat by evaporative cooling) or into the abdomen (where moths and bumble bees lose heat by radiation and expiration of warm air).

Food is pumped backward by the pharynx through the esophagus and into the crop. From the crop, food moves through the proventiculus into the midgut, where the major events of digestion take place. In most insects the food particles are confined within the peritrophic membrane. Digestion products are absorbed largely in the anterior midgut.

Insects must conserve water while they eliminate metabolic wastes. The bulk of the waste nitrogen is in uric acid. Potassium is secreted from the hemolymph into the malpighian tubules, and water follows. As the fluid flows backward, ions, metabolites, and water are reabsorbed. The rectum is a major site of water and solute recovery. Both the malpighian tubules and the rectum are under hormonal control.

Selected Readings

Anderson, S. O. 1979. "Biochemistry of insect cuticle." *Annual Review of Entomology*, vol. 24, pp. 29–61.

Burnett, A. L., and T. Eisner. 1964. *Animal Adaptation*. New York: Holt, Rinehart and Winston. 136 pp.

Bursell, E. 1970. *An Introduction to Insect Physiology*. New York: Academic Press. 276 pp.

Heinrich, B. 1979. *Bumblebee Economics*. Cambridge, Mass.: Harvard University Press. 245 pp.

———, ed. 1981. *Insect Thermoregulation*. New York: Wiley. 328 pp.

———, and G. A. Bartholomew. 1972. "Temperature control in flying moths." *Scientific American*, vol. 226, no. 6, pp. 70–77.

Hepburn, H. R., ed. 1976. *The Insect Integument*. Amsterdam: Elsevier. 571 pp.

Maddrell, S. H. P. 1981. "The functional design of the insect excretory system." *Journal of Experimental Biology*, vol. 90, pp. 1–15.

Mordue, W.; G. J. Goldsworthy; J. Brady; and W. M. Blaney. 1980. *Insect Physiology*. New York: Wiley. 108 pp.

Neville, A. C. 1975. *Biology of the Arthropod Cuticle*. New York: Springer-Verlag. 448 pp.

Wall, B. J.; J. L. Oschman; and B. A. Schmidt. 1975. "Morphology and function of malpighian tubules and associated structures in the cockroach *Periplaneta americana*." *Journal of Morphology*, vol. 146, pp. 265–306.

Wigglesworth, V. B. 1972. *The Principles of Insect Physiology*. Seventh edition. London: Chapman and Hall. 827 pp.

Development and Reproduction

evelopment includes both growth in size and changes in form. As an insect develops, the most dramatic increases in body size and the major changes in form take place when the insect molts. The study of insect development is particularly concerned with the preparations for each molt and the hormonal regulation of molting. Reproductive function is also regulated by hormones and therefore falls within the scope of this chapter.

The Events in Development

The development of every insect involves three major stages: the embryo, the immature, and the adult. Embryonic development occurs within the egg, which is well supplied with yolk and surrounded by a delicate outer shell, or **chorion.** Following hatching, or **eclosion,** the insect feeds and grows, molting several times until the adult reproductive stage is reached. In insects with complete metamorphosis, emergence of the adult from the pupa is often also referred to as eclosion. Before we examine the endocrine control of development, let us briefly review the stages in development.

Embryonic development begins after fertilization as the zygote nucleus passes to the center of the yolk mass and begins to divide (Fig. 4–1). In *Drosophila*, there are 12 to 13 successive division cycles, each of which takes between 5 and 10 minutes, eventually yielding about 5000 **cleavage nuclei.** Each cleavage nucleus migrates out to the surface of the egg, hence the term **superficial cleavage.** Cell membranes form to segregate each nucleus within a distinct cell. These cells comprise the **blastoderm,** a thin cellular monolayer that surrounds the yolk. A portion of the blastoderm thickens and becomes the **germ band.** In most insect species the germ band then develops an infolding, which folds down into the yolk to produce a two-layered embryo. **Mesodermal** structures (muscles, fat body, hemocytes, gonads) arise from the inner layer, while the outer layer is transformed into the **ectodermal** structures, including the foregut and the hindgut. The midgut (**endoderm**) grows inward from rudiments at either or both ends of the body (Fig. 4–1). The germ band grows and continues to develop as it sinks into the yolk. The embryo becomes segmented, and the major organ systems are formed. In successive stages many of the segments fuse, and their boundaries become obscured in such regions as the head. The embryo goes through several molts, and finally an immature insect hatches from the egg.

Metamorphosis.   The postembryonic life of an insect is divided into **instars.** Insects progress from one instar to the next by making a new cuticle and shedding the old one. The hard exoskeleton requires molting to allow growth, and successive instars differ from each other in form and size. Such changes of form are termed **metamorphosis,** and metamorphic development is characteristic of arthropods. It is a vivid illustration of the basic biological phenomenon of differential gene action in development.

Figure 4–1
Embryonic development in an insect. (a) Diagrams illustrating the formation of the blastoderm. (b) Diagrams illustrating the development of the midgut. (R. F. Chapman, 1969, reproduced by permission of the publishers, Hodder and Stoughton Ltd., from *The Insects: Structure and Function.*)

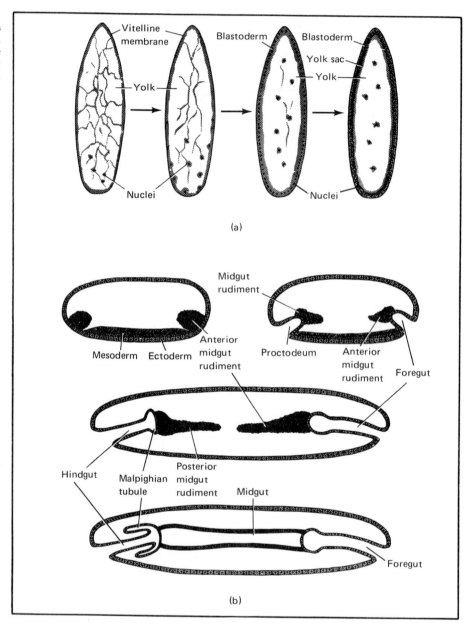

As every student has learned in general biology, not all genes are expressed simultaneously. As the development of any tissue proceeds, some genes are turned on while others are turned off. Insect metamorphosis reflects such a progression through a developmental sequence of genetic programs. It is a particularly dramatic example for three reasons: (1) The changes in all tissues are coordinated; (2) the times of change are marked by the overt act of molting; and (3) the morphological differences between immatures and adults may be very pronounced.

Primitive insects (Apterygota) change little in form when they molt from one instar to the next. These insects are referred to as **ametabolous.** The first instar is a small version of the adult except for the fact that the reproductive organs and external genitalia are not mature (Fig. 4-2, left column), and change in form throughout development is very slight.

The paleopterous orders (Ephemeroptera and Odonata), as well as exopterygote orders such as the Orthoptera, Dictyoptera, and Hemiptera, undergo **incomplete,** or **direct, metamorphosis** (Fig. 4-2, center column). These insects are termed **hemimetabolous.** Larvae of hemimetabolous insects are sometimes called **nymphs.** In these orders the immatures are usually similar to the adult in appearance, food habits, and habitat. This is not strictly true of orders such as Odonata and Ephemeroptera, which are aquatic as immatures; but in every case development involves a gradual increase in size of the wing pads until the final molt, after which the wings and reproductive organs are fully formed. For the most part the same cells and tissues that make larval structures change their pattern of gene expression and go on to make adult structures at the final molt. For example, the abdominal epidermis of *Rhodnius* produces five larval cuticles in succession and then lays down adult cuticle.

In the Endopterygota, which includes many of the most successful orders of insects, such as Diptera, Coleoptera, Lepidoptera, and Hymenoptera, the life stages are egg, larva, pupa, and adult. These insects are termed **holometabolous** and their development is described as **complete,** or **indirect, metamorphosis** (Figs. 2-5, 4-2, right column).

In holometabolous insects the larvae and the adult, or **imago,** differ greatly in body form and habits, and a pupal stage intervenes between larva and adult. The **pupa** is characteristically sedentary, but it is actually metabolically very active as old larval tissues and organs are remolded or replaced to form adult organs. Some adult organs, such as the malpighian tubules of Lepidoptera, are formed by transformation of the larval equivalent. Others are derived from **imaginal discs,** which are special groups of adult cells that were set aside in the embryo. Imaginal discs grow but do not differentiate during larval development and are activated to differentiate at metamorphosis. Many of the discs are invaginations of the epidermis. At pupation these invaginations fold out to form adult structures. In *Drosophila* there are paired discs for legs, wings, eyes, and antennae, and an unpaired genital disc (Fig. 4-3).

In many holometabolous insects, there is a quiescent stage in the last larval instar, immediately preceding pupation. This is called a **prepupa.** An insect that has molted but is still enclosed within its old exoskeleton is said to be in a **pharate** stage.

Molting.
Molting is a complex process, which is summarized in Fig. 4-4. Over the course of a molt cycle, the epidermal cell demonstrates an impressive capacity for synthetic versatility. Seven major steps take place in a molt: (1) apolysis, (2) epicuticle formation, (3) new procuticle deposition, (4) ecdysis, (5) procuticle expansion, (6) hardening and darkening, and (7) intermolt endocuticle deposition.

1. **Apolysis** is the retraction of the epidermal cells from the inner surface of the old endocuticle. Between the cells and the cuticle, a subcuticular space is formed. Into this space molting gel is secreted, and then beneath the molting gel a new cuticle is laid down. For some time the building blocks for the cuticle of the next instar continue to be secreted into the subcuticular space.

2. **Epicuticle formation** starts when dense plaques appear above the tips of fine microvilli that project from the epidermal cells. The first layer to be laid down, the cuticulin, is extensively wrinkled and folded. The cuticulin will determine the

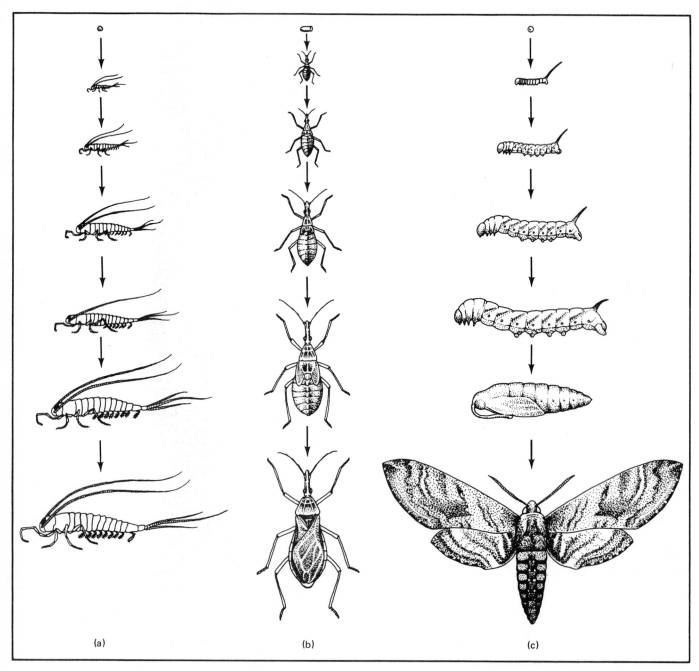

Figure 4–2
Basic types of development in insects. (a) *Machilis*
(Archeognatha); (b) *Rhodnius* (Hemiptera); (c)
Manduca (Lepidoptera).

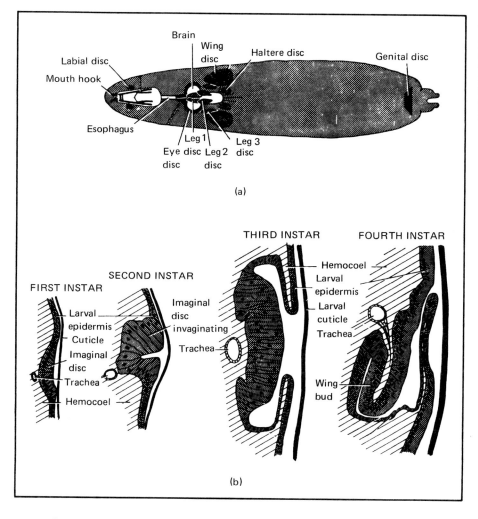

Brain

Wing
disc

Labial disc

Haltere disc

Genital disc

Mouth hook

Esophagus

Leg 1
disc

Leg 3
disc

Eye disc   Leg 2
disc   disc

(a)

THIRD INSTAR   FOURTH INSTAR

Hemocoel

Larval
epidermis

Larval
cuticle

Trachea

SECOND INSTAR

FIRST INSTAR

Imaginal
disc
invaginating

Larval
epidermis

Cuticle

Imaginal
disc

Trachea

Trachea

Hemocoel

Wing
bud

(b)

Figure 4–3
Imaginal discs in holometabolous insects. (a) Imaginal discs of a mature *Drosophila* larva seen from the ventral surface; (b) sections through the developing wing bud in the first four instars of the caterpillar of the imported cabbageworm. (R. F. Chapman, 1969, reproduced by permission of the publishers, Hodder and Stoughton Ltd., from *The Insects: Structure and Function.*)

surface area and surface patterning of the new cuticle. Next, the inner protein epicuticle is deposited just inside the cuticulin. By some mechanisms that probably include the action of a peroxidase, the cuticulin and inner epicuticle are chemically stabilized and rendered insoluble.

3. **Procuticle deposition** occurs by formation of chitin microfibrils in the subcuticular space beneath the inner epicuticle. Small sugar precursors are linked together to form the linear chains of chitin, and 18 to 20 of the chitin molecules are aligned in parallel in the paracrystalline array that is the microfibril. At the same time as the new procuticle is being formed, the inner endocuticular layers of the old cuticle are digested by activation of the molting gel. Molting gel contains digestive enzymes, which are inactive when first secreted. But when activated, the digestive enzymes are confined to the compartment above the new cuticulin and beneath the old exocuticle. Thus the newly deposited procuticle is protected from hydrolysis.

Figure 4–4
The events of the molt cycle.

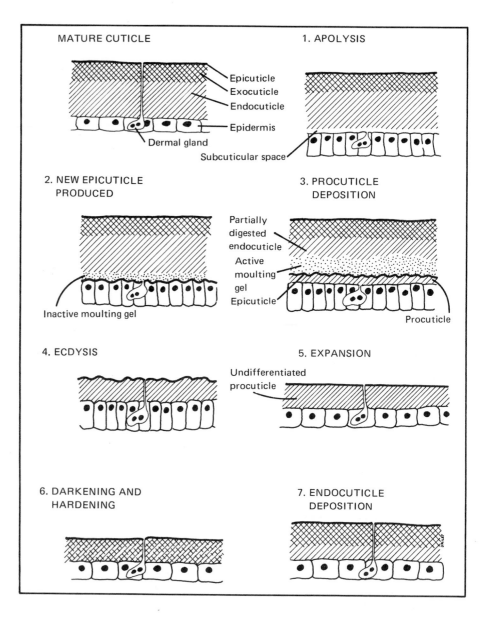

4. **Ecdysis** occurs when the old cuticle splits along a middorsal ecdysial suture. The cast skin consists of epicuticle and exocuticle and includes indigestible lipid, protein, and chitin. The endocuticular chitin and proteins have been recovered and largely recycled into the new procuticle.

5. **Expansion** of the soft, whitish new cuticle takes places as the insect swallows air to inflate and smooth out many of the wrinkles in the cuticulin. At this stage much of the procuticle is stretched to allow increase in surface area. Apparently the proteins slide over one another to allow the cuticle to expand, and there is no significant elastic recoil because the proteins are not yet cross-linked.

6. **Hardening and darkening** stabilize the new procuticle in its expanded state. Quinones, formed by oxidation of diphenols, cross-link cuticular proteins. The quinones react especially heavily in the outer layers of the procuticle, and thus the exocuticle is formed.

7. **Endocuticle deposition** includes laying down of both chitin and protein and continues for some time after ecdysis. In many insects a lamina of endocuticle is formed every 24 hours; thus the age of the cuticle can be measured by counting the growth lines.

Epidermal Cells and Molting.   Most of these molting events involve the sheet of epidermal cells. The cells define the environment within the subcuticular space —the low crypt within which all the precursors are assembled and polymerized into a multimolecular three-dimensional network. Some of the molecular components of these networks are derived from small precursors that polymerize in the subcuticular space, whereas others merely use the epidermal cells as a conduit to reach the cuticle-forming site. It is the epidermal cell that regulates the flow of molecules into the subcuticular polymerization mixture, and it is the epidermal cell that ultimately mediates each step in the complex process of laying down the new exoskeleton.

Over the course of cuticle deposition, the epidermal cells develop an extensive rough endoplasmic reticulum and other cellular adaptations for protein synthesis and export. Once new cuticle has been laid down, this intracellular machinery is somewhat superfluous. It appears that the organelles are "aged" by the extreme demands of the molt cycle. A typical vertebrate solution to such a problem of worn-out cytoplasm would be to discard the cells and, by mitosis, to replace the old cells with new ones. Some insect cells also die and are replaced; but for the most part, the insect solution is cell-conservative. The insect cell lives on while its cytoplasm is rejuvenated. The cytoplasm is reorganized by localized self-digestion. Tired organelles are wrapped in special membranous sacs, and the contents of these sacs are broken down by lysosomal hydrolytic enzymes. Thereafter the epithelial sheet remains thin as the cells maintain the cuticle during intermolt, but cells grow and become packed with rough endoplasmic reticulum as a new cycle of cuticle formation begins.

The Control of Molting

For a successful molt, all epidermal cells must act together. All the epidermal cells must be committed to making a particular kind of new cuticle for a larval, pupal, or adult instar, and then they must act out that commitment in synchrony. The decision to molt, the choice of which instar, and many of the metabolic events in the molt cycle are regulated by circulating hormones.

Many of the hormones were discovered because insects can survive massive surgical insults. A headless cockroach grooms itself and responds to touch for days, and an isolated abdomen of a moth pupa will continue to breathe for months. Such drastic surgery has been used routinely to investigate the role of endocrine centers (Fig. 4–5) in insect development. The brain, the thorax, the ventral nerve cord, and glands associated with the nervous system produce hormones that influence molting and metamorphosis.

Production of Prothoracicotropic Hormone (PTTH) by the Brain.
The pioneer experiments of insect endocrinology were carried out between 1915 and 1920 by Stefan Kopeć, a brilliant experimentalist working at the Agricultural Research Station in Pulawy, Poland. Kopeć suspected that the insect brain might some-

Figure 4–5
A diagram of the major neuroendocrine centers in the heads of insects. (Modified from R. F. Chapman, 1969, *The Insects: Structure and Function,* with permission of the publishers, Hodder and Stoughton Ltd.)

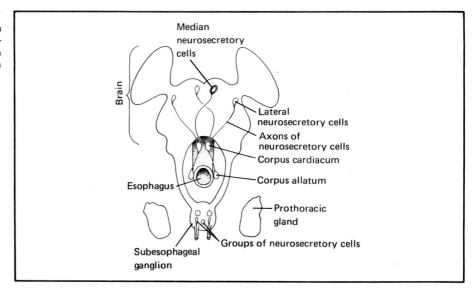

how control metamorphosis in the gypsy moth. He carefully placed two ligatures of silk thread around the middle of gypsy moth caterpillars of known ages and then cut the caterpillar in two between the ligatures (Fig. 4–6). Then he waited to see whether the front half and the back half of the caterpillar could pupate. As long as the ligature was placed at 7 days or later (after the caterpillar had fed), the front half could pupate, but the back half remained larval when ligatures were applied at less than 7 days. Both halves pupated when the ligatures were applied at 10 days. Kopeć hypothesized that something important was happening between 7 and 10 days and that this important event required the presence of the brain. The front half, with a brain, could pupate only if the brain had acted before the ligature was applied. Kopeć then removed brains from caterpillars at 7 days and 10 days and waited for pupation. Caterpillars debrained at 7 days did not pupate, whereas those debrained at 10 days pupated normally (Fig. 4–6). These results led Kopeć to draw two conclusions: (1) a hormone, secreted by the brain, is necessary for pupation, and (2) the brain hormone acts between the seventh and tenth day of the instar.

Kopeć's conclusions were confirmed 10 years later by V. B. Wigglesworth, working with the blood-sucking bug, *Rhodnius prolixus. Rhodnius* will molt only after it has been able to fill its crop with blood. Normally ecdysis occurs 10 days after the blood meal. When larval *Rhodnius* are decapitated just after feeding, the body survives for months, but it does not molt. When decapitation is delayed for 5 days, the headless fed body molts on schedule at 10 days. The experiments of both Wigglesworth and Kopeć emphasize the importance of timing. The brain must be present for a *critical period;* during this critical period the brain secretes its hormone. Decapitation *after* the critical period fails to prevent molting because the brain has finished its secretory role.

Molting is triggered by activation of the brain. In *Rhodnius,* feeding causes stretch receptors in the abdomen to send nerve impulses up to the brain, and these nerve impulses lead to secretion of the brain's hormone. In cecropia silkmoths, Carroll Williams, of Harvard University, showed that cold and an appropriate day length are necessary to activate the brain. Like many insects, cecropia overwinters in a state of

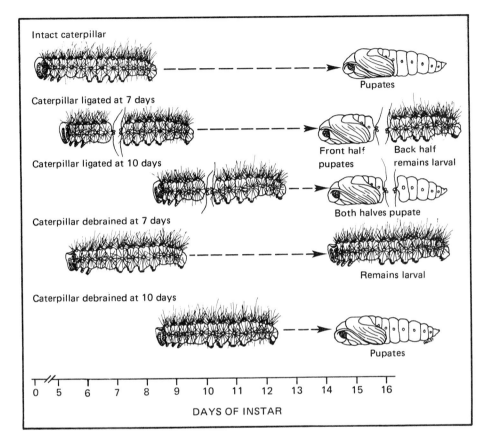

Intact caterpillar

Pupates

Caterpillar ligated at 7 days

Front half   Back half
pupates   remains larval

Caterpillar ligated at 10 days

Both halves pupate

Caterpillar debrained at 7 days

Remains larval

Caterpillar debrained at 10 days

Pupates

0 5   6   7   8   9   10 11 12 13 14 15 16

DAYS OF INSTAR

Figure 4–6
A diagrammatic summary of the experiments of Stefan Kopeć on caterpillars of the gypsy moth. The last larval instar normally lasts 16 days. When the caterpillar is ligated at midbody (between thorax and abdomen) on the seventh day, only the front half pupates, and the back half remains larval. When ligated 3 days later, both halves pupate. A caterpillar debrained at 7 days remains larval, but one debrained 3 days later pupates normally. From these results, Kopeć inferred the existence of a brain-produced hormone that acts between 7 and 10 days.

arrested development called **diapause.** When caterpillars complete their last larval instar, they pupate and enter diapause. Diapause persists until the brain has been activated by a natural or artificial (refrigerator) winter. Williams transplanted an activated brain from a developing pupa into another pupa deep in diapause, and the transplanted brain caused its host to begin adult development. As we shall see later, Williams showed that the target of the brain hormone is an endocrine organ called the prothoracic gland. The brain hormone is now known as the prothoracicotropic hormone (PTTH).

PTTH is made in specialized nerve cells of the brain, which are called **neurosecretory cells.** In the tobacco hornworm, there are a total of 30 neurosecretory cells, arranged in four symmetrical groups on each side of the brain. Axons from these cells transport the neurosecretory products downward and backward toward the **corpora cardiaca** and the **corpora allata.** In 1980 Larry Gilbert and his co-workers (now at the University of North Carolina) dissected out individual neurosecretory cells from freshly pupated tobacco hornworms and used homogenates of each isolated cell to stimulate prothoracic glands in organ culture. PTTH activity was localized to a *single* neurosecretory cell in each half of the brain (another example of the cell-conservative bias of insects). One axon from each of the two PTTH cells runs down to the corresponding corpus allatum where PTTH is stored before its release into the blood. In 1981 Williams and his colleagues at Harvard confirmed Gilbert's results and showed that the corpora cardiaca are also sites of PTTH release. Storage-release structures such as the corpora allata and the corpora cardiaca are known as **neurohemal organs.** The corpora allata and the corpora cardiaca are the neurohemal organs for PTTH and for other neurosecretory products such as those controlling metabolism, ion balance, and heart rate, which were discussed in Chapter 3.

In spite of the efforts of many fine laboratories, the chemical structure of PTTH is still not fully established, but it appears to be a peptide. As is common with vertebrate peptide hormones, PTTH apparently acts on its target tissue (the prothoracic gland) by increasing the concentration of the intracellular biochemical messenger, cyclic AMP.

Production of α-Ecdysone by the Prothoracic Glands.

Twenty-four years after Kopeć's work with gypsy moths, S. Fukuda, of the University of Tokyo, used ligature techniques to show that a thoracic endocrine center was required for pupation in oriental silkmoths. The thoracic center is the **prothoracic gland,** a diffuse, translucent secretory mass associated with tracheal trunks (Fig. 4–5). Its function was clearly established by Carroll Williams.

A similar sequence of hormone peaks occurs before each molt. Williams demonstrated the presence of both the thoracic and cephalic centers with a modification of Kopeć's experiment. He cut off the abdomens of newly molted cecropia pupae and sealed them shut with a coverslip glued over the wound. These isolated abdomens did not develop toward adulthood. He then implanted an activated prothoracic gland into such an abdomen, and it developed onward to yield a furry adult abdomen. Some of the adult abdomens were so perfect in morphology and function that they emitted sex attractant and laid eggs.

The newly molted cecropia pupa is in diapause. Williams showed that the brain alone cannot initiate development in a diapausing isolated abdomen, for no molt followed after an activated brain was implanted. Likewise, no molt occurred after a prothoracic gland from a diapausing pupa was implanted into the isolated abdomen. But when Williams put an activated brain from a chilled pupa and also an *inactive* prothoracic gland into the same isolated diapausing abdomen, diapause was broken. The abdomen began adult development. The activated brain from the head secreted PTTH, which acted on the implanted prothoracic gland. The prothoracic gland then secreted a second hormone called **ecdysone,** or molting hormone.

Ecdysone is a steroid, related to cholesterol and to vertebrate sex hormones. Two major forms of ecdysone are found in insects. α-Ecdysone is made in the prothoracic gland by hydroxylation of cholesterol from the diet. α-Ecdysone is secreted into the blood and converted to the true hormone in the target tissues. The true hormone is **β-ecdysone,** formed by adding a hydroxyl group to α-ecdysone. This final hydroxylation occurs in almost all tissues (epidermis, fat body, malpighian tubules, gut, and so on) except for the prothoracic gland itself.

How do the concentrations of ecdysone change as an insect develops? The titer of ecdysone can be measured by extracting an insect and injecting the extract into a diapausing abdomen. There are a score of other bioassays, but the most sensitive method is a radioimmunoassay similar to those widely employed by vertebrate endocrinologists. Radioimmunoassays use antibodies produced in a mammal to distinguish ecdysones from other similar steroids. With radioimmunoassays, ecdysone content has been measured recently in over a dozen species. The ecdysone surges occur at different times in development, and in some instars there is more than one peak of ecdysone. Furthermore, these reliable and sensitive assays have shown that the small peaks may last only a few hours.

α-ecdysone

β-ecdysone

In hemimetabolous insects, there is a major ecdysone peak two days before adult ecdysis (Fig. 4–7a). In holometabolous insects, such as the tobacco hornworm, there are two distinct peaks in the last larval instar (Fig. 4–7b). The first peak at day 4 is associated with a change in behavior when the caterpillars stop feeding and begin to search for a place to pupate. A second peak at about 8 days initiates apolysis and synthesis of pupal cuticle. Following pupation, a large sustained surge in ecdysone triggers adult development.

Ecdysones are rapidly removed from circulation. Free ecdysones are metabolized to hormonally inactive derivatives, which are excreted. The blood is a pipeline for ecdysone delivery; it is not a reservoir for ecdysone storage.

Ecdysones alter the pattern of gene expression in target cells. At the cell level, β-ecdysone appears to act like a typical vertebrate steroid hormone. β-Ecdysone enters a target cell; it binds to a cytoplasmic receptor protein, and then the hormone-receptor complex is translocated to the nucleus. The nuclear localization can be well seen in the salivary glands of *Drosophila* larvae. These glands contain the giant polytene chromosomes so beloved by geneticists. During the molt cycle, the polytene chromosomes acquire swellings called "puffs," which appear and disappear according to a specific developmental schedule. Puffs are sites of intense synthesis of ribosomal and messenger RNAs. Recent evidence shows that radiolabeled ecdysone molecules are concentrated at the precise sites of puffing where the hormone-receptor complex is presumed to act directly on chromatin.

Some puffs follow the addition of ecdysones to salivary glands in culture. In order to mimic the pattern of puffing seen in *Drosophila*, ecdysones must be added to the cul-

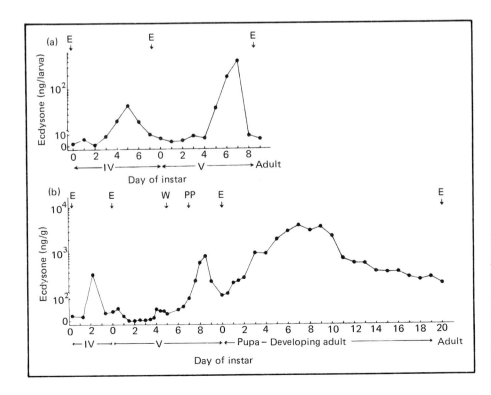

Figure 4–7
Ecdysone concentrations through larval growth and metamorphosis of (a) a hemimetabolous insect, *Schistocerca gregaria,* and (b) a holometabolous insect, *Manduca sexta.* Ecdysone was extracted from whole animals and measured by radioimmunoassay. The break and slash in the Y axis indicate a scale change from linear to log. *E,* ecdysis; *W,* wandering larva; *PP,* prepupa. (L. M. Riddiford and J. W. Truman, 1978, from *Biochemistry of Insects,* edited by M. Rockstein, with permission of the publishers, Academic Press, Inc., and the author.)

tures at just the right concentrations and for just the right periods of time. As in vertebrate hormones, the concentration and patterns of rise and fall are important to the proper action of ecdysones. The output of the prothoracic gland is precisely matched to the blood volume of the insect so that the concentration of circulating hormone is correct. We know that PTTH activates the prothoracic gland and that the prothoracic gland degenerates at the final (adult) molt. But we know almost nothing about the ways in which subtle and apparently important differences in circulating ecdysone concentrations are regulated.

Triggering of Ecdysial Behavior by Hormones.   The shedding of the old cuticle is the culmination of the molt cycle. As the time of ecdysis approaches, the insect begins a sequence of stereotyped preecdysial movements that loosen the old cuticle. For the adult **eclosion** of cecropia (ecdysis from pupa to moth), the process begins with a slow rotation of the abdomen. After 30 minutes of rotation and another 30-minute period of rest, the animal begins vigorous movements. Peristaltic waves sweep forward from the abdomen into the thorax, and the wing bases are pulled free from the old cuticle in a series of shrugging movements. The cuticle ruptures over the head and thorax; it is peeled off the legs and wings; and it is pulled out of the gut and tracheae. Then the moth gradually wiggles forward and out of the pupal skin. All insects have an analogous program of ecdysial behavior.

In several species ecdysial behavior is released by a hormone. This **eclosion hormone** was discovered in giant silkmoths by James Truman, of Harvard University and the University of Washington. Truman began by studying eclosion behavior. For each species of moth there is a characteristic daily rhythm of eclosion. When the light:dark cycle is 16 hours of light and 8 hours of darkness, moths of **Hyalophora cecropia** eclose in the morning, and moths of **Antherea pernyi** eclose over the 4 hours just before dark (Fig. 4–8). When Truman removed the brains of the developing pupae, cuticle deposition continued, and the animals eventually shed their pupal skins. But the eclosions no longer occurred on the usual daily schedule (Fig. 4–8), and the ecdysial behaviors were irregular and out of the normal rigid sequence. Truman then took brains from animals that were scheduled to eclose on the next day and implanted them into abdomens isolated from similarly aged animals. Isolated abdomens with these floating

Figure 4–8
The brain contains the clock that determines the time of eclosion in two species of giant silk moths, *Hyalophora cecropia* and *Antherea pernyi*. The black horizontal bar indicates 8 hours of darkness. Intact *H. cecropia* eclose after dawn, while intact *A. pernyi* eclose in the late afternoon. Debrained moths are irregular. Implantation of a brain into the abdomen of a debrained moth restores the normal schedule.   (J. W. Truman, 1973. Reprinted by permission, *American Scientist,* 61: 703, Journal of Sigma Xi, The Scientific Research Society, Inc.)

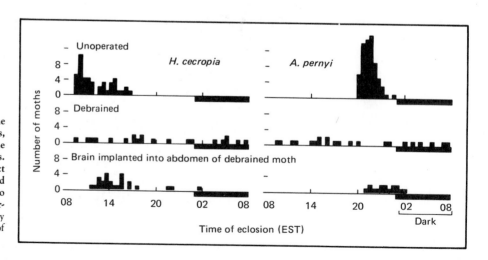

brains eclosed on the normal daily rhythm (Fig. 4–8), whereas control abdomens (without brains) eclosed only much later in an arhythmic and uncoordinated manner. The eclosion clock apparently resides in the brain. The brain recognizes the time of day and somehow triggers eclosion behavior. Since no direct nervous connections link a floating brain with the ventral nerve cord, the brain apparently triggers eclosion by liberating a hormone.

The eclosion hormone of cecropia moths has been isolated by Truman and found to be an acidic polypeptide. It is produced by neurosecretory cells in the brain of the preecdysial adult and stored in the corpora cardiaca. In the last instar larva, hormone is also synthesized and stored in the ganglia of the abdomen and the thorax and then released from these centers before pupal ecdysis. Eclosion hormone acts directly on the ventral nerve cord to release ecdysial behavior for pupal, adult, and probably larval molts. It also affects physiology. Just after adult eclosion, the hormone induces degeneration of the special muscles needed for molting and increases the plasticity of cuticle to allow wing expansion. At pupal ecdysis, eclosion hormone stimulates the dermal glands to produce the cement layer that helps to waterproof the cuticle. Truman has recently reported that eclosion hormone activity is found in five orders of insects. It seems quite likely that eclosion hormone controls ecdysis in many insect species.

For endocrine regulation, there must be an endocrine gland, its hormone product, and a target cell that is responsive to that hormone. The ventral nerve cord is the target for eclosion hormone, and its sensitivity to the hormone varies with the age of the animal. The ventral nerve cord of a young pupa is insensitive to eclosion hormone, and the cord becomes responsive only on the day before ecdysis. Control of responsiveness is a matter for future study.

Control of Cuticular Tanning by Bursicon.
Blow flies pupate underground. In 1936 Gottfried Fraenkel, now at the University of Illinois, reported that the newly emerged fly was shrunken and dull colored. Only after the fly had dug up to the surface of the ground did the wings expand, the cuticle become smooth, and the flies acquire their characteristic iridescent color. A quarter-century later Fraenkel and his coworkers, Katherine Hsiao and also Kip Cottrell of Cambridge University (working independently), simultaneously reported that tanning and expansion are controlled by a hormone, now called **bursicon,** which is produced by neurosecretory cells of the brain. Bursicon is a polypeptide that has been found in many insect orders. It is usually released from neurohemal organs associated with the thoracic or abdominal ganglia. Blow flies do not stretch their cuticle nor do they tan it until bursicon is secreted after the fly has reached the surface.

Bursicon affects many postecdysial processes. It promotes the synthesis of precursors of tanning quinones. It plasticizes cuticle so that expansion can take place. It triggers cell death in the wing epidermis. It leads to a thickening of the endocuticle in abdominal sclerites. The stimuli that cause bursicon to be released vary with the biology of the species. At least in moths, one of the prerequisites is prior secretion of eclosion hormone.

Secretion of Juvenile Hormone by the Corpora Allata.
A molt can lead to a larva, a pupa, or an adult. The choice among these alternatives is largely determined by **juvenile hormones (JHs)** from the **corpora allata,** as was first demonstrated by Wigglesworth's classic experiments with early-instar *Rhodnius*. When third-instar *Rhodnius* were decapitated immediately *after* the critical period for the brain, the headless body molted, but the result was not a fourth-instar larva (Fig. 4–2). Instead it

was a small precocious adult. Apparently the head of the intact third instar produced factors that prevented adult development. Headless animals, freed of that inhibition, skipped the remaining larval instars and molted to adults. Wigglesworth showed that endocrine glands behind the brain, the corpora allata, prevented adult development. When active corpora allata were implanted into the headless, well-fed body of an early-instar larva, the result was another normal larval instar rather than a precocious adult. Implantation of active corpora allata into the fifth-instar larva, which would normally molt to an adult, produces instead an unnatural stage—a giant sixth-instar larva.

Juvenile hormone (JH) from the corpora allata maintains the juvenile stage and prevents metamorphosis. JH is neither a polypeptide nor a steroid. Rather, it is a terpene, derived from the condensation of three isoprene units. There are three juvenile hormones (JH1, JH2, JH3), which differ only in the nature of the side chains. One or another JH predominates in different species, but no general pattern seems to be widespread. All three JHs are relatively insoluble in water. Their transport from the corpora allata to the target tissue is aided by JH-binding proteins in blood plasma. Circulating JH in plasma is rapidly degraded by enzymes called JH-specific esterases.

Juvenile hormone modulates the effects of ecdysone. When JH is present at high titer, ecdysone surges trigger molts to another instar. Throughout most of larval life, the corpora allata continue to secrete much JH. But in the last larval instar, the corpora allata become inactive. In hemimetabolous insects this inactivity persists through the ecdysone peak that causes the final molt to the adult. As we have said, there are two ecdysone peaks in the last larval instar of the holometabolous tobacco hornworm (Fig. 4–7). The first small ecdysone peak comes when JH is below detectable levels. This ecdysone peak affects the commitment of the epidermis and shifts epidermal synthetic pathways from larval cuticle toward pupal cuticle. But no cuticle is made yet. When it acts in the absence of juvenile hormone, ecdysone primes the epidermal cells so that the next time cuticle is made, it will be pupal cuticle. The corpora allata are again reactivated late in the last larval instar. A brief rise in JH coincides with the large ecdysone peak that triggers cuticle deposition. If no JH were present at this point, some parts of the animal might try to skip the pupal stage and proceed directly to the adult. JH levels fall again and remain low during the pupal stage.

What controls the activity of the corpora allata? Some evidence suggests that neurosecretions from the brain inhibit and/or activate the corpora allata. How does JH act on its target cells? Probably JH binds to a cytoplasmic receptor protein in the epidermis and somehow affects gene transcription.

Pupariation. Many flies, including blow flies and house flies, begin metamorphosis in an unusual manner. The cuticle of the last-instar larva contracts into a barrel shape and tans to form a **puparium** (Fig. 2–5). The true pupa lies within that puparium.

Gottfried Fraenkel showed that **pupariation** (formation of the puparium) of blow flies is triggered by a substance liberated in the anterior half of the animal. He ligated larvae at midbody before pupariation. When surgical trauma was slight, the front half tanned but the rear did not. But when ecdysone-laden blood was injected into the rear part, it would tan. This experiment forms the basis of the "*Calliphora* test," a biological assay for detection of ecdysone activity that long preceded the modern radioimmunoassay. Ecdysone acts indirectly by causing the release of several pupariation neurohormones that influence larval behaviors and changes in shape, as well as the tanning of the puparium.

JH1 — C_2H_5 ... C_2H_5 ... $COOCH_3$ (O, H)

JH2 — C_2H_5 ... CH_3 ... $COOCH_3$ (O, H)

JH3 — CH_3 ... CH_3 ... $COOCH_3$ (O, H)

Diapause.   In a previous section we discussed diapause in the cecropia moth, which occurs in the pupal stage inside a cocoon. In this instance diapause, defined as a state of arrested development, cannot be broken until the pupa has been exposed to a period of cold and the brain activated to produce PTTH. If the brain of a pupa that has been activated is implanted into a pupa deep in diapause, development ensues. These and other experiments demonstrate that diapause is regulated by the endocrine system. Diapause is a very widespread phenomenon, particularly among insects occurring in unstable environments or in areas where there are long periods of cold. In a diapausing state, insects are often able to survive extreme cold or other unfavorable situations that they could not survive in a nondiapausing state.

Depending on the species, insects may diapause in any stage: in the egg, in any larval instar, or as pupae or adults. Diapause differs from simple quiescence (for example, during a brief period of suboptimal temperature) in being more prolonged and requiring a particular stimulus before it can be initiated or broken. Diapause tends to anticipate unfavorable climatic conditions; for example, many insects anticipate the winter by entering diapause in the late summer or fall in response to decreasing day length. In places where there is a prolonged dry period in the summer, or when the normal food is not available during the summer, diapause may occur during the warmer months of the year. Or it may embrace both summer and winter, as in species specializing on plants that flourish only in the spring. Some insects spend as much as 10 months of the year in diapause. Diapause is usually preceded by the accumulation of nutrient reserves and by behavioral adaptations, such as seeking concealment or spinning a cocoon.

Insects that have a single generation a year are said to be **univoltine,** while those having two are **bivoltine,** and more than two **multivoltine.** Univoltine species usually have a long diapause that is **obligate;** that is, it is intimately tied to certain environmental factors that influence the endocrine system in specific ways and always occurs when these factors are present. However, in many insects diapause is **facultative;** that is, the insect may or may not enter diapause, depending on external conditions. Bivoltine and multivoltine species frequently go through two or more generations not interrupted by diapause, but in the autumn they enter a diapausing stage. Such insects can usually be reared continuously in the laboratory if supplied with uniform conditions.

Considering the fact that diapause may occur in any life stage and in response to many factors (photoperiod, temperature, food quality, and others) and that it may be either obligate or facultative, it is not surprising that its endocrine basis is not fully understood and not necessarily alike in all species. The corpora allata are clearly involved in many cases, a specific titer of JH being required to prolong that particular instar and to regulate metabolism. A **diapause hormone** has been described in the silkworm; it is produced by the female pupa and causes her developing eggs to diapause after they are laid. It is usually assumed that PTTH and consequently ecdysone are not produced during larval and pupal diapause; however, cases are known of larvae that molt (but do not grow) during diapause, so ecdysone must be produced at certain times in these cases. Ecdysone is clearly involved when the insect molts following the termination of diapause (unless, of course, it has diapaused as an adult).

Production of Insect Growth Regulators by Plants.   Plants produce substances that mimic or antagonize insect endocrine secretions. Such **insect growth regulators** appear to cause changes in the hormone balance of insects feeding on the plants. These changes produce abnormal development that usually leads to death. Some insects do, however, feed and develop normally on plants rich in hormone mim-

Figure 4–9
Diagram of testis follicle (a) and egg tubes (ovarioles) (b, c). (b) A panoistic ovary that lacks nurse cells; (c) a meroistic ovary with nurse cells. (Modified from V.B. Wigglesworth, 1972, *The Principles of Insect Physiology.* London: Chapman and Hall. 827 pp.)

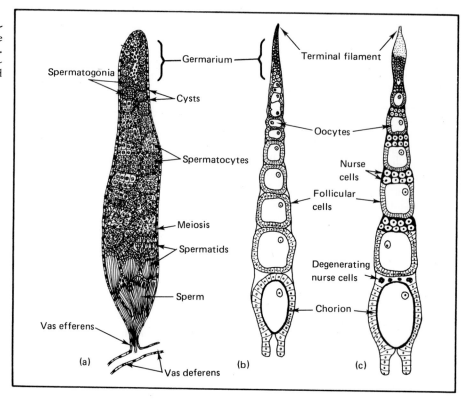

Figure 4–10
Drawing of the reproductive system of a male mealworm beetle. The bean-shaped gland produces a solid product (black) that forms the spermatophore.

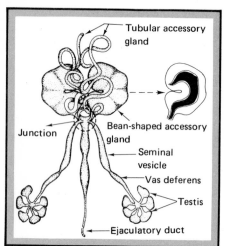

ics. Perhaps they are able to detoxify the plant-produced hormones and excrete inactive metabolites. The subject of insect growth regulators is pursued further in Chapter 10, dealing with the defenses of plants against insect attack.

Reproduction

Reproductive maturation and reproductive function are regulated by ecdysone, juvenile hormones, and neurosecretions from the brain.

Primary gonads (ovaries and testes) are biological factories that produce gametes. Testes spew out multitudes of small spermatozoa, while ovaries produce many fewer but much larger oocytes. In both the testis and the ovary, the sequence of developmental stages from youngest to oldest is laid out along the length of the tubular components (follicles and ovarioles). Stem cells are at the apex, and mature gametes are at the base. In both sexes the fully developed eggs or sperm move from the gonad into mesodermal ducts and finally exit through cuticle-lined invaginations of the body wall (Fig. 1–16c,d).

Hormones and Sperm Maturation.   The testes contain one or more sperm follicles (Figs. 4–9, 4–10). At the tip of each tubular follicle is the **germarium.** Within the germarium, **spermatogonia** divide, and some of the cellular progeny become invested with a coat of somatic cells. The coated germ cells then divide mitotically five or six times to form a cyst of 64 to 256 **spermatocytes.** Meiotic divisions follow and yield rounded **spermatids,** which eventually mature into elongate **spermatozoa.** As they are produced, the cysts of sperm accumulate at the base of the follicle and are shed into an inflated portion of the vas deferens called the **seminal vesicle.**

Ecdysone and JH indirectly affect the rate of spermatogenesis by controlling the rates of mitotic division in the follicle. Ecdysone accelerates germ cell mitoses, and juvenile hormone inhibits its action. The interplay between ecdysone and JH means that mitoses are slow in the larva and speed up in the last preimaginal instar. Germ cells will mature only after they have undergone a certain number of cell divisions. When an experimenter removes the corpora allata (an operation called **allatectomy**) from a third-instar male *Rhodnius*, the insect will molt into a precocious adult that is almost perfect except for the testes. The sperm follicles of the testes contain spermatogonia and spermatocytes but lack spermatids and spermatozoa.

Spermatozoa are stored in the seminal vesicles, where they are mixed with secretions of the mesodermal ducts and the accessory glands. The differentiation of these glands is triggered by the preimaginal surges in ecdysone. In many hemimetabolous insects the corpora allata are required for maintenance of accessory glands in adults, but in some species, such as blow flies, growth of the adult accessory glands seems independent of hormones.

Spermatophores.

In many insect species, spermatozoa are delivered to the female within a protective sac. This sperm sac, fashioned from secretions of the male tract, is called a **spermatophore.**

Male mealworm beetles have two pairs of accessory glands that contribute their products to the spermatophore (Fig. 4–10). The tubular glands pass soluble proteins into the seminal fluid that bathes the sperm, while secretions from the bean-shaped gland form the wall of the spermatophore. The spermatophore is formed during the act of copulation. Within a minute after the male has inserted his aedeagus, ejaculation begins. Vigorous contractions of the strong muscular coat of the vas deferens and of the ejaculatory duct force fluid semen and tubular gland products into the pasty secretion oozing out of the bean-shaped accessory glands. The mass of sperm and secretions is pumped backward along the ejaculatory duct. Within minutes after copulation has started, a multilayered elongate spermatophore pops from the gonopore at the tip of the aedeagus. The spermatophore then swells and explosively ruptures to liberate the spermatozoa (Fig. 4–11).

A male machilid (Archeognatha) spins out fine silken threads to which he affixes several spermatophores. During a "mating dance," he maneuvers the female into position so that she walks over the spermatophore, and as her genital opening passes over the sperm sac, it is engulfed (Fig. 4–12). In some crickets, grasshoppers, and meloid beetles, the spermatophore is assembled in the female. Secretions from various parts of the male tract pass out in succession into her special sperm receptacle. In some grasshoppers these copious secretions may be nutritive to the female, and thus the male is investing nutrients toward egg production as well as transferring his genetic information (Fig. 7–22). In some species, such as the honey bee, sperm and seminal fluids are transferred directly without formation of a spermatophore.

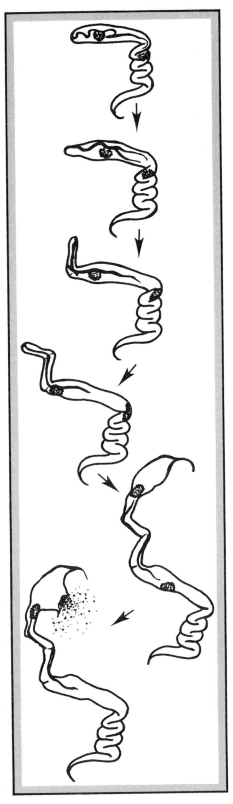

Figure 4–11
The spermatophore of the mealworm beetle is compact when the male places it into the female median oviduct. Within five minutes it undergoes several expansions and bursts to release the sperm.

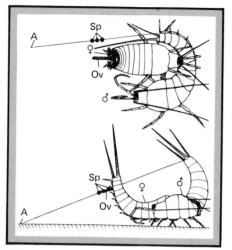

Figure 4–12
Courtship in *Machilis* (Archeognatha). The male places spermatophores (Sp) on a thread and guides the female, causing her to pick them up with her ovipositor (Ov). (From F. Schaller, 1968, *Soil Animals.* By permission of The University of Michigan Press.)

Ovaries, Ovarioles, and Vitellogenesis.

The ovary is a cluster of several egg tubes, or ovarioles (Figs. 1–16d, 4–9). The apical zone of each ovariole is the **germarium,** a site where **oogonia** divide and give rise to **oocytes.** Each oocyte becomes surrounded by an epithelial sac of **follicle cells** that is homologous to the cyst that coats the primary spermatocytes. Oocytes grow many times in volume as they proceed single-file toward the base of the ovariole. In some insects, oogonia also produce **nurse cells** that either accompany the oocytes down the ovariole or remain in the germarium, connected to the descending oocytes by cytoplasmic strands. These ovaries are termed **meroistic;** those that lack nurse cells are called **panoistic.**

Both nurse cells and follicle cells pass nutrients into the growing oocyte. When present, the nurse cells contribute RNA and proteins for the early stages of oocyte growth. **Vitellogenesis** (yolk formation) follows later and requires the cooperation of the fat body, the follicle cells, and the oocyte itself. **Vitellogenins** are precursors of egg yolk proteins. They are synthesized in the follicle cells and also in the fat body. In cockroaches, locusts, flies, butterflies, and many other insects, JH stimulates the fat body to synthesize the messenger RNA for the vitellogenic proteins. Once produced, vitellogenins are secreted into the hemolymph and thus pass to the ovary. In some species JH is also required for vitellogenin uptake by the ovarian follicle.

JH, ecdysone, and a neurosecretory substance from the brain are required for vitellogenesis in the mosquito *Aedes aegypti* (Fig. 4–13). Shortly after ecdysis, the corpora allata secrete JH, which primes the system. JH initiates follicle growth and makes the ovary responsive to the egg-development neurosecretory hormone (EDNH), which will be produced later by the brain. JH also makes the fat body responsive to ecdysone. Once the female has obtained a blood meal, her brain liberates EDNH, which acts on the previtellogenic ovary. Henry Hagedorn, of Cornell University, has reported that the ovary responds by synthesizing and secreting ecdysones. (Remember that the prothoracic gland degenerated at the last molt.) Ecdysone from the ovary acts on the JH-primed fat body to promote vitellogenin synthesis.

Choriogenesis.

As each oocyte receives its proper quota of yolk, the follicle cells turn to the synthesis of the egg shell. First, the follicle cells secrete a thin vitelline membrane that coats the yolk mass, and next they produce the thicker **chorion,** or egg shell. The chorion is a multilayered structure, composed of 20 to 100 different proteins that are impregnated with waxy materials. Each chorion protein is synthesized by the follicle cell, and all are secreted in a fixed sequence. Unlike vitellogenesis, choriogenesis takes place in the absence of hormonal modulators. As shown by Fotis Kafatos and his co-workers at Harvard University and the University of Athens, the first proteins are assembled into a loose scaffolding, and then the interstices are filled. The whole mass is stabilized and cross-linked by the action of oxidative enzymes.

Egg shell surfaces are rough and have a variety of plaques and protrusions (Fig. 4–14). Sperm fertilize the oocyte by traversing one or more tiny pores called **micropiles,** which run through the layers of the chorion and through the vitelline membrane (Fig. 4–14b). The various surface processes of the egg shell help to anchor the egg at the site of laying and to form an **aeropile.** The aeropile is a respiratory plastron that traps air and allows the egg to continue to breathe even when it is submerged by rains and temporary floods.

Fertilization.

As eggs mature, they pop out of the follicles and collect in egg sacs at the bases of the ovarioles. Here they usually remain until after mating. At copulation, sperm from the male are received by the female and then are transported into a special cuticular storage chamber called the **spermatheca.** Sperm can remain viable in the spermatheca for weeks, months, or even years. Following mating, eggs slip from the egg sac, are forced back through the lateral oviducts, and enter the median oviduct where fertilization takes place. The spermatheca opens into that oviduct, and the micropile of each egg is presented to the opening from the spermatheca. Spermatozoa are extruded; several penetrate the micropile(s), and one of the sperm nuclei fuses with the oocyte nucleus.

Accessory Glands of the Female.   Female accessory gland secretions coat the eggs and protect them from dessication and mechanical damage. Some accessory glands require JH for development and maintenance. The **colleterial glands** of female cockroaches (Fig. 1–16d) provide an especially clear illustration of the function of these accessory glands.

Many cockroaches deposit their eggs within a hard protective structure called an **ootheca,** which is composed of tanned protein (Fig. 4–15). After fertilization the eggs are lined up, side by side, in the expandable median oviduct. The secretion of the left colleterial gland is poured over, around, and between the egg mass. The left colleterial gland produces three major products: soluble structural protein, a phenyl-glucoside, and the enzyme phenol oxidase. This mixture is stable and does not tan itself until secretions of the right gland are added. The right gland adds another enzyme—a glucosidase. Tanning occurs only when the products of the two glands are mixed together. The result is hard ootheca, which encloses the eggs.

Unusual Modes of Reproduction.   Although the vast majority of insects lay eggs and are said to be **oviparous,** exceptions do occur. Some flies, for example, retain the eggs internally until they hatch and deposit small first-instar maggots. They are said to be **larviparous,** or **ovoviviparous.** Truly **viviparous** insects are few in number. These insects retain the larva internally through much of its development and pass nourishment to it. An especially dramatic example is provided by the tsetse fly, vector of trypanosome diseases such as African sleeping sickness. Only one tsetse oocyte matures at a time, and it is fertilized in the median oviduct. The female nourishes the single offspring in her uterus (a modified median oviduct) and lays a fully grown larva, which pupariates within one to two hours. At "birth" (larviposition), the larva weighs as much as the mother. The pregnancy cycle repeats itself every 10 days, and coincident with it are recurrent cycles of secretory activity in the uterus.

There are also some insects (especially in the order Coleoptera) in which the male sex is absent and reproduction is by **obligate parthenogenesis.** Since there is no genetic recombination, all offspring are alike. Parthenogenetic species are regarded as evolutionary "dead ends," since they lack the ariation on which natural selection must work. In aphids, parthenogenesis is **cyclic.** There are usually no males during the summer months, and females produce living offspring parthenogenetically; but in the fall males are produced, mating occurs, and the females lay eggs that survive the winter. Aphids thus take advantage of the rapid buildup of populations made possible by viviparity and asexual reproduction without sacrificing the advantages of sexual reproduction.

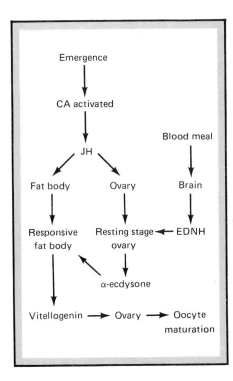

Figure 4–13
Scheme for hormonal control of oogenesis in the mosquito *Aedes aegypti.* (L. M. Riddiford and J. W. Truman, 1978, from *Biochemistry of Insects,* edited by M. Rockstein, with permission of the publishers, Academic Press, Inc., and the author.)

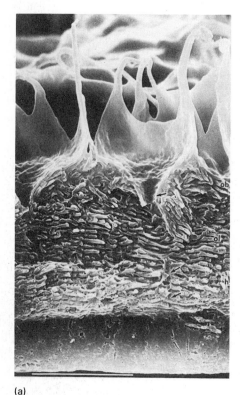

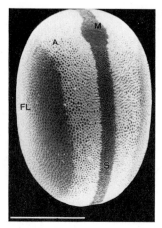

(b)

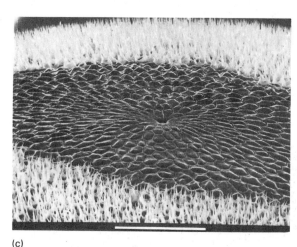

(c)

(a)

Figure 4-14
Scanning electron micrographs of the egg shell of a
silk moth. (a) Composite of three micrographs
through aeropile crown of egg shell. Arrow points
to the aeropile channel. ob, oblique layer; ol, outer
lamella; h, holey layer. (b) The whole egg, showing
four surface regions: FL, flat; A, aeropyle crown;
S, stripe; M, micropile (bar = 1 mm). (c) Detail of
anterior pole of same egg (white bar = 100 μm).
(Grace D. Mazur, Harvard University, from *Developmental Biology*, vol. 76, 1980, with permission of
Academic Press, Inc., and the author.)

In some Hymenoptera, parthenogenesis is obligate and males are absent. But in
the majority of Hymenoptera (and a few other insects), it is facultative. Mating occurs
and sperm are stored in the female spermatheca; however, the female is able to control
a sphincter muscle that permits the release of sperm. Unfertilized eggs are haploid and
produce males; fertilized eggs produce females. This is termed **haplodiploidy.** The ability of females to determine the sex of their offspring was an important prerequisite for
the evolution of social life in ants, bees, and wasps (Chapter 8).

An unusual type of asexual reproduction occurs among a few Diptera and Coleoptera. In certain gall midges (Cecidomyidae), for example, the ovaries become functional in mature female larvae, resulting in the body of the female filling up with daughter
larvae that break out, killing the mother. Reproduction by immature larviform individuals is called **paedogenesis.** It is usually cyclic, sexual reproduction by true adults
occurring at other times in the life cycle.

Summary

Insects grow and develop by molting. The final molt is to a reproductively competent
adult. In most insects adults, but not juveniles, are winged. The changes in form that
occur at molting are called metamorphosis. In primitive insects the metamorphosis is
not marked (ametabolous). Hemimetabolous insects with incomplete metamorphosis
molt from a last-instar larva (or nymph) to a winged reproductive adult. Holometabolous insects with complete metamorphosis have a pupal stage between larva and adult.

New cuticle is laid down by a carefully controlled sequence of biosynthetic steps,
most of which involve the epidermal cells. The first step in molting is apolysis, and its
climax is ecdysis.

By the late 1950s, it was generally thought that insect development was controlled by three classical hormones: juvenile hormone from the corpora allata, ecdysone from the prothoracic gland, and PTTH from the brain. Bursicon and eclosion
hormone are recent additions to that list. Thus the molt cycles of most insects are controlled by PTTH, ecdysone, perhaps eclosion hormone, and bursicon, each acting one
after the other. A tentative sequence of the hormonal signals and the molting events
follows:

Hormonal signal	Molting event

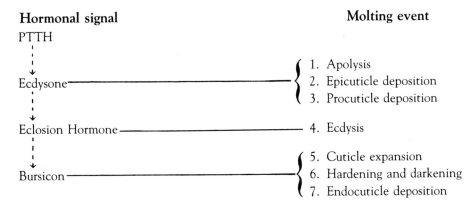

Figure 4–15
A female giant Florida cockroach, *Eurycotis floridana*, about to deposit her ootheca in a depression she has dug in the sand. Some cockroaches merely drop their oothecae without protecting them in any way; a few retain the ootheca internally and supply water and nutrients to the developing embryos, giving birth to living young. (Photograph by Thomas Eisner, Cornell University.)

When high titers of juvenile hormone act in concert with ecdysone, the new instar is a larva. When juvenile hormone is absent, an adult is produced. Rises and falls in hormone titers are synchronized with changes in the responsiveness of the target tissues. Disturbance of the normal patterns may lead to developmental arrest, to death, or to production of monsters that are partly adult and partly immature. Diapause, defined as a state of arrested development, is also under hormonal control.

Sperm and oocytes mature in sperm follicles and ovarioles, respectively. Ecdysone promotes cell division that allows the progressive differentiation of germ cells; juvenile hormone acts to counter this. Spermatozoa collect in the seminal vesicle and are passed to the female. Often they are enclosed in a spermatophore produced by the male accessory glands.

Oocyte maturation may require contributions of RNA and protein from nurse cells, and it always requires sequestering of yolk protein (vitellogenesis) and addition of the egg shell (choriogenesis). JH, ecdysone, and a neurosecretory hormone (EDNH) are required for vitellogenesis in some species. Oocytes are fertilized after spermatozoa enter the micropile. Accessory gland secretions form a protective coat around the eggs.

While the majority of insects lay eggs (oviparous), a few retain the larvae internally and supply nourishment to them, giving birth to larvae that are fully grown (viviparous). Parthenogenesis is not uncommon among insects. It may be obligate or facultative; in the latter case males are produced from unfertilized eggs and are haploid.

Selected Readings

Chippendale, A. M. 1977. "Hormonal regulation of larval diapause." *Annual Review of Entomology*, vol. 22. pp. 121–38.

Lawrence, P. A., ed. 1976. *Insect Development*. Symposium of the Royal Entomological Society of London, no. 8. 230 pp.

Locke, M., and D. S. Smith, eds. 1980. *Insect Biology in the Future: VBW '80*. New York: Academic Press. 977 pp.

Menn, J. J., and M. Beroza, eds. *Insect Juvenile Hormones: Chemistry and Action*. New York: Academic Press. 341 pp.

Miller, T. A., ed. 1980. *Neurohormonal Techniques in Insects*. New York: Springer-Verlag. 340 pp.

Riddiford, L. M., and J. W. Truman. 1978. "Biochemistry of insect hormones and insect growth regulators," in *Insect Biochemistry* (M. Rockstein, ed.), pp. 308–17. New York: Academic Press.

Tauber, M. J., and C. A. Tauber. 1976. "Insect seasonality: diapause maintenance, termination, and postdiapause development." *Annual Review of Entomology*, vol. 21, pp. 81–107.

Truman, J. W. 1973. "How moths 'turn on': a study in the action of hormones on the nervous system." *American Scientist*, vol. 61, pp. 700–6.

Wigglesworth, V. B. 1959. *The Control of Growth and Form*. Ithaca, N.Y.: Cornell University Press. 140 pp.

———. 1970. *Insect Hormones*. San Francisco: Freeman. 159 pp.

Willis, J. H. 1974. "Morphogenetic action of insect hormones." *Annual Review of Entomology*, vol. 19, pp. 97–115.

Pheromones and Allomones

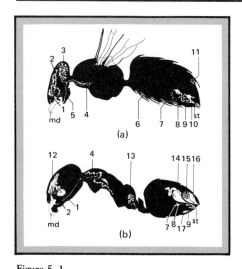

Figure 5–1
Exocrine gland systems of (a) a honey bee worker and (b) a worker ant (*Iridomyrmex humilis*). Glands of the two species are labeled with the same number if they are homologous. (1) Mandibular glands; (2) hypopharyngeal glands; (3) head labial glands; (4) thoracic labial glands; (5) hypostomal gland; (6) wax glands; (7) poison glands; (8) reservoir of poison gland; (9) Dufour's gland; (10) Koschevnikov's gland; (11) Nasanov's gland; (14) hindgut; (15) anal gland; (16) reservoir of anal gland; (17) Pavan's gland. (Reprinted by permission of the publishers from *The Insect Societies*, by E. O. Wilson; Cambridge, Mass.: The Belknap Press of Harvard University Press, copyright 1971, by the President and Fellows of Harvard College; modified from C. R. Ribbands, 1953, and M. Pavan and G. Ronchetti, 1955.)

C hemical senses, such as taste and smell, are much more important to insects than to humans. Insects are repeatedly bombarded with meaningful chemical signals that tell of food, of nest sites, of prey, of predators, or of a suitable mate. Many of these signals produce a rapid and specific behavioral response. The study of the chemical signals that act between organisms constitutes a new discipline called chemical ecology. This field has been the subject of several recent books and innumerable research articles, and is one of the most exciting frontiers in entomology.

Chemical signals include hormones, pheromones, and allomones. Endocrinologists study hormones, which we considered in Chapter 4. **Hormones** are internal signals that are produced by an endocrine gland and that are carried through the blood to act on target tissues within the animal. Chemical ecologists study external signals that are secreted into the environment by exocrine glands and that act between different organisms. **Pheromones** act between individuals of the same species. Pheromones attract the sexes, aid in courtship, announce the presence of danger, mark trails and home territories, and control many other intraspecific interactions. E. O. Wilson and W. H. Bossert, of Harvard University, divided pheromones into two general subclasses: (1) **releasers,** which act rapidly to produce a behavioral response, and (2) **primers,** which act to modify the physiological state. **Allomones** act between different species to benefit the producer. Allomones repel predators, confuse prey, and mediate symbiotic interactions.

Although the terms *pheromone* and *allomone* are of recent origin, our knowledge of chemical signals began in the nineteenth century for pheromones and even earlier for defensive allomones. Yet it is only since 1950 that the analytical power of organic chemistry has developed to the point that identification of the molecules can be accomplished repeatedly from very small samples. With the explosion of information on the molecules themselves, it has become possible to ask precise questions about their impact on behavior and physiology. The molecular structures are of some intrinsic interest as well, but the most fascinating aspect of chemical ecology is the delicate intercourse among organisms that is mediated by the chemical signals.

Chemical communication requires a source of signal molecules, the chemical signal itself, and a receiver. The source is most often an **exocrine gland,** which opens on the surface of an insect. Glands may be associated with mouthparts, intersegmental membranes, sclerites, legs, and other body parts (Fig. 5–1). Somehow the products of these glands must cross through the impervious insect exoskeleton. Type I secretory cells lie well below the epidermis itself and pass their secretions to the outside through elongate ductules that are invaginations of the epicuticle (Figs. 5–2, 5–3). Type II secretory cells have cuticles that are riddled with tiny perforations so that the secretion percolates up from the secretory cell to the outside (Figs. 5–2, 5–4). The molecular signals and their receptors will be discussed below.

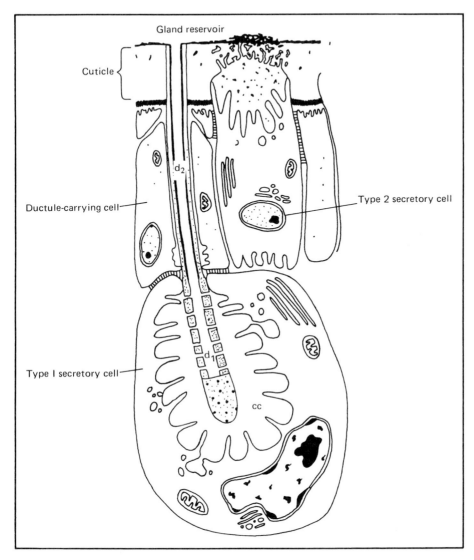

Figure 5–2
Secretory epithelium of an insect exocrine gland, showing two cell types. This gland regulates the growth of symbiotic fungi that inhabit the cuticular pouch that is the gland reservoir. Type 1 cells pass their products into an enclosed extracellular cavity (*cc*), within which lies the perforated tip of a cuticular ductule (d_1). The ductule (d_2) is surrounded by a ductule-carrying cell as it runs to the reservoir. Type 2 cells secrete their products into a pocket of subcuticular space, and the secretions then ooze through the overlying porous cuticle to reach the reservoir. (Diagram based on the mycangium of a female southern pine beetle, from Happ, Happ, and Barras, 1971, from *Tissue and Cell*, vol. 3, with permission of the publishers, Longman Group Ltd.)

Pheromones

From among a noisy jumble of competing chemical stimuli, a pheromone must stand out and convey an unambiguous message. Not all molecules make good pheromones. The signal must be novel, and in a terrestrial environment it must be volatile. Glucose and glycogen, for example, are poor candidates on both counts. In theory, the larger the molecule, the greater are the possibilities for a unique structure that conveys a clear message. But in practice, the requirement for volatility restricts pheromones to molecular skeletons of 20 carbons or less. Most are derived from common biochemicals like fatty acids or amino acids.

Throughout the class Insecta, the molecular structure of most known hormones varies little. In contrast, many thousands of different molecules have been identified as

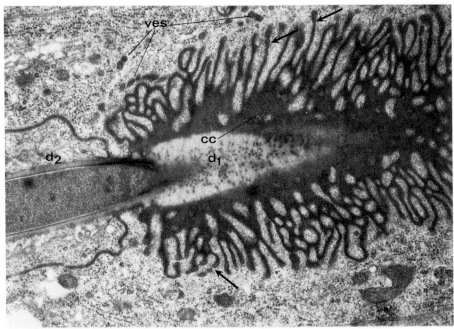

Figure 5–3
In the fungal transport pouch of a southern pine beetle (Fig. 5–2), a cuticular ductule carries the products of the type 1 secretory cells to the pouch. The secretion is produced in small vesicles (*ves*), which pass their contents into the central cavity (*cc*). The wall of the end apparatus (d_1) is perforated with fine channels through which the dense secretion flows from the central cavity into the end apparatus and out along the ductule (d_2). (Happ, Happ, and Barras, 1971, from *Tissue and Cell*, vol. 3, with permission of the publishers, Longman Group Ltd.)

Figure 5–4
In the fungal transport pouch of a southern pine beetle (Fig. 5–2), the cuticle above a type 2 secretory cell contains many irregular canals through which secretion oozes to reach the space above the cuticle where the fungi are maintained. (Happ, Happ, and Barras, 1971, from *Tissue and Cell*, vol. 3, with permission of the publishers, Longman Group Ltd.)

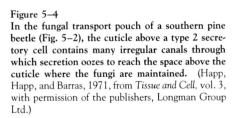

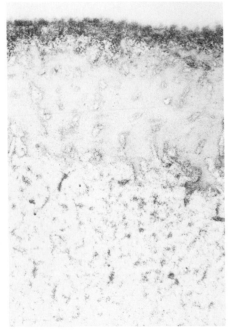

pheromone components. Their diversity among species helps to avoid signal confusion. To increase the information content even more, most pheromones are mixtures of several chemical substances so that the actual signal is a medley of several components.

Active Spaces.   Pheromone signals occupy space and persist for some time. Many of the special features of the communication systems are related to the expansion of the signal and to its fade-out in space and in time. Imagine an insect sitting in the middle of an empty basketball court and emitting pheromone. Assume that there are no breezes. As the pheromone begins to evaporate from the insect, its concentration rises in the air nearby. As times goes on, the molecules diffuse outward to the edge of the center circle and beyond. There will be a hemispheric cloud of pheromone, with its concentration highest at the source and falling off in all directions.

To trigger a response, the pheromone must be perceived by another insect. In most cases the response is fairly stereotyped, once the pheromone has exceeded a threshold concentration. Far away from the source, pheromone concentrations are extremely low, and no insects respond. As one moves closer to the pheromone source, concentration in the hemispheric cloud reaches threshold. And at lesser distances, concentration is above threshold. Thus there is a zone within which other insects respond and a surrounding environment where they do not. Wilson and Bossert called this zone of response the **active space.** In Wilson's words, "The signal is the active space."

How large is an active space? Its size is determined by both sender and receiver. When an animal has given off pheromone for a long period, the active space is likely to be at its maximum size. An active space will be larger for a receiver with a low threshold of response than for one with a high threshold. When the emission rate of the pheromone is high, the active space will expand faster and achieve a larger volume than when the emission rate is low. By altering the rate of emission, the threshold, and the chemical signal itself, the communication system can be tuned to optimum performance for a particular task. The active space for a sex pheromone may be large, but that for an alarm substance should be quite small.

A basketball court is an artificial environment, but the lessons learned there about our hypothetical insect can be applied to the natural world. The major differences between the gymnasium and the forest are winds and obstacles. In nature, breezes act on odor clouds. The cloud of pheromone drifts downwind and forms a plume, or aerial odor trail (Fig. 5–5). The plume will rarely be perfectly symmetrical, since breezes increase and decrease unpredictably. The resulting turbulence gives the plume ragged edges (Fig. 5–6), and these edges are further distorted by flowing over hills and into valleys, by going around and through trees, and by changes in temperature and air pressure. Every plume is a little different from all the others. Although there remains much debate about the detailed shapes of these clouds and about the effects of wind and weather, the general concept of an odor plume serves well to give us a feel for the chemical signal.

Sex Pheromones.

Some pheromones attract one sex to the other over long distances, while others act at close range to trigger some of the elements in a chain of courtship behaviors. These pheromones are not produced continuously by all members of a species, but only by animals that are reproductively mature and often only at a particular time of day. Because disruption of reproduction is an appealing strategy for control of insect pests, a large number of species have been studied in university laboratories and agricultural research stations throughout the world. The research began as a descriptive search for the molecules, and as more and more have been identified, the field is shifting toward careful analyses of the mechanisms by which pheromones act.

Early on (around 1960) it seemed that one sex of each species might produce a single, unique substance that automatically triggered a rigid sequence of stereotyped behavior in the opposite sex. That view of pheromones was attractive because of its simplicity. However, natural selection has left much more interesting and complex phenomena for the entomologist to analyze. Most pheromones are complex medleys of scents, and these odor blends have many subtle shades of meaning. Rather than list all the molecules, or even a representative few hundred, we shall describe the biology and chemistry of the sex pheromones used in the reproduction of two groups of insects: moths and queen butterflies.

Moth Attractants.

In the nineteenth century, the French naturalist J. H. Fabre demonstrated that male peacock moths are attracted to a paper that previously held a female but pay no attention to a female under a glass bell jar. He correctly concluded that a residual sex scent on the paper was responsible. But it was not until 1959 that Adolf Butenandt, of the University of Munich, Germany, became the first to determine the chemical structure of a pheromone. That was the culmination of a 30-year effort with *Bombyx mori*, the oriental silkworm moth, which has been cultured for many centuries in Japan and China.

When a female *Bombyx* moth emerges from her pupal cocoon, nearby males become highly excited and will copulate immediately with her if given the opportunity. Males do not even need to see the female. They exhibit characteristic excitement (wing fluttering, flying upwind, abdominal movements, attempting copulation) when exposed to female scent. That scent is normally emitted from two glands that

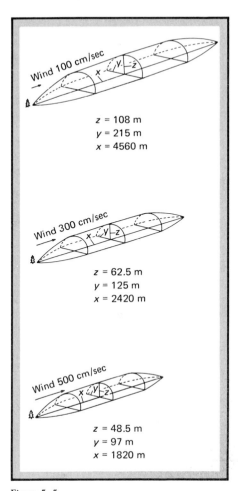

Figure 5–5
The active space created by a single female moth releasing sex attractant downwind. So long as the female continues to emit pheromone, the space will persist. If wind velocity increases, the active space shrinks. (E. O. Wilson, 1970, from *Chemical Ecology*, edited by E. Sondheimer and J. B. Simeone, with permission of Academic Press, Inc., publishers.)

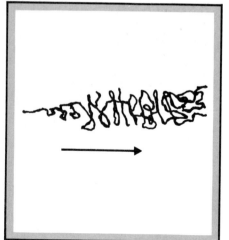

Figure 5–6
The irregular filamentous structure of an odor plume formed by continuous emission of a pheromone into air moving in the direction of the arrow. (Modified from H. H. Shorey, 1976.)

open in the intersegmental membrane at the tip of the female abdomen (Fig. 5–7). Butenandt and his co-workers painstakingly snipped off the tips of the abdomens of 500,000 virgin female *Bombyx*. These tips were then extracted with ethanol-ether, and the active components were carefully purified from the crude mixture and then characterized by brilliant microchemical techniques that were devised by Butenandt for this project. He obtained about 12 mg of pure attractant and identified it as a 16-carbon alcohol, which he called bombykol.

Male *Bombyx* are very sensitive to female scent, and the male fluttering response is used as a bioassay to compare different concentrations of pheromone. For each test a sample of dilute pheromone is placed on a filter paper cartridge and blown by an airstream to resting males. A typical dose-response curve for pure bombykol is shown in Fig. 5–8. Examine it closely. Such curves may seem dry and abstract, but they neatly summarize the data and show how they fall into a pattern. Below 10^{-6} μg the response is no different from background activity. Between 10^{-6} μg and 10^{-4} μg there is an abrupt rise as an increasing percentage of males respond. Above 10^{-4} μg, essentially all the males respond.

For almost 20 years it was believed that bombykol is the only component emitted by female *Bombyx*. But in 1978 K. E. Kaissling and his colleagues at the Max Planck Institute in Seewiesen, Germany, found that the female also produces the corresponding aldehyde in the ratio (bombykol:bombykal) of 10:1. When males are stimulated with pure bombykal at moderate doses, there is *no* behavioral effect. But when the

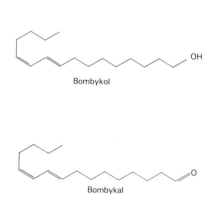

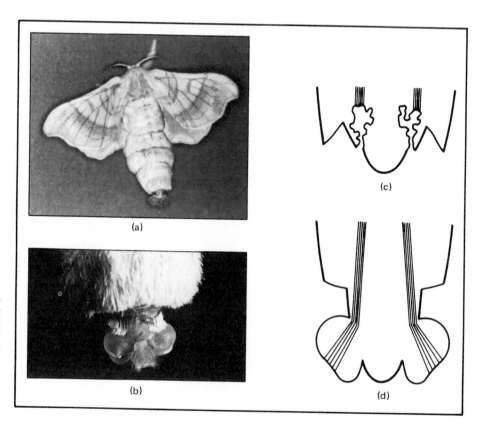

Figure 5–7
(a and c) The gland of the female silkworm moth that secretes the sex pheromone. The secretory cells open into a pouch formed by the infolded intersegmental membrane. When the female is "calling" (emitting pheromone), the pouch is folded out so that the pheromone can evaporate, as shown in the lower figures (b and d). (R. A. Steinbrecht, 1964, from *Zeitschrift mikr. Anat.*, vol. 64, pp. 227–261, with permission of the publishers, Springer-Verlag, Inc., and the author.)

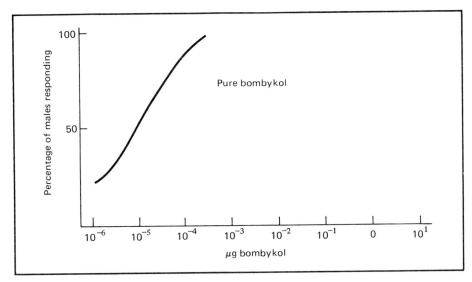

Figure 5–8
Fluttering response of male silkworm moths when exposed to increasing concentrations of bombykol. (Modified from K. E. Kaissling, 1979, "Recognition of pheromones by moths, especially in Saturniids and Bombyx mori," in *Chemical Ecology: Odour Communication in Animals,* F. J. Ritter, ed. Amsterdam: Elsevier, pp. 43–51.)

alcohol and aldehyde are blended together as in the natural pheromone, bombykal partially inhibits the response of the males to bombykol. The dose-response curve for this mixture (which presumably approximates the natural pheromone) is shown in Fig. 5–9. Note that relative to pure bombykol, the curve is shifted to the right, and its rise is much more gradual as the dose increases. The curve shown for the blend reflects a much more graded response to pheromone concentration.

Pheromone detectors are found on the antennae of male *Bombyx.* The large, feathery antennae are studded with about 64,000 sense hairs, of which about 80% are specialists that respond only to sex pheromone. The structure of these sensilla and the way they function are described in Chapter 6 (p. 151). Each sensillum has two **sensory**

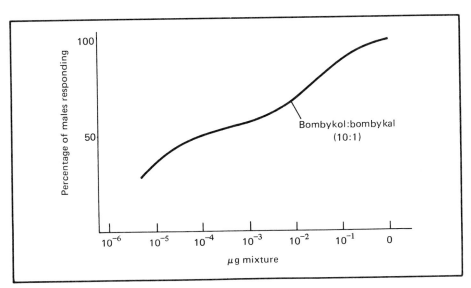

Figure 5–9
Fluttering response of male silkworm moths when exposed to increasing concentrations of a bombykol:bombykal mixture (10:1). (Modified from K. E. Kaissling, 1979.)

neurons (Fig. 6–11), one sensitive to bombykol, the other to bombykal, and each sends a signal independently to the brain. The brain computes the relative concentrations and makes the decision whether or not to flutter.

On any summer evening many different species of moths emit pheromones simultaneously, and somehow males must distinguish conspecific female scents from all the others. Wendell Roelofs and his co-workers at Cornell University have shown that the pheromone blends put out by some species are very precisely regulated. For example, two species of the genus *Archips* (leafroller moths) share the same four components in their pheromones. But the two species differ in the relative proportions of the components—60:40:4:200 for *A. argyrospilus* and 90:10:1:200 for *A. mortuanus*. Males of each species have specialist receptors for each of these molecules. Their sensilla have at least four neurons, and apparently each neuron recognizes one of the components. Signals from these sense cells are sent in parallel to the brain, where the relative proportion of the components must be computed. In this way males are able to recognize the brand of perfume emitted by females of their own species.

How well can moths discriminate between a signal, such as bombykol, and other molecules of similar size and with similar properties? The fluttering response can be used to compare similar molecules, such as geometric isomers of bombykol. Bombykol has two double bonds and thus four geometric isomers. Butenandt synthesized all four isomers, and experiments showed that the threshold for bombykol itself is 100 to 1000 times lower than for any other isomer. Many other experiments have supported the conclusion that precise molecular shape is recognized by pheromone receptors.

How many males can a single female attract? The pheromone gland of each female *Bombyx* contains about 164 ng of bombykol, in theory enough molecules to excite 10^{12} males if each received a threshold dose. But such calculations ignore the impact of bombykal in the blend, and even if so adjusted, they would not be biologically meaningful. The important advertising signal is the active space. Production of large amounts of pheromone is largely an adaptation to maximize the volume of active space. By casting her net wider, the female increases the likelihood of attracting males.

Once stimulated by pheromone within the active space, a male requires orientation cues in order to find the female. For long-range orientation, he appears to use mostly **anemotaxis**—movement based on the direction of the wind.

Insects flying upwind are moving toward the source of wind-borne scent. It is difficult to observe sustained oriented movement in free-flying moths, but Ilse Schwinck, of the University of Munich, Germany, found a clever alternative system. She used a flightless *Bombyx* mutant. When excited by female pheromone, flightless males wave their useless wings and try to run to the female. Schwinck watched the males as they ran along a table in the laboratory. When exposed to female scent, males ran upwind in a zig-zag course irrespective of the actual origin of the scent (Fig. 5–10).

Later experiments have shown that the "Schwinck effect" also applies to free-flying moths. J. S. Kennedy and D. S. Marsh, of Imperial College, London, confined male moths to a special wind tunnel and placed a "calling" female upwind. Males flew upwind in a series of diminishing, irregular zig-zags. When the calling female was suddenly removed, males flew back and forth across the wind with frequent left-right reversals (Fig. 5–11). In some other field experiments, the males actually blundered out of the plume. To find the female, they flew back downwind to reenter the odor trail and began anemotaxis again (Fig. 5–12).

Moths might orient also by **chemotaxis**, detecting slight differences in pheromone concentration that would permit them to follow a chemical gradient upstream toward its highest concentration where the female sits. Some evidence suggests that moths

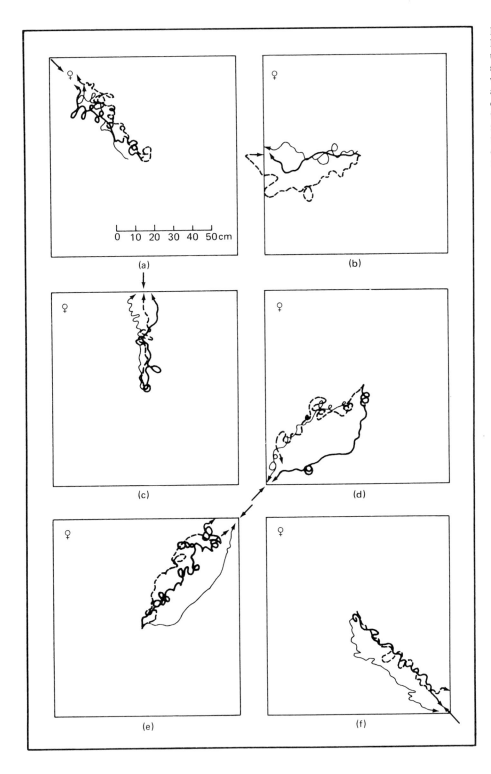

(a)

0 10 20 30 40 50 cm

(b)

(c)

(d)

(e)

(f)

Figure 5-10
Flightless male silkworm moths run across a table to reach the source of female scent. Each panel shows three records of runs after the male moth was placed in the center of the table. The female is always placed at the upper left corner. The wind direction, indicated by the small arrows, was systematically moved around the table. It is clear that males move *upwind* rather than in the direction of the female. (I. Schwinck, *Zeitschrift für Vergleichende Physiologie*, vol. 37, p. 36, with permission of the publisher, Springer-Verlag, Inc.)

can detect and follow chemical gradients in still air, but it has yet to be shown that they can tell upstream from downstream. Chemotaxis certainly is used for short-range orientation in insects, but it remains to be demonstrated as an important mechanism in long-distance orientation of moths.

When males get close to a female, their behavior changes. They hover or land and run about until they collide with her. How do the males know that a female is nearby? In some species it appears that the male sees a silhouette against the sky or sees the female herself and then alights. In other species a hierarchy of active spaces may be involved. In the redbanded leafroller moth, Roelofs and his colleagues found that the pheromonal blend included components that acted at a distance as well as one, dodecyl acetate, that served in close-range orientation and elicited landing responses.

Some arctiid moths might use yet another feature of their unusual signal—its pulselike character. William Conner and Thomas Eisner, of Cornell University, have recently shown that a female arctiid moth emits its sex lure in short pulses (100 times per minute). At distances of several meters from the source, the pulses would probably be smeared together by turbulence, but close in, they remain discrete (Fig. 5–13). When males are but a few centimeters downwind, their antennal receptors show corresponding pulsations in electrical potential. It is possible that when the continuous signal transforms into an on/off pattern of pulses, males switch from upwind anemotaxis to short-range search.

Queen Butterfly.
Visual cues attract males to females in most diurnal butterflies, such as the queen (*Danaus gilippus berenice*). The courtship sequence, a chain of signals between male and female, was described in detail by Lincoln and Jane Van Zandt Brower while they were at Amherst College in Massachusetts (Fig. 7–17).

Male queen butterflies have a pair of large organs, called **hair pencils,** which are really tufts of fine hairlike processes of cuticle (Fig. 5–14). When a male queen sees a female in flight, he rapidly overtakes her. As he passes over her dorsal surface, he extrudes the large hair pencils from the tip of his abdomen and bobs up and down over the female's head and antennae. As shown by Thomas Pliske and Thomas Eisner, of Cornell University, the hairs are covered with dustlike particles that are brushed onto the antennae of the female. These numerous particles are actually microscopic spherical bits of epicuticle that were formed during pupal development. The particles clump

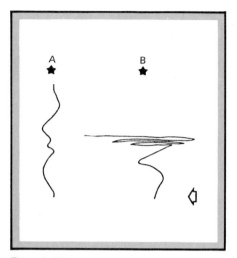

Figure 5–11
Two 3-second flight tracks of the same male moth as it came within 0.5 m of a source of female scent (star). The wind is blowing from top to bottom in the diagram. (a) The source remained in place throughout the male's approach; (b) the source was removed as the male reached the level of the arrowhead so that the male reached unscented air as he flew 0.5 m upwind. (J. S. Kennedy and D. Marsh, *Science*, vol. 184, p. 100, copyright 1974 by the American Association for the Advancement of Science.)

Figure 5–12
Male moths approaching a pheromone source often hover near it before alighting. If they accidentally leave the odor trail, the males fly downwind and start over. (J. Murliss and B. W. Bettany, 1977. Reprinted by permission from *Nature,* vol. 268, no. 5619, pp. 433–35, copyright 1977 Macmillan Journals Ltd.)

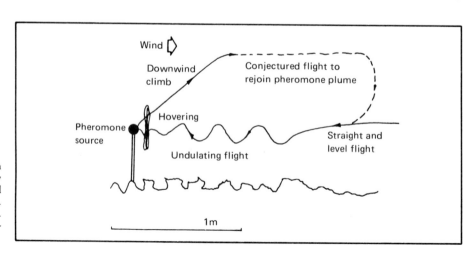

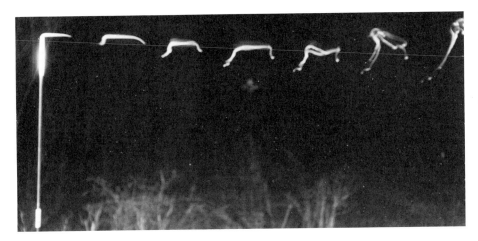

Figure 5–13
Simulation of a temporally patterned chemical signal. Air laden with a visible marker (titanium tetrachloride) is pulsed at a rate of 2 puffs per second from the glass capillary shown at the left. The pulses remain discrete over a distance of 60 cm. (W. E. Conner et al., 1980, from *Behavioral Ecology and Sociobiology*, vol. 7, p. 62, with permission of the publishers, Springer-Verlag, Inc.)

together and stick to the hairs and to the antennae because they are coated with an oil film. The oil contains the real signal, and the dust is just a means for its transfer. The oil is apparently produced by a secretory cell at the base of each hair. Jerrold Meinwald and his team at Cornell identified two major components in the secretion: a ketone and a viscous terpenoid diol.

In response to the hair penciling, the female lands. The male continues to hair pencil vigorously, and the female folds her wings. The male then alights and attempts copulation. Pliske and Eisner deprived males of their hair pencils and found that, although they courted actively, the courtship very rarely led to copulation. Males raised indoors on cuttings of the major food plant failed to mate successfully, although males raised outdoors, mostly on the same intact food plant, courted successfully. Comparison of these two male groups showed that the indoor males had the diol but lacked the ketone, whereas the outdoor males had both components. The outdoor males obtained the chemical precursor of the ketone from another plant on which they spend very little time and of which they eat little. But without that snack, they could not court females.

When artificial ketone in an oily vector was applied to the deficient males, they began to court successfully. The behaviorally active aphrodisiac pheromone is apparently the ketone. The diol provides the oily vehicle for its spread and adherence.

Aggregation Pheromones.

Many kinds of insects form aggregations for feeding, mating, shelter, oviposition, or some other purpose. The gregarious behavior may wax and wane with the passing seasons, perhaps occurring only in the cold months for hibernation. It may be confined to a certain time of the day, as in sleeping aggregations of butterflies. It may be associated with a developmental stage, as in tent caterpillars, which are gregarious as larvae but not as adults. It is commonly associated with opportunistic behavior, as when lycid beetles form large aggregations to feed on sweet clover and to mate. Once the temporary or restricted resource is exhausted, the aggregation breaks up. The majority of insect aggregations form in response to pheromones. The most intensively studied species are bark beetles and ambrosia beetles, both members of the family Scolytidae.

Bark beetles and ambrosia beetles aggregate for feeding, mating, and oviposition. Because the beetles and the fungi that they carry from tree to tree do such enormous

Figure 5–14
Hair pencils on the abdominal tip of a male queen butterfly. (Photograph by Thomas Eisner, Cornell University; from T. Pliske and T. Eisner, *Science*, vol. 164, p. 1171, copyright 1969 by the American Association for the Advancement of Science.)

damage to pines of our forests and to the American elm, the aggregation behavior has been much studied. Bark beetles' pheromones illustrate well the complexity and natural history of a chemical signal system.

Shortly after their final ecdysis, the adult bark beetles emerge from their host tree and begin a dispersal flight to find a new host. Depending on the species of beetle, either males or females (but never both) are the pioneers. When they reach a new host, the pioneers burrow through the bark and construct galleries. As they chew their way along, the beetles seed the galleries with fungal spores, which germinate to form the mycelial mass on which beetle larvae will feed.

The colonization of a new tree begins with pioneer beetles that orient to a particular host tree by several cues, such as its shape and odor. In the genus *Dendroctonus*, females are the pioneers; in the genus *Ips*, males are the first to colonize. Once they land, the beetles burrow into the bark. Healthy trees produce a flow of resin or pitch that may eject the burrowing beetle from its partially completed hole or may drown it. But when a few pioneers manage to complete the entrance tunnel and to establish themselves, they release aggregation pheromone. More beetles are attracted, and the balance begins to shift toward the intruders. The mass attack causes mechanical damage to the phloem, and fungal hyphae block the conducting vessels of the tree. The tree is overwhelmed and dies. Since dead timber dries rapidly, it is advantageous for many beetles to exploit the resource quickly before it becomes unsuitable.

All of the bark beetle pheromones identified thus far are monoterpenes, structurally similar to those produced by the host tree. The aggregation is a response to a medley of molecules that enhance each other's effectiveness. Such molecular components are described as **synergistic.** Each of the synergists may be somewhat attractive when tested alone, but the attractiveness of the blend is much greater than we would expect from merely adding up the sum of all the contributions from each separate component.

The first bark beetle pheromone to be identified was that of *Ips confusus*, the California five-spined ips. As the beetles chew their way into ponderosa pine trees, the wood passes through their gut and is deposited in pellets called frass. David Wood, an entomologist at the University of California, Berkeley, and Robert Silverstein, an organic chemist now at Syracuse University, isolated three terpene alcohols, ipsensol, ipsdienol, and *cis*-verbenol, from the frass of attacking male engraver beetles. In a laboratory bioassay with walking female beetles, ipsenol alone was excitatory as a single component, and it was synergized by mixing with one or both of the others. In field tests only the combination of all three synthetic compounds proved attractive. When the synthetic mixture was later compared with "natural" scent from male-infested pines, that natural scent was much more effective than a synthetic blend of ipsenol, ipsdienol, and *cis*-verbenol. Apparently the natural pheromone contains other components that are not yet identified.

Upon arriving at the male-infested tree, a female runs about, searching for entrances to a male burrow. Outside its entrance, she stridulates by rubbing the serrated edge of her elytra against a scraper on her abdomen. In response to the female chirps, the male may allow her to enter his burrow; mating follows. Each male will allow only three females to enter the nuptial chamber, and, after mating, each female constructs an egg gallery in the phloem and outer xylem. If any more females stridulate or attempt to enter the burrow, the male blocks the entrance with his body.

Several species of *Ips* have overlapping geographic distributions. Both *Ips paraconfusus* and *Ips pini* attack ponderosa pine in California, but the two species rarely attack

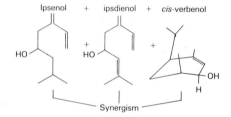

Ipsenol + ipsdienol + *cis*-verbenol

Synergism

exactly the same tree. Martin Birch and his colleagues at the University of California, Davis, have shown that resource competition is minimized because each species responds only to its own attractant blend and because each blend includes an inhibitory scent for the other species. The pheromones emitted by males include the following:

Ips paraconfusus	*Ips pini*
100% S-ipsenol	
90% S-/10% R-ipsdienol	100% R-ipsdienol
1S,4S,5S-*cis*-verbenol	

S-ipsenol inhibits females of *I. pini*, and pure R-ipsdienol inhibits the attack by females of *I. paraconfusus*. One component of the aggregation pheromone of each species is an inhibitory allomone.

As mentioned earlier, in bark beetles of the genus *Dendroctonus* it is the females that are pioneers. This genus includes many serious pests of coniferous forests, such as the southern pine beetle and the western pine beetle. From the frass of the latter species, Wood and Silverstein isolated and identified the unusual terpene *exo*-brevicomin. Pure *exo*-brevicomin is quite attractive to males, but many more males were attracted to a synthetic mixture of *exo*-brevicomin and myrcene, a terpene produced by wounding the host ponderosa pine. After responding to these substances, males arrive at a tree laden with females and give off a pheromone of their own, frontalin, which attracts other females. Like *exo*-brevicomin, frontalin is synergized by myrcene. The combination of these substances is responsible for the mass attack of both sexes. Following mating, both sexes begin to release two additional pheromones, *trans*-verbenol and verbenone, which inhibit further arrivals.

Inhibitory pheromones are one device to control density so that the tree is not overloaded with larvae. Another device to control density is high attractant concentration, as demonstrated in trapping experiments. Beetles are readily caught in artificial traps baited with a mixture of brevicomin, frontalin, and myrcene. However, when pheromone concentration gets very high, the catch in the trap declines. At very high concentrations of pheromone the incoming beetles turn away from the baited tree and attack nearby trees. The result is a cluster of infected trees that produces a "hot spot" of brown foliage in the forest.

The ability of beetle receptors to recognize differences between the pheromone components and similar molecules is impressive. Geometric isomers are readily discriminated; thus *trans*-verbenol is an effective inhibitor for attacking western pine beetles, while *cis*-verbenol is not.

We do not know exactly how the chemicals in bark beetle aggregation pheromones are manufactured. No discrete exocrine glands seem to be responsible. The aggregation scent is not produced if the host terpenes are absent, and many of the pheromones seem to be oxidized derivatives of the host terpenes.

For example, male *Ips paraconfusus* convert myrcene to ipsenol and ipsdienol. David Wood and his associate, John Byers, have shown that the oxidation of myrcene to ipsenol and ipsdienol is inhibited by the antibiotic streptomycin. They suggest that symbiotic bacteria, restricted to the male, are responsible for myrcene oxidation. If bacteria play such a role, the beetle depends on the tree for the starting material and the bacteria for the metabolic conversion in order to produce an insect aggregation pheromone.

Figure 5–15
After an attack by a predatory bug (Hemiptera, Nabidae), an aphid secretes a droplet of fluid that contains the alarm pheromone. (Photograph by L. R. Nault, Ohio Agricultural Research and Development Center; from W. S. Bowers et al., *Science*, vol. 196, p. 680, copyright 1977 by the American Association for the Advancement of Science.)

β-farnesene

4-methyl-3-heptanone

Advertising an aggregation carries a risk. Predaceous carabid beetles, which feed on the bark beetles, are also attracted by the bark beetle aggregation pheromones. When the carabid arrives at the infested tree, it consumes the source of the signal.

Alarm Pheromones.   Wherever insects congregate, a predator has an easier time finding prey. Upon approach of the predator, the group may disperse or it may invoke a colonial defense response. In some hemipterans and in social insects, an alarm pheromone triggers these behaviors.

Aphids form dense feeding aggregates. When a predator attacks, the victim secretes a droplet of fluid from a large gland opening called a **cornicle** (Fig. 5–15). Within that fluid is an alarm pheromone, of which the most common is farnesene. Nearby aphids respond by falling to the ground, walking away, or leaping from the plant.

Alarm Pheromones of Ants.   Everyone who has stepped on an ant nest has seen the outrush of defenders, which open their mandibles and swarm over the attacking foot. This alarm behavior is released by a pheromone emitted by the mandibular glands and by glands associated with the sting.

Wilson and Bossert predicted that an alarm signal should be localized, blatant, and brief. The pheromone should spread rapidly to coordinate colonial defense and should fade out rapidly to avoid prolonged false alarms. For the site of the disturbance to be easily located, the active space should remain small. To test these predictions, they investigated the size and time-course of the active space for alarm in a harvester ant. The major component of this ant's mandibular secretion is 4-methyl-3-heptanone. Wilson and Bossert quickly crushed the head of a worker ant to release all of its mandibular gland contents at once. They watched the response of nearby workers, and by repeating the experiment many times, they were able to deduce the properties of the active space. The active space expanded to its maximum radius (6 cm) in 13 seconds, and as the pheromone continued to diffuse, the active zone shrank to nothing by 35 seconds. In a natural situation with a serious invasion of the nest, other workers, excited by the first secretion, would have encountered intruders and added their own pheromone. This added pheromone would amplify and prolong the signal and expand the defensive frenzy. The size of the group response is matched to the severity of the attack.

Alarm pheromones are very volatile, and most are of low molecular weight, containing 12 carbons or fewer. Those from the mandibular glands usually have a keto- or aldehyde group, while those from Dufour's gland (near the sting) are most often hydrocarbons. Many are mildly toxic and distasteful; they thus serve the simultaneous functions of alarm and defense.

African weaver ants construct an elaborate nest by sewing leaves to one another with thread from the silk glands of their own larvae. Both the nest itself and the territory surrounding it are vigorously defended by the major workers. Any insect that happens to wander into that territory is rapidly attacked and often becomes prey for the ants. The defensive behavior is a sequence of four components: (1) alerting, in which ants raise their heads and open their jaws; (2) attraction toward the disturbance; (3) stopping near the source of the disturbance; and (4) biting and holding for several minutes.

Alarm behavior follows the expulsion of a fragrant but musty odor from the mandibular glands of a major worker. J. W. S. Bradshaw, R. Baker, and P. E. Howse, of the University of Southhampton, England, have shown that at least 33 volatile components are released by crushing the head of a major worker. Four of these seem to trigger specific aspects of the alarm response. These molecules differ in their volatility from hexanal (most volatile) to 2-butyl-2-octenal (least volatile). Their mechanism of action can be best understood if we imagine that an excited ant deposits a droplet of mandibular gland secretion at the center of the concentric circles shown in Fig. 5–16. In still air each component diffuses away from the site of deposition. Because they differ in volatility, some fill space more quickly than others. After a short time, hexanal defines the largest active space, indicated by the outermost circle. When ants enter this space, the alerting behavior is characteristic. As they move into the next zone (hexanol), the major workers are attracted toward the site of its deposition. In the innermost zone 2-butyl-2-octenal is the biting marker, and 3-undecanone contributes to biting and is also a short-range orientation signal. When the secretion has been allowed to evaporate for some minutes, no effects of hexanal or hexanol are detectable, but when the workers come close to the deposition site, biting behavior is released. Thus a chain of behaviors is sequentially triggered by the discrete chemical components of the alarm pheromone that are localized on the intruding object.

Minor workers of the weaver ant emit a considerably different set of components from their mandibular glands when they are disturbed. If other minors are exposed to that secretion, they do not defend the nest but instead flee from the site of disturbance. In contrast, major workers are attracted and arrested by high concentration of at least one component in the minor's perfume. Thus major workers come to the defense of a violated minor worker and of the nest.

How private are the alarm signals? Are they specific for each ant species or for each colony? Apparently they are much less so than for sex pheromones. Often the same components are shared among many species within a genus and among genera within a subfamily. However, there are subtle and biologically important differences between similar species, as shown by studies in the ant genus *Myrmica*. Four species of this genus that coexist in similar habitats in England and Belgium have been found to have similar blends of several components in their alarm pheromones, but their relative proportions differ. In effect, each species has a different dialect.

Trail and Territorial Markings.   Both food and habitat are limited. Territorial and trail pheromones enable efficient resource exploitation. Many foods, like blueberries or caterpillars, come in discrete packages. Female insects that place their eggs in such packages may mark them with a pheromone that is like a sign that says "Occupied." Some parasitic wasps include such a territorial pheromone with each egg injected into an insect host. Female apple maggots mark each fruit after ovipositing in it. After she has laid an egg, the female drags her ovipositor over the surface of the fruit and thus deposits pheromone. The size of the mark is matched to the future needs of one larva from hatching to maturity. A small fruit, like a cherry, is completely coated after one oviposition and thus will be avoided by subsequent females. A large fruit, like an apple, is marked on only a portion of its surface. It will collect several eggs until its surface coat of pheromone reveals that it is loaded to capacity with larvae.

Efficient resource utilization requires choices among alternatives. Many social insects use recruitment trails to direct nestmates to the richest food sources. Recruitment trails are continuous directional signals that precisely guide the foragers to abundant goodies. Stingless bees use secretions of their mandibular glands to lay recruitment trails. Like honey bees, stingless bees have large colonies with many workers, use wax

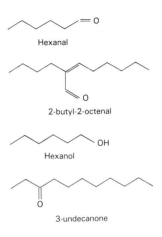

Figure 5–16
The mandibular gland secretions of major workers of weaver ants include four components with specific behavioral effects. These four differ in their volatility, and at 20 seconds after it has been deposited, the mixture creates a set of concentric active spaces. (J. Bradshaw, R. Baker, and P. Howse, reprinted by permission from *Nature*, vol. 258, no. 5532, pp. 230–31, copyright 1975 by Macmillan Journals Ltd.)

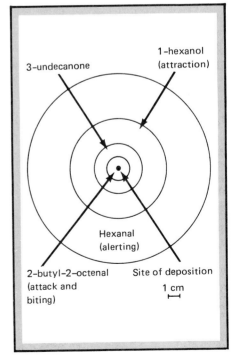

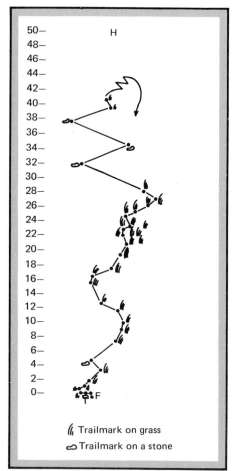

Figure 5-17
The record of a trail-marking flight of *Trigona postica,* a stingless bee. Each point indicates one of the short landings on her return flight, during which she deposited an odor trail mark. *H,* hive; *F,* feeding table. (Reprinted by permission of the publishers of *Communication Among Social Bees,* by M. Lindauer; Cambridge, Mass.: Harvard University Press, copyright 1961 by the President and Fellows of Harvard College.)

as a building material, and collect and store honey and pollen. Martin Lindauer, of the University of Munich, showed that stingless bee foragers recruit one another very well. When a pioneer forager finds a new nectar source, she makes a direct flight back to the nest with food. If the nectar source is still copious after several trips, she returns to the nest in a different manner. Instead of a quick flight home, she flies in small stages—stopping repeatedly to wipe secretions from her mandibular glands onto rocks, plants, or dirt along the route (Fig. 5-17). At the nest she arouses a number of her sisters, and with the pioneer as leader and pilot, the group flies along the trail of scent marks. The foragers stop along the way to check the marks, and eventually they reach the rich nectar source.

The recruitment tactics of these stingless bees require both the pilot bee and an odor trail. In many ant and termite species an odor trail alone is sufficient to guide a follower to a food source.

Ants and termites lay trails by dragging their abdomens along the substrate and depositing streaks of chemicals. The active space of the trail is a narrow corridor and extends upward in a half-cylinder. Followers run along within this active space, weaving from side to side and turning back to the center as their waving antennae break out of the active space into an unscented zone. The followers do not have to touch the odor marks but have only to stay within the active space. They can even follow a trail drawn overhead on the ceiling if ceiling and floor are close together.

The exact roles of these trails vary with the biology of each species. In harvester ants, Bert Hölldobler, of Harvard University, has shown that some are fixed trunk trails while others are more temporary recruitment trails. Trunk trails, laid by secretions of Dufour's gland, are like spokes radiating out from the nest entrance into foraging territory. Trunk trails are fairly permanent and are used mostly by ants that are returning to the nest. Recruitment trails, laid by the poison gland, branch off from trunk trails. They are laid by a successful forager as she returns from a food source. These trails entice other foragers to follow the chemical signals and thus to help retrieve the food. As food-laden foragers come back to the nest, they lay recruitment trails that reinforce the previous signals. Foragers that do not find food and return hungry to the nest lay no recruitment trails. The recruitment trail has a short half-life, and soon after a food resource has been exhausted, the recruitment trail fades away.

It is relatively easy to show that a given species of ant or termite follows a chemical trail, but the molecular components of the trail signal have proved difficult to isolate in most cases. To demonstrate the presence of a chemical trail, a researcher need only paint a line of a chemical sample along a substrate, such as a piece of filter paper, and then watch to see whether foraging ants or termites follow the line when they chance to cross it. Extracts of whole insects or their glands or chemical fractions can be readily assayed.

Less than a dozen trail pheromone components have been identified, and most of these are not very volatile. Neocambrene A, a complex terpene, is the trail substance of nasute termites, and a pyrazine is the trail substance in several species of myrmicine ants. At least in these ants, the true trail pheromone is not species specific. This pyrazine is found in all seven species of British *Myrmica,* and it is found also in a related genus of tropical leaf-cutter ants. The power of modern microchemistry is such that only 50 *Myrmica* workers provided enough pyrazine for the definitive chemical identification of the trail substance.

The trail pheromone of *Myrmica* is produced by the poison gland. When Marie-Claire Cammaerts, of the University of Brussels, Belgium, drew lines with poison gland extracts, workers that blundered across the lines turned to follow them, but foragers were not attracted to the trail. Further experiments revealed that the volatile components of Dufour's gland are attractants and excitants. Both glands have a common opening at the tip of the abdomen and both contribute to the trail.

In nature, trails are not public thoroughfares that are shared commonly by all species. Each species, and usually each nest, follows its own trails. An artificial trail, laid with a combination of the pyrazine and volatile components of the Dufour's secretion of one species of Myrmica, is followed quite well by foragers of the same species and equally well by foragers of other species. There is no apparent species identification. However, when Cammaerts added the remainder of the Dufour's secretion, which is an oily mixture of low volatility, she produced trails that were species specific. The oils differ among the species and allow each species to identify its own trail.

The oils from Dufour's glands are a **territorial pheromone** for Myrmica. Pioneer foragers, seeking new food sources far from the nest, move forward slowly and tentatively. As they walk ahead onto new ground, each forager repeatedly marks the ground with her Dufour's gland. The oils persist for long periods, and subsequent foragers will run quickly across the same ground that was so slowly crossed the first time. The oily marks distinguish home territory from alien territory.

The natural foraging trails of Myrmica have three kinds of components, each of which serves a distinct purpose. Secretions of the poison gland define the trail itself. Volatile substances from Dufour's gland attract and increase running speed, but these signals are short-lived. Finally, oily substances from Dufour's gland remain and specify the home territory.

Pheromones and Individual Differences.

Not all members of a species emit the same complement of pheromones. Differences between life stages (larva versus adult) and between sexes are obvious. In addition, genetic strain, physiological state, and nest and caste identification in social insects are communicated by pheromones.

For successful mating of the fruitfly, Drosophila melanogaster, the female must emit pheromones that stimulate the male to court, and the male must emit pheromones that cause the female to accept. Both sexes prefer to mate with "strangers" that are genetically quite different, and they recognize the differences between related and unrelated partners by their differing pheromones. Males and females from the same inbred line do not spontaneously court one another unless they are fooled by being brought together in an environment filled with alien pheromones.

Many halictine bees are colonial and nest together in burrows. At the entrance to each colony, a guard bee inspects newcomers and rejects those that are not residents (Fig. 8–5). Charles Michener and his associates at the University of Kansas wondered how the guards of one species, Dialictus zephyrus, distinguished nestmates from aliens. Bees were brought to the nest opening in a piece of plastic tubing and poked out into the entrance with a pipe cleaner. Of the 228 nonresident bees introduced into other nests, 50% were attacked by guards before contact and 92% on contact. Fifty residents were introduced in a similar manner, and all were accepted. Next, Michener killed 20 residents and 32 nonresidents by freezing and presented them, after thawing, one by one to the guard bees. Only 10% of the frozen residents were rejected, while 97% of the nonresidents were attacked. Nonresidents were also rejected when the room was illuminated with a red light to which bees are blind. These experiments and others established that touch, vision, and hearing are not used for rejection. Odor must account for recognition of nestmates. Does the guard remember the scent of a nestmate, or does she merely compare the odor of the entering bee with that of the nest? When guards are placed on duty in a foreign nest, they accept only their *own* nestmates and reject the owners of that nest. It is the odor of the bee, not that of the nest, that these guards use as a criterion for rejection.

The colony odor has a strong genetic component. Nestmates are sisters and are genetically similar. Guard bees of *Dialictus zephyrus* can recognize individual odors that are variants on the pheromone blends produced by their nestmates, and they will thus admit genetically related bees that were reared in other nests. In honey bees, as well, colony odor has a genetic basis. When a honey bee queen is suddenly presented to workers of another colony, she is usually attacked. But she is much more likely to be accepted by these workers if she is closely related to their own queen. Michael Breed, of the University of Colorado, Boulder, has shown that inbred sister queens are accepted in 35% of the exchanges, while nonsisters are invariably rejected.

Learning plays a part in the recognition of individuals or classes of individuals. In laboratory colonies, an halictine guard bee, placed in a "foreign" nest, can habituate to the odors of that nest and then will accept a bee with that odor, but such a situation is probably rare in nature. The response of male halictines to females is more meaningful. Upon first scent, a male responds avidly to the novel odor of a new female. But if she is unwilling to be courted, he becomes unresponsive to her now familiar odor. When exposed to a second new female with yet a different odor, he immediately attempts to court again. It appears to be an advantage to learn not to carry the torch for a lost cause.

From what glands is that individually distinctive body odor derived? Probably from many. The surface layers of insect cuticles contain odorous molecules that originate in the underlying epidermal cells. Other contaminants are left on the surfaces around the exocrine glands and dissolve into the coat of epicuticular and surface lipids. Each exocrine gland has many components, and their proportions vary slightly between individuals. The rich mix of secretions from many glands, the scents from the environment, food odors, and the intrinsic odors of the epicuticle combine to produce pheromonal polymorphism among individuals and thus allow individual recognition.

Primer Pheromones.   Primers affect the physiological state of the recipient. It is usually assumed that primer effects are mediated by the endocrine system. Thus pheromones influence endocrine glands, and hormones directly regulate the physiological state. Primers are found in subsocial and social situations where the persistence of the group allows stimuli produced by one individual to act on another.

Many primers influence reproductive capacity. A pheromone from male desert locusts increases the rate of development in immatures of both sexes. The scents of male and female mealworm beetles accelerate egg development and other aspects of reproductive maturation in young female beetles. Honey bee queen substances inhibit the growth of ovaries in the workers.

Even more dramatic are the primers that switch development from one end point to an alternative. Desert locusts have such a primer, locustol, which triggers development from the solitary form to the gregarious form. The gregarious form is responsible for the devastating locust swarms that have ravaged crops in Africa since biblical days (Fig. 16–10). As more and more locusts feed together, they begin to produce locustol in their feces. Locustol causes a gradual transformation of the light-colored solitary types into much darker gregarious forms that gather together and migrate, eating everything in their path.

Many termite colonies have only two functionally reproductive individuals at any one time. If an accident or an entomologist removes the male or female reproductive,

Locustol

one of the nymphs of the same sex develops into a "supplementary reproductive." The presence of a functional reproductive inhibits the maturation of others of the same sex. It is assumed that each reproductive individual produces an inhibitory pheromone, which is passed around by mutual grooming to workers and among workers. In at least one species of termite, it is possible to inhibit sexual maturation of nymphs by introducing extracts of the functional reproductives. Soldier production is probably limited by an analogous soldier-specific inhibitory primer, but the experimental evidence is not yet conclusive.

Defensive Allomones

Some insects avoid being eaten because they are distasteful. Foul-smelling and evil-tasting chemicals in both animals and plants confer protection from attackers.

Perhaps the simplest chemical defensive behavior for an insect is regurgitation—vomiting gut contents at an attacker, as do many grasshoppers when they "spit tobacco." The regurgitate may be enriched in unpleasant chemicals derived from the host plant. For example, sawfly larvae that feed on pine trees sequester terpenoid resins from the needles in two pouches of the foregut. The pouches, like the rest of the foregut, are lined with a cuticle that prevents diffusion of the resins into the blood of the insect. When attacked by an ant or a spider, the sawfly rears up its front, regurgitates a droplet of resinous fluid, and wipes the regurgitate onto the attacker. The regurgitate is rich in the resins of the pine tree, which are repulsive and distasteful to birds and to arthropod predators such as ants and spiders.

Blood-Borne Allomones. Many distasteful insects are aposematic (warningly colored), such that predators can readily associate the appearance of the animal with its unpalatability. For example, monarch butterflies are strongly aposematic with their striking orange and black wing patterns. When blue jays try to eat monarch butterflies, they vomit up the partially eaten carcass (Fig. 14–6). Lincoln and Jane Van Zandt Brower, then at Amherst College, Massachusetts, offered monarchs and other butterflies to captive jays. The birds quickly learned to associate the monarch color pattern with vomiting, and thereafter the butterflies were refused. Monarch butterflies are unpalatable because their blood and tissues contain cardiac glycosides. These glycosides were sequestered by the caterpillars from milkweed in their diet.

When the defensive allomone is contained in blood, it may be liberated by a special localized bleeding, or autohemorrhage. Blood bubbles out from weakened sites at leg joints, antennal joints, or between sclerites when the insect is attacked. Rough handling of a blister beetle will cause it to release clear droplets of blood from its leg joints, and lady beetles exude yellow drops of blood from similar sites (Fig. 5–18). The blood of the blister beetle contains cantharidin (spanish fly), while that of the lady beetle carries the alkaloid precoccinelline. In contrast to the cardiac glycosides used by monarch butterflies, both cantharidin and precoccinelline are manufactured by the insect itself rather than sequestered from a food plant.

Cantharidin and precoccinelline are distasteful to many vertebrate predators, but not to all. Thomas Eisner reports that toads will repeatedly consume blister beetles. Small predators are further deterred by the mechanical properties of the droplet. Ants become entangled in the sticky blood. As it coagulates, one leg is glued to another, antennae become stuck to mouthparts and ants adhere to each other. Ants break off the attack and go through a frenzy of cleaning activity.

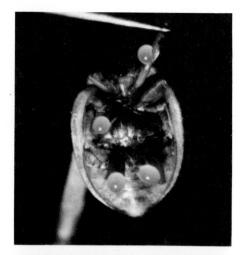

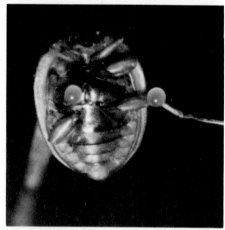

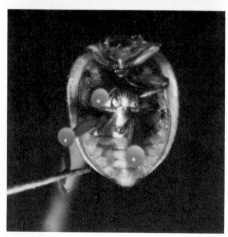

(b)

(a)

Figure 5–18
When roughly handled, lady beetles (a) and blister beetles (b) release droplets of blood from their leg joints. The droplets contain distasteful chemicals. Note that when a leg is probed with forceps, lady beetles release droplets from that leg. (Photographs by Thomas Eisner, Cornell University; a from G. Happ and T. Eisner, *Science*, vol. 134, p. 130; b from J. Carrel and T. Eisner, *Science*, vol. 183, p. 755; copyright 1961, 1974 by the American Association for the Advancement of Science.)

Allomones Produced in Exocrine Glands. Defensive allomones are frequently sequestered in cuticular reservoirs (as were those of the sawfly). Most often the reservoirs are storage compartments for secretions made within exocrine glands. The reservoir may be everted and wiped on the site of attack, as seen in the larvae of swallowtail butterflies (Fig. 5–19). The defensive fluid of these caterpillars contains isobutyric and 2-methyl butyric acid. Alternatively, the secretion may ooze from the gland opening, much as blood oozes from the knee joints of blister beetles. Finally, the secretion may be ejected in a jet of acrid mist. The spraying glands are dramatic in their impact and often extremely accurate. The defensive fluid of a Florida walking stick (*Anisomorpha buprestoides*) is aimed directly at the site of attack (Fig. 5–20) and effectively repels both vertebrate and invertebrate predators.

Cuticular reservoirs broaden the possibility of potential defensive allomones, for within a sac of impervious cuticle, small reactive molecules can be safely stored. In contrast, these toxins would be very difficult to store in the blood.

Defensive chemicals operate in a very different context from that of chemical signals such as pheromones. The potential prey is at an advantage when the signal is so blatant that the predator cannot ignore it. The predator is at an advantage when it can ignore the blatant insult or when it is insensitive to the defensive substance. Any insect has a host of potential predators, and one might expect natural selection to favor defensive allomones that are nonspecific, that is, are repellent to many species. The defensive secretions of many insects contain small reactive molecules that are broadly toxic to most living tissue. Small acids, like formic or acetic, readily coagulate protein and irritate excitable tissues. Formic and acetic acids are used in the chemical defenses of ants, beetles, and caterpillars. Reactive aldehydes, which bind readily to proteins, are painful tear gases to humans and animals. Molecules such as these are found in the defensive secretions of walking sticks (*Anisomorpha*), cockroaches, beetles, and true bugs. Benzoquinones, which also bind proteins and "tan" them, are common in the defensive secretions of beetles and millipedes. Many predators find such small reactive substances unpleasant and spit out the prey that sprays them. In addition, these defensive allomones are effective as antibiotics against microorganisms.

Storage of reactive defensive molecules could pose a problem for the insect producing them. How does such an animal avoid being poisoned by its own secretions? The answer is twofold. First, the secretions are stored within reservoirs of cuticle that have special permeability properties that prevent any leakage of the toxic fluids. Effectively the secretion is stored outside the animal. Second, many of the secretions are produced in a stepwise manner, such that the cytoplasm of the secretory cells contains only harmless precursors rather than the toxic end product. The sequence in the production machinery is best seen in the adaptation for quinone production by bombardier beetles and darkling beetles.

Bombardier beetles are named for their defensive behavior. When disturbed, they emit a jet of hot secretion at the site of attack, and the ejection of secretion is accompanied by a clearly audible explosion. The glands open at the tip of the abdomen into movable turrets that are aimed when the tip of the abdomen is bent toward the attack (Fig. 5–21). The spray contains a mixture of *p*-benzoquinones.

Quinones are *not* stored in the reservoir of the gland but rather are produced at the instant of attack. The gland reservoir consists of two compartments (Fig. 5–22). As shown by H. Schildknecht, of the University of Heidelberg, Germany, the inner compartment contains hydroquinones (diphenols), which are precursors to the final defensive quinones. The hydroquinones are mixed with hydrogen peroxide. When the beetle is attacked, some of the content of the inner compartment is passed into the outer compartment. The outer compartment is a reaction chamber. The small outer compartment contains a mixture of enzymes (catalases and peroxidases) that oxidizes and decomposes the hydrogen peroxides to give water and oxygen and also oxidizes the hydroquinones to yield benzoquinone. The oxidation reactions liberate heat and gaseous oxygen. Thomas Eisner has shown that the temperature of the mixture reaches 100°C. The gaseous oxygen provides the propellant that accounts for the audible "pop" when the glands discharge. Eisner has also used high-speed photography to analyze encounters between attacking ants and bombardier beetles. The entire sequence, from ant bite to beetle discharge to ant release, takes place in less than one second.

Darkling beetles, such as flour beetles and mealworm beetles, also use quinones in their defensive secretions. Unlike the bombardier beetles, these darkling beetles store the actual quinones, not a precursor, in the reservoir of the glands. But like the bombardier beetles, the darkling beetles protect their secretory cells from quinoid products by a strategic sequence of localized reaction steps. The secretory cells of these defensive glands pass their products to the reservoir via fine cuticular ductules. In these defensive glands there are two distinct populations of secretory cells (Fig. 5–23). The secretory cells that are of particular interest to quinone production are types 2a and 2b, which are arranged in pairs. Each pair is strung in series on a cuticular ductule that leads to the reservoir. Each cell of the pair surrounds a central cavity within which lies the efferent ductule. The secretory product of the cell passes into that extracellular cavity and thence through the wall of the efferent ductule and to the reservoir.

The system of cell cytoplasm, cavity, and ductule can be viewed as reaction compartments, arranged in series. The two final steps in quinone production are hydrolysis and oxidation. The reaction steps for quinone production apparently occur in the se-

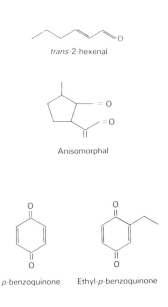

trans-2-hexenal

Anisomorphal

p-benzoquinone   Ethyl-*p*-benzoquinone

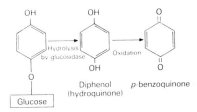

Phenyl glucoside

Diphenol   *p*-benzoquinone
(hydroquinone)

Figure 5–19
Caterpillar of a swallowtail butterfly extruding its osmeteria in response to a disturbance. (Photograph by Allan Hook, Colorado State University.)

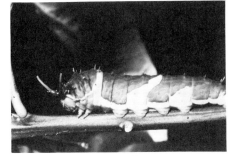

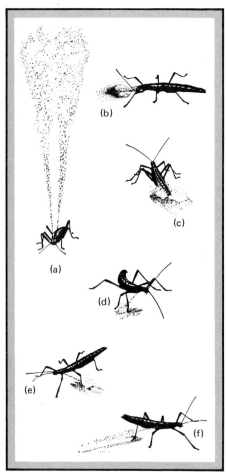

Figure 5–20
Anisomorpha, a walking stick, spraying in response to stimuli applied in various ways. (a–c) Bilateral discharges elicited by tapping the dorsal
thorax (a), touching both antennae with a heated
probe (b), or pinching the rear of the abdomen (c).
(d–f) Unilateral discharges elicited by pinching the
right foreleg (d), left middle leg (e), and right hind
leg (f).   (T. Eisner, *Science,* vol. 148, p. 967, copyright 1965 by the American Association for the Advancement of Science.)

quential compartments. In the cytoplasm of the distal cell, phenyl glucoside is stored.
The glucoside is hydrolyzed as it is secreted into the cavity of the distal cell, and the resulting diphenol can then pass along through the wall of the ductule. Within the ductule, the diphenol is oxidized to produce the final quinone. The oxidative enzymes
(phenolases and peroxidases) are apparently built into the wall of the cuticular ductule.
Thus the secretory cells avoid being poisoned by quinones because they are never actually exposed to the final poison.

A third technique for avoiding self-poisoning is gradually to secrete appropriate
enzymatic catalysts and precursors of the allomones into the reservoir, and then to allow the final steps to take place in the reservoir. Reservoir fluids often consist of a
watery phase and an oily phase. Murray Blum and his colleagues report that in the two-
phase secretion of a true bug (*Leptoglossus*), the enzymes are in the aqueous phase and
the reactive defensive allomones are in the oily phase.

Like pheromones, defensive secretions are usually mixtures of several substances.
The components often differ sharply in their polarity, as in the darkling beetles, where
several relatively polar quinones are mixed with nonpolar hydrocarbons. The mixture
is therefore well suited to dissolve and disrupt a variety of permeability barriers at the
surface of the integument. Furthermore, a mixture of several different repellent substances contributes to the nonspecific effectiveness of defensive secretions in repulsing
many different kinds of predators. In some species, 20 to 40 distinct components make
up the defensive fluid.

Defensive Allomones of Social Insects.

Whenever the nest of a social insect
is invaded, members of the colony pool their resources for collective defense. In some
ants and many termites, a special caste, the soldiers, is dedicated to nest defense. The
secretions they use are often dual purpose: repellent to the attackers and alarm signals
for nestmates. As shown by André Quennedey at the University of Dijon in France,
the major source in higher termites is the frontal gland of the soldiers (Fig. 5–24). In
Cubitermes, the strong mandibles pinch the attacker (usually an ant), and the frontal
gland secretions pour out on the sites of contact. The secretions are somewhat repellent, but in this genus the mandibular pinch is the major defensive act. In *Rhinotermes*
there are two categories of soldier: a major caste with enormous mandibles and a minor
caste with tiny mandibles. The minor caste has a huge frontal gland reservoir that
extends back from the head all the way to the abdomen. It secretes a mixture of ke-
tones toxic to many insect species. In the third style of defense, found in nasute termites, the chemical weapon is paramount. The soldiers are remarkable in that their
mandibles are reduced in size and their heads are huge, elongated snouts—squirting devices for the frontal gland secretion. When the nest wall is broken, soldiers advance
outward in parallel or in a staggered line, precisely spaced by just the distance of antennal contact. A soldier sprays the frontal gland contents when it is bumped. The secretion contains volatile terpenes that are alarm releasers for termites and irritants to the
attacker, and other unusual terpenes that are sticky and entangle ants, which are common enemies of these termites.

Allomones Promoting Associations

Defensive allomones terminate interspecific associations between potential prey and a
diverse array of predator species. In contrast, there are a host of less-studied allomones
that promote intimate associations between two particular species. Many of these associations are mutualistic, with both partners deriving substantial benefits, while others
are parasitic, with benefits accruing to only one of the partners.

A bizarre group of specialized insect species are "guests" in the nests of social insects. The alien insect enters an ant nest or termite nest without being attacked and is sometimes even assisted by a host worker. In some cases the volatile body lipids of guest and host are very similar in composition so that the intruder appears to mimic the body odor of the hosts. In other cases the guests placate their hosts with allomones that E. O. Wilson has termed **appeasement substances.** Jacques Pasteels, of the University of Brussels, Belgium, has shown that the appeasement substances of rove beetles are produced in a battery of special exocrine glands that are found only in species that are social parasites. Bert Hölldobler has observed the process by which a rove beetle guest gains entry to an ant nest (Fig. 5–25). When a forager of *Myrmica* approaches a beetle, the beetle promptly bends its abdomen forward and presents a secretion to the ant. The ant feeds on the appeasement secretion. Next, the ant licks an "adoption gland" on the lateral margins of the beetle's abdomen, and finally it carries the beetle into its nest.

Once within the nest some guest species are fairly benign, merely soliciting regurgitated food from the workers and exploiting the shelter and stable habitat provided by the hosts. Other guests return the hospitality with a bad turn. Kenneth Hagen, of the University of California, Berkeley, has reported that some lacewing larvae eat their termite hosts. The lacewing larvae move about freely in the termite galleries. At certain developmental stages the lacewings approach termite workers and wave the tips of their abdomens in front of the termites. The termites are paralyzed by a gaseous allomone, and the larvae then feed on the immobile workers.

Allomones also promote symbiosis between insects and microorganisms. Leaf-cutter ants, certain termites, and bark beetles transport and consume symbiotic fungi. The associations are quite specific—each insect species cultures only one or two fungal species. As the fungus grows in ant nests, the cultures remain clear of any contaminating species. Leaf-cutter ants weed their crop by mechanical and chemical means. Alien spores are picked out of the fungus garden, and allomones, produced by the metapleural gland of the thorax, regulate the species composition of the growing fungi. H. Schildknecht and V. Maschwitz reported that the metapleural secretion has three components: phenylacetic acid, β-indolylacetic acid, and myrmicacin (β-hydroxyde-

Figure 5–21
Bombardier beetle (Carabidae) ejecting spray toward its left front leg, which is being pinched with forceps. For photographic purposes, the beetle has been fastened to a wire hook attached to its back with cement. (Photograph by Thomas Eisner and Daniel Aneshansley, Cornell University.)

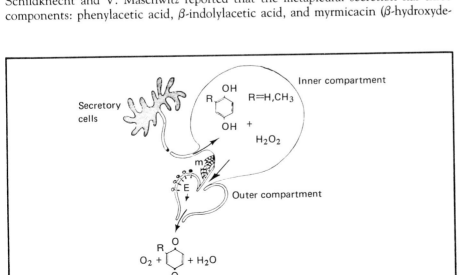

Figure 5–22
Diagram of one of the paired defensive glands of the bombardier beetle. The secretory cells produce a mixture of hydroquinones and hydrogen peroxide, which are stored in the reservoir. When a muscle (m) opens the valve, some of the contents of the inner compartment flow into the outer one, where catalases and peroxidases catalyze an explosive exergonic reaction. (Modified from T. Eisner, 1970, "Chemical defense against predation in arthropods," *Chemical Ecology,* E. Sondheimer and J. B. Simeone, eds. New York: Academic Press. pp. 157–217.)

Figure 5–23
Secretory cells in the defensive glands of darkling beetles (Tenebrionidae). In the type 1 secretory units, a single secretory cell is drained by a cuticular ductule. In the type 2 secretory units, two secretory cells are arranged in series along a cuticular ductule. In an entire gland there are hundreds of each type of secretory unit, all of which pass their products to the reservoir. (Modified from G. Happ, 1968, "Quinone and hydrocarbon production in the defensive glands of *Eleodes longicollis* and *Tribolium castaneum* [Coleoptera, Tenebrionidae]," *Journal of Insect Physiology*, vol. 14, pp. 1821–1837.)

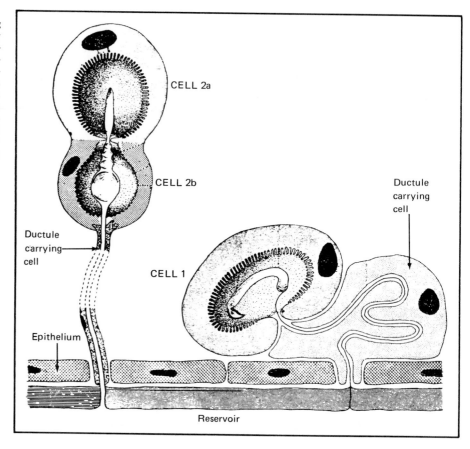

canoic acid). Phenylacetic acid is a general-purpose antibiotic that prevents the growth of bacteria and of some weed fungi. Indolylacetic acid, a plant hormone, is thought to promote growth of the symbiotic crop. Myrmicacin inhibits the generation of fungal spores. The species of fungus normally cultured in colonies of these ants is not affected by phenylacetic acid or myrmicacin.

Bark beetles and ambrosia beetles transport symbiotic fungi within cuticular pits or in large cuticular pockets. The pits and pockets are associated with secretory cells (Figs. 5–2 to 5–4), and cycles of secretory activity correlate with the peak of fungal proliferation in the pocket. In at least the southern pine beetles, the secretions apparently nourish the symbiotic fungi and restrict their growth to yeastlike budding rather than production of long hyphal strands. The yeastlike progeny drop out of the transport pocket as the beetle passes through the tunnels and galleries and start new growths of symbiotic fungi that will nourish beetle larvae. In addition, the microflora in the pocket is restricted to two species. Presumably the secretions favor the growth of these two symbionts while excluding all the weed fungi that contaminate the integument of the beetle. Unfortunately we know nothing of the chemical nature of allomones that play these roles.

The microsymbiont of the association may be an aid to digestion. As shown by the late L. R. Cleveland, of Harvard University, intestinal protozoans (flagellates) help the primitive cockroach *Cryptocercus* to digest its meal of cellulose. The protozoans live in the anaerobic hindgut of the cockroach. In the presence of high concentrations of oxygen, the protozoans die. To grow, the cockroach must molt its entire cuticle, including that of the hindgut. If abruptly exposed to the atmosphere together with the shed skin, the protozoans would not survive, and the cockroach would lose its vital digestive partners.

The protozoans survive over a molt because they "anticipate" it and enclose themselves in a cyst that excludes poisonous oxygen. The allomone that triggers en-

Figure 5–25
A rove beetle (Staphylinidae) (white) being taken into the nest by a worker ant (black). The beetle presents its appeasement gland to a *Myrmica* worker that has just approached it (a). After licking the gland opening (b), the worker moves around to lick the adoption glands (c–d), after which it carries the beetle into the nest (e). In (f), ag = adoption glands; dg = defensive glands; apg = appeasement glands. (Reprinted by permission of the publishers from *The Insect Societies*, by E. O. Wilson; Cambridge, Mass.: The Belknap Press of Harvard University Press, copyright 1971, by the President and Fellows of Harvard College; after B. Hölldobler, 1970, *Zeitschrift vergl. Physiologie*, vol. 66, pp. 215–250.)

Figure 5–24
Drawings of the use of the frontal gland for defense in three species of higher termites. In *Cubitermes*, the secretions supplement the strong mandibular pinch. In *Schedorhinotermes*, the minor soldiers have a very large defensive gland extending back into the abdomen. The defensive secretions of nasute soldiers, like *Trinervitermes*, entangle the attacker. (A. Quennedey, 1975, from *La Recherche*, vol. 6, no. 54, with permission of the author and the publisher.)

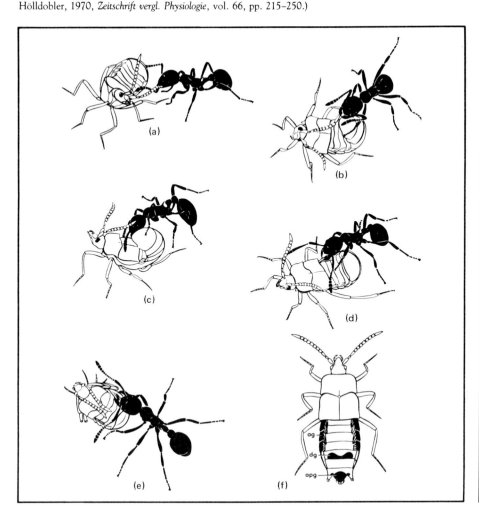

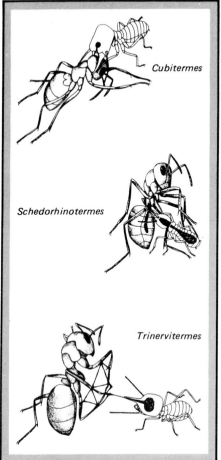

cystment is the insect hormone that induces apolysis: the familiar steroid ecdysone. The newly molted cockroach consumes its old cuticle, and once within the anaerobic gut environment, the protozoans become active again.

An Overview of Chemical Signals

Chemical signals have acted between cells since cells first appeared on the earth, and intercellular chemical communication is a characteristic of all major phyla. Hormones, pheromones, and allomones share some common origins back in the evolutionary past. Exocrine secretions for diverse purposes often have metabolic byproducts, and it is not surprising that these products, which are produced only in certain physiological states, might communicate that state to animals of the same species. For pheromones and those allomones that promote associations, selective pressures act on both the producer and the target organism to evolve an unambiguous signal that is detected at low levels. For defensive allomones, selection operates on the producer to make the signal blatant and insulting, while the potential target organisms are at an advantage if they can ignore the insult. Thus many defensive allomones are broad-spectrum reactive protein poisons that are very difficult for a living system to ignore.

Similar carbon chains form the skeletons of the molecules that function as allomones, pheromones, or hormones. The same molecule may play more than one role, as does ecdysone in *Cryptocercus*. In theory structural diversity of chemical signals should be almost unlimited, but in practice only a few families of carbon skeletons predominate among pheromones, hormones, and allomones. These include aliphatic straight chains, terpenes and the related steroids, and small ring compounds. In addition, a significant number of hormones are peptides, and a significant number of defensive allomones are complex and unusual poisons. The explanation of the widespread molecular conservatism is probably rooted deep in the metabolic pathways common to living systems. All cells have the enzymes for manufacturing certain skeletons. To develop a signal molecule, it is simpler to amend a preexisting biosynthetic pathway and a preexisting receptor surface than to evolve totally new ones.

Each class of molecular signals has common characteristics. Most pheromones and many allomones are volatile. Many defensive allomones are simple poisons with quite reactive functional groups, while others are complex and unusual toxins (such as prococcinelline) that irritate sensory receptors and are difficult to degrade metabolically. Most pheromone components are much less reactive, but their skeletons tend to be rigid and to have asymmetric carbons that determine precisely the shape of the molecule. Both defensive allomones and pheromones are medleys, but for very different reasons. Defensive allomones are medleys so that the producing animal covers as many bets as possible when it plays its defensive hand. Pheromone medleys allow the construction of less ambiguous signals.

Summary

Chemical communication between two insects requires an exocrine gland, its chemical product, and an individual that receives the signal. Pheromones are chemical signals that act between members of the same species. Allomones act on members of different species.

Most known pheromones are releasers that elicit a prompt behavioral response. The chemical signal diffuses out from the source to create an active space—a zone within which other organisms respond. The size and lifetime of the active space vary with different types of signals.

Sex pheromones include attractants and courtship signals. The attractants of moths are usually multicomponent blends emitted by the female, which attract males. Active spaces are often large. When pheromone concentration is above threshold, males fly upwind until they come close to the calling female and then they alight. Most moth attractants are aliphatic alcohols, acetates, or esters. Species are distinguished from one another by differences in the signal molecules and in the proportions of each within a blend. Males detect the species-specific molecules with their feathery antennae, which are effectively molecular sieves. Queen butterflies begin to court in response to visual stimuli. The courtship pheromone of the male is dusted on the antennae of the female. In response, she alights and allows copulation.

Pheromones establish and maintain aggregations for feeding, oviposition, mating, hibernation, and other activities. In bark beetles the oviposition aggregations form after pioneer beetles have begun to burrow into a tree and emit an attractant for the other sex. When the mates arrive, they in turn emit a pheromone for more beetles of the pioneer sex. The host trees give off odors that are synergistic with the insect ones. All bark beetle pheromones are monoterpenes, many of which are derived from terpenes of the host tree. To terminate the attack, many mated beetles give off inhibitory terpenes that inhibit the approach of more colonists. The antennal receptor of the beetles distinguishes between geometric isomers and enantiomers. Every species has a precise blend of molecules, each of which has a precisely defined geometry.

Alarm pheromones of ants, bees, wasps, and termites coordinate colonial responses to intruders. The molecules are volatile, and the active spaces are usually small. Various aspects of defensive behavior may be individually triggered by particular components in the alarm-producing pheromone blend. Related species may have a common set of chemicals—a common vocabulary—in their alarm blend, but often their dialects are different.

Trail and territorial substances distinguish known or exploited resources from unfamiliar or virgin ones. Stingless bees, ants, and termites organize their food-gathering activities by movement along trails. The trail pheromones may have components that increase locomotion, that distinguish home territory from foreign areas, and that specify the route itself.

Individual insects differ in their body odors. These individual scents give information on genetic strain, physiological state, caste, and home nest. Both development and genetic constitution play a part in defining individual scents and in distinguishing among them.

Primer pheromones influence physiological state. They accelerate maturation and control the production of the various castes in social insects.

Defensive allomones may be contained in blood and tissue or may be stored within cuticular reservoirs associated with secretory cells. Some of these molecules are complex in structure, but most are small, reactive, broad-spectrum toxins. Many of the unpleasant mixtures are forcibly ejected at an attacker. Production of the reactive mixtures might cause self-poisoning. Such is not the case because the final steps in the biosynthesis of the molecules often take place in extracellular cuticular compartments.

Some allomones promote intimate interspecific associations. Social parasites in insect colonies placate their hosts with allomones. Certain insects harbor and culture certain species of symbiotic fungi with the aid of allomones.

Similar carbon skeletons are found in allomones and pheromones. The basic skeletons are modified and functional groups are added so that the chemical signal is appropriate for its role. Most chemical signals are medleys of several components that complement one another.

Selected Readings

Blum, M. S. 1981. *Chemical Defenses of Arthropods*. New York: Academic Press. 562 pp.

Eisner, T., and J. Meinwald. 1966. "Defensive secretions of arthropods." *Science*, vol. 153, pp. 1341–50.

Happ, G. M. 1973. "Chemical signals between animals: allomones and pheromones." In *Humoral Control of Growth and Differentiation*, J. Lobue and A. S. Gordon, eds., vol. 2, pp. 149–90. New York: Academic Press.

Jacobson, M. 1972. *Insect Sex Pheromones*. New York: Academic Press. 382 pp.

Roelofs, W. L. 1978. "Chemical control of insects by pheromones." In *Biochemistry of Insects* (M. Rockstein, ed.), pp. 419–65. New York: Academic Press.

Shorey, H. H. 1976. *Animal Communication by Pheromones*. New York: Academic Press. 167 pp.

Sondheimer, E., and J. Simeone, eds. 1970. *Chemical Ecology*. New York: Academic Press. 336 pp.

Weaver, N. 1978. "Chemical control of behavior—intraspecific." In *Biochemistry of Insects* (M. Rockstein, ed.), pp. 359–89. New York: Academic Press.

———. 1978. "Chemical control of behavior—interspecific." In *Biochemistry of Insects* (M. Rockstein, ed.), pp. 391–418. New York: Academic Press.

The Nervous System

I t is hard to catch a cockroach and almost impossible to snatch a dragonfly out of the air. Many insects avoid attacks by evasive reactions that succeed because of the rapid information processing within their nervous systems. Like the endocrine system, which we discussed in Chapter 4, the nervous system responds to stimuli by triggering appropriate responses, but the nervous system reacts much faster than does the endocrine system. Before we can understand behavior, we must study the nervous system that controls that behavior.

Nervous systems detect, decide, and react. They include **input elements,** like eyes and ears, which sense changes in the surroundings; **decision centers,** which evaluate the many inputs in order to choose a response; and **output elements,** which direct that response. Somehow the nervous system must select and filter the important facts from a snowstorm of signals coming from the environment. It must transmit this information to its decision centers, which integrate input messages; then select the appropriate behavioral response; and finally coordinate that response by directing the contractions of muscles.

The nervous system of every animal is an organized network of individual nerve cells, or **neurons.** The neurons are of three functional types: **sensory neurons,** the input elements; **interneurons,** the collecting and computing elements within the central nervous system; and **motor neurons,** the output elements. Each individual neuron has several functional zones, including specialized processes for input, called **dendrites,** and for output, called **axons** (Fig. 6–1). The system makes decisions as the cells interact with one another. The sites of cellular interaction, called **synapses,** can be thought of as places where neurons talk to each other.

The nervous system of an insect is both fundamentally similar to and significantly different from that of mammals. The individual nerve cells of an insect are rather like ours, but there are many fewer cells in the insect. The nervous system of a large cockroach has at most a few hundred thousand neurons, while that of a comparably sized mouse has over a billion. In the nervous system, as in the other physiological systems, cellular economy is a hallmark of arthropods.

Both insects and vertebrates have neurons clustered together in a **central nervous system** (brain and nerve cord), into which sensory nerves run and from which motor signals are sent. In addition, both have a specialized **visceral nervous system** associated with their internal organs. But the detailed anatomy is very different. Although both insects and vertebrates have their brains in their head capsules, the nerve cord of a vertebrate is dorsal (within the backbone), whereas that of an insect is ventral (Figs. 1–16a, 6–2).

The basic anatomical unit of the insect central nervous system is the **ganglion;** almost every major segment in the embryo begins with a mass of nerve cells in an embryonic ganglion. Several ganglionic masses fuse together to form the insect brain, and other ventral abdominal segmental ganglia are often merged together as well. In each

Figure 6–1
A bipolar sensory neuron (left), which synapses with another neuron (right). Excitation is conducted from the dendrite through the cell body and along the axon to its tip. Vesicles of a chemical neurotransmitter are released at the synapse and cause a depolarization in the dendrite of the postsynaptic cell.

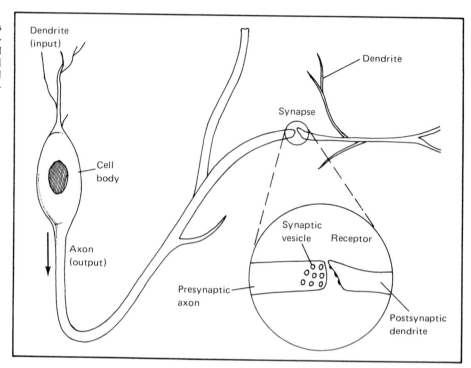

Figure 6–2
(a) Diagram of the central nervous system of a cockroach, showing the dorsal brain (br), subesophageal ganglion (so), thoracic ganglia (t_1, t_2, t_3), a motor axon to the metathoracic legs, the abdominal ganglia, and a sensory fiber running from a cercus (ce) to the sixth abdominal ganglion (a_6). (b) Diagram of the main nerve elements of the evasive response (gf, giant axons in the abdominal nerve cord; mf, motor fibers supplying the leg muscles). (Reprinted by permission of the publishers of *Nerve Cells and Insect Behavior*, by K. Roeder; Cambridge, Mass.: Harvard University Press, copyright 1967 by the President and Fellows of Harvard College.)

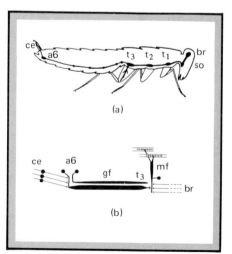

ganglion the nerve cells are distributed around the outside, forming a **cortex** of cell bodies. Each nerve cell sends one or more processes into the center of the ganglion. This central portion, called the **neuropile** (Fig. 6–3), is a closely packed mass of narrow cell processes, each of which divides like the root system of a plant. Despite the apparent tangle, each process goes to a specific place in the mass. Points of contact between the fibers are the sites of synaptic interaction. It is difficult for us to trace nerve cell processes into that tangle; it is *very* difficult to map all the synapses; and it is almost impossible for us to know how every single one of them functions during a complex behavior. But we can trace some of the processes, we can monitor some of the interactions, and from these clues we can generalize. In the last few decades these data about the function of identified neurons have been brilliantly correlated with the events in certain insect behaviors. But before we consider the ways in which nerve networks control behavior, we must begin with the basic properties of nerve cells.

Neurons and Synapses

Neurons talk to one another in an electrical and chemical code. The inside of the resting nerve cell has a slight deficiency in positive ions so that it is negatively charged relative to the surrounding extracellular fluids. This slight charge imbalance can be measured as a voltage, or potential difference. The **resting potential** of an unstimulated neuron is about −70 millivolts (mV), negative in comparison with the outside. The voltage difference means that the cell is polarized like a battery. As in any battery, electrical charges will flow only if the electrical circuit is complete. In a resting neuron the circuit is incomplete and there is no net flow of charges.

The circuit is incomplete because not all ions flow equally well across the nerve cell membrane. When a neuron is at rest, potassium (K^+) and chloride (Cl^-) can flow freely into and out of the cell, but the resting membrane has a very low permeability to sodium (Na^+). When that neuron is stimulated, the cell membrane is modified so that electrical current can flow. Membrane channels, which admit mostly sodium, suddenly open and the positive sodium ions rush in. The inrush of positive charges **depolarizes** the cell. This depolarization, called a spike or an **action potential,** lasts for only a few thousandths of a second. The voltage difference across the membrane changes from -70 mV (the resting potential) to 0 mV, and it even goes slightly beyond to perhaps $+25$ mV. This is the rising phase of the action potential. Then the sodium gates slam shut. Positive potassium ions flow out during the falling phase of the action potential. The loss of these positive charges makes the inside of the cell negative once again and thus restores the original resting potential. In the milliseconds thereafter, a special membrane pump exchanges sodium inside for potassium outside and restores the original ion distribution. These various steps are indicated in Fig. 6–4. The details are quite complex and are beyond the scope of this text, but it is important to remember the electrical and ionic bases of the resting potential and the action potential.

An action potential is a wave of depolarization that moves across the surface of a nerve cell and passes quickly down the axon. By means of action potentials, a state of

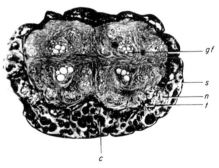

Figure 6–3
Cross section of an abdominal ganglion of the American cockroach, *Periplaneta americana*. The cell bodies (c) of the motor neurons and interneurons are in the outer cortex and their processes in the central, lightly staining neuropile. Giant axons (*gf*) are cut in cross section in the neuropile (*s*, sheath or neural lamella surrounding the ganglion; *n*, nerve about to enter the neuropile; *t*, trachea). (Reprinted by permission of the publishers of *Nerve Cells and Insect Behavior*, by K. Roeder; Cambridge, Mass.: Harvard University Press, copyright 1967 by the President and Fellows of Harvard College.)

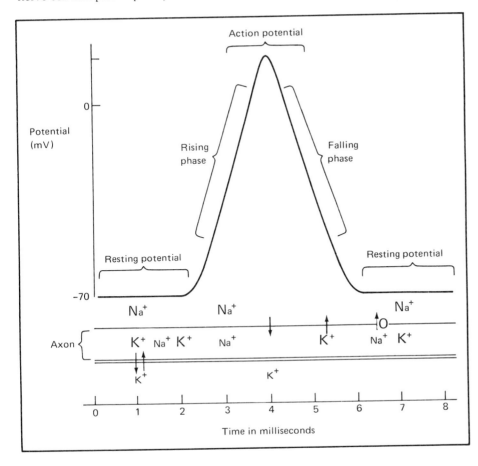

Figure 6–4
An axon and its basic electrical features. At rest (far left) the resting potential is stable. The inside of the axon is rich in potassium and poor in sodium; the outside is the reverse. When an action potential takes place (center), sodium rushes in during the rising (depolarizing) phase and potassium rushes out in the falling (repolarizing) phase. After the action potential (right) the sodium pump exchanges sodium inside the neuron for potassium outside it.

Acetylcholine

Glutamic Acid

GABA
(Gamma amino butyric acid)

Octopamine

Dopamine

5-HT
(5-Hydroxytryptamine)

excitation is conducted from a peripheral sense organ into the central nervous system or from the ventral nerve cord to a muscle. Muscles also have action potentials. A spike sweeps along the muscle cell membrane, and thereafter the contractile proteins begin to pull.

Synapses are the points of interaction between a nerve cell and another neuron or a muscle (Fig. 6–1), a gland, or a specialized output organ (such as a firefly light organ). Almost all nerve cells give off secretions from the ends of their axons. You already know that the neurosecretory cells that make neurohormones such as PTTH (Chapter 4) release their products into the blood. The secretions of most neurons are released into the very narrow (about 100 nm) synaptic space. Information flows in one direction, from the upstream, or **presynaptic,** cell to the downstream, or **postsynaptic,** cell. The information is transmitted by chemicals called **neurotransmitters,** which are stored within spherical **vesicles** that accumulate in large numbers at the tip of the axon in the presynaptic cell. There are several neurotransmitters reported from insects, such as acetylcholine, glutamic acid, GABA, octopamine, dopamine, and 5-HT.

When an action potential sweeps down the axon and reaches its tip, several of the presynaptic vesicles fuse with the plasma membrane and release their chemical contents. These chemicals diffuse across a very narrow space (about 100 nm), act on receptor proteins in the postsynaptic cell, and cause ions to flow across the postsynaptic membrane.

At many synapses the postsynaptic cell produces a spike after it has been exposed to enough neurotransmitter, but at others the downstream cell is inhibited. Each nerve cell releases only one kind of transmitter as its axons synapse with other cells. But a neuron may be acted on by several neurotransmitters at its input dendrites. For every different neurotransmitter, there is a different kind of receptor protein on the postsynaptic cell membrane. For every different neurotransmitter–receptor interaction, there can be a different effect on the membrane potential of the postsynaptic cell. Part of the computing capacity of the central nervous system stems from the many kinds of results that can follow synaptic transmission.

Many nerve cells will produce a spike only when they are acted on by another cell (at a synapse) or by some other outside energy input (at a sensory receptor). But some nerve cells and some muscles are inherently unstable. When left in isolation, they fire off an action potential from time to time. Such nerve cells are called **spontaneously active,** and the pattern of spikes can be given off at a regular rhythm. The cells of the insect hindgut or the mammalian heart are spontaneously active and continue to beat rhythmically even when no nerve impulses are being sent. When the physiological rhythm is due to a spontaneously active muscle, it is termed **myogenic,** and when it is driven by spontaneously active neurons, it is termed **neurogenic.** As we find out more and more about the functioning of insect nervous systems, we are discovering that these built-in rhythmic discharges are an important aspect of the oscillating nerve circuits that control behavior.

Among the simplest of behaviors are those that involve only a few synapses to yield a quick response. One of these is the cockroach evasive response, which was beautifully dissected in the classic studies of the late Kenneth Roeder, of Tufts University. A slight puff of air near the hind end of a cockroach causes it to run forward quickly in a response time of less than a tenth of a second. The behavior is a **reflex,** like the reflex that causes you to blink when an object is thrown at your face. The neural circuitry responsible for the simple response has just three elements (Fig. 6–2): the sensory neurons, an interneuron in the nerve cord, and the motor neurons that lead to the leg muscles. The signals sweep along nerve processes by action potentials and across synapses by means of chemical transmitters. We shall return to this simple behavioral example as we describe each of the three components. We shall look first at sensory input, then at the signals that flow outward to muscles, and finally at the transmission and computing that takes place between input and output.

The Sensory System

In contrast to motor neurons and interneurons, **sensory neurons** have their cell bodies outside the central nervous system, near the stimuli themselves. Each sensory neuron is a part of a **receptor,** which responds to stimuli by sending action potentials into the central nervous system. The axon of a sensory neuron joins other axons in a peripheral nerve and runs directly to a central nervous system ganglion. In the ganglion the axon passes through the peripheral region into the neuropile, where it branches out and forms synapses with dendrites of interneurons or motor neurons.

A dendrite of a sensory neuron is either much branched (**multiterminal**) or a single process (**uniterminal**). Multiterminal sensory neurons fire spikes when parts of the body are stretched. These stretch receptors are **proprioreceptors,** organs that detect the relative position of parts of the animal's own body. One or more dendrites of each proprioreceptive neuron make numerous points of contact with some organ inside the insect. In a thorough study of the blow fly larva, M. Osborne, of the University of Birmingham in England, has shown that the soft integument is underlain by a complex of branching multiterminal sensory neurons, 24 in the prothorax, 28 in both the mesothorax and the metathorax, and 30 in each abdominal segment.

Other multiterminal neurons send dendrites to a variety of internal organs, including other nerves, muscles, and connective tissue strands. Apparently all insects possess stretch receptors in the dorsal longitudinal muscles or in dorsal connective strands. In the bug *Rhodnius*, these receptors are stretched when the bug fills its gut with a blood meal, and the signal thus sent to the central nervous system leads to the release of PTTH and thus initiates a molt (Chapter 4).

Most sensory neurons in an insect's body are uniterminal, with a single dendrite. The dendrite usually inserts into a cuticular structure that is specialized to receive a certain stimulus. The sensory neuron is called **bipolar** because two extensions (the axon and the single dendrite) sprout independently from the cell body. The combination of neuron, accessory cells, and cuticular structure is called a **sensillum** (plural, sensilla) (Fig. 6–5).

Each sensory receptor is especially sensitive to a particular kind of stimulus. Movements of shapes and changes of illumination are detected by the simple and compound eyes, odors by sensilla that are usually concentrated on the antennae, tastes by sensilla on the mouthparts and tarsi, and mechanical disturbances by hairs at various sites all over the body (touch) or by ears (sounds). Each sensillum is a **transducer,** which detects environmental changes and transforms that information into an electrical signal that flows into a ganglion.

Although all bipolar sensory neurons have common features, the sensilla can be highly modified to receive different sensory modalities. The importance of sensilla cannot be overemphasized; the cuticle forms a nonliving, almost impermeable outer coat for the insect, and the sensilla provide the only "window" by which the insect can perceive the outside world (Fig. 1–2).

All of the cells that make up each sensillum are descended from ancestors that were epidermal cells of the integument. The ancestral epidermal cell divided two or more times to yield granddaughter or great-granddaughter cells that differ from one an-

Figure 6–5

Scheme of cells of an insect sensillum: ax, axon; bm, basement membrane; cut, cuticle; ep, epidermis; nl, neurolemma cell (glial cell); rlc, receptor lymph cavity; sc, sensory cell; th, thecogen cell; to, tormogen cell; tr, trichogen cell. (Drawing by T. Keil, from U. Thurm and J. Küppers, 1980, from *Insect Biology of the Future*, edited by M. Locke and D. S. Smith; with permission of the publishers, Academic Press, Inc.)

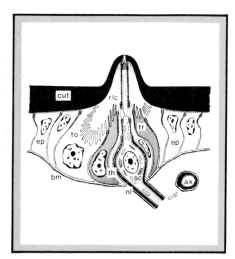

other. The descendent cells wrap around one another so that the neuron is at the center of a concentric cluster of cells. As one proceeds outward from the neuron in a "typical" sensillum, one finds in succession: the **glial cell,** which surrounds the axon as it travels into the central nervous system; the **thecogen cell,** which surrounds the glial cell and secretes a sheath around the outer segment of the dendrite; the **trichogen cell,** which secretes a cuticular process (such as a hair); and the **tormogen cell,** which connects that process to the rest of the body cuticle. Working together, the tormogen and trichogen cells secrete a fluid, the **receptor lymph,** which bathes the ciliary dendritic process (Fig. 6–5).

All sensilla of insects are variants of this general theme. For convenience we shall consider the receptors in five functional categories: **mechanoreceptors** (touch, position, sound), **chemoreceptors** (taste and smell), **hygroreceptors** (humidity), **photoreceptors** (sight), and **thermoreceptors** (heat).

Mechanoreceptors.   The simplest mechanoreceptors are cuticular hairs, sometimes called **trichoid sensilla.** The trichogen cell produces a long hairlike process, the **seta,** and the tormogen cell secretes a socket of flexible cuticle around the hair so it can bend. When the hair is touched, it rocks in the socket. The dendrite sheath attaches to the base of the hair, and when the hair moves, the compression on the dendrite causes the neuron to fire one or more spikes.

In specialized mechanoreceptors called **campaniform sensilla** (Fig. 6–6g), the seta is replaced by a flexible dome, or bubble, of cuticle. The dendrite sheath attaches to the inside of the dome. When the cuticle is deformed by body movements or by pressure, the bubble bulges out or is pulled inward, and thus the dendrite is stimulated.

Mechanoreceptors are scattered over the surface of the body. Some are dotted at intervals across the sclerites, while others are clustered together to form larger sense organs.

Campaniform sensilla are especially numerous in the bases of wings (700 in the fore wing of a honey bee), whose movement they monitor during flight. The remarkable **halteres,** which take the place of hind wings in flies, are knoblike sensory organs with several groups of aligned campaniform sensilla in their bases (Fig. 1–10). The halteres beat passively 180° out of phase with the wings but in a vertical plane, so they act as gyroscopes. Any deviation from straight flight is perceived as strain in the bases of the halteres by the campaniform sensilla, and the flying fly can use this information to control stability in all planes.

One of the clearest demonstrations that tactile stimulation controls behavior is seen in the evasive response of the cockroach. The cerci, on the tip of the abdomen, are sensitive to puffs of air. As Jeffrey Camhi and his colleagues at Cornell University have recently shown, each cercus has about 440 tactile setae on its underside. The hairs are organized into nine columns that run longitudinally along the cercus and many rows that run across the cercus. Are these setae wind detectors? To test that possibility, Camhi coated the bottoms of the cerci with wax so that the hairs could not bend. As expected, the waxed cockroach no longer responded to the wind.

The rows and columns of setae are not all identical. Each hair bends back and forth in just one direction; all hairs in a column bend the same way, and the preferred direction changes from column to column. A puff of wind from a fixed direction will bend some columns of hairs more than others. It is possible to insert a fine electrode into the cercal nerve (Fig. 1–16a) and to record the spikes from a single seta. Such recordings show that each column responds to a fairly narrow range of wind directions, but one column or another will respond to wind from any angle. These sensory axons

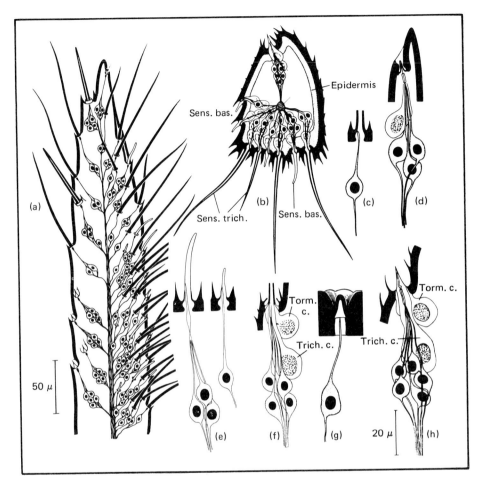

Figure 6–6
Representative sense organs and their arrangement on the antennae of *Bombyx*. (a) Tip of a branch showing the positions of different sensilla; (b) the same in transverse section; (c) tactile seta; (d) conical peg; (e) cylindrical peg; (f) tactile seta; (g) campaniform sensillum; (h) pit-peg organ. Torm c, tormogen cell; trich c, trichogen cell. (D. Schneider and K. E. Kaissling, from *Structure and Function in the Nervous System of Invertebrates*, edited by T. H. Bullock and G. A. Horridge; copyright 1965, W. H. Freeman and Co.)

are "labeled lines," wired into the central nervous system so that the cockroach can tell which column of hairs is stimulated, and from that information the insect can judge the direction of the attacker. When a large predator, such as a toad, strikes at the cockroach, its lunge produces a puff of wind. Because the cercal hairs are directionally sensitive, the cockroach runs in exactly the opposite direction from the attacker.

Insects frequently have beds of tactile setae called **hair plates** in joints that perceive the relative position of the two components that form the joint, and thus they are proprioreceptors. Hair plates in the neck of a praying mantid enable it to determine its head–thorax angle as it visually follows a prospective prey insect, so it can accurately strike with its forelegs. Similar hair plates in a honey bee are used as gravity detectors because they perceive the pull of gravity on the head relative to the thorax. The accurate detection of "down" is not as straightforward for an insect as you might at first think, and a variety of mechanoreceptors are used in different insects. Burrowing roaches use plumb-bob-shaped setae on their cerci. A final use for tactile setae is the detection of sound, as we shall discuss below.

Hearing is a special kind of mechanoreception. Sound waves may be transmitted through air, water, or solid substrates. Ears detect either pressure oscillations or particle vibrations in air. Arthropods and vertebrates are the only animals that can hear, and among the insects we find an astonishing diversity of independently evolved ears, located on nearly every part of the body. Perhaps the simplest "ears" are the vibration detectors that provide the remarkable ability of some caterpillars to hear the approaching of predatory wasps. Jürgen Tautz and Hubert Markl, of the University of Konstanz in Germany, have recently shown that the larva of the cabbage moth *Barathra brassicae* has eight tactile setae on its thorax that are vibrated by the frequency of yellow jacket wasp wingbeat. The caterpillar can hear the approaching predator 70 centimeters away and either "freeze" or drop from the plant. Larvae with intact setae are attacked 30% less than are larvae whose sensilla are removed.

The sound of a female mosquito attracts males of the same species. During most of the day and night, the males are at rest, and the long hairs on their antennae droop against the flagellar shaft. But at dusk these setae are erected by changing the pH in the cells at their bases. When a female is nearby, the erectile hairs and indeed the entire antennal flagella vibrate in resonance with the female hum.

An additional sense organ at the base of the flagellum detects the vibrations of the antenna. This structure, called a **Johnston's organ,** is an ordered cluster of many individual sensilla called **chordotonal sensilla,** or **scolopidia.** Each (Fig. 6–7) is a variant of the generalized sensillum we saw in Fig. 6–5. It is as if the entire sensillum had been retracted inward until all of the unit was just beneath the epidermis. Each scolopidium contains two neurons, the outer segments of whose dendrites are both enclosed in a **scolopale,** an intracellular structure of the surrounding scolopale cell. The dendrite tips are covered by an extracellular **cap** that attaches to the cuticle (Fig. 6–7a). Studies of other scolopidia have shown that they may contain from one to four neurons, and that an attachment cell generally connects the cap to the epidermal cells (Fig. 6–7b). The trichogen and tormogen cells of a typical sensillum are represented by the scolopale and accessory cells, respectively, and the dendrite sheath is represented by the cap. The neurons fire when the sensillum is stretched, which occurs when the integument to which it is attached moves.

Many thousands of these scolopidia are in the elaborate Johnston's organs of mosquitoes and black flies (7000 to 7500 in the mosquito *Aedes aegypti*). As you can see from Fig. 6–8, the antennal flagellum (bearing the erectile hairs) extends upward from a circular plate of cuticle that lies at the bottom of a cuplike inpocketing of the second antennal segment (pedicel). The circular plate is freely suspended, surrounded by a ring of flexible cuticle and cushioned underneath by the pressure of the blood. A circle of prongs radiates out from the base plate, and the distal ends of the scolopidia are anchored to these prongs. The flagellum tilts, vibrates, and oscillates easily, and precise information on its movements is passed into the antennal lobe of the brain. The shaft and hairs provide a frequency filter, resonating to the pitch of female flight sounds. Somehow the overall flagellar movements provide directional information for the male to "home-in" on female sounds. In flies and bees, the Johnston's organ allows the animals to measure their own flight speed by air currents moving the joint. Surface-dwelling whirligig beetles detect irregularities in the water surface and detect vibrations from other beetles by means of their Johnston's organs.

Many ears, such as those in the abdomens of grasshoppers (Fig. 1–13a), the forelegs of crickets, and the thoraces of moths, are **tympanic organs,** with cavities formed from tracheal air sacs that resonate to sounds like the inside of a drum. Scolopidia are usually in these cavities and attach to a movable tympanic membrane. When sound strikes the membrane, its movements trigger a chain of action potentials that sweep into the central nervous system.

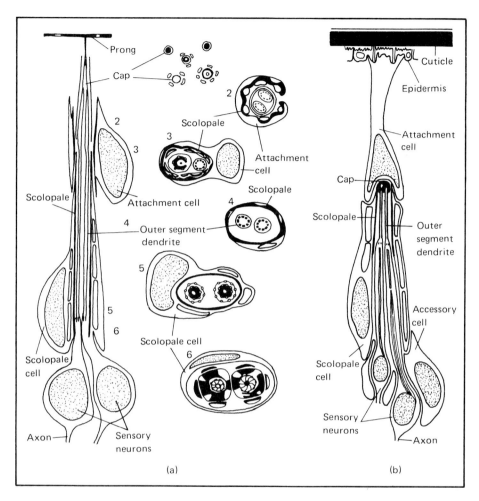

Figure 6–7
Scolopidia from the Johnston's organ of the antenna of a male mosquito, *Aedes aegypti*. (a) Predominant type of scolopidium, with transverse sections at numbered levels to right; (b) Scolopidium associated with antennal blood vessel, illustrating attachment cell. Note that each scolopidium contains two sensory neurons. (Reprinted with permission from *International Journal of Insect Morphology and Embryology*, Vol. 4, K. S. Boo and A. G. Richards, "Fine structure of the scolopidia in the Johnston's organ of male *Aedes aegypti*." Copyright 1975, Pergamon Press, Ltd.)

The best-studied insect ears are the tympanic organs of noctuid moths (Fig. 6–8b). As shown by the elegant studies of Kenneth Roeder, these ears are bat-detectors that allow flying moths to sometimes avoid bats, their voracious predatory enemies (Fig. 14–11). Each ear consists simply of a tympanic cavity and membrane and three neurons in scolopidia. When stimulated by ultrasonic vibrations, such as the cries of a hunting bat, a train of spikes flows along the tympanic nerve. The simplicity of the system allowed Roeder to record from the nerve and draw precise conclusions about the information-filtering and transmitting capacity of a single sensory cell. Only one neuron, the A_1 cell, is sensitive to low intensities, and it has been the most studied. It fires a train of spikes while a bat is calling some distance away (up to 40 meters). When the sound stops, the spikes stop, and thus we can "listen" to a bat cry (at frequencies far beyond our hearing capacities) by recording the trains of spikes coming in from a moth ear (Fig. 6–9). The ears are bilateral—one on each side of the metathorax. As the moth's wings flap up and down, the sound from above flickers as the ears are briefly shadowed on the downbeat. By balancing the intensities of the bat cry as heard in the left and right ears and by minimizing the flicker, the moth can aim away from the bat

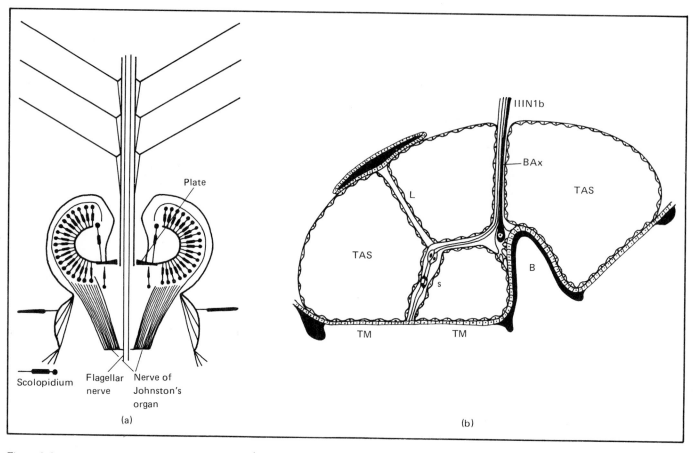

Figure 6–8
(a) Johnston's organ of a male mosquito. The antennal hairs are shown erected. The flagellum ends in the plate, which is supported by flexible cuticle. The chordotonal sensilla (scolopidia) of the Johnston's organ are arranged radially around the extensions of the plate. (b) The tympanic organ (ear) of a noctuid moth. The scolopidia (s) contain two acoustic receptors, or *A* cells. They are attached to the tympanic membrane (TM) and suspended in the tympanic air sac (TAS) by a ligament (L) and by the axons passing to the skeletal support, B. Here the *A* fibers are joined by the *B* cell and run together with the *B* nerve fiber (BAx) to form the tympanic nerve (*III N1b*). Activity was recorded from the tympanic nerve after it had left the tympanic air sac. (a: Modified from T. H. Bullock and G. A. Horridge, 1965. b: Reprinted by permission of the publishers of *Nerve Cells and Insect Behavior*, by K. Roeder; Cambridge, Mass.: Harvard University Press, copyright 1967 by the President and Fellows of Harvard College.)

and can compute an optimum flight path to avoid it. If a bat approaches closely, within about 3 meters, the A_2 sensory neuron, sensitive to high intensities, fires. The moth then makes quite different responses—abrupt dives or quick, erratic loops.

A characteristic group of 10 to 40 scolopidia occurs in the tibiae of many insects, where it forms the **subgenual organ.** The subgenual organ is sensitive to substrate vibrations transmitted through the legs.

Figure 6–9
Tympanic nerve responses of a noctuid moth to the cries of a bat. (a) The approach of a cruising bat; (b) a bat flying near the preparation. (Reprinted by permission of the publishers of *Nerve Cells and Insect Behavior*, by K. Roeder; Cambridge, Mass.: Harvard University Press, copyright 1967 by the President and Fellows of Harvard College.)

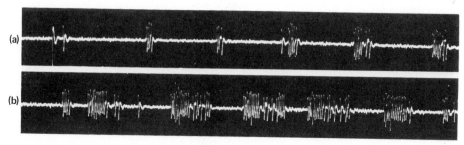

Chemoreceptors. Chemoreception includes both taste and smell. For an insect, chemical cues are among the most important in locating food and finding a mate and in social interactions. Functionally, chemosensory sensilla tend to fall into two general classes: (1) **specialists,** which respond to only one or a few specific molecular signals and (2) **generalists,** which respond to a broad spectrum of stimuli. We shall first describe two well-studied sensilla, one from each class, and then make some generalizations concerning the remarkably diverse chemosensory sensilla found in insects.

Specialist receptors are especially important in the perception of pheromones, whose importance in mating was described in Chapter 5 (p. 117). They have been best studied in moths, in which the male's large feathery antennae (Fig. 6–10) perceive the sex attractant released by the female's exocrine glands (Fig. 5–7). In a classic study of the silkworm moth, Dietrich Schneider, of the Max Planck Institute for Behavioral Physiology, near Münich, West Germany, has described how the antennae filter the pheromone molecules out of the air. Each antenna has several types of olfactory sensilla, of which 17,000 are long hairs that respond only to the sex pheromone (Fig. 6–11). These long hairs have some important differences from the tactile setae described on p. 146. First, the walls of the hair have thousands of tiny pores, from which several times as many inner tubules run to the fluid-filled center of the hair. Second, the sensillum contains two sensory neurons, whose dendrites do not terminate at the base of the hair but instead leave the dendrite sheath and extend into the fluid-filled lumen of the hair, where they adjoin the tubules. The cell membrane of one neuron responds to bombykol, and that of the second responds to bombykal, the two components of the sex pheromone (p. 118).

When a molecule of bombykol or bombykal is absorbed on the antenna, it diffuses into a pore and through a tubule, where a receptor site on the appropriate dendrite apparently conforms to its molecular structure and triggers a discharge in the neuron. Both neurons have very low thresholds; in fact, only *one* molecule of pheromone is effective. Bombykol and bombykal act independently on their respective specialist neurons, which send parallel action potentials to the deutocerebrum along parallel axons. The brain acts as a computer that determines the concentration of each component of the pheromone and makes the decision as to whether the moth should fly upwind in search of the female. The antenna as a whole is a highly efficient sieve for a low concentration of molecules. About 200 sensilla must be stimulated to cause a male to respond. An air current that contains about 1000 molecules of pheromone per cubic centimeter may be sufficient to activate that number of receptors.

Generalist receptors are often important in discriminating what is or is not acceptable food. Vincent Dethier, now at the University of Massachusetts, has described the physiology of the hairlike chemosensory sensilla that form a fringe around the labella of a blow fly's proboscis (Figs. 1–5g and 6–12). There are 245 to 257 of these setae on the proboscis and also over 3000 on the tarsi. The blow fly also has about 65 shorter peglike sensilla on its proboscis. In aggregate these receptors allow the fly to taste and react to a multitude of substances (nectar, fatty juices in carrion, feces, esters, and so on) routinely encountered in nature.

Each hairlike sensillum at first glance closely resembles a typical tactile seta. However, it has five neurons, of which one is a mechanoreceptor that inserts at the base of the seta (Fig. 6–13). The other four are chemoreceptive, and their dendrites extend within the dendrite sheath to the very tip of the seta. The dendrites do not branch, and they are bathed in a common pool of receptor lymph. Unlike the pheromone receptor that we just described, the walls of the seta have no pores, but the apex of the hair does have one pore. When the fly puts its proboscis (or its tarsus) into some food,

Figure 6–10
Head of a male cecropia moth, showing the greatly developed, feathery antennae. (Photograph by Howard E. Evans.)

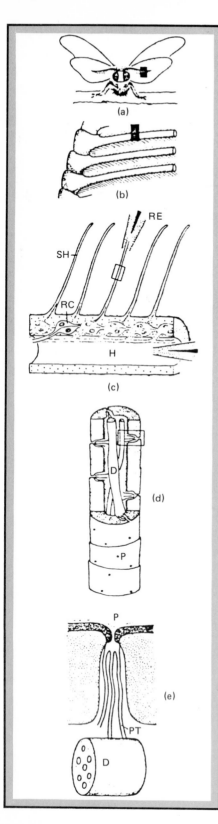

(a)

(b)

RE

SH

RC

H

(c)

D

P

(d)

P

D

PT

(e)

chemicals diffuse into this pore, into the receptor lymph, and reach the receptor sites on the dendrites. When they are exposed to pure substances, the four neurons can be divided into one water, one sugar, and two salt receptors. However, experiments with mixtures (salts with sugars, plant juices, and the like) show that each receptor cell reacts to more than one stimulus but shows a preference toward some particular substances. In other words, the receptors are heterogeneous. Some are spontaneously active; they fire spikes even in the absence of an appropriate stimulus. When stimuli are presented, the trains of spikes in these neurons accelerate or slow down. These neurons are especially sensitive to concentration changes. In addition, the response spectra of the four cells differ from one sensillum to the next. The insect can detect subtle differences in stimuli by comparing the patterns among the various incoming axons.

Helmut Altner, of the University of Regensburg in West Germany, has recently proposed a classification of sensilla based on their fine structure, which we shall follow here. The taste receptors on the blow fly proboscis belong to the **uniporous** category. These are usually variously shaped hairs and pegs, characterized by a single apical or subapical pore. The pore may be a simple opening, have cuticular channels or ridges, or even be equipped with a closing apparatus. From 1 to 10 (usually 3 to 6) chemosensory neurons are in each sensillum, whose dendrites are unbranched and extend within a dendrite sheath to the pore. Many have flexible sockets and double as mechanoreceptors, as described for the blow fly.

While some are olfactory receptors, the majority of uniporous sensilla are contact chemoreceptors, or organs of taste. We tend to think of taste in connection with our mouths, but insects are not so limited. Insects have uniporous sensilla on their mouthparts, especially the palpi; on the terminal segments of the antennae; on the ovipositor; and especially on the tarsi. Indeed, a fly or butterfly first tastes its food by stepping on it!

The pheromone receptors on the moth antenna belong to the **multiporous** category. These assume a wide variety of shapes, from pointed, flattened, or club-shaped hairs to elongate, conical, or spherical pegs to disclike plate organs to the weird "circumfilar sensilla" on gall midge antennae. While most project from the surface, some are sunk in shallow or deep pits (pit-peg organs). All are characterized by numerous pores in the walls of the sensilla, through which chemicals enter to reach the dendrites.

Multiporous sensilla are olfactory receptors—they perceive air-borne chemicals. As you might expect, they are most abundant on an insect's antennae, although some occur on other appendages like the palpi and the ovipositor. Most have from one to six neurons, although a few unusual types, like the pitted plate organs on hymenopteran antennae and the circumfilar organs on gall midge antennae, have hundreds of neurons, perhaps being composite structures. The plate organs on honey bee antennae, incidentally, perceive several pheromones: the queen substance (which doubles as a sex pheromone when perceived by drones), the alarm pheromone, and an attractant pheromone.

Figure 6–11
The sex pheromone receptors of the male silkworm moth. The area enclosed in a box is enlarged in the next figure. (a) Moth; (b), branches of antenna; (c) sensory hairs, each containing two sensory neurons, one sensitive to bombykol, the other to bombykal; (d) longitudinal section of a hair, showing dendrites and pores in the cuticle; (e) cuticular pore, with tubules extending into the lumen of the hair to adjoin the dendrites. D, dendrite; H, lumen of antennal branch with inserted electrode; P, pore; PT, pore tubule; RC, receptor cell (sensory neuron); RE, recording electrode inserted over tip of sensory hair; SH, sensory hair. (With permission from R. A. Steinbrecht and D. Schneider, 1980, *Insect Biology of the Future*, edited by M. Locke and D. S. Smith. Copyright: Academic Press Inc. [London] Ltd.)

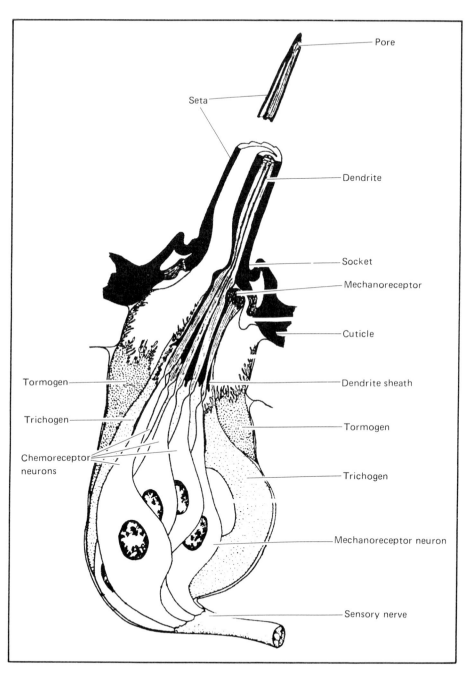

Pore

Seta

Dendrite

Socket

Mechanoreceptor

Cuticle

Tormogen

Trichogen

Chemoreceptor neurons

Dendrite sheath

Tormogen

Trichogen

Mechanoreceptor neuron

Sensory nerve

Figure 6–12
Oral view of the opened labellar lobes of a blow fly, showing the peripheral chemosensory setae. (Photograph by Vincent G. Dethier, University of Massachusetts.)

Figure 6–13
Hairlike chemoreceptor sensillum from the labella of a blow fly. (Modified from V. G. Dethier, 1976.)

Photoreception.   The **compound eyes** on the heads of adult insects and larval exopterygote insects are image-forming photoreceptors, with the same function as our own eyes. Insect and vertebrate eyes are similar in photochemistry but strikingly different in their optics and their physiology. In both animal groups, light is detected when it strikes membrane elaborations that are rich in **rhodopsin,** which is a protein containing retinene, related to vitamin A. In mammals a large lens projects a single image on the retina, but in all compound eyes a group of receptors behind each facet accepts light from only a small fragment of the entire visual field in the same way as a rod or cone in our eyes accepts light from only a tiny part of the outside world. This is done by a converging lens in the convex surface of each facet or within the cornea or cone. There are hundreds of facets (up to 30,000).

The functional unit of insect eyes is the **ommatidium,** one behind each facet, with, three components: the optical system, the pigment cells, and the retinula cells (Fig. 6–14). The optical system consists of a transparent patch of cuticle, the **cornea;** below it the **crystalline cone;** and finally the **rhabdom,** which absorbs the light. The **retinula cell** is a monopolar sensory neuron with one axon. From 6 to 12 (basically 8) retinula cells are packed together in a tall cylinder. In cross section each retinula cell is wedge-shaped, like a piece of pie, and all contribute to the rhabdom that runs down the center of the retinular cylinder. Each retinula cell contributes a **rhabdomere** as its share of the rhabdom. In flies and some bugs the rhabdomeres are separate and therefore look in different directions from behind each facet, but in other insects the rhabdom is compact. Each rhabdomere is a wedge of stacked tubular membranes (microvilli) that are laden with photosensitive molecules of a protein-retinene complex called rhodopsin. Light on the optical axis is focused by the optics on the rhabdom, which acts as a light guide. The narrow (about 0.03 μm) microvilli that compose the rhabdomeres are perpendicular to the light's path, and their membranes are stabilized by a central filament with side arms and by bridges between them. The light energy causes a change in the rhodopsin in the microvilli, releasing an amplification process that eventually increases the permeability of the retinula cell membrane. The resulting ion flow triggers a depolarization in the retinular axon.

If light rays were to bounce about at random within the eye, the insect would not resolve contrasts, which is the job an eye is adapted to do. The optics described ensure that each ommatidium is sensitive in a different direction. It has an optical axis different from other ommatidia. To cut out glare, each ommatidium is shielded by **primary pigment cells** that surround the cone and also by **secondary pigment cells** around the cone and retinular cluster. In diurnal eyes of the apposition type (see below) light cannot cross from one rhabdom to the next. Ommatidia sometimes have a sleeve of tracheae, and the silvery walls of these air-filled tubes help keep the light within each ommatidium. In nocturnal eyes of the superposition type (see below), a basal layer of tracheae (the **tapetum**) doubles sensitivity by reflecting light a second time through the rhabdom and causes the reflected glow seen when a flashlight is shone into a moth's eye.

What can an insect see? For most insects, the world is colorful. Within a single ommatidium the retinula cells differ in their sensitivity to colors because of differences in the proteins forming the rhodopsin molecules. Randolph Menzel, of the Free University of Berlin, for instance, has identified the nine retinula cells in the ommatidium of the worker honey bee: four are sensitive to green, two to blue, and three to ultraviolet. Like us, the bee has trichromatic vision, but with a surprising difference: it is blind to red but sees in the ultraviolet, a color to which we are blind but which is present in sunlight (see Chapter 11, p. 253, for a discussion of how UV vision affects foraging). Because of scatter, the sky at 90° to the sun is relatively rich in ultraviolet, and the scattered UV light, which is used in navigation by the bee (see below), is polarized. Most insects show this shift of visible range toward shorter wavelengths (ultraviolet) and of blindness toward longer wavelengths (red), but there are exceptions. Most

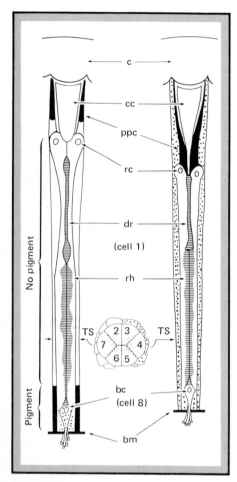

Figure 6–14
An ommatidium of the night eye (dark adapted) (left) and the day eye (light adapted) (right) of a whirligig beetle (Gyrinidae), with correct proportions in the vertical scale but enlarged more sideways. The central inset shows the transverse section (TS) of the rhabdom region with cells 2–7. Bc, basal retinula cell; bm, basement membrane; c, cornea; cc, crystalline cone; dr, distal rhabdomere; ppc, principal pigment cell; rc, retinula cell soma; rh, rhabdom. Most nocturnal insects do not have the distal rhabdomere (dr) crossing the clear zone, and some have an extension of the cone instead, to carry the light in the day eye. (G. A. Horridge, The Australian National University, Canberra, Australia.)

butterflies have tetrachromatic vision: they see red as well as ultraviolet and indeed have a visual spectrum from 300 nm to 680 nm, the broadest in the animal kingdom.

Insects and other arthropods share another capability that we lack: they can discriminate the plane of polarized light. When light from the sun passes through the earth's atmosphere, it is scattered; at greater angles to the sun it is more polarized, in a plane at right angles to the directions of the sun and the observer. When viewed through a polarizing filter, the blue of the sky is no longer uniform in intensity, and this pattern changes during the day with the angle to the sun. An animal with a polarizing filter can thus use the pattern to navigate even with only a small patch of blue sky. The compound eye is such a filter, thanks to the oriented rhodopsin molecules in the microvilli, which absorb a maximum of light when the direction of the polarization is parallel to the microvillar axis. Menzel considers that in honey bees the polarized light detector is a short, basal, ultraviolet-sensitive rhabdomere (like cell 8 in Fig. 6–14), oriented in different directions in different ommatidia; but Rudiger Wehner at the University of Zurich, Switzerland, considers that polarization detection is done by typical long UV-sensitive cells along the dorsal edge of the honey bee eye.

What actually is in the image that an insect sees? To answer this question, we must begin with the basic fact that an insect brain receives 8 to 10 axons that are labeled in color and polarization from each ommatidium. How are these many samples of the entire visual field used to detect shape, position, and contour? The classical theory of image formation, stemming largely from the nineteenth-century study of Johannes Müller, states that the insect forms a **mosaic image.** Really, this simply says that the ommatidia look in different directions; it says nothing about reconstituting an image of the visual world in the brain.

To understand insect vision, think of the compound eye as an aggregate of cylinders (ommatidia) that radiate out like spines of a startled porcupine. Every ommatidium is always pointed (aimed) toward a slightly different view from its neighbors. The lens system of an ommatidium, with the cornea and crystalline cone, brings light rays to a focus at the base of the cone. In day-active insects like the honey bee, this focal point is also the top of the rhabdom. The array of ommatidia is a solid fan of sampling devices. Only those light rays that fall along or near the axis of each ommatidial cylinder are effective, by virtue of the optical system and the screening pigment. Each ommatidium looks at only a very small part of the total view. The visual field of each ommatidium is about 2 degrees wide in diurnal insects, with axes separated by 2 degrees, but fields of adjacent ommatidia overlap more in most insects that are active in dim light. The mosaic theory simply says that the eye sees contrasts and the relationships between them by assembling the fragments side by side. The subtlety lies in the subsequent processing of the angle-sampled data.

Not all ommatidia of an eye are identical. In hunting day-active species such as mantids, dragonflies, and wasps, one cluster of forward-looking ommatidia have smaller angles between them, so that this sector of the visual field is sampled more densely. These particular ommatidia are also larger, so each has greater resolving power; thus fields are narrower and the field overlap is constant. This region is analogous to the **fovea** of vertebrates—it is the area that best sees detail. Prey objects are centered in this fovea before the predator strikes.

About 100 years ago S. Exner divided eyes into two general anatomical classes that he correlated neatly with day and night activity. Day-active insects, such as bees, dragonflies, and butterflies, have **apposition** eyes, in which the cone abuts directly against the top of the rhabdom and each ommatidium is optically isolated by a shield of pigment (Fig. 6–14, right figure; Fig. 6–15). Many nocturnal insects, especially

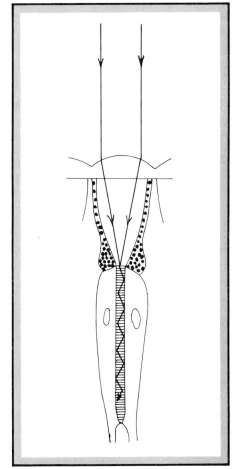

Figure 6–15
The apposition eye of a day-flying insect. The crystalline cone is shielded by pigment and abuts directly on the tip of the rhabdom. Only light near the axis, as shown, stimulates the rhabdom.

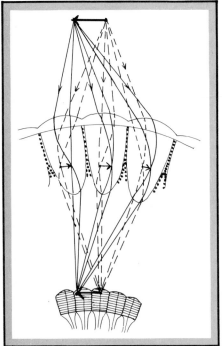

Figure 6–16
The essential optics of a superposition eye. The rhabdom receptor layer (cross-hatched) is far below the crystalline cone. Light rays are bent within the crystalline cone and cross in the clear zone from one ommatidium to the next.

moths and some beetles, have **superposition** eyes (Fig. 6–14, left figure; Fig. 6–16), and some diurnal insects like skipper butterflies have them too. Between the cone and rhabdom in superposition eyes is a **clear zone** that acts like the space in a camera or telescope where rays come together to form a **real image.** Parallel rays from a distant point pass through a patch of facets in this type of eye and are bent by the cornea and cones so that they converge to a real image on the rhabdom layer (Fig. 6–16). The crossing of rays between ommatidia is essential to form the superposition image. Rays from one point pass through many facets to land on one receptor. The superposition eye loses some resolution by this complicated summation of rays but is more sensitive than an apposition eye because the collecting area is larger. By day the screening pigment migrates from between the cones (where it lies at night) into the clear zone, where it surrounds long, narrow tracks for light. The superposition eye by day can therefore still see, but each ommatidium is now separate as in the apposition eye. The light guides across the clear zone are sometimes called **crystalline threads.**

The interesting point about mosaic vision is how the image is processed by the brain. Here insects seem to differ from ourselves. They seem to see only contrasts that move across the eye. It is the movement in the visual field that is seen, not a still photograph. Actually, in ourselves a stabilized image on the retina soon fades. An insect eye detects the movement of an object as it passes from one ommatidial viewing field to the next. One can imagine the process as a change from one angle-labeled line to the next. When an edge or a spot moves across the visual fields of adjacent ommatidia, the pattern of electrical activity changes, and obviously these changes carry the visual message. When a stationary object is present in front of an immobile insect, there is no flicker. It is almost as if the stationary object were invisible. Now, imagine the inflow of information as a dragonfly or mantis approaches directly toward an insect. The prey remains centered in the fovea, and the foveal ommatidia flicker because the angles between them are small. As the prey gets closer, it looms larger and moves off the fovea when very close. That is when the prey is struck. You will remember that a complementary strategy is used by noctuid moths. Minimal sound flicker told the moth it was flying directly away from the bat.

Insect eyes are not as good in discriminating shapes. The visual acuity of a honey bee (the reciprocal of the angle between ommatidial axes) is only 1/60 to 1/100 that of a human. However, the compound eyes are good at perceiving movement, and they are fast. The "flicker fusion" of a honey bee is 300 flashes per second, compared to our 15 to 20. This means that fluorescent lights, which we see as a steady glow, flash 100 to 120 times per second to a bee.

Simple eyes are of two kinds: **stemmata,** which are found in larval holometabolous insects, and **dorsal ocelli,** the accessory eyes of most adult insects as well as larvae of exopterygotes (Fig. 6–17). Form perception in stemmata is very poor. Stemmata consist of a few large ommatidia with more retinula cells than those of adults. In beetle and mosquito larvae, the lens system is mostly the corneal cuticle, but in Lepidoptera, Neuroptera, and Trichoptera, there are crystalline cones as well. Light is detected by a dozen or more sensory cells that form a rhabdom. Demonstration of the effectiveness of stemmata is difficult for lack of good behavioral signs, but at least they distinguish light and dark and seem to enable the animals to orient themselves.

Dorsal ocelli are best developed in flying insects. They have a corneal lens and beneath it 500 to 1000 sensory cells, which form rhabdomlike structures in groups of 2 to 5. Dorsal ocelli do not form images, but they are very sensitive to changes in light intensity and to low levels of light intensity. Ocelli are important in regulation of daily rhythms in some species, and they may be involved in maintaining flight stability; but their functions are poorly understood.

Thermoreception, Hygroreception, and Magnetic Sensitivity. Mosquitoes and other blood-feeding insects are attracted to the warmth of their warm-bodied hosts. In tsetse flies, heat is the major stimulus that induces probing. The major heat-sensitive receptors are on the fly's antennae, but in addition there are heat-sensitive hairs on the tarsi of the prothoracic legs. Some parasitic wasps detect their concealed insect prey by the latter's body heat, which is perceived as infrared radiation by plate organs on their antennae. The remarkable beetle *Melanophila* is attracted to forest fires so that it can lay eggs in newly burned trees; sensory pits on its thoracic sternum enable the beetle to orient to the heat.

We also know that humidity is sensed by insects because many species choose a specific relative humidity in a humidity gradient. In many insects the perception of a humidity gradient has been traced to **poreless** sensory pegs (sometimes sunk as pit-peg organs) on the antennae. Of the three or four neurons, one responds to dry air, one responds to moist air, and one responds to cold—these sensilla double as thermoreceptors. Exactly how such a sensillum can perceive relative humidity remains a mystery.

Finally, recent studies by Martin Lindauer and Herman Martin, of the University of Würzburg, have indicated that honey bees can perceive the orientation of the earth's magnetic field. Exactly how bees accomplish this feat is still unknown, but James Gould, of Princeton University, has located crystals of magnetite, the natural magnetic mineral also called lodestone, in the bee's abdomen.

Motor Nerves and the Neuromuscular Junction

The cell bodies of motor neurons lie in the cortex of ganglia of the central nervous system. The best-studied ones are those monopolar motor neurons that control walking or flying muscles. There are only a few hundred such cell bodies in each thoracic ganglion, and it is possible to map important ones in a very precise way so that the same kind of neuron can be studied repeatedly in many different animals (Fig. 6–18). A large process runs from the cell body into the neuropile of the ganglion and gives off several branches, or **collaterals.** Some branches are dendrites that synapse with axons of other neurons. The main process is the axon that continues out of the ganglion in the company of other axons that together form a peripheral nerve. The axon runs directly to a muscle, where it branches to end in a series of junctions or synapses with the muscle cell (Fig. 6–19). Except at these **neuromuscular junctions** and at the synapses in the ganglion, the motor neuron is surrounded by specialized nonneuronal glial cells that isolate the neuron electrically and also nourish it.

In the jumping (extensor) muscle within the grasshopper metathoracic femur (Fig. 1–6c), there are just four axons, and each plays a different role. The first acts quickly, producing a fast, or **phasic,** contraction of the tibia; the second acts slowly, producing a gradual, or **tonic,** contraction; and the other two have a more subtle function. These neuromuscular junctions are accessible to fine electrodes and are considerably easier to study than are synapses in the neuropile.

When a spike travels down the fast axon, it causes releases of glutamic acid at the junction. Less than a tenth of a second later, there is an action potential in the postsynaptic muscle cell. The muscle proteins slide over one another, and the muscle as a whole contracts quickly.

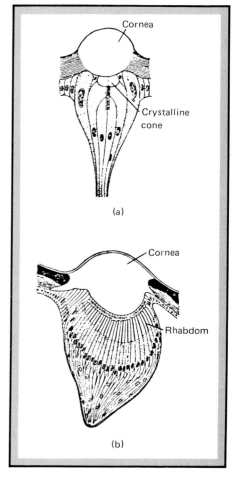

Figure 6–17
Simple eyes: (a) the lateral eye (stemmatum) of a caterpillar; (b) the dorsal ocellus of a male ant. (After R. E. Snodgrass, 1935, *Principles of Insect Morphology,* Figs. 279F and 281D; reproduced with permission of McGraw-Hill Book Co.)

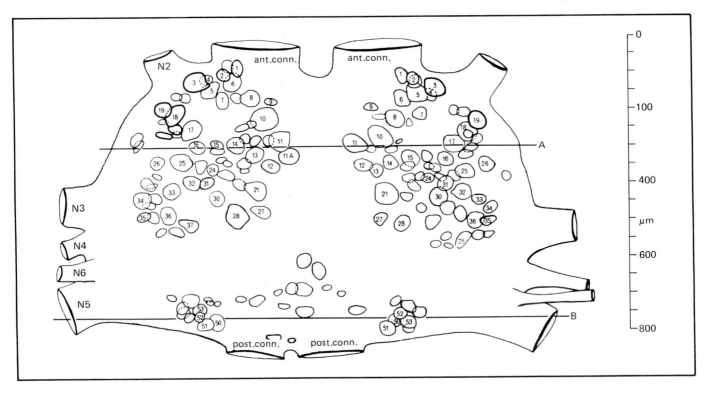

Figure 6–18
Motor neurons in the metathoracic ganglion of a late-stage immature American cockroach. Cells that are specific landmarks or associated with particular nerve trunks are numbered. N 2–6, nerve trunks, ant. conn., anterior connectives; post. conn., posterior connectives. (M. J. Cohen and J. W. Jacklet, 1967, *Philosophical Transactions of the Royal Society of London*, vol. 252, used with permission.)

The importance of glutamic acid can be confirmed in a simple experiment: When you use a micropipette to add glutamic acid in the vicinity of the neuromuscular junction, the muscle contracts. Glutamic acid is a common amino acid in almost all proteins, and as you know, the products of protein digestion in the gut include amino acids, which are absorbed as such into the hemolymph (Chapter 3). Why doesn't glutamic acid, absorbed in that way, stimulate the muscle? We don't yet have a complete answer, but two factors contribute. First, insect blood plasma contains little glutamic acid. By some mechanism that is not yet understood, the hemolymph is kept largely free of this neurotransmitter in many species. Second, the glial cells may help to prevent any glutamic acid that contaminates the hemolymph from reaching the postsynaptic membrane.

When stimulated once, the second, or slow, axon produces only slight postsynaptic depolarization and a slight increase in muscle tension. After repeated stimulation, the small depolarizations add up and build on each other to yield finally a large potential change and a full contraction. These slow contractions seem to be used when the grasshopper is merely standing about or walking slowly. As in the fast axon, the transmitter is glutamic acid. Why does glutamic acid have two different effects? The different speeds of muscle contraction with the slow and fast axons could occur because slow and fast axons release differing amounts of the neurotransmitter or perhaps because the distinct postsynaptic receptor proteins are very different at the two different types of synapses.

The third axon produces no response when stimulated alone, but it inhibits the response of the muscle to both the slow and the fast axons. The neurotransmitter for this common inhibitory axon is GABA.

The fourth axon is the most difficult to understand. It originates in a dorsal unpaired neuron, called a DUM cell, which is found in the top of the metathoracic ganglion (Fig. 6–23b). Its transmitter, octopamine, has three distinct effects. First, it enhances contraction when the third axon is stimulated, and it is therefore called a neuromodulator. Second, it affects a myogenic rhythm. The tibial extensor is a heterogeneous muscle, and some portions of it are spontaneously active. They show muscle action potentials and contractions even when all neuronal input has been abolished. The axon from the DUM cell modulates that rhythm by reducing its frequency, while the slow axon modulates the rhythm in the opposite direction. Third, octopamine helps to mobilize food reserves (much as adrenalin does in vertebrates), which provide more energy for muscular work.

Let us now compare the control of insect leg muscles with the analogous control in mammals. Mammals have many hundreds of motor neurons that send axons to the many functional subunits of each leg muscle. Every one of these vertebrate motor neurons produces a muscle twitch akin to the action of the fast axon in insects. Each vertebrate axon terminates in a single complex end plate at one site on a muscle cell. In contrast, insects usually have only a few (two to four) axons leading to each muscle. Once again we see that insects are cell conservatives. But each insect axon branches extensively, and the processes wind across the surface of the muscle to form many simple neuromuscular junctions.

The intensity of muscle contraction in vertebrates is graded because of the multitude of subunits, each controlled by a distinct motor axon. By recruiting one, a few, a majority, or all of the subunits, the speed and strength of the contraction can be adjusted. Insects have many fewer cells to work with, but the neurons are of four kinds: fast, slow, common inhibitor, and DUM cell. The insect solution allows enough fine control to permit a host of complex behaviors, as will be discussed in the next chapter. But the insect solution does not permit the complexity and precision that is possible in vertebrates.

Interneurons and the Ventral Nerve Cord

All interneurons are located entirely within the central nervous system. Like a motor neuron, the interneuron is monopolar. Its cell body is located in the periphery of a ganglion, and its axon enters the neuropile where it branches. The axon may terminate there or may pass through the connectives to synapse with neurons in the neuropiles of other ganglia. Messages to the interneuron from the terminal branches of sensory neurons or other interneurons reach the dendrites and are transmitted along the axon. In most cases the action potentials do not reach back into the cell body. Unlike the bipolar cells that we saw in so many sensilla, these monopolar interneurons and the monopolar motor neurons confine the action potentials to their processes in the neuropile. The cell body is almost silent. Messages from the interneuron are sent down the axon to the terminal branches, where they are transmitted to other interneurons or to motor neurons. The amount of information an interneuron may process depends on the number of synapses that it forms with other neurons.

Let's step back from our examination of a single interneuron and look at the structure of an entire ganglion in the ventral nerve cord (Figs. 6–3, 6–20). The ganglion is surrounded and mechanically supported by a fibrous, noncellular **neural lamella,** which is secreted by the underlying layer of cells, the **perineurium.** The neural lamella is actually a thick (up to 5 μm) version of the basement membrane that coats the peripheral nerves and indeed all organs in the hemocoel, including the epidermis. The

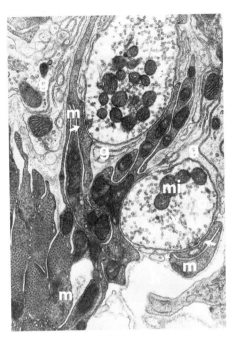

Figure 6–19
Neuromuscular junction in the body wall of a stick insect, × 30,000. Two axons (a) make synaptic contact with projections from the surface of the muscle cell (m) at points marked by arrow. Unapposed areas of the axons are capped by the glial cell (g). The axons contain synaptic vesicles (v) and mitochondria (mi). In some places (*) the axons are poorly isolated from the hemolymph (h). (M. P. Osborne, 1970, from *Symposia of the Royal Entomological Society of London,* no. 5, with permission of the publishers, Blackwell Scientific Publications Ltd.)

Figure 6–20
The cellular and extracellular components in the last abdominal ganglion of the American cockroach. In this figure, glial cytoplasm is indicated by light stippling; extensive extracellular space, by dark stippling; and neuronal cytoplasm is unstippled. The ganglion is surrounded by a noncellular neural lamella (NL) and a cellular layer, the perineurium (PN). The cortex, or outer layer, of the ganglion (OG) contains the cell bodies of the neurons, encapsulated by glial cell processes. The axon at the left (ax_1) is surrounded by a glial sheath, as are the axons (2–10) in the neuropile (NP). Arrows within axons indicate potential sites for synaptic interaction. Tracheolar cells (tr) also penetrate the neuropile. (D. S. Smith and J. E. Treherne, 1963, from *Advances in Insect Physiology*, vol. 1, with permission of the publishers, Academic Press, Inc.)

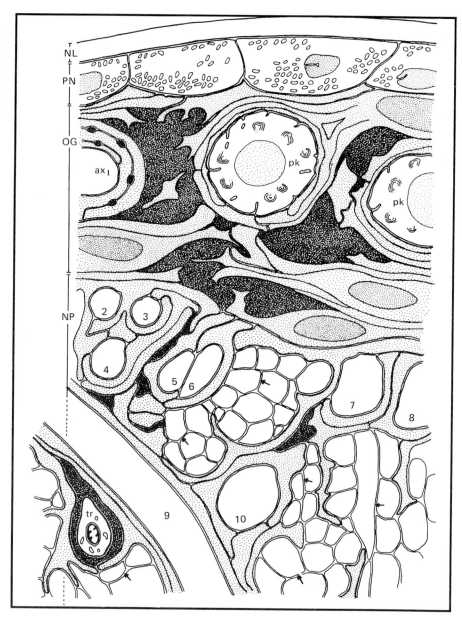

perineurium also is responsible for transporting nutrients from the hemocoel to the ganglion cells and, as John Treherne, of Cambridge University, has demonstrated, for maintaining an ionic differential with the hemolymph, analogous to the "blood-brain barrier" of vertebrates. The high extracellular concentration of sodium necessary for action potentials is thus maintained within the ganglion. The peripheral cortex underlying the perineurium contains all the cell bodies of the motor neurons and interneurons of the ganglion, plus their associated glial cells. The central neuropile contains only glial cells and neuronal processes, and it is the site of synaptic interactions. Many transmitters, such as octopamine, dopamine, 5-HT, and acetylcholine, are found in the neuropile, but we have yet to sort out the roles of each one.

Sensory Filtering.   When the spikes from the sensory axons flow into the ventral cord, the information is transferred to interneurons. As a result of synaptic interaction within the neuropile, the information is filtered and punctuated before being passed on to output centers. One of the clearest examples of this process is provided by the auditory interneurons in noctuid moths.

As we already noted, bat cries are given in short ultrasonic pulses that follow each other in trains. For *Myotis lucifugus*, a common bat in Massachusetts, where Kenneth Roeder did his research, the individual pulses vary in length (duration), loudness (intensity), pitch (frequency of sound waves), and pitch modulation. Trains of pulses vary in the length of the interpulse silence and in the length of the overall train. When the bat is cruising and looking for prey, it calls at a low pulse repetition rate, but when it has discovered prey and is moving in for the kill, the pulse repetition rate increases to a buzz. Which of these parameters are used by the moth? To answer this question, Roeder recorded the action potentials from the sensory nerve running from the ear to the metathoracic ganglion and from the ganglion itself. For this purpose he used fine wires that could be bent into hooks to pick up the nerve or sharpened into fine needles and stuck into the ganglia. When the moth was mounted and electrodes were implanted, Roeder presented a bat cry and recorded the result.

Let us first consider individual pulses as they are transmitted along the tympanic nerve. One actual recording is shown in Fig. 6–9. By changing intensity, duration, and pitch, Roeder was able to find which of these parameters passed the primary filter—the ear itself. He discovered that although the bat's ultrasonic frequencies varied from 25,000 to 60,000 Hz, the ear took no note of that fact. The moth does not hear differences in pitch; it is tone deaf. Therefore the moth also fails to detect pitch modulation. In contrast, intensity differences are detected and transmitted centrally (Fig. 6–21). Intensity is encoded in several features of the spike flow, including the spacing between individual spikes and the delay between the arrival of the sound and the first action potential. If a sound arrives from the left, it will be heard by both ears but will be louder at the left ear. Action potentials from the left ear will arrive first in the ventral cord and will follow each other more closely than those from the right ear. The moth can tell whether the bat is to the right, to the left, or (if intensity is equal in both nerves) at the midline.

The interneurons are more radical in their transformation. Refer to Fig. 6–22 to see what each of these units does to the bat cry. Of the various classes, the most commonly found and most subtle filter is the **repeater interneuron.** The repeater passes on information about intensity and duration of a single bat cry as well as train frequency and duration, but the repeater fails to fire an action potential every time it receives the presynaptic signal. It fires quickly at the first stimulus in a series and then becomes less responsive. In other words, it emphasizes the leading edge of each pulse, in a way analogous to emphasis on certain beats in music. The repeater makes the pulses easier to distinguish from one another. **Pulse marker interneurons** produce a radical transformation of the acoustic information. They produce only one spike for every bat cry, no matter how great its loudness. By listening in at the pulse marker, we can easily determine the pulse repetition rate. The **train marker interneuron** fires at a fixed rate as long as a bat continues to emit regular cries. No interneuron that would compare the intensities in the two ears has yet been found.

Figure 6–21
Tympanic nerve response in a noctuid moth to a pure tone of 40 KHz. The occasional large spikes originate in the B cell. (a) Response to sound intensity close to the threshold of the A_1 cell. The sound intensity in (b) is 7 db, in (c) 15 db, and in (d) 23 db above that in (a). Time line in (d), 100 msec. (Reprinted by permission of the publishers of *Nerve Cells and Insect Behavior*, by K. Roeder; Cambridge, Mass.: Harvard University Press, copyright 1967 by the President and Fellows of Harvard College.)

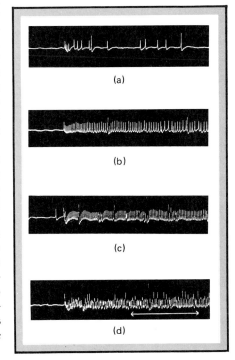

(a)

(b)

(c)

(d)

Figure 6–22

The response of the nervous system of a noctuid moth to the cries of a bat. (a) Diagrammatic picture of the bat cry and the corresponding spikes in the primary sensory cell (A_1 cell) and the repeater, pulse marker, and train marker interneurons of the ventral cord. (b) A hypothetical depiction of the filtering at various levels in the sense cells (A) and interneurons. Single pulses of bat cries vary in pitch, intensity, and duration, and trains of pulses vary in pulse frequency and train duration. Pitch is discarded at the level of the receptor. The repeater transforms the signal, with the loss of some information about intensity and duration of the individual pulse. The pulse marker discards information on individual pulse variables, and the train marker only signals the presence of bat cries. (See text for further information.)

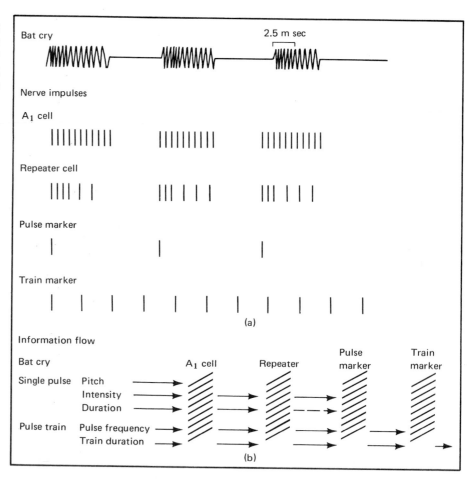

Roeder's data show us what is important to the moth. Intensity must be encoded so the moth can localize the bat in space. The ears certainly pass this information centrally, but we do not yet know how central processing compares right and left. The repeater punctuates the signal, making each pulse stand out. The pulse maker discards information on intensity and merely tells the moth whether the bat is cruising and searching or buzzing in for the kill. The **train marker** is a kind of continuous alarm, which reminds the ventral cord that a bat is around even while it is silent between pulses.

Giant Axons. For escape behavior there is another aspect of interneuron physiology that is very important: the speed with which the interganglionic axons can transmit a signal. Vertebrate axons that transmit high-speed messages are myelinated—they are wrapped in the layered membranes of specialized glial cells that insulate the axon and permit the speed of transmission to be increased some 5 to 10 times. For the most part, insects and other invertebrates lack true myelin. An alternative way to increase the speed of transmission is to increase the diameter of the axon. Consequently a few interneurons in elongated insects (as well as crustaceans, worms, and some molluscs) have a few **giant axons** of large diameter. Many originate in the terminal abdominal ganglion and transmit messages to locomotory centers in the thorax. These giant axons appear to be especially concerned with triggering escape behavior, where speed may be a life-or-death matter.

The giant axons in the cockroach, *Periplaneta,* have been analyzed by Kenneth Roeder, Jeffrey Camhi, and their respective colleagues. The roach has seven pairs of giant axons. The largest is about 30 μm (0.03 millimeters) in diameter and thus occupies an area roughly 100 times that of a typical axon (Fig. 6–3). The cell bodies for these giant axons lie in the terminal abdominal ganglion, where their collaterals synapse with sensory axons from the cerci. When the cercal sensilla are stimulated by a puff of air, a message is relayed up the ventral cord axons toward the head. Giant axons carry that message. Synapses occur in the metathoracic ganglion with motor neurons of the legs, and the roach is stimulated to begin running.

Transmission across a synapse takes time, and time is short when a predator is striking. Delays at the fastest synapses are at least several milliseconds. At least three neurons and three synapses (don't forget the neuromuscular synapse) are involved in this response. Incredibly, it takes only 44 *milliseconds* for the roach to begin running from the time the cercus is stimulated. The giant axons transmit signals at 6 to 7 meters per second, 10 or more times faster than in a normal axon. It takes only 2.8 milliseconds for a message to travel the length of such an interneuron and to reach the thorax.

However, this simple, attractive interpretation of the role of the giant interneurons in *Periplaneta* has come into question in the last decade. Several scientists electrically stimulated giant axons and were unable to show any excitation in the leg motor neurons. Part of the explanation for the inconsistent results may lie in physiological diversity among giant fibers. This controversy will be resolved only as we map each of the neurons involved and work out the details of their synaptic interactions. The neuronal control of insect behavior is not so simple as we once thought, and it becomes much more interesting as we come to appreciate the subtleties of interneuron physiology.

Insects pay a price for the speed gained by giant axons. They lose the additional information that could be transmitted by the many interneurons that were potentially displaced by one giant interneuron. The size of the ganglia, and consequently the number of motor neurons and interneurons, is limited not only by the size of the insect but also by the problem of transporting nutrients into and wastes out of the ganglia across the perineurium. Although tracheae do penetrate the ganglia, the high oxygen demand of neurons must also limit the diameter of ganglia. Insects thus have a nicely intermediate number of neurons, not the hundreds of millions that make the tracing of neuronal pathways in vertebrates a nightmare, but also not the very few neurons that make sea slugs such behaviorally uninteresting animals. The mesothoracic ganglion of a roach, for instance, has about 300 motor neurons, 200 interganglionic interneurons, and 1500 intraganglionic interneurons.

Nervous Circuitry and Rhythm Generators.   Recent techniques with intracellular dyes have provided a true breakthrough in insect neurobiology, one that allows us to see all the processes of a single neuron and makes possible the dream of tracing the entire neural circuitry involved in functionally important behavior. First, the function of a particular neuron is determined by following its axon to a muscle or by penetrating its process in the neuropile with a fine electrode and recording electrical discharges that are correlated with behavior. Then the process is injected

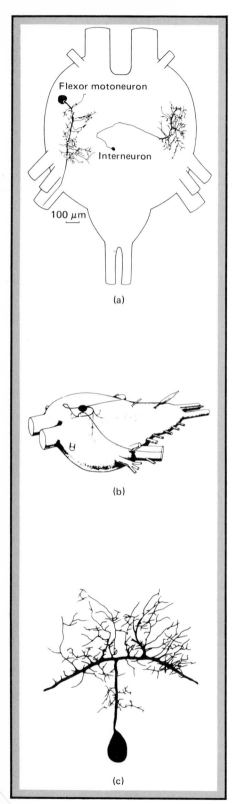

Flexor motoneuron

Interneuron

100 μm

(a)

(b)

(c)

through the electrode with procion yellow, Lucifer yellow, or a cobalt dye, or the cut end of the peripheral axon is dipped in the dye. The dye remains within that one cell. It diffuses through much of the neuron including the silent cell body but does not penetrate other neurons. Then the ganglion is fixed and prepared for histology. The behaviorally identified neuron can thus be located and its connections traced when the nerve or ganglion is sectioned or dissected (Fig. 6–23). Perhaps all central nervous system neurons will ultimately be traceable in this fashion, although the smaller brain interneurons will pose a technical challenge. These studies show that the cell body and processes of each motor neuron and interneuron occupy a characteristic location in the ganglia. Its axon passes into **fiber tracts,** ordered groups of axons that run through the connectives and the ganglionic neuropiles. These locations are constant from insect to insect of the same species. For instance, the positions of each of the 14 giant axons illustrated in Fig. 6–3 will be the same in the connectives of any *Periplaneta americana* you section, as will the positions of their cell bodies in the terminal ganglion. The importance of "identified neurons" in neurobiological research is obvious—they can be numbered and mapped, and the results of neural identification in one specimen can be directly applied to another.

Early students of neurophysiology and behavior assumed that no rhythmic activities, such as walking, could take place in the absence of sensory feedback. Much to the astonishment of many biologists, experiments in the past two decades have shown that locomotor rhythms persist even when the sensory nerves are cut in the flying locust.

You will remember that the indirect flight muscles consist of the dorsal longitudinals and the dorsoventrals (Chapter 1). When the late Donald Wilson, of Stanford University, cut the sensory nerves in flying migratory locusts and stimulated the nerve stumps with random electrical impulses, the insects continued to flap their wings up and down in a coordinated manner. Somehow the motor neurons were continuing to discharge in a correct sequence without receiving feedback signals from stretch receptors. Wilson concluded that the interneurons must form a **central rhythm generator** that orders the discharge of motor neurons in a correct sequence that oscillates repeatedly from longitudinals to dorsoventrals and back again. Input from the sensory neurons helps to maintain a heightened central excitatory state so that the rhythm generator continues to oscillate, but it does not direct the switching from one muscle group to the other. Input from the brain or from other ganglia can also serve to fine tune the motor patterns. The sensory input from sensilla on the wings, pressure-sensitive hairs on the head, abdominal nerves, and others are fed into multimodal interneurons, like those discussed earlier, and thus the insect makes small adjustments so as to steer in flight.

What is the central rhythm generator? For cockroach walking, it may be some of the interneurons in the thoracic ganglia that were discovered by Keir Pearson and Charles Fourtner, of the University of Alberta. These interneurons have collaterals

Figure 6–23
The detailed structure of the processes of individual motor neurons and interneurons as visualized by the injection of cobalt salts or fluorescent dyes into the cells. The large circular structure is the cell body. (a) Slow motor neuron and nonspiking interneuron in the locust metathoracic ganglion. (b,c) DUM neuron in the metathoracic ganglion of a locust. The cell body is located dorsally on the midline. Axons go to both sides. (a: M. J. Burrows and M. V. S. Siegler. Reprinted by permission from *Nature*, vol. 262, p. 222. Copyright © 1976, Macmillan Journals Limited. b: G. Hoyle, 1974, from *Journal of Experimental Zoology*, vol. 187, reproduced by permission of Alan R. Liss, Inc. c: with permission from C. W. Heitler and C. Goodman, 1978, from *Journal of Experimental Biology*, vol. 76. Copyright: Academic Press Inc. [London] Ltd.)

that excite or inhibit the slow motor neurons that innervate the leg muscles. A small network of these interneurons oscillates in a reciprocal manner, so that flexor and extensor motor neurons are alternately excited (Fig. 6–24). A similar control mechanism seems to control many other behaviors, ranging from breathing rhythms to the stereotyped set of movements required for ecdysis from the old cuticle. The sequences of movements are largely programmed within the central nervous system, but the detailed orientation and timing of the movements are influenced by sensory feedback. Sensory information does not dictate the behavioral repertoire, but it may trigger its start and modulate its expression.

The Brain

The ventral nerve cord ganglia directly receive and integrate all sensory information received by an insect except that coming via the eyes and antennae, and they are involved in all movements except that of the antennae. Even the mouthparts are innervated by a portion of the ventral nerve cord, the subesophageal ganglion. However, an insect does have a brain. What role does the brain play in coordinating an insect's movements? Almost none, surprisingly; except for movement of the antennae, the actual patterning of motor neuron impulses that results in a coordinated sequence of muscular contractions does not require the brain. A headless insect (with the neck sealed so the hemolymph does not leak out) can live for days and conduct coordinated walking, swimming, flying, grooming, copulating, and oviposition movements. Indeed, judicious cutting of nerve cord connectives can demonstrate that movements of a pair of appendages are coordinated solely by the ganglion of that segment—so flying, swimming, and walking muscular contractions are coordinated solely by the thoracic ganglia. This means that one part of an insect can be doing something quite independently of what another part is doing, as a female mantid demonstrates when she eats the head of her mate, whose genitalia, controlled by his terminal abdominal ganglion, continue to carry out copulatory movements. (This gory topic will pop up again in the next chapter.)

Since the ventral nerve cord coordinates rhythmic movements, can an insect be said to truly possess a brain? Indeed, Aristotle said that an insect's brain is situated somewhere between its head and tail, and Linnaeus partially defined insects by the absence of a brain! Of course, an insect does have a brain, located in its head above the foregut. But an insect's brain is quite small, especially considering that much of the interior of the head is occupied by the muscles that operate the mouthparts. The largest insect brain is about 7.5 millimeters wide, while the diameter of an ant's brain is about 150 μm (0.15 millimeter). To say that a friend has the "brain of a fly" is the beginning of the end of the friendship. Yet some scientists have spent much of their professional lives unraveling the secrets of the brains of house flies and blow flies. One such eminent scientist is Nicholas J. Strausfeld, of the European Molecular Biology Laboratory in Heidelberg, who in 1976 prepared a remarkable atlas of the brain of a fly. A house fly's brain weighs 0.35 (male) to 0.42 (female) milligrams, with respective volumes of 0.28 and 0.31 cubic millimeters. A female fly brain contains about 338,000 neurons, compared with about 850,000 in a honey bee brain and the estimated 10 billion neurons in a vertebrate brain. The brain is certainly important to the fly, however. It receives and integrates visual and olfactory information; initiates, directs, or inhibits motor activities; and is the principal center of learning in the nervous system.

You may recall that an insect's brain consists of three pairs of ganglia (Fig. 6–25). The smallest of these, the **tritocerebrum,** is actually a pair of ventral nerve cord gan-

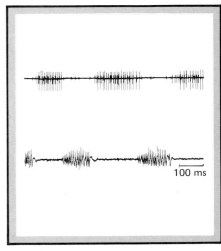

Figure 6–24
Simultaneous electrical recording of the spikes in levator motor neurons (top) and depressor motor neurons (bottom) in the metathoracic ganglion of a cockroach during walking. (K. G. Pearson and C. R. Fourtner, 1975, from *Journal of Neurophysiology,* vol. 38, with permission of the American Physiological Society.)

Figure 6–25
The insect brain and associated parts of the nervous system. (After R. E. Snodgrass, 1935. Reprinted with permission of Macmillan Publishing Co., Inc. from *Insects in Perspective* by M. D. Atkins. Copyright © 1978 by Michael D. Atkins.)

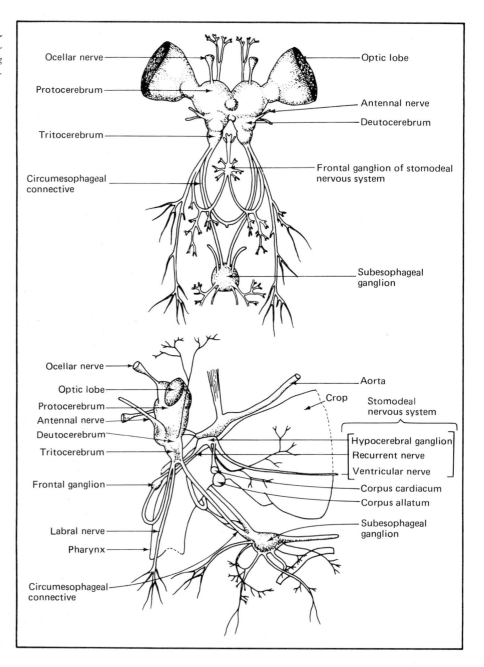

glia that during development migrate above the gut to join the rest of the brain. Their connectives to the next pair of ganglia in the ventral nerve cord (the mandibular part of the fused subesophageal ganglion) pass on either side of the foregut to unite the brain with the remainder of the central nervous system. The tritocerebrum sends nerves to the labrum and to the stomodeal nervous system.

The remaining two parts of the brain, the deutocerebrum and protocerebrum, develop dorsally, separate from the ventral nerve cord. Their potential homology with segmental ganglia is a controversial topic in embryology. The **deutocerebrum** controls movements of the antennae and also receives messages from the sensilla located on these important sensory organs. The sensory neurons synapse with interneurons in the neuropile of the antennal lobes in the deutocerebrum. These interneurons in turn mostly synapse with other interneurons in the protocerebrum.

The **protocerebrum** is the largest and most anterior of the parts of the brain and most closely fits our human concept of what a brain should be. One important function is to receive signals from the visual receptors, and most of the protocerebrum is occupied by two massive **optic lobes** that join the compound eyes (Figs. 6–25, 6–26). The optic lobes contain 76% of the neurons in a fly's brain. Each optic lobe contains three masses of neuropile that progressively filter and integrate visual information. The optic ganglia are successive tiers of interneurons through which the primary visual fibers run. The interneurons filter and abstract information on shape and movement, reacting to such general stimulus features as vertical or horizontal movement of edges or dots. The details of stimulus filtering in these lobes are not yet clear, but with the possibility of studying morphologically identifiable interneurons (Fig. 6–27), we may soon be able to make sense of the information-processing scheme. Smaller nerves transmit information from the simple eyes, the dorsal ocelli.

The functional organization of the protocerebrum is still little understood, except for the optic lobes. The neurosecretory cells that produce prothoracicotropic hormone (PTTH) (see Chapter 4) are located anteriodorsally, on each side of the midline. Their axons leave the brain as nerves that run to the corpora cardiaca and corpora allata.

Within the neuropile of the protocerebrum are several conspicuous fiber tracts, of which we shall consider two. The **mushroom bodies** (corpora pedunculata), complex fiber tracts with hundreds of thousands of specialized interneurons contained wholly

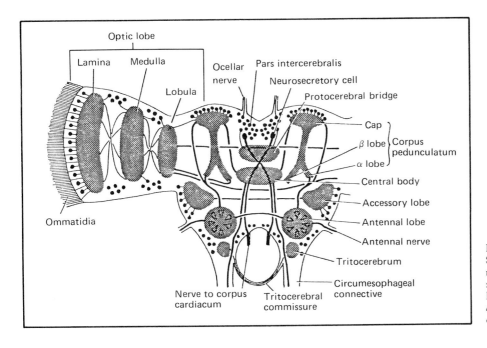

Figure 6–26
Schematic diagram of an insect brain, showing more important neuropile regions (hatched) and zones of cell bodies (black dots). (Modified from R. F. Chapman, 1969, from *The Insects: Structure and Function,* with permission of the publisher, Hodder & Stoughton Ltd.)

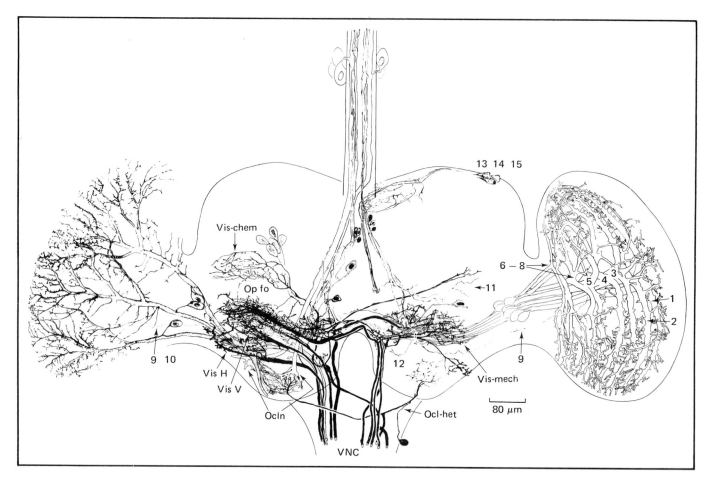

Figure 6–27
Visual interneurons in the brain of a house fly. The ocellar nerve runs in from the top; the optic ganglia (lobula) are shown on both sides. Each neuron was demonstrated by injection of cobalt chloride. Cells 1–8 (right) are giant interneurons that have their dendritic fields oriented in the vertical in the lobula. The three large interneurons shown on the left have their dendritic fields oriented in the vertical. Descending interneurons that run to the ventral cord include those that respond to visual cues (*vis H, vis V*) as well. Others respond to several kinds of sensory input (*vis-chem*, left, and *vis-mech,* right). Cells 13, 14, and 15 are lateral neurosecretory cells. *VNC* = ventral nerve cord. (N. J. Strausfeld, 1976, *Atlas of an Insect Brain,* reprinted with permission of the publishers, Springer-Verlag, Inc.)

within the mushroom bodies, occur on either side of the PTTH neurosecretory cells (Fig. 6–26). The cell bodies of the mushroom body interneurons form one or two caps over the stalk of the mushroom; the stalk then branches into at least two lobes, where synapses occur with interneurons from other parts of the central nervous system. The mushroom bodies are largest in insects with a well-developed sense of smell, and their synapses are mostly with interneurons coming from the deutocerebrum and from the optic lobes. Their principal function appears to be to synthesize sensory information. They have long been thought to be the seat of learning in an insect, largely because they are best developed in worker social insects, the "smartest" of all insects. These, however, are also the insects that have the greatest need to integrate diverse chemical signals. Mushroom bodies are also large in the "dumber" moths and roaches, which also must perceive odors, but are relatively small in visually oriented flies and dragonflies.

This does not, however, exclude a memory role for the mushroom bodies. Stephen and Ruth Bernstein, a husband-wife team at the University of Colorado, found that worker *Formica* ants that are more efficient in learning to run a maze also have, on the average, broader caps in the mushroom bodies. Indeed, Martin Lindauer's laboratory at the University of Würzburg has actually transferred memory, in the form of a learned time of feeding, from trained honey bees to naive bees by implanting mushroom bodies from the former into the head (but not the brain) of the latter! Randolf Menzel, Jochen Erber, and colleagues at the Free University of Berlin further investigated the cellular basis of learning in honey bees. They presented an odor to a bee just before giving it food and found that after one training session the bee learned the odor and then extended its proboscis whenever the odor was presented, even if given no reward. They confined the bee in a tube, exposed its brain, and blocked neuronal activity in a small area with a thin, cold needle. With this technique they demonstrated that the mushroom bodies are the essential structures in the transfer of odor information from short-term to long-term memory.

The other fiber tract we shall identify in the brain is the **central body,** located deep in the center of the protocerebrum. It is well developed in all insects, and it receives input from most regions of the brain and also from the ventral nerve cord. It appears to be the most important part of the brain for integrating different types of sensory information. It is also an activating center, initiating the body movements that are coordinated by the thoracic ganglia. The central body interneurons are apparently "command" interneurons, whose axons enter the ventral nerve cord and go to the thorax. The interaction of these neural centers in controlling cricket chirping is described in Chapter 7.

Sensory signals trigger behavior, and sensory feedback influences the duration and intensity of motor patterns. The brain is primarily a center for the sorting, evaluating, and blending of sensory information that flows in along axons from the eyes, the antennae, and other cephalic receptors. The brain contributes to the selection of particular behavioral responses, but we do not understand exactly how that selection takes place. Most motor patterns are encoded in the thoracic ganglion of the ventral cord. The control of an entire behavioral sequence requires a selection of particular motor patterns in a useful sequence and their expression in a timely manner. The intensity of the output is precisely modulated. Locusts use the same muscles and movements in kicking as in swimming, and the motor pattern for leg flexion and extension is the same whether for a weak swimming stroke or for a powerful jump. Only the strength of the contractions differs.

With the advent of techniques for tracing the processes of individual identified neurons, for recording cellular interactions, and for reconstructing small networks, we are gaining significant insights into the neural bases of behavior. We still have much to learn about the "executive function" of the brain and the control centers of the ventral nerve cord, but exciting studies are being published at a high rate. Readers can be sure that parts of this chapter are already out of date.

Summary

Nervous systems are composed of sensory neurons, motor neurons, and interneurons. All the cells are electrically polarized and conduct a state of excitation by electrical potential changes. Each neuron has an input portion, the dendrite, and an output portion, the axon. The cell bodies of motor neurons and interneurons are in the cortex of each ganglion, while their processes run into the central neuropile. The cell bodies of sensory neurons are peripheral, associated with tactile hairs, eyes, ears, and so on.

Neurons interact with one another at synapses, all of which are in the neuropile. The transmission across synapses is brought about by release of neurotransmitters, which include acetylcholine, glutamic acid, GABA, and octopamine.

Sense organs respond to visual stimuli, sound, touch, pressure, scents, tastes, heat, and moisture, among other things. The basic sensory unit is a sensillum, composed of a sensory neuron and various accessory cells that include the glial cell, the trichogen cell, and the tormogen cell, the latter two secreting the cuticular portion of the sensillum.

Mechanoreceptors include tactile setae and ears. In the latter, chordotonal organs detect the vibrations of a larger structure, such as an antenna or eardrum. The sensory neuron in the ear and the interneurons in the central nervous system filter and process the signal until essential information stands out sharply. For example, a moth evades a bat merely by minimizing the flicker and balancing the sound levels in both ears.

Chemicals are perceived by a wide variety of specialist and generalist sensilla that are provided with pores through which the chemicals can reach the sensory neuron dendrites. Uniporous sensilla usually are taste receptors and occur especially on the mouthparts and tarsi. Multiporous sensilla are odor receptors, concentrated on the antennae.

The compound eye contains many ommatidia. The light detector in an ommatidium is the rhabdom, composed of processes from several visual sensory neurons called retinula cells. The ommatidium sees only a small portion of the visual field; the whole field is communicated as a kind of mosaic image. The important features for insect vision seem to be *moving* stimuli that pass from one visual field to another. The eye does not work by producing snapshots; rather it gathers information from the flickering of moving patterns.

Motor neurons synapse with muscles at neuromuscular junctions. The motor axons are of several kinds: fast and slow excitatory axons (glutamic acid is the usual transmitter), inhibitory (GABA is the transmitter), and modulatory (octopamine is one such transmitter). The motor control of insect muscles uses many fewer axons and muscle subunits than are found in vertebrates.

Interneurons are confined to the central nervous system. They occur in small networks that are responsible for rhythmic trains of impulses along motor axons. These networks are central pattern generators that oscillate spontaneously between alternative outputs. Sensory input is integrated and filtered by sensory interneurons, which then pass that information to the central pattern generators. The bulk of the complex movements that insects carry out are "hard-wired" in central patterned circuits, which lie mostly in the ventral nerve cord. Behavior can be controlled by multilayered systems of such oscillators.

The brain is primarily a center for filtering, processing, and storing visual and olfactory information, and for initiating or inhibiting behavior that is coordinated by the ventral nerve cord. It consists of the protocerebrum, the largest portion, which includes two large optic lobes and the mushroom bodies and central body; the deutocerebrum, the site of antennal motor neurons and for filtering of antennal input; and the tritocerebrum, which is actually part of the ventral cord.

Selected Readings

Camhi, J. M. 1980. "The escape system of the cockroach." *Scientific American*, vol. 243, no. 6, pp. 158–72.

Dethier, V. G. 1963. *The Physiology of Insect Senses.* New York: Wiley. 266 pp.

———. 1976. *The Hungry Fly.* Cambridge, Mass.: Harvard University Press. 489 pp.

Horridge, G. A. 1968. *Interneurons: Their Origin, Action, Specificity, Growth and Plasticity.* San Francisco: Freeman. 436 pp.

———. 1977. "The compound eye of insects." *Scientific American*, vol. 237, no. 1, pp. 108–20.

Howse, P. E. 1975. "Brain structure and behavior in insects." *Annual Review of Entomology*, vol. 20, pp. 359–79.

Hoyle, G. 1970. "Cellular mechanisms underlying behavior-neuroethology." *Advances in Insect Physiology*, vol. 7, pp. 349–444.

Pearson, K. 1976. "The control of walking." *Scientific American*, vol. 285, no. 6, pp. 72–86.

Schneider, D. 1974. "The sex-attractant receptor of moths." *Scientific American*, vol. 231, no. 1, pp. 28–35.

Strausfeld, N. J. 1976. *Atlas of an Insect Brain*. Berlin and New York: Springer-Verlag. 214 pp.

Wehner, R. 1976. "Polarized-light navigation." *Scientific American*, vol. 235, no. 1, pp. 106–15.

In his award-winning book *The Lives of a Cell*, Lewis Thomas remarks that a person who writes about insects often takes pains to caution that their behavior is "absolutely foreign, totally unhuman, unearthly, almost unbiological. They are more like perfectly tooled but crazy little machines. . . ." He then goes on to enumerate some of the accomplishments of these most successful and ubiquitous of creatures. Clearly they are unhuman—but unbiological, never. That they are "perfectly tooled"

Behavior

P A R T

and often machinelike in their actions can hardly be questioned. The brains of the minute gnats called no-see-ems measure only 0.2 mm, yet these insects are able to create misery in animals with 10 billion cells in their brain, namely, ourselves. Obviously there is something rather special about insects. Can we hope to understand them? The answer is: we had better, if we hope to coexist with them and to render them less competitive with ourselves.

T H R E E

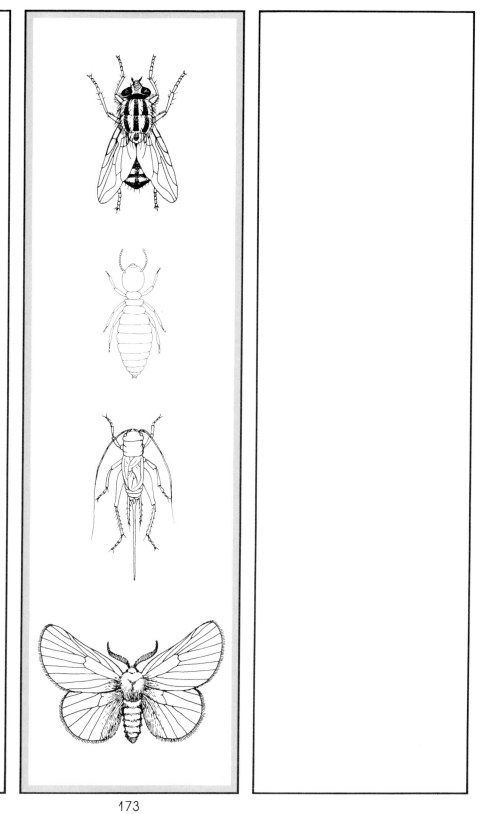

Fundamentals of Behavior

Insects are often said to exhibit **programmed behavior,** which means that they are born with the capacity to behave in certain set patterns on receipt of appropriate stimuli. They are also **specialists;** that is, they are programmed to do certain things with great efficiency; other things, not at all. The gardener's potatoes may be defoliated by Colorado potato beetles, and perhaps the eggplants too, but the gardener may feel confident that these insects will not touch the carrots or the strawberries. Female moths of the fruittree leafroller, a pest of apples and other fruits, produce a volatile pheromone that attracts males of their own species from a considerable distance—yet the males do not respond to the pheromones of closely related species even though they differ only slightly in molecular configuration. A tiny parasitic wasp called *Aphytis holoxanthus,* when introduced from Hong Kong to Israel, virtually wiped out the red scale insect of citrus in that country—a specialist *par excellence,* programmed to find and consume a specific host insect and no other. That is not to say that all insects are highly specialized feeders or that all behavior is fully programmed —matters we shall consider shortly—only that these are two of the more striking features of insect behavior—features that we may sometimes turn to our own advantage.

Innate Behavior

Behavior is the result of interactions between an organism and its environment by way of receptors (eyes, sensory setae, and the like) and effectors (muscles, exocrine glands). What an insect does at a given time is the result not only of the cues it is receiving from the outside but also of such internal factors as patterns inherent in the nervous system, physiological states (such as hunger or fatigue), kind and level of hormones circulating the blood, and learned information.

One of the striking features of insect behavior is that much of it is performed without previous experience and without interaction with other members of the species. Such behavior is inborn, or **innate.** Innate behavior may be fairly simple and straightforward, as when a mated female imported cabbageworm butterfly, responding to specific cues, lays her eggs on a farmer's broccoli. Or it may be complex and capable of being switched from one behavior pattern to another, as when a worker yellowjacket chews wool pulp from a fence post, mixes it with saliva, and applies it to the covering of a paper nest, all the time coordinating its behavior with that of other workers and with the needs of the colony. There are several questions we may ask of such behavior—none of them easy to answer, but all of them the subject of much current research:

1. How is innate behavior inherited?
2. How is such behavior stored so as to be called on when needed?
3. To what extent can behavior be modified in response to environmental variables?
4. Can insects learn, and if so, what is the relationship between innate behavior and learning?

Inheritance and Coding of Behavior.

Students of behavior often use the term **fixed-action pattern** for segments of behavior that are performed in a stereotyped, species-characteristic manner. Much of the performance of insects consists of fixed-action patterns, often following one another in series, guided by internal and external cues. These patterns are in large part coded in the genes and expressed by specific nervous pathways that link receptors and effectors. Much behavior is polygenic, that is, under the control of numerous genes, each of which may code other aspects of development.

The effects of single genes on behavior are known in a few cases. For example, the mutant *yellow* in the fruit fly *Drosophila melanogaster* causes males to vibrate their wings in courtship at a lower frequency than normal and thus reduces their success in mating. These same mutants are also more strongly attracted to light than the wild type. Even complex behavior has sometimes been shown to have a rather simple genetic basis. Certain strains of honey bees are resistant to the bacterial disease called American foulbrood. The workers of these strains uncap cells containing diseased larvae and pupae, remove them, and clean the cells. By appropriate crossings, Walter Rothenbuhler, of Ohio State University, showed that the switch from "hygienic" to "nonhygienic" behavior was controlled at two loci, one affecting uncapping and the other removal of diseased offspring. The underlying behavioral mechanisms may well be under the influence of many genes, but in this case two pairs of alleles control the expression of hygienic behavior.

Cricket Acoustic Behavior.

When a male field cricket reaches sexual maturity, he begins to sing a "calling song," a series of chirps familiar to everyone who has lived in the country. This song serves to attract females, who approach the male and mate with him. Males of each species of cricket produce a pattern of chirps characteristic of that species, and in each case the female is programmed to respond to the appropriate pattern; in this way the integrity of each species is maintained. When male crickets of a given species are reared from the egg in isolation, or with exposure only to the songs of alien species, they nevertheless produce their own species-specific song when they reach maturity. Thus these songs are the output of fixed action patterns.

Crickets of different species can sometimes be induced to mate in the laboratory, and in some cases the hybrids are fertile and can be backcrossed with the parental stock. As might be predicted, first generation males produce songs intermediate in pattern between the two parental species, and first-generation females are most attracted to these intermediate songs. With appropriate backcrossing, it can be shown that the song patterns shift in a manner corresponding to the proportion of genes inherited from each species (Fig. 7-1). The different elements (number of chirps, chirp interval) do not segregate independently, suggesting that many genes are involved in determining the neural basis of song production.

These studies were carried out by David Bentley, of the University of California at Berkeley, and Ronald Hoy, of Cornell University, who went on to ask another fundamental question of their crickets: How are sound patterns generated in the nervous system? Cricket songs are produced by movements of the fore wings, in which a "file" on one wing is moved over a "scraper" on the other wing and sounds are amplified by a resonating membrane on the partially elevated wings. It is possible to insert very small electrodes into the nerves and muscles of a cricket through small holes in the cuticle, without interfering in any important way with the cricket's behavior. It was found that only the first two thoracic ganglia are required for generation of the calling song, but the song is elicited by a "command neuron" emanating from the central body of the

Figure 7-1
Song patterns of two related species of crickets (a and f) and of F_1 hybrids (c and d) and backcrosses (b and e). Patterns shift in proportion to the ratios of genes inherited from each species. (D. R. Bentley, *Science*, vol. 174, p. 1140; copyright 1971 by the American Association for the Advancement of Science.)

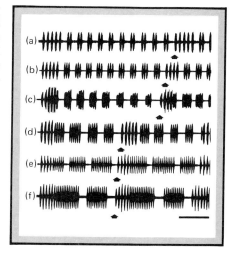

protocerebrum (see Chapter 6). This interneuron has complex interconnections with interneurons and motor neurons within the ganglia, and the latter produce appropriately rhythmical contractions of the longitudinal and dorsoventral muscles—the same muscles that under a different nervous regimen produce flight (Figs. 1–8c, 7–2).

The circuitry necessary for production of the calling song develops gradually through the several molts of the immature cricket, and premature activation of any part is prevented via inhibition from the brain. During development, new cells are added and new dendritic and synaptic connections are developed. A single motor neuron of an adult insect may have an enormously complex dendritic branching within the ganglion. The construction of this neuronal programming network is genetically determined, much as an electronic transistor might be built from blueprints and templates.

Inhibitory Centers. The importance of central inhibitory centers has been demonstrated in many insects, and it is probable that much behavior is initiated by removal of inhibition rather than by stimulation. Destruction of inhibitory centers often

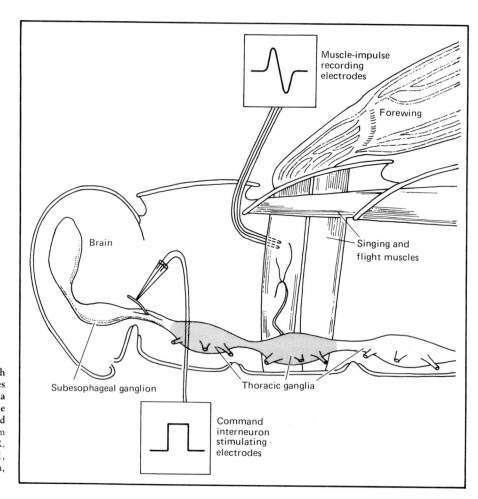

Figure 7–2
Anterior part of a cricket nervous system with electrodes implanted to elicit and record impulses responsible for the song. The two shaded ganglia are sufficient for generating the song, via the thoracic muscles. Stimulation of the command interneuron induces singing. (Modified from "The neurobiology of cricket song," by D. R. Bentley and R. R. Hoy, *Scientific American*, vol. 231, pp 34–44, copyright 1974 by Scientific American, Inc. All rights reserved.)

results in abnormally high activity levels. Lesions in the protocerebrum of the brains of many insects produce an increase in locomotor activity that may last for hours. When a male cockroach is decapitated and electrodes placed in the nerve trunk to the genitalia, activity of the neurons climbs steadily and remains high for several hours. Such cockroaches show much movement of the tip of the abdomen, including the genitalia. In the praying mantid, this behavior may have strong survival value, for the female mantid commonly attacks her mate and begins feeding on him from the anterior end. Kenneth Roeder and his colleages at Tufts University have shown that removal of inhibitory centers in the head of the male mantid produces greatly increased firing in the nerves supplying the genitalia (Fig. 7–3). The result is that when the female has eaten the head of the male, the abdomen of the latter undergoes more vigorous movements than occur in the intact male. These result in prolonged copulation and implantation of a spermatophore in the female.

Endogenous Rhythms. Many insects show an increase of activity at a certain time of day. Such peaks of activity are often dependent on external factors, such as light intensity, but in other cases they result from rhythms inherent in the nervous system. Such rhythms persist for several days even when the insect is placed in constant light or darkness (Fig. 7–4). However, the "biological clock" must first be set by expo-

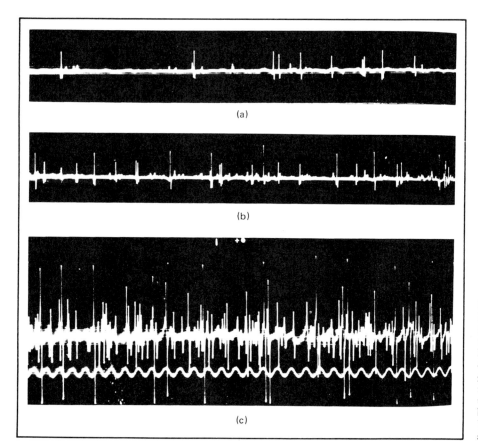

(a)

(b)

(c)

Figure 7–3
Pattern of impulses in the nerves supplying the genitalia of the male mantid: (a) before the connection with the remainder of the nervous system has been severed; (b) three minutes after the nerve cord has been cut; (c) four minutes later, showing greatly increased nervous activity following the removal of inhibition. (Reprinted by permission of the publishers from *Nerve Cells and Insect Behavior*, by Kenneth D. Roeder, Cambridge, Mass.: Harvard University Press, copyright 1967 by the President and Fellows of Harvard College.)

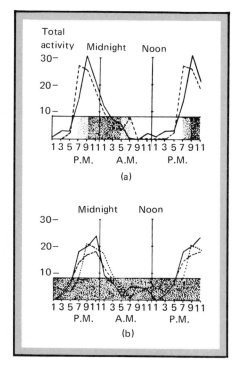

Figure 7-4
Activity of a male cricket under normal day—night conditions (a) and when later placed in constant darkness (b). (From K. D. Roeder, 1953, *Insect Physiology,* with permission of the publishers, John Wiley and Sons, Inc.; after F. E. Lutz, 1932.)

sure to the normal cycle of light and darkness. Since most such activity peaks occur on about a 24-hour cycle, they are termed **circadian rhythms,** from the Latin words *circa* and *dies,* meaning "about a day." Some insects exhibit different daily peaks for different activities: *Drosophila* flies, for example, are active in mating during the day, but the females lay most of their eggs in the evening. Endogenous rhythms are normally independent of temperature, at least within the normal temperature range of the species. They are also largely independent of hunger and other physiological states.

Indeed, little is known regarding the control of endogenous rhythms. Cells in certain parts of the brain have been shown to possess daily rhythms of spontaneous firing. Through synaptic connections with neurosecretory cells, these may provide the generators of rhythmic activity. But for the present these matters remain in the realm of hypothesis.

Hormones and Behavior.   In Chapter 4 we reviewed the major endocrine organs of insects and the hormones they produce. Hormone balance intimately controls all aspects of insect development, including that of nervous circuitry, receptors, and effectors. A precocious adult, produced by removal of the corpora allata, also *acts* like an adult, within certain limits, demonstrating the importance of hormones in maintaining functionality of all body parts at specific stages of development.

The importance of hormones in controlling various aspects of vertebrate behavior has long been appreciated, but it is only within the last few decades that researchers have come to appreciate the many ways that hormones may influence insect behavior. In fact, it has been argued that because of the simplicity of the insect nervous system, hormones may be of special importance in supplying the basis for greater behavioral diversity. That is, they may alter the "mood" of an insect in such a way that different stimuli assume importance and different responses are elicited. Being blood borne, hormones reach every cell in the body. However, changes in hormone titer cannot take place instantaneously. Hence hormones do not usually trigger a quick response, but much more often supply the background for a gradual change to a different type of behavior.

For example, female *Gomphocerus* grasshoppers, after the last molt, at first repel approaching males by kicking and fleeing. In the course of time, the corpora allata of the female begin producing juvenile hormone, causing the female to stridulate in answer to the male and eventually to allow him to mate. This same effect can be produced by treating newly molted adult females with this hormone. In many grasshoppers, secretions of the corpora allata are required for the maturation of the male sexual behavior. When the corpora allata are excised, males do not exhibit sexual behavior, but implantation of active corpora allata into such males restores this behavior. Allatectomized males of *Schistocerca* locusts do not produce the pheromone that hastens sexual maturation in other males. Many other examples are known of interactions between hormones and pheromones.

A particularly good example is supplied by the silkworm moth *Antheraea pernyi.* As in other moths, females of this species produce a volatile sex pheromone that attracts males from a distance. J. W. Truman and L. M. Riddiford, a husband—wife team now at the University of Washington, showed that in the absence of eclosion hormone (see Chapter 4) males remain totally unresponsive to the female pheromone. They discovered this by removing brains from pupae, an operation that if properly performed does not prevent molting to the adult. But when brains were implanted into the abdomens of debrained pupae, the resulting adults underwent intense flight activity when exposed to female sex pheromone. Electrodes implanted in the antennae of males lacking eclosion hormone showed normal response to the pheromone, demonstrating that the "behavioral block" was not in the receptors but in the central nervous system.

Very often the effect of hormones is not to facilitate or inhibit a nervous response totally, but to alter the **threshold of response.** For example, "readiness to lay eggs" implies a spectrum of possible responses. If only slightly motivated, a female may respond to stimuli only if they are strong and persistent; the threshold of response is said to be high. On the other hand, a female long under the influence of internal stimulation but deprived of suitable external stimuli may undergo a lowering of threshold of response, such that the eggs are laid in a wholly inappropriate place. This is only one of several sources of behavioral variation and unpredictability, a matter to which we must now turn our attention.

Behavioral Plasticity. We have seen examples of highly stereotyped behavior, endogenous in the sense that it involves nervous and hormonal mechanisms that are under genetic control. Such fixed-action patterns are generally concerned with mating, egg laying, feeding, and other fundamental aspects of growth and reproduction. Stereotypy is essential to an insect when learning or practice is disadvantageous or impossible, as is often true of short-lived animals. Fixed-action patterns are triggered by environmental **releasers,** but they can also be modified to fit particular situations, again guided by environmental cues. Oviposition is typically quite stereotyped, but a moth about to lay her egg may be guided by leaf shape and texture, exposure to sun or wind, and the like.

A good example of complex stereotyped behavior capable of modification involves the building of cases by the aquatic larvae of caddisflies. As a rule, each species makes a case of characteristic form by combining small items in the environment with silk from its labial glands. But building behavior is sufficiently plastic that cases can often be built under unusual circumstances. Several experiments were performed with a species that normally builds its case of bits of leaves, using its front legs to seek leaf pieces and its middle legs to manipulate them. When larvae are removed from their cases and one pair of legs is amputated, they are able to build reasonably normal cases in only slightly more time than usual; another pair of legs assumes the role of the amputated pair. When decased larvae are given no access to leaves but are given other, dissimilar kinds of plant materials, they are also able to build fairly adequate cases. However, when they are presented with sand grains, they are unable to build a case, even though other species normally use sand grains in case construction.

Obviously the capacity to adapt fixed-action patterns to suit the situation has high survival value, for no insect can be assured a perfectly constant environment. There are also times in the life of an insect when searching behavior must be performed; that is, a caddisworm must find suitable particles from which to build a case; a caterpillar, a place to hibernate; and so forth. During such behavior the organism is constantly monitoring its environment. Such highly variable searching activities are often called **appetitive behavior.** In fact, appetitive behavior and fixed-action patterns are simply two extremes, and many aspects of behavior fall somewhere in between, depending on the degree to which the insect is being guided by external cues toward "finding" something or is being guided by internal cues toward performing a specific act. In appetitive behavior **orientation** with respect to certain environmental cues is of overriding importance, and this is a subject we must now consider.

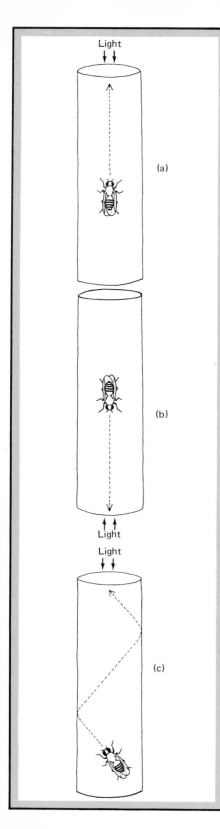

Orientation

Insects are programmed to respond directionally to such environmental cues as light, gravity, odors, and other stimuli. The German-American biologist Jacques Loeb spoke of such directional adjustments as "forced movements," believing that an animal receiving a stimulus unequally on the two sides of the body will move with machinelike precision and predictability to attain equal stimulation on the two sides—and thus will be oriented toward the stimulus. He called these movements "tropisms," borrowing a word used by botanists to describe the bending of plants toward a stimulus, and more properly used in that sense. Loeb's writings were influential during the early part of this century, but both his theories and his terminology have since undergone many modifications.

The present tendency is to recognize two major classes of orientation movements among animals, **kineses** and **taxes** (singular, kinesis and taxis). To these words may be attached prefixes denoting the sensory modality (for example, geotaxis) or the type of behavior exhibited (for example, orthokinesis) (see Table 7–1). In addition, one may add the words *positive* or *negative* to denote whether the insect faces toward or away from the source of stimulation (for example, positive phototaxis, or simply photopositive).

This complex terminology is useful in describing directional components in behavior, though it does nothing to explain them. Taxes and kineses are usually regarded as **reflexes,** that is, simple, built-in responses to simple stimuli. Much depends, of course, on the insect's sense organs and on the precise nature of the stimuli, so that a term such as *chemotaxis* may conceal a variety of different situations. Orientation is much less mechanistic than Loeb believed. Response is subject to threshold effects, to changes in internal state, to behavioral context, to stimulus quality, and to many other factors. Taxes are often quickly reversible; for example, virgin queen ants are photopositive when they emerge from the parental nest, but following mating they are photonegative. Hungry caterpillars of several kinds are photopositive, but after feeding they become photonegative. When their container is jarred, mosquito larvae exhibit a negative phototaxis and a positive geotaxis—adaptive since the change in response constitutes an escape reaction from surface predators. Circadian rhythms often involve a change in response to physical factors. An insect may turn toward a weak stimulus but will turn away when the same stimulus is intensified. Orientation movements may be modified by learning, as we shall see in the next section.

Phototaxis has been especially well studied in *Drosophila*, where it has been shown to be influenced by temperature, time of day, time since feeding, age, sex, and several other variables. These flies often show individual variation in response to light and to gravity, and artificial selection experiments have often succeeded in producing photopositive or photonegative stocks within a few generations. Normal *Drosophila melanogaster* are photopositive and geonegative, but phototaxis overrides geotaxis when the two are in conflict, and blinding of one eye modifies the response to light (Fig. 7–5).

Figure 7–5

(a) A normal *Drosophila melanogaster,* being geonegative and photopositive, moves upward in a tube in darkness or if lighting is from the top. (b) But if lighting is from below, orientation is downward, since phototaxis overrides geotaxis. (c) If one eye is blinded and illumination is from the top, the fly moves in a spiral as it attempts to achieve equal illumination in the two eyes (tropotaxis). (Parts a and c after Hotta and Benzer, 1970, "Genetic dissection of the *Drosophila* nervous system by means of mosaics," *Proceedings of the National Academy of Sciences,* vol. 67, pp. 1156–1163.)

Table 7–1. Terminology of Orientation Behavior in Animals

Classified by Type of Stimulus	Classified by Type of Behavior

Kineses: Undirected movements in which the speed of movement or the frequency of turning depends on the intensity of stimulation

Hygrokinesis: with respect to humidity or surface moisture

Photokinesis: with respect to a gradient of light

Stereokinesis: with respect to contact with surfaces

Chemokinesis: with respect to a chemical gradient

Orthokinesis: speed of movement depends on intensity of stimulus

Klinokinesis: frequency of turning depends on intensity of stimulus

Taxes: Movements directed toward or away from a source of stimulation

Hygrotaxis: with respect to moisture

Geotaxis: with respect to gravity

Chemotaxis: with respect to taste, odors (= osmotaxis)

Thermotaxis: with respect to temperature

Anemotaxis: with respect to air currents

Rheotaxis: with respect to water currents

Phototaxis: with respect to light

Phonotaxis: with respect to sound

Astrotaxis: with respect to sun, moon, or stars (Also: Orientation with respect to plane of polarization of light)

Klinotaxis: comparison of intensity of stimulus by symmetrical deviations from a straight path

Tropotaxis: achievement of equal stimulation in bilaterally symmetrical sense organs

Telotaxis: fixation on a stimulus without bilateral scanning

Menotaxis: maintenance of a constant angle with stimulus direction (= compass orientation)

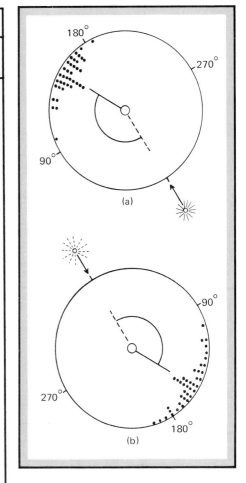

Figure 7–6
Santschi's experiments to demonstrate sun-compass orientation in ants. (a) The ants are maintaining about a 150° angle with the sun's rays. (b) The sun is covered and reflected from the opposite side by a mirror; the ants turn around and proceed in the opposite direction. (H. Markl and M. Lindauer, 1965, from *The Physiology of Insects*, edited by M. Rockstein, with permission of the publishers, Academic Press, Inc.; after R. Jander, 1957.)

The ability of insects to find the optimum environment during various periods of their development, feeding, mating, and dispersal is intimately dependent on proper orientation to physical cues. Their navigational capacities are commonly dependent on their ability to maintain a constant angle with the direction of light rays from the sun (menotaxis, or light-compass orientation). One of the first persons to demonstrate this relationship was Felix Santschi, who in 1911 performed experiments with ants with the use of a mirror (Fig. 7–6). In nature, insects are of course not often presented with such an abrupt change in the direction of light rays as in this experiment, and many of them are able to compensate for the slow movement of the sun across the sky and to navigate successfully despite a gradual change in the direction of light.

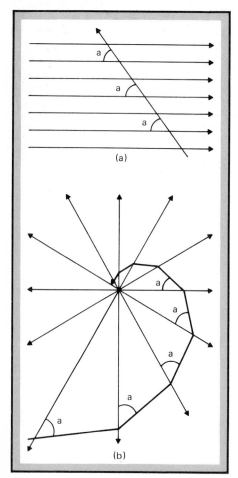

Figure 7–7
(a) A moth maintains a straight line when navigation is via rays of light that are from a distant source and thus essentially parallel. (b) But when the source is close by, maintenance of a constant angle with the light rays causes the moth to circle toward the source.

Orientation in night-flying insects is not well understood, but moon-compass orientation has been demonstrated in a few cases, and the use of starlight cannot be entirely discounted. In all cases of orientation by a distant light source, rays striking the insect are approximately parallel. When the light source is close by—as in a lantern or a light trap—the rays diverge such that an insect maintaining a constant angle is drawn ever closer to the source (Fig. 7-7); thus the common sight of moths circling a light and often being drawn to their death.

Insects may be guided by taxes in concert or in rapid succession. A homing bee may use the angle of the sun's rays or may be guided by the pattern of polarization when the sun is obscured; it may also respond to certain landmarks; close to the nest odor cues may play a major role. It should not be forgotten that some insects are able to provide their own orientation cues (as in the chemical trails of ants); and others convey information on·distance and direction to be traversed, even transposing from one sensory modality to another (the honey bee; see Chapter 8).

Learning and Memory

It is probably safe to say that all animals have some capacity to modify their behavior as a result of experience. Animals with short life cycles and relatively simple nervous systems, such as insects, cannot be expected to learn a great deal or to retain learned patterns for very long. It is obviously advantageous for a larva to begin feeding quickly or for an adult to find a mate quickly. Nevertheless there are situations where the capacity to learn is valuable: for example, in remembering the appearance of bad-tasting prey or in finding one's nest after a foraging flight. In general, learned behavior in insects is short term and subserves the performance of innate behavior.

Types of Learning.   Simple forms of learning can be demonstrated in most insects, and for that matter sometimes in individual ganglia. In a now classic series of experiments, G. A. Horridge, of the Australian National University, trained headless cockroaches and locusts to flex their legs to avoid shock (Fig. 7–8). The insects were suspended in pairs over containers of electrified saline solution, such that one of them received a shock whenever the leg fell below a certain level. In the other (the control) a shock was received but without regard to position of the leg. After half an hour the experimental animal changed its behavior so that fewer shocks were received. When the two were then compared, both receiving a shock when the leg was lowered to a certain position, the experimental animal received appreciably fewer shocks than the control. Further studies have shown that learned leg flexion can be mediated by a single isolated ganglion. Such behavior is perhaps most simply termed **shock avoidance learning.**

Of a somewhat similar nature is **habituation,** defined as learning not to respond to a stimulus that has no effect on the organism. For example, it was shown long ago that a spider will drop to the ground at the sound of a tuning fork; but after repeated exposure to the stimulus the response lessens, until finally the spider ceases to respond at all. Body lice prefer a rough, woolen surface to silk, and on crossing from wool to silk will show increased turning until they are back on the wool (a **kinesis**). However, after prolonged exposure to a smooth, silken surface their rate of turning decreases greatly. It is well known to students of insect behavior that their subjects often become used to being handled and manipulated. In nature such behavior may be highly advantageous, permitting the insect to remain alert to unusual disturbances and yet able to adapt to repeated stimuli that produce no harm.

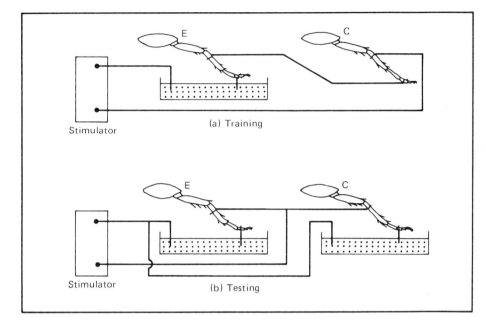

(a) Training

Stimulator

(b) Testing

Stimulator

Figure 7–8
Experiment demonstrating shock avoidance learning in headless cockroaches. In (a) the experimental animal (E) receives a shock whenever its leg touches the electrified saline solution, but the control (C) is shocked without regard to leg position. After a period of time both are wired to receive a shock when the leg is lowered, but the experimental animal has learned to receive fewer shocks. (After G. A. Horridge, 1963, "Learning of leg position by the ventral nerve cord in headless insects," *Proceedings of the Royal Society of London,* series B, vol. 157, pp. 33–52.)

It is only a short step from this to **associative learning** or **conditioning;** that is, the capacity to associate a stimulus with a reward or punishment. For example, if cockroaches are placed in a cage that is half dark and half lighted, they will congregate in the dark portion. However, if the floor of the dark portion contains an electric grid through which a shock can be administered, after a time the roaches learn to remain in the lighted area. In nature, learning is perhaps more often linked with reward than with punishment. A hunting wasp, for example, may learn that an abundance of suitable prey is associated with a particular clump of bushes and may return repeatedly until its behavior is no longer rewarded. Karl von Frisch, a Nobel laureate in 1973 in recognition for his work on honey bee communication, showed many years ago that honey bees are able to associate color of the feeding place with the presence of food.

It is a remarkable recent discovery that conditioned responses may be retained through metamorphosis despite the extensive reorganization of the nervous system during the pupal stage. Mealworms, the larvae of grain beetles (*Tenebrio molitor*), when placed in a chamber with a floor partly of lucite and partly of electrified copper, quickly learn to remain in the lucite section. When these larvae are allowed to pupate and then tested as adults, they remain in the lucite section much more consistently than control groups that were not trained as larvae or that received shocks irrespective of their substrate (controls A and B, respectively; see Table 7–2).

It has also been shown that responses to odors established by conditioning during the larval stage may sometimes be retained through metamorphosis. Such **preimaginal conditioning** may have practical significance, since it implies that adults of species that have more than one host plant or animal may "remember" the odor of the host species they developed on and show a preference for that host. This may favor the development of "host races" and conceivably may play a role in the evolution of host-specific species (Chapter 15). When the parasitoid wasp *Nasonia vitripennis* is reared on its normal host, the pupae of blow flies, and the offspring are given a choice of blow fly

Table 7–2. Performance of adult grain beetles trained as larvae to avoid a copper surface.			
Group	Number tested	Mean time to first entry to copper surface	Mean number of trials to learn
Trained as larvae	25	30+ minutes	0
Controls (A)	25	41 seconds	3
Controls (B)	31	86 seconds	3

Source: After Somberg, Happ, and Schneider, 1970, "Retention of a conditioned avoidance response after metamorphosis in mealworms," *Nature*, 228: 87–88.

or house fly pupae, nearly 100% select blow fly pupae for oviposition. But when they are reared on house fly pupae, only about 75% of the offspring, on the average, show a preference for blow flies. This percentage does not increase materially in subsequent generations reared on house flies, indicating that a small measure of preimaginal conditioning has been imposed on a genetically based preference for blow fly pupae.

Maze Learning.   Considering the fact that associative learning is widespread in insects, it is not surprising that some insects are capable of learning to run a maze containing a terminal reward or an initial punishment. One of the first persons to demonstrate this behavior was C. H. Turner, in 1913. Turner's maze was a simple, homemade one, built of copper strips over a pan of water. An overhead light served as a negative stimulus; a jelly glass "home" at the end of the maze, as a reward. After only a few trials most cockroaches greatly reduced the time required to complete the maze.

Turner also performed some of the earliest experiments on maze learning in ants. These studies were extended by the late T. C. Schneirla and his colleagues at the American Museum of Natural History. When ants were compared with white rats in the same maze, all of the rats ran the maze with only a few errors after 6 trials; and with no errors at all, after about 12 trials. However, the ants continued to make a few errors after 30 trials (Fig. 7–9). Nevertheless the contrast between the two groups is less than one might have predicted, considering the vastly more complex nervous system of the rat. The contrast is greater when the animals are asked to run the maze in reverse. Rats show a considerable transfer of learning from the original running, whereas ants perform as if it were a wholly new problem. Similarly rats quickly accommodate to a small change in the maze, while ants again act as if it were a different maze. These and other experiments reveal that insects have little capacity for **transfer learning;** that is, learned patterns cannot be readily transposed to a different situation.

Locality Learning.   Insects that have a nest in which their brood is reared must be able to leave the nest to hunt and to return quickly and with a minimum of errors, even though there may be small changes in the environment during their absence (for example, a change in the angle of sunlight or disturbance to landmarks by wind or other factors). Such insects perform an "orientation flight" or "locality study" on leaving the nest, during which they learn a configuration of cues at, surrounding, and often some distance from the nest entrance. Such **locality** or **exploratory learning** is a form of **latent learning,** in the sense that the insect is not rewarded until somewhat later, on the return flight.

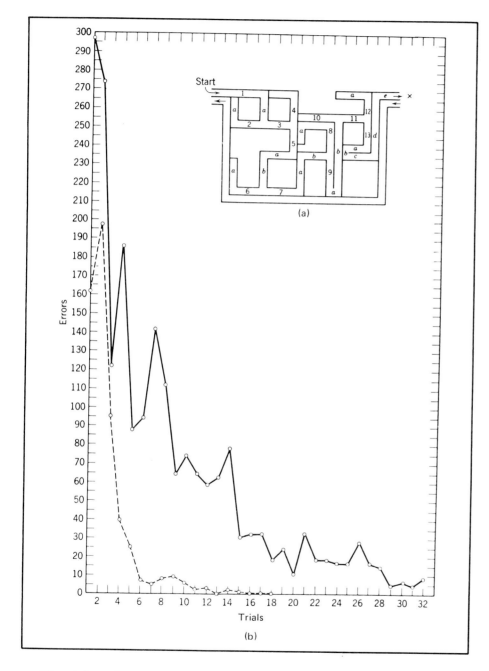

Figure 7–9
(a) Maze used for testing ants and hooded rats, the ants being rewarded with their nest at *X*; the rats, with food. (b) Performance of the ants (solid line) compared with that of the rats (broken line). (From T. C. Schneirla, 1953, "Modifiability in insect behavior," in *Insect Physiology*, K. D. Roeder, ed., with permission of the publishers, John Wiley and Sons, Inc.)

The ability of bees and wasps to learn a complex configuration of landmarks quickly and to return without error from a considerable distance is a continual source of amazement. In *Philanthus* digger wasps, an orientation flight of 6 to 10 seconds is sufficient to ensure safe return to the nest. Many wasps prepare a series of nests in the course of the season, each time learning the necessary configuration of landmarks.

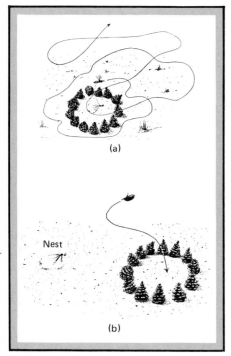

Figure 7–10
(a) Locality study of the digger wasp *Philanthus triangulum*. A ring of pine cones was placed around the entrance. When the wasp left, the ring was displaced 30 cm. (b) Upon returning, after 90 minutes, she at first approached the nest directly, then was attracted to the center of the ring of cones. This was repeated several times with similar results. (N. Tinbergen, 1951, from *The Study of Instinct*, with permission of the publishers, Oxford University Press.)

Most remarkable are certain species of *Ammophila* that maintain three to five nests at one time, each at a different stage of development of the egg or larva in the nest. In this instance the female visits each nest in the morning (remembering the precise location of each), inspects the cell contents, then supplies each nest with the needed number of prey or closes the nest permanently if the larva is mature. In the latter case she no longer visits that site but prepares a new nest and remembers its location until its larva has reached maturity.

The nature of homing behavior varies considerably depending on the terrain and whether the insect returns over the ground or in the air. Species of *Ammophila* digger wasps that bring their caterpillar prey to the nest by walking over the ground appear to have an intimate knowledge of the area in which the nest is located. In the bee-wolf, *Philanthus triangulum*, a species that flies with its prey, the Dutch behaviorist Niko Tinbergen and his co-workers found that cues close to the nest entrance are all-important, and returning females can be misled by displacement of landmarks (Fig. 7–10). Further experiments showed that the wasps do not learn individual landmarks as such, but their configuration; the displacement of one or a few objects may cause only temporary disorientation. Members of a related genus, *Bembix*, often nest in broad expanses of bare sand yet return successfully to their nests. They do so by learning subtle cues in concentric circles from the nest, including even the profile of the distant horizon. Thus they usually return successfully even when there are major disturbances to a portion of their nesting area. Wasps have been known to forage as much as 3 km from their nest and to return successfully after many hours delay. Honey bees, when their hive is displaced, will remember its previous location as much as 12 days later.

Homing is by no means entirely dependent on memorization of landmarks. Many ants, for example, lay odor trails, which colony members follow to food sources and back to the nest. As we have seen, many insects are able to navigate by maintaining a constant angle with the direction of sunlight, or even with the plane of polarization of light if they cannot see the sun directly. Even these basic mechanisms of orientation may involve learning. For example, honey bees have an excellent **temporal memory;** that is, they are able to compensate for the passage of time and consequent change in angle of the sun's rays while they are foraging. Honey bees can be taught to fly to a food source at a certain time of day; in fact, they are able to remember over a period of days the location and time of presentation of several different food sources. Accurate time compensation for sun-compass orientation may always require experience. Ants tested early in the spring, without much experience with the sun, maintain a constant angle with the sun regardless of the passage of time; but later in the season they are able to adjust their angle of progress in accordance with changes in the sun's position.

Insect Memory.   There is evidence that insects readily forget learned behavior when the training period is quickly followed by performance of a different task. For example, when cockroaches are trained to avoid shock and then immediately forced to run a treadmill, they tend to lose their learned ability to avoid shock. But if they are forced to remain quiescent for a period right after training, there is no such loss of memory. Thomas Alloway, of the University of Toronto, showed that when yellow mealworm beetles were trained to run a maze, both learning and retention were enhanced when the beetles were exposed to cold (1.7°C) between training sessions and during retention periods of 1 to 10 days. Evidently reduced activity improves the entrenchment of learned behavior. Some learned behavior is, however, deep-seated and less readily extinguished by performance of other activities. Even after ether anesthetization, honey bees are able to find their hive readily, and as we have seen honey bees remember over several days both the location and the time of presentation of food sources. Such learning is evidently more firmly fixed in the memory and not readily extinguished. Perhaps the existence of two "levels" of memory, short term and long term, helps to explain why some learned responses are easily extinguished, while others persist much longer, as in insects with a permanent nest (such as the honey bee), which are able to remember over many days not only nest location but place and time of food availability. Evidence of the role of the mushroom bodies in memory was reviewed in the previous chapter.

Mental Capacities of Insects.   In summary, one may say that insects are able to learn many things, from simple reflex behavior (such as shock avoidance) to the complex homing behavior seen in various social insects. It is not surprising to learn that some persons have maintained that insects are capable of **insight learning,** that is, the ability to combine processes learned from previous experience to meet a new problem. One example that is often cited is "tool using" in the digger wasp *Ammophila:* When the female has completed a nest, she fills the burrow and then pounds the soil in place with a small pebble held in her mandibles and later discarded (Fig. 7–11). Recent students of *Ammophila* have concluded that this behavior is by no means "intelligent," as sometimes claimed. Rather these wasps are performing inherited patterns capable of the usual fine tuning to the precise environmental situation. In some species of this genus, "tool using" is optional—that is, this behavior is a latent pattern that in some situations is elicited, in some situations not. (For another example of the use of foreign objects as "tools," see Chapter 15, p. 330.) Similarly, the ability of many Hymenoptera to repair their nests is readily explained on the basis of stimulus-response behavior. In sum, there is no clear evidence that insects are capable of insight learning.

Perhaps the most widely cited example of higher mental capacities in insects relates to the "dances" of the honey bee, which we shall describe more fully in the following chapter. Not only are worker bees able to indicate direction and distance of a food source by angle and duration of elements of the "dance," but the angle of flight with respect to the sun is translated in the darkness of the hive to a vertical angle with respect to gravity. This has often been cited as an example of symbolism, and Donald Griffin, in his provocative book *The Question of Animal Awareness,* cites it as an example of the "evolutionary continuity of mental experience." Be that as it may, there is no real evidence that "insight" plays an appreciable role in the performance of this re-

Figure 7–11
Two examples of "tool use" in insects: (a) an *Ammophila* digger wasp pounding the soil in her nest entrance with a stone (after G. W. and E. G. Peckham, 1898); (b) worker weaver ants using their larvae as "shuttles" to weave together leaves to make their nest (the larvae spin silk, but the adults lack silk glands). (W. M. Wheeler, 1910, from *Ants, Their Structure, Development, and Behavior,* with permission of Columbia University Press; after Dolfein, 1905.)

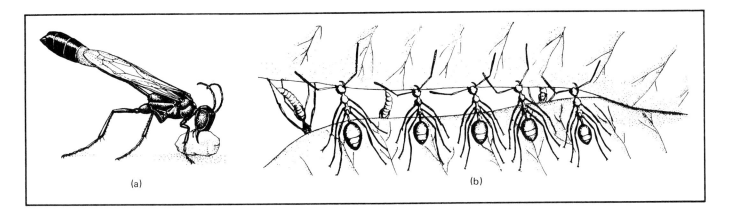

(a)   (b)

"There's one advantage in our inability to make decisions—we never make the wrong ones."

Drawing by Bill Clark; from *To Know a Fly*, by Vincent Dethier; reproduced with permission of Holden-Day, Inc., San Francisco.

markable behavior. Rather, it is a supreme example of the processes of evolution achieving startling results with a nervous system of limited capacity.

In general, insect learning is highly adaptive but limited to matters directly related to their specialized, programmed lives. What and how well they learn is related to the capacities of their sense organs and their nervous system, and ultimately to their genes. Honey bees learn to fly to feeding stations with blue or yellow markers in only 2 or 3 trials; but 7 to 12 trials are required if the markers are blue-green or black. They can be trained to arrive at a food source 24 hours later, but not at some other time interval. Perhaps it is unfair to ask what "human" traits insects cannot perform, when it is clear that they can perform a great many insectan traits that are far beyond our capability.

Response to the Environment

We have been looking at the insect from the "inside out," asking questions about its equipment for dealing with the environment. We now change our focus slightly, asking what environmental cues are important, how they are received and processed, and how the insect responds on the basis of its internal state.

Much insect behavior can be thought of in terms of **stimulus-response,** although this simple statement conceals a multitude of variables. Response is influenced by many factors in the internal environment, as we have seen. **Stimuli** are those elements in the environment capable of eliciting a response. A stimulus for one species may be totally ignored by another—for example, a species-specific sex pheromone. Or a stimulus attractive to members of one species may repel members of another; for example, mustard oils repel many plant-feeding insects but are stimulants for larvae of the imported cabbageworm. Failure to respond is often an indication that a stimulus is not being received. A medieval control for fleas and bed bugs in churches was to declare them excommunicated. That they failed to leave was not a reflection of their religious beliefs; they simply were unable to detect the modulations of the human voice. On the other hand, certain insects are supremely capable of detecting and responding to certain sounds—female crickets to the songs of the males, moths to the ultrasonic cries of bats. In these and many other cases a built-in response mechanism, often called a **releasing mechanism,** is essential for survival and reproduction. The stimulus itself is in this case called a **releaser.**

Insects often respond to stimuli that we cannot perceive except with special equipment: volatile pheromones, secondary plant substances, ultrasound, ultraviolet, and so forth. Of course they often also respond to images and sounds within our own perceptual field. But any one species is likely to respond to only a few key stimuli; other stimuli elicit no response and often are not received at all. **Sensory filtering** is a term applied to the many instances in which only a few environmental cues pass through the "filter" provided by the insect's sense organs and releasing mechanisms. Moths' specialized receptors for ultrasound—and their deafness to sounds having no biological meaning for them—provide a superb example (Chapter 6, pp. 149–150).

Stimulus Quality.   Not all insect receptors are as "finely tuned" as the moth tympanic organs to the ultrasound of bats. Yet most behavior is triggered by certain **sign stimuli** to which the insect is specially tuned to receive and to respond. Response will vary depending on the internal state and on the precise quality of the stimulus. In a species with an appropriate releasing mechanism, and in an individual with a high level of motivation, a suboptimal stimulus may produce a weak response or none at all. It is of considerable theoretical interest to know the precise qualities of stimuli that are most effective in eliciting a response—and there are times when this information may have practical value in manipulating the behavior of a pest species. Stimulus quality is perhaps best appreciated in the case of visual cues, since we ourselves are "visual animals."

Some years ago Dietrich Magnus, of the Zoological Institute at Darmstadt, West Germany, made a study of sexual releasers in a fritillary butterfly, *Argynnis paphia*. Males of this species, beginning two or three days after emergence and under suitable conditions of temperature and light, begin searching for females. Having recognized a female, the male pursues her in flight, but courtship and mating ensue only on receipt of specific odor cues. With a series of experiments, Magnus showed that the orange coloration of the wings served as a releaser, but the precise pattern on the wings was of no significance. Shape was also unimportant: Dummies in the form of circles, triangles, or squares were as effective as those shaped like butterflies, so long as they were colored orange. Size was, however, quite important. Models smaller than the usual size of the female (surface area about 22 square centimeters) produced fewer approaches. But those larger than the female, in fact up to four times her surface area, were still more effective in eliciting a response (Fig. 7–12). Magnus also showed that speed of wing flutter was important. When given a choice between a slow alternation of orange and black and one twice as fast, males showed a preference for the faster speed of flicker—up to about 75 alternations per second, which taxed the resolving power of the butterflies' eyes. But in nature females do not move their wings anywhere near this rapidly. Thus Magnus had discovered two ways in which visual stimuli could be made *supernormal:* by increasing the size or by increasing the speed of flutter well beyond the normal. One might expect evolution to produce larger females with more rapid wing movements, since these qualities are more successful in attracting males. However, there are many other selection pressures operating to maintain the present size and speed of wing movement.

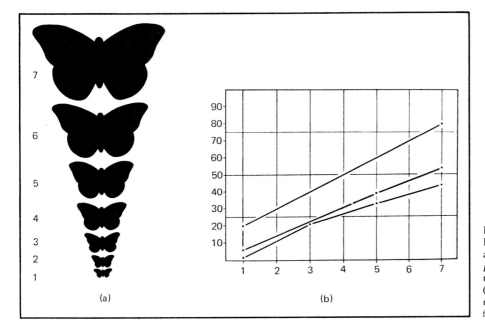

(a)   (b)

Figure 7–12
Experiments on the effect of size of females on approaches by males of the butterfly *Argynnis paphia:* (a) relative size of the seven models used, the third from the top being the normal size; (b) percentages of approaches by male (ordinate) to models of the seven sizes (abscissa). (Modified from D. Magnus, 1958.)

Magnus noted that fritillaries of both sexes approach objects colored blue or yellow, since these are the colors of the flowers they visit for nectar. Niko Tinbergen and his colleagues, working on another species of butterfly, the grayling, found that in this species color is of little importance in sexual pursuit, but hungry individuals of both sexes show a strong response to blue and yellow. Similarly the imported cabbageworm butterfly selects blue and yellow flowers for feeding, but the female selects green objects for oviposition. Obviously stimulus quality cannot be evaluated except with reference to the internal state of the animal.

A striking example of the practical importance of understanding stimulus quality has recently been provided by Ronald J. Prokopy, now at the University of Massachusetts in Amherst. His studies concern the apple maggot, a major cause of "wormy" apples throughout many parts of the United States and traditionally controlled by a schedule of carefully timed chemical sprays. Apple maggot flies are highly responsive to visual stimuli. Prokopy found that flies of both sexes are attracted to rectangles hung in apple trees (covered with tanglefoot to capture the flies)—but especially if colored yellow, most particularly daylight fluorescent saturn yellow. They are also attracted to spheres 7.5 centimeters in diameter (about apple size) when painted tartar dark red. If the size of the sphere is increased, the response decreases if they are red but shows a slight relative increase if they are yellow (Fig. 7–13). There is also a tendency for red spheres to be relatively more effective than yellow rectangles later in the season. Yellow evidently attracts flies because it is indistinguishable from green to the flies but has

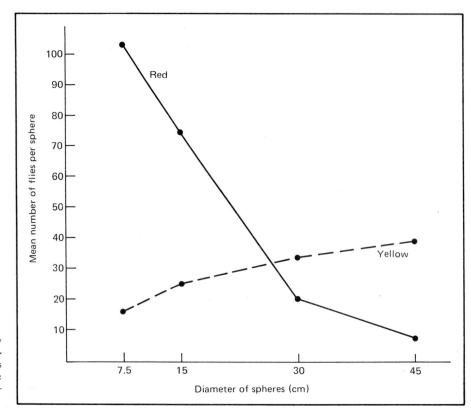

Figure 7–13
Response of apple maggot flies to red and yellow spheres of various sizes, under orchard conditions. (After R. Prokopy, 1968, "Visual responses of apple maggot flies, *Rhagoletis pomonella* (Diptera: Tephritidae): orchard studies," *Entomologia Experimentalis et Applicata*, vol. 11, pp. 403–422.)

higher reflectance. Thus it attracts them to apple foliage, which is a source of honey-dew for the flies, while the red spheres present possible oviposition sites for the females and mating sites for the males.

The possibilities of using these supernormal stimuli, either singly or in combination (red spheres on a yellow rectangle), for monitoring or for reducing apple maggot populations, is intriguing. In his own orchard, Prokopy placed 469 sticky-coated spheres in 81 unsprayed apple trees. He captured about 10,000 apple maggot flies on these spheres and discovered that the infestation of apples by maggots that season was 1% to 3% as compared to 97% to 98% in nearby unsprayed and untrapped orchards. While conditions in this isolated orchard may have been somewhat unusual, there is reason to hope that these techniques may be further refined and may someday be employed against some of the other serious pests in the fruit fly family (Tephritidae).

Reaction Chains.   In nature, insects do not often perform isolated bits of behavior but rather perform acts in continuous series, each act commonly dependent on completion of the preceding one and on receipt of appropriate stimuli. A comparison might be made with a car passing along a street with a traffic light at each intersection. Each block must be traversed before the next one can be started; at each corner one must have the appropriate stimulus—in this case a green light; and eventually one reaches a goal—in this case perhaps a supermarket. The goal in an insect might be copulation, oviposition, return to the nest with food, or some other essential act. Orientation will be especially important in the appetitive behavior preceding attainment of the goal; and as we have seen, even taxes may be sequential. Use of the term *goal* does not, of course, mean that the insect purposefully proceeds to a perceived goal. Rather, goal orientation is a result of natural selection, which has produced a series of responses that, over time, have resulted in maximum survival and reproduction.

Such chains cannot often be short-circuited by presenting a stimulus out of place. For example, if suitable prey is presented to a digger wasp while it is digging its nest, the wasp will show no reaction—such a reaction would of course be nonadaptive at that stage. If prey is presented right after a temporary closure but before the initiation of the hunting flight, it may also be rejected—but not necessarily so if the wasp's larva is large and hungry, for then the wasp's threshold of response may be low. If prey is removed from a fully stocked nest, before final closure is begun, the female may be induced to bring in additional prey to fill the nest before the closure is made. Fig. 7–14 is an attempt to outline the major steps in the nesting cycle of a typical solitary wasp, but in fact each step is capable of further analysis. Consider, for example, "prey found."

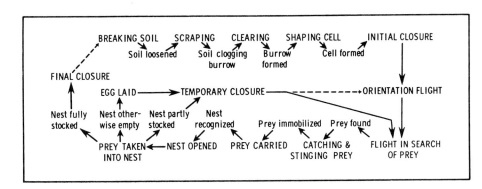

Figure 7–14
Model of the reaction chain of a nesting female digger wasp. Fixed-action patterns are shown in capitals; presumed releasers, in small letters. (H. E. Evans. Reproduced with permission from the *Annual Review of Entomology*, Volume 11. © 1966 by Annual Reviews Inc.)

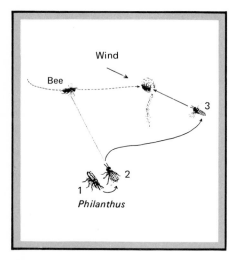

Figure 7–15
Sequence of hunting behavior of the bee-wolf,
Philanthus. (N. Tinbergen, 1935, from *Zeitschrift für*
vergleichende Physiologie, vol. 21, with permission of
the publishers, Springer-Verlag, Inc.)

Figure 7–16
Some terms used in the study of animal communi-
cation. (From *Animal Communication*, Second Edition, Revised and Enlarged, by Hubert and Mable
Frings. Second, revised edition copyright 1977 by
the University of Oklahoma Press.)

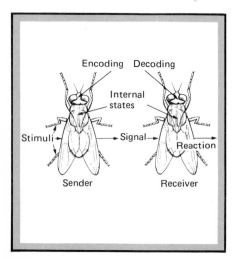

Tinbergen has shown that in the bee-wolf this in itself is sequential and involves three different sensory modalities. The first response is to a visual stimulus, which must be moving and of about the right size; at this stage there is no response to the odor of the normal prey even when it is presented close to the wasp. Having spotted potential prey, the wasp flies downwind of it, and at this point odor becomes the dominant stimulus; dummy prey are not approached unless supplied with bee odor. Finally the wasp flies to, seizes, and stings the prey, evidently employing tactile stimuli (Fig. 7–15).

It is convenient to think of reaction chains as of two general types: **insect–environment** (including both physical and biological factors, other than members of the same species) (Fig. 7–14), and **insect–insect** interactions (between members of the same species, as in Fig. 7–17). Examples of the first include the nest preparation and hunting behavior of the digger wasp just considered; the search for and discovery of a host plant, and subsequent oviposition, by a moth or butterfly; and the like. Exchange of signals between members of the same species is properly termed **communication** and is considered in greater detail below. To be sure, many reaction chains include interactions both with conspecifics and with the environment—for example, the mating swarms of midges, in which individuals come and stay together by means of common responses to landmarks and to air currents and other physical factors, at the same time responding to one another in complex ways. It is nevertheless convenient to consider communication a separate topic, since it involves a precise, sequential exchange of signals between two or more individuals.

Communication

Communication is defined as the production of a signal by an individual that influences the behavior of another individual and that is mutually beneficial (Fig. 7–16). Commonly both individuals are of the same species, but not necessarily so. For example, a person may communicate meaningfully with a dog, or ants with aphids that they guard and from which they obtain honeydew. But interactions that are not mutually beneficial (such as predator–prey) are not usually said to be communicative.

Insects communicate in all sensory modalities, and members of any one species may employ several modalities. The queen butterfly employs chemical, visual, and probably tactile signals; crickets employ acoustic, tactile, and in some species chemical signals; water striders communicate by patterned sequences of surface waves produced by leg movements. The messages conveyed are diverse: alarm, attraction, recruitment to food sources, information on conditions within the nest, readiness to mate, and so forth. Reproductive behavior includes some of the finest examples of elaborate, species-specific behavior. We have already mentioned examples among crickets, silkworm moths, and butterflies. We consider here a few somewhat different examples as well as a few general concepts.

Mating Behavior.   Effective communication between members of the opposite sex may serve several functions:

1. Signals may be used to draw members of the opposite sex from a distance.
2. Mates must recognize one another as members of the same species, thus avoiding wastage of gametes, as well as time and energy, by inappropriate matings.
3. Signals may be used to bring the partner to a state of readiness or to cause it to remain quiescent during copulation.

4. Females may select males producing the most effective stimuli, thus ensuring that their offspring receive superior genes.
5. Mating should be accomplished in such a way that minimal predation occurs during courtship and copulation.

In the queen butterfly, discussed in Chapter 6, the male detects a flying female visually, pursues her, and brushes her with hair pencils that are extruded from his posterior end (Figs. 5–14, 7–17). Similar behavior, involving both visual and chemical cues, occurs in many butterflies; but moths, being largely nocturnal, tend to use mainly chemical cues.

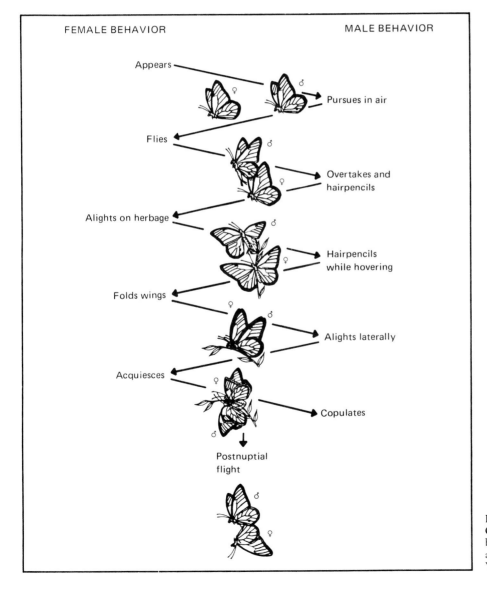

Figure 7–17
Courtship of the queen butterfly. (From an article by L. P. Brower, J. V. Brower, and F. P. Cranston appearing in *Zoologica*, 1965, published by the New York Zoological Society.)

A very different type of mating behavior is exhibited by hangingflies (*Bittacus* and related genera of the order Mecoptera). These curious nocturnal insects receive their name from the fact that they hang from vegetation by their front legs. The males produce a volatile sex attractant from protrusible exocrine glands on the abdomen, but they do this only after they have caught a small insect, which they hold with their prehensile hind legs as a "gift" for the female (Fig. 7–18). A female attracted by the pheromone alights beside the male and tastes the prey; if she accepts it, copulation follows. At the end of copulation, the pair may struggle for what is left of the prey, and if the male is able to abscond with a sizable piece, he may use it to entice another female.

Randy Thornhill, at the University of New Mexico, has shown that when a female is presented with small or undesirable prey, she breaks off copulation within a few minutes and flies away. If copulation lasts less than 5 minutes, no sperm are transferred, and between 5 and 20 minutes the amount of sperm transferred is proportional to the time (Fig. 7–19). Thus it is to the male's advantage to present the female with a relatively large insect (the size of a house fly or larger), and it is to the female's advantage to choose a mate with large and desirable prey: In this way the male maximizes his contribution of genes to the next generation, and the female obtains nourishment that permits her to devote less of her energies to obtaining food and more to laying her eggs. This is a particularly fine example of **sexual selection** involving female choice, a widespread phenomenon that does much to explain female "coyness" and the development of vigorous courtships and optimum signaling devices on the part of males.

Another example of sexual selection, though of a somewhat different nature, is provided by certain dragonflies. These insects have large eyes and small antennae, and as might be surmised communicate almost entirely by visual signals. Differences in wing pattern and body color serve as species and sex recognition features. In one of the common North American species, *Plathemis lydia*, sexually mature males have the up-

Figure 7–18
A copulating pair of hangingflies, *Hylobittacus apicalis*. The male (left) is holding a "nuptial gift" for the female. (Photographed at night with electronic flash by Nancy W. Thornhill, University of New Mexico.)

Figure 7–19
The number of sperm transferred as a function of the duration of copulation in the hangingfly *Hylobittacus apicalis*. (After R. Thornhill, 1976, "Sexual selection and nuptial feeding behavior in *Bittacus apicalis* (Insecta: Mecoptera)," *American Naturalist*, vol. 110, pp. 529–548, with permission of the University of Chicago Press.)

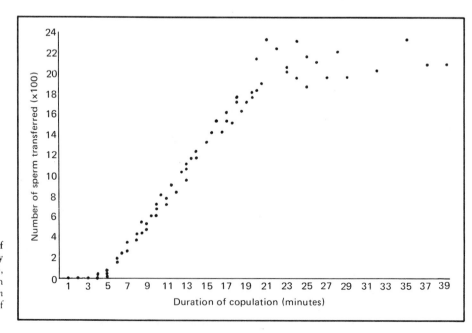

per side of the abdomen brilliantly bluish-white. Such males select small areas along the margins of ponds that provide suitable oviposition sites for females. They patrol these areas with a slow flight and much hovering, often flipping their conspicuously colored abdomens up and down while doing so. If another male intrudes, he is pursued and sometimes attacked, whereupon the intruder flees with his abdomen depressed until in his own territory, when he may become the pursuer. In this way territory boundaries are maintained (Fig. 7–20). Males with their abdomens blackened artificially are less successful in defending territories and ultimately less successful in obtaining mates.

Researchers at Syracuse University have found that in this species territories may at times be occupied by several males; but in this case one male assumes a dominant role by superior aggressive interactions, particularly in the middle of the day, when females are most likely to enter the territory. The dominant male enjoys the greatest mating success and is most able to guard the female while she is laying eggs. Although males tend to be faithful to the same territory over a period of days, when they are recently emerged or after they have mated several times, they are less likely to be dominant. Thus there are changes in the hierarchy from time to time in any one territory.

Sexual selection in dragonflies is largely **intrasexual**—that is, male–male interactions determine which is dominant on a territory—and the most vigorous males and those at the peak of sexual arousal mate most frequently. Female selection doubtless relates to a suitable oviposition site, and whichever male is dominant there becomes her mate. In contrast, in the hangingflies we considered earlier, selection is **intersexual** (between sexes), since the female selects a mate on the basis of his superior courtship.

Territoriality is more widespread among insects than was realized just a few years ago, especially in Hymenoptera. In solitary bees and wasps, males of many species defend an area advantageous for mating—for example, a group of flowers likely to be visited by females or a plot of soil where females are nesting or are likely to emerge.

Figure 7–20
Two male dragonflies (*Plathemis lydia*) displaying at the edges of their territories. Note the elevated abdomens, which are brilliant white on the upper surface. (Drawing by Peter Eades, after M. E. Jacobs, 1955, "Studies on territorialism and sexual selection in dragonflies," *Ecology*, vol. 36, pp. 566–586.)

Figure 7–21
A male bumblebee-wolf, *Philanthus bicinctus,* **on his territorial perch.** (Photograph by Darryl T. Gwynne, Colorado State University.)

Such males commonly pursue other insects that enter their territories, sometimes even small birds or stones that are thrown. If the intruder is a male of the same species, he is chased away, sometimes with bodily contact. Although only one male normally occupies a territory at one time, there is often competition for superior sites, that is, those most likely to contain females. An individual male, such as that of the bumblebee-wolf *Philanthus bicinctus*, may return to the same perch day after day, tolerating no other males within a radius of a meter or more and attempting to mate with females within his territory (Fig. 7–21). In certain Hymenoptera that parasitize the eggs of other insects, the first male to emerge from a group of parasitized eggs patrols the egg mass and defends it from other males. As females emerge from the egg mass, they are successively mated.

Alternative Mating Strategies.   By and large the mating behavior of insects is stereotyped and proceeds to a rapid conclusion, as indeed it must in such short-lived animals. But recent studies have revealed a number of instances in which males may employ more than one strategy, depending on certain variables in the population or in the environment. In the case of *Centris pallida*, a ground-nesting bee, John Alcock, of Arizona State University, has found that males fall into one of two classes when searching for females. Large males are patrollers, cruising over the ground in the nesting area, detecting virgin females emerging from their cocoons in the soil, and digging in to mate with them just as they emerge from the soil. Smaller males are hoverers, poising in peripheral areas or at flowering plants well away from the nesting area. Here they pursue other insects, especially females of their own species; evidently they mate with virgin females that escape the attentions of the larger males. Even though small males are generally less successful in obtaining copulations, they continue to be produced in each generation. Alcock has made several suggestions as to why this is so. For example, at low population densities, larger males may have difficulty in finding females. Also, these bees have certain nest parasites, and when these are abundant it may be advantageous for the females to prepare a large number of small cells (containing fewer provisions and producing smaller offspring); that is, by distributing her efforts over more offspring, the mother may ensure that more will survive the onslaughts of the parasites.

The importance of similar factors in controlling alternative strategies is well shown in the case of the field cricket *Gryllus integer*. The males of this species attract females with a calling song similar to those we have described earlier. At times these crickets are extremely abundant in the southern United States and may do much damage in cotton fields. William Cade, then at the University of Texas, found that when populations are high, many males do not call but are "satellite males," apparently entering territories of calling males by stealth, detecting females by odor, and achieving some success in mating. By not calling, they avoid being attacked by the aggressive, calling males, and they also do not attract certain tachina flies that orient to the songs of the males and deposit lethal maggots on them. (It has been shown that domestic cats also orient to singing Orthoptera.) Evidently the lessened mating success of satellite males is compensated for by greater survival as a result of reduced intraspecific aggression and reduced incidence of parasitism by tachina flies.

In scorpionflies of the genus *Panorpa* (order Mecoptera), Thornhill has described no less than three ways in which males may achieve copulations. Unlike the hanging-flies we discussed earlier, these scorpionflies are scavengers, feeding on dead insects in woodlands of the eastern United States, where they are often very common. Males locate dead insects and defend them from other males, sometimes using their large genital bulb as a "club." These male insects produce a pheromone that attracts females, which then feed on the "nuptial gift" and are mated. Males may, on other occasions, produce a mass of salivary material on which a female will feed after being attracted to the site by a pheromone. The third method does not involve nuptial feeding, the male merely seizing a female, clamping down her wings, and mating with her as a form of "rape." In Thornhill's words, "the availability of food over both ecological and evolutionary time is probably the major factor determining mating behavior in *Panorpa*."

Parental Behavior

For both males and females, the act of mating may be only part of their investment of time and energy in the perpetuation of their genes. **Parental investment** is defined as behaviors that increase the probability of some offspring surviving to reproduce at the cost of the parent's ability to produce more offspring. Minimal behavioral investment tends to be compensated for by maximum physiological capacity to produce offspring (for example, in house flies). Conversely, maximum investment (such as brood care) tends to be offset by reduced production (but higher survival) of offspring. As a general rule, the investment of the female is much greater than that of the male; she not only converts most of her nutrients to the production of eggs, but she seeks a suitable site for oviposition and in some cases guards her eggs or even feeds her offspring. The male, on the other hand, produces large numbers of much smaller sperm, with less energetic investment, and attempts to use these to inseminate as many females as possible. But it is not always quite as simple as this, and it will pay us to look briefly at male parental investment before returning to that of the female.

Male Investment and Assurance of Paternity.   Male sperm, as we have seen, are stored in the female's spermatheca and released at the time of oviposition. When females mate more than once, it is generally the sperm from the last mating that are released to fertilize the eggs. This has been termed **sperm precedence.** In instances of sexual selection involving male territoriality or nuptial feeding, it is to the male's advantage, once he has mated, to ensure that the female does not mate again before she lays her eggs. Thus a male dragonfly guards the female while she lays her eggs, ensuring his paternity of the resulting offspring. Indeed, Jonathan Waage, of Brown University, showed that the penis of certain damselflies serves a dual function: By means of a scooplike extension of the penis, any sperm present in the female genital tract is removed before the male introduces his own sperm.

Male field crickets, having attracted a female via a calling song, switch to a "courtship song;" then, after mating, to a "staying together" song, which ensures that she will not mate again before laying eggs. In the case of the hangingflies we discussed earlier, the male has invested efforts in a "nuptial gift," and it is to his advantage to ensure that the resulting offspring carry his genes; if the gift is large and the female fully inseminated, she enters a refractory period and will not mate again until her eggs have been laid.

The male hangingfly, in fact, contributes indirectly to egg production by supplying nutrients to the female. It has recently been shown by the use of radioactive tracers that male butterflies of certain species contribute to egg production via nutrients supplied in the spermatophore. The spermatophores of some of the Orthoptera are particularly large. In the Mormon cricket 20% of the male's weight is lost in a single mating. Much of the spermatophore consists of a mass of proteinaceous material

Figure 7–22
A freshly mated female Mormon cricket feeding on proteinaceous material in the spermatophore deposited by the male. (Photograph by Darryl T. Gwynne, Colorado State University.)

Figure 7–23
A male giant water bug bearing eggs, in this case in part hatching to produce striped offspring known to have been fathered by another individual. (Photograph by Robert L. Smith, University of Arizona.)

that is consumed by the female following mating (Fig. 7–22). Females compete vigorously for calling males, and males reject smaller females, who have been shown to be less fecund. Since rejection usually occurs after the female has mounted the male, it appears that it is at this time that the male assesses the weight of the female. This unusual example of "sex reversal"—competition among the females for males and male selection of females likely to produce the most offspring—has recently been elucidated by Darryl Gwynne, then at the University of New Mexico.

Sex reversal in which the male cares for the eggs is also known in a few instances. In giant water bugs there are repeated bouts of copulation interspersed with egg laying; these bouts are dominated by the male, who receives several eggs on his back following each mating. On one occasion a pair were seen to copulate over 100 times in 36 hours, resulting in the transfer of 144 eggs to the back of the male. While carrying the eggs, the male subjects them to necessary aeration, and he assists the young as they emerge from the eggs. Such a system ensures that the male carries eggs that bear his genes. But a male can be "cuckolded" if he mates with a female who carries sperm from a previous mating. Robert Smith, of the University of Arizona, vasectomized a male and mated him to a female that had previously mated to a male homozygous for a dominant genetic marker, a dorsal stripe. The vasectomized male received 75 eggs as a result of this mating; most were infertile, but several were fertile and gave rise to striped offspring (Fig. 7–23), demonstrating that this male had been "cuckolded" by another male. Thus, despite this elaborate mating system, a male does run a risk when he mates with a nonvirgin female.

Brood Care by Females.   Since, in general, it is the female that invests the most in terms of physiological commitments to her offspring, it is not surprising that when brood care occurs, it is usually the female that is involved. Brood care has evolved in a variety of insects that have developed strategies opting for maximum protection of the offspring as opposed to production of large numbers of offspring that are "left to their own devices." Extreme examples are provided by certain cockroaches and by tsetse flies, which retain and nourish their larvae internally, in a uteruslike structure. More commonly eggs are "brooded" by the female, although brooding does not usually imply the transfer of heat as it does in birds. Rather it serves to protect them from predators, egg parasites, mold, or other factors that might destroy them. Female stink bugs of several species cover their eggs and small larvae much as a hen will cover her chicks. Such eggs suffer high mortality from ants and other predators if the female is removed.

A more advanced type of parental care occurs in certain crickets in which the female prepares a burrow that she guards vigorously from intruders. The eggs are laid in a cell at the bottom of the burrow, and when the young hatch, they are fed by the mother, at first with small, infertile eggs that she lays, later with food brought in from the outside. Beetles of several families exhibit brood care that in its initial stages may involve both sexes. Dung beetles (Scarabaeidae) prepare balls of dung, roll them to a suitable site, and bury them as food for the larvae (Fig. 13–15). Often male and female work together to make the dung ball and bury it. In some cases the female remains in the burrow after laying her eggs, standing guard and keeping the pellets clean and well formed. She may remain with her offspring until they are fully developed, then emerge from the ground with them. In carrion beetles (Silphidae), male and female work together to bury a dead animal, then prepare a ball of decaying flesh in which the eggs are laid. As the larvae grow, they are fed by the female with regurgitated food, much as a bird feeds its nestlings. (Fig. 7–24).

Brood care reaches its greatest development among the Hymenoptera, particularly in social groups such as the ants and some of the bees and wasps, which we shall consider in the next chapter. In solitary wasps—from which the social Hymenoptera are believed to have evolved—the female commonly prepares a nest, provisions it with paralyzed insects or spiders, and after oviposition seals it off in such a way that it is well protected against predators and physical factors in the environment. We have already discussed some of these wasps briefly. Of special interest are those species in which the nest cell is not sealed off immediately but is visited repeatedly by the female, who brings in prey over a period of days as the larva grows. **Progressive provisioning,** as this is called, involves much contact between parent and offspring as well as further protection of the larva as a result of the mother's continued presence in the nest. Of even greater interest are cases of **communal nesting,** that is, instances in which several females are active in the same nest, preparing and provisioning individual cells without aggression with other females in the nest. In many cases these females are sisters or mother and daughters, and the "extended family" they represent is perhaps an important progenitor for the colonies of social species. In both cases—progressive provisioning and communal nesting—there is much evidence that the selective advantages relate to a reduction in the opportunities for entry by various nest parasites and predators.

Figure 7-24
Female carrion beetle (or burying beetle) feeding her larvae. (Drawing by Sarah Landry. Reprinted by permission of the publishers of *The Insect Societies,* by Edward O. Wilson, Cambridge, Mass.: The Belknap Press of Harvard University Press, copyright 1971 by The President and Fellows of Harvard University.)

Summary

The behavior of an insect results from the interaction of environmental cues and of internal factors such as inherited patterns, physiological states, and learned behavior. Although insect behavior is often described as "programmed," it is capable of considerable plasticity dependent on variables both in the internal and in the external environment.

Certain aspects of behavior—for example, the calling of male crickets—are innate and genetically determined, as evidenced by the fact that interspecies hybrids produce songs intermediate in pattern. In this instance the required nervous circuitry is inborn but remains under inhibition until the adult stage. Many insects show increased activity at certain times of day, the result of inborn rhythms, which, however, must be "set" by exposure to rhythms in nature. Hormones influence behavior by altering the "mood" of the insect—for example, by rendering it sexually responsive. This influence has been demonstrated by removal or implantation of endocrine organs at critical periods. Behavior may also often be modified by learning—for example, learning to avoid an unfavorable stimulus or learning landmarks that will enable the insect to return home successfully. There is no evidence that insects are able to combine learned processes to meet new problems (insight or intelligence).

Much insect behavior can be thought of in terms of stimulus-response, stimuli being those elements in the environment that are capable of eliciting a response. In most cases only a few "key stimuli" are "filtered" from the environment. Quality and intensity of these stimuli must be considered in relation to the internal state at time of reception. External cues are of major importance in guiding the orientation of insects, and an elaborate terminology has been developed to describe orientation with respect to cues such as light, gravity, sound, and so forth.

Production of signals by one individual that influence the behavior of another individual is termed communication. Insects communicate in all sensory modalities, and their signals have evolved to produce specific adaptive responses, as in courtship and mating, in recruitment to food sources, and so forth. Mating behavior involves a precise interchange of signals between prospective mates. However, several instances are known in which alternative strategies are available to males, depending on such variables as male size, distribution and abundance of females, and so forth.

Sexual selection, in premating behavior, may be intrasexual or intersexual. In the former case, males compete among themselves and may establish and defend territories where females are likely to occur; in the latter, females select males that provide them with the strongest stimuli. In those instances in which males make a major contribution to egg development, via nutrients in the spermatophore, the role of the sexes may be reversed, competition for mates occurring among the females, mate choice among the males.

Males may perform postcopulatory or parental behavior when they have invested much time and energy in mate procurement and it is thus to their advantage to ensure that the offspring bear their genes. More usually, it is the female that is involved in brood care where it occurs, as it does in diverse insects that produce relatively fewer offspring but ensure their greater survival from physical and biotic factors in the environment. Brood care in solitary Hymenoptera may have been the precursor of the elaborate brood care in social bees, ants, and wasps.

Selected Readings

Alcock, J. 1979. *Animal Behavior: An Evolutionary Approach.* Second edition. Sunderland, Mass.: Sinauer Associates. 532 pp.

Alloway, T. M. 1972. "Learning and memory in insects." *Annual Review of Entomology,* vol. 17, pp. 43–56.

Atkins, M. D. 1980. *Introduction to Insect Behavior.* New York: Macmillan. 237 pp.

Baker, R. R. 1983. "Insect territoriality." *Annual Review of Entomology,* vol. 28, pp. 65–89.

Barton Browne, L., ed. 1974. *Experimental Analysis of Insect Behaviour.* New York: Springer-Verlag. 366 pp.

Blum, M. S., and N. A. Blum, eds. 1979. *Sexual Selection and Reproductive Competition in Insects.* New York: Academic Press. 463 pp.

Eickwort, G. C. 1981. "Presocial insects," in *Social Insects,* vol. 3 (H. R.

Hermann, ed.), pp. 199–280. New York: Academic Press.

Fraenkel, G., and D. L. Gunn. 1940. *The Orientation of Animals.* New York: Oxford University Press. (Reprinted in paperback, with additions, by Dover, 1961, 376 pp.)

Lloyd, J. E. 1983. "Bioluminescence and communication in insects." *Annual Review of Entomology,* vol. 28, pp. 131–60.

Matthews, J. R., and R. W. Matthews, eds. 1982. *Insect Behavior: A Sourcebook of Laboratory and Field Exercises.* Boulder, Colo.: Westview Press. 324 pp.

Matthews, R. W., and J. R. Matthews. 1978. *Insect Behavior.* New York: Wiley. 507 pp.

Roeder, K. D. 1967. *Nerve Cells and Insect Behavior.* Revised edition. Cambridge, Mass.: Harvard University Press. 238 pp.

Thornhill, R., and J. Alcock. 1983. *The Evolution of Insect Mating Systems.* Cambridge, Mass.: Harvard University Press. 547 pp.

Social Behavior

When entomologists speak of "social insects," they have in mind a somewhat more restricted use of the word *social* than is common. Many persons would regard any form of interactive group behavior as social—a herd of deer, perhaps, or a winter roost of monarch butterflies. Students of vertebrates often consider parental care—for example, among song birds—as social behavior. As we saw in the preceding chapter, there are many examples of parental care among insects that are usually regarded as "solitary." In some cases the mother even feeds the offspring progressively, resulting in much mother–offspring contact. In instances of communal nesting, several females share a nest, each preparing and provisioning her own cells. But all of these are examples of what entomologists call **presocial behavior.** The highest level of cooperative behavior, termed **eusocial behavior** (literally, "truly social") is found in only four groups: termites, ants, and some bees and wasps. Eusocial insects have the following attributes in common:

1. Brood care is cooperative; that is, individuals often feed offspring that are not their own.
2. There is a **caste system** involving a reproductive division of labor, such that many colony members are sterile.
3. There is an overlap of generations, some offspring assisting the parental generation in the rearing of further offspring.

The question of how this unique form of behavior may have evolved is a topic of much lively discussion among entomologists. The diversity of lifestyles among presocial insects suggests that there have been several routes to eusociality. In this chapter we shall be concerned with types of behavior especially characteristic of eusocial insects, but we shall also take occasion to consider the advantages of presocial and of eusocial behavior and to ask why (considering the success of ants, termites, and the like) not all insects have evolved eusociality.

Social Communication

The behavior of the many individuals in the colonies of social insects must be integrated in such a way that there is cooperation in defense, building, foraging, brood rearing, and the production of reproductive individuals at appropriate times. This can only be accomplished by broadcasting messages throughout the colony. Since visual signals would be difficult to transmit within the darkness and complexities of the nest, social signals are more often acoustic, tactile, gustatory, or especially olfactory. Some of the social pheromones were reviewed in Chapter 5. We present here an outline of the types of messages conveyed with a few examples of the mechanisms that have evolved in the various groups of wasps, bees, and ants (Hymenoptera) and termites (Dictyoptera). In the words of E. O. Wilson, of Harvard University, "the modes of

Figure 8–1
A queen honey bee surrounded by a retinue of workers. (Photograph by Norman E. Gary, University of California, Davis.)

communication found in the social insects are awesomely diverse." Wilson's book *The Insect Societies* (1971) provides a much fuller exposition of the subject than is possible here.

The nests of social insects present a rich source of food for vertebrate predators (such as skunks or bears) and for arthropod predators (such as other social insects). Not only are they often filled with thousands of larvae and pupae, but there may be large amounts of food in storage. All social insects have evolved methods of dealing with predators, either by attacking en masse or by fleeing to a place of safety. But first a message must be transmitted quickly throughout the colony concerning the danger. Such **alarm signals** are often highly volatile pheromones, which fade quickly unless renewed. Rapid movements within the colony, with much tactile stimulation, may also convey alarm. Termites, when alarmed, vibrate their bodies against the substrate, producing a sound audible to humans, and some ants are able to stridulate by rubbing together specialized ridges on parts of the abdomen. Alarm usually mobilizes defense, but in some cases the alarm signals themselves may also serve a defense function.

Attractants also play a major role in social life, especially with respect to the queen. In most social insects the queen rarely if ever leaves the nest and must be fed and groomed by the workers. In a honey bee hive, for example, the queen is usually accompanied by a circle of attending workers (Fig. 8–1). Indeed, the term *queen* was first suggested by the retinue surrounding the mother of the colony, who of course "reigns" only in the sense that she produces chemical signals causing workers to forego reproduction and to attend to the queen. In the honey bee the signal is the continued production of ketodecenoic acid (often called "queen substance") from her mandibular glands.

The most prevalent attractants are the subtle combinations of odors usually called "nest odor"—apparently compounded of pheromones on the cuticle of individuals; volatile pheromones in the colony; and the odor of the brood, food, substrate, and other unknown odor sources. Individual ants, termites, and other social insects orient toward these odors when close to the nest and are able to identify their own nest among others of the same species. Individuals also recognize one another as members of the same colony by these same cues, which may be said to be not only attractants but also **recognition signals.**

Another major group of social signals serves in **recruitment** to food sources or new nest sites. One of the simplest forms of recruitment is shown by ants of the genus *Leptothorax*. When a foraging worker is successful, she returns to the nest and regurgitates food to nestmates. She then raises her abdomen, extrudes her sting, and discharges a droplet of fluid (Fig. 8–2). This serves to attract other workers, one of whom touches the abdomen or hind legs with her antennae and proceeds to follow the forager to the food source. The leader then lowers her abdomen, but if the follower becomes lost, the leader once again elevates her abdomen and initiates calling behavior. **Tandem running,** as this is called, involving only two or at most a very few individuals, appears an inefficient system for recruitment to a food source, but it is interesting as an apparent evolutionary precursor of the more common and efficient method of laying **odor trails.** As E. O. Wilson has shown, when a fire ant worker locates a source of food, she returns to her nest with her abdomen lowered, dragging her sting lightly over the substrate and depositing small quantities of a trail-marking pheromone. When she arrives at the nest and regurgitates to some of the workers, they are able to follow the trail to the food source, reinforcing it on the way back as long as the food lasts (see Chapter 5).

Termites also lay odor trails from abdominal glands—a remarkable case of evolutionary convergence, since ants and termites belong to quite unrelated orders of insects. In more primitive termites, odor trails are used in recruiting workers to breaches in the nest wall. In more advanced termites, which forage outside the nest, odor trails serve in recruitment to food sources, much as they do in ants.

In groups in which the workers are winged (wasps and bees), recruitment is accomplished quite differently. Some tropical social wasps as well as some stingless bees (tropical relatives of the honey bee) deposit pheromone on twigs, stones, and other objects between the food source and the nest or between a potential new nesting site and the old nest (Fig. 5–17). The amounts of pheromone deposited are much larger than in the case of trail-laying ants and termites, and in some cases the odor can be detected by human observers. Workers are able to detect the odor from a distance of a few meters and are able to proceed along the trail by flying from odor spot to odor spot, even high into the trees in some cases.

Honey Bee Communication.   Finally, a word must be said about the unique recruitment behavior of the honey bee, the so-called dances described many years ago by Karl von Frisch. A worker bee, returning to the hive after having found a rich source of nectar, conveys information to other bees concerning both the direction and the distance of the food source. If she is standing on a horizontal surface (as would rarely occur in nature), *direction* is indicated by pointing the straight run of the dance directly toward the food source (Fig. 8–3). On the vertical surface of a comb, in the dark of the hive, direction is indicated by pointing the straight run to conform to the angle of flight in relation to the sun, as if the sun were directly above, thus substituting gravity for a visual cue.

These statements apply to the **waggle dance,** which is performed if the food source is more than 35 to 80 meters away (different races of the honey bee differ in this regard). In this dance more precise information on *distance* is indicated by the duration of the straight run: The greater the distance, the greater the duration of the straight run. If the food source is close to the hive, however, a different dance having no information on direction or distance is performed, the so-called **round dance.** This consists of a circle, at the end of which the worker turns around and repeats the circle facing in the other direction. Workers reading this message fly out in all directions near the hive and search for odors like those carried on the body of the dancing bee. Odors of the food source are also important in the waggle dance, which in itself serves only as an approximation of distance and direction.

The dances are decoded within the hive by other bees, using tactile, olfactory, and acoustic cues. Bees cluster about the dancer; contact her with their antennae; and having read the message, fly to the vicinity of the food source. At times there may be several workers dancing on behalf of several food sources. In this case the most persistent dancers tend to recruit the most workers. These same dances are also used for recruitment to new nesting sites at the time of swarming.

In recent years there have been a number of criticisms of von Frisch's research. Possibly he underestimated the importance of odor of the food source clinging to the body of the dancers and the information in the wing sounds they produce; and it is true that different races of honey bees have different "dialects." But recent novel experiments have tended to confirm von Frisch's findings. James Gould, of Princeton University, established two feeding stations equidistant from the hive but in opposite directions, both similarly scented or unscented. In one the concentration of sugar was twice that in the other. Most of the foragers danced to the direction and distance of

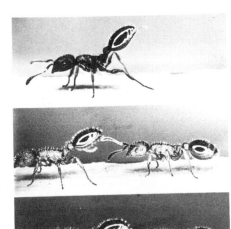

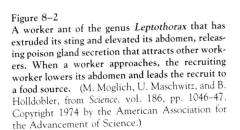

Figure 8–2
A worker ant of the genus *Leptothorax* that has extruded its sting and elevated its abdomen, releasing poison gland secretion that attracts other workers. When a worker approaches, the recruiting worker lowers its abdomen and leads the recruit to a food source. (M. Möglich, U. Maschwitz, and B. Hölldobler, from *Science,* vol. 186, pp. 1046–47. Copyright 1974 by the American Association for the Advancement of Science.)

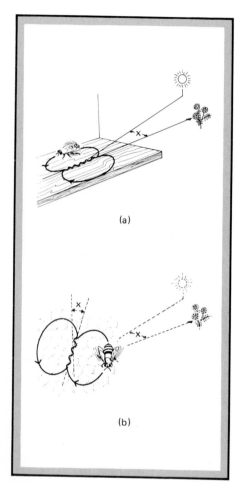

Figure 8–3
Waggle dance of the honey bee. On a horizontal surface (a) the straight part of the dance points directly toward a food source, but on the vertical surface of a comb (b) the straight part points to an angle with the sun as if the sun were directly above. (From Curtis, *Biology*, 3rd ed. Worth Publishers, New York, 1979, page 918.)

the station with the highest concentration of sugar, and nearly all recruitment was to that station. Factors such as site-specific odors, wind direction, and so forth were controlled, and it was also shown that in the absence of dancing almost no bees found the food sources. In a later experiment Gould developed a sophisticated design in which some of the bees responded to the information in the dances even though these were performed by bees coming from a different food source than that specified in the dance. That is, the recruits responded to distance and direction cues in the dances rather than to odors conveyed by the dancers.

The communication signals of social insects are parsimonious in that the same signal is often used in different contexts to transmit different messages. "Queen substance" of the honey bee inhibits worker ovarian development and the initiation of queen cells, stimulates grooming and feeding of the queen by the workers, maintains colony cohesion during swarming, and attracts males during the queen's nuptial flight—a single substance conveying different information in different circumstances.

This by no means exhausts the subject of social communication. For example, larvae of paper wasps may indicate hunger by hitting their heads against the cell walls, producing an audible sound; queens on the nests of these same wasps may signal their dominance by tail wagging and other visual signals. Interchange of food and secretions between adults and between adults and larvae conveys many messages on conditions within the colony; for example, the proper balance between the castes may be maintained by the amount of caste-specific pheromone being circulated in the colony. Studies using radioactive tracers have shown that in *Formica* ants substances imbibed by a single worker are rapidly spread throughout the colony, and within 27 hours every member of the colony has received some of it. The behavior of individual social insects is intimately guided by these messages. Since colonies may contain many thousands of individuals, it is difficult to imagine how they would function coherently without such a system of mass communication.

Division of Labor in Social Insects

The most unique feature of social insects is the existence in each species of several **castes**: at least two (queen and workers), often a third (soldiers), and sometimes several subcastes differing in appearance and function. Wasps and bees lack a distinct soldier caste; in the ants soldiers are simply large and specialized workers; but in the termites the soldiers constitute a caste quite different from the workers. Worker and soldier termites may be either male or female, and workers may be either immatures or adults. In contrast, worker Hymenoptera are always adult females. Workers and soldiers do not mate; they represent the worker force and defense of the colony, and reproduction becomes the sole function of the queen and males. In termites, a male reproductive ("king") is a permanent attendant of the queen, but in Hymenoptera males die soon after mating.

Members of different castes of one species often differ radically in appearance, a striking case of **polymorphism** that has, for the most part, an environmental rather than a genetic basis. Members of different castes and subcastes also exhibit very different behavior, a phenomenon called **polyethism**. A soldier termite is a specialist in defense and often cannot even feed itself. But in other cases polyethism has a temporal element, as in the subcastes of worker ants shown in Fig. 8–4. In the case of the honey bee, division of labor among the workers is wholly temporal. There is no polymorphism within the worker caste, but workers perform different tasks depending on their age (see Chapter 11, p. 260).

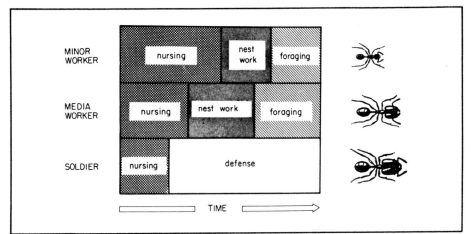

Figure 8–4
Polymorphism and polyethism in an ant of the genus *Pheidole*. Smallest workers, called minors, are at first nurses, feeding and tending larvae; later they do other kinds of work within the nest; and finally they may leave the nest to forage for a period. Larger workers, or medias, spend less time as nurses and more in other activities. The largest workers, called majors, may also be termed soldiers, since a major part of their time is spent in colony defense. The increasingly disproportionate size of the head is a result of allometric growth: larger workers have received more food as larvae. (Reprinted by permission of the publishers of *The Insect Societies*, by Edward O. Wilson, Cambridge, Mass.: The Belknap Press of Harvard University Press, copyright 1971 by the President and Fellows of Harvard College.)

How can all these diverse behavior patterns be built into one individual? And how is one species able to produce several kinds of individuals that differ so much in appearance and in behavior? If workers and soldiers do not reproduce, how are their genes conveyed to the next generation? If natural selection favors those individuals that produce the most surviving offspring, how does one account for the existence of sterile castes? Obviously the social insects present a host of questions that cannot easily be answered. Perhaps some will seem less intractable if we look briefly at certain species that seem not to be "fully eusocial"—that is, that appear to be on the verge of acquiring a worker caste.

Caste Determination in "Primitively Eusocial" Hymenoptera. Charles D. Michener and his students at the University of Kansas have for some years been studying a small ground-nesting bee called *Dialictus zephyrus*. These bees form aggregations in earthen banks, provisioning their nests with pollen and nectar gathered from flowers in the vicinity. In midsummer, nests contain several cooperating females, all superficially alike (Fig. 8–5). But closer inspection reveals that a division of labor does occur. One of the females proves to be the major egg layer, while the others act as guards or foragers. (We discussed the pheromones that enable guards to recognize nestmates in this species in Chapter 5, pp. 129–130.) The major egg layer ("queen") nudges the other females periodically, causing them to move into guarding positions, or she leads them down into the burrow where there are stimuli for building or provisioning. When the subordinates lay eggs, as they do occasionally, the queen eats them and replaces them with her own. If the queen is removed, one of the subordinates assumes her role. Ovarian development in the subordinates is believed to be retarded as a result of the queen's behavior, but all the workers have the potential to mate and become queens (males are present throughout the summer).

A somewhat different situation prevails among bumble bees, which live in small colonies in cavities in the soil. Inseminated females overwinter and start colonies in the spring, and the workers and the new crop of males and queens are all the foundress's offspring. The queen dominates the workers by pheromones transmitted on contact, aided by aggressive behavior, and the workers come to assume various roles in the colony, the smaller ones mostly as nurses, the larger ones mostly as foragers, although

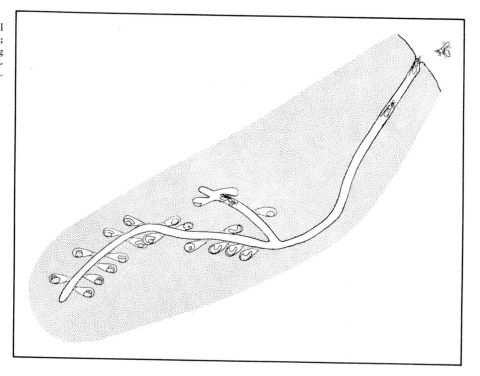

they may change their role as a result of age or the needs of the colony. If the queen is removed, one of the larger workers will assume her role. Workers tend to be larger late in the season, and eventually a new crop of large females destined to be queens is produced (as well as males). Late in the season there are more workers present in the colony in relation to the number of larvae. The increase in size of individuals may simply be the result of the increasing worker: larva ratio; larvae receive more food and become larger adults.

Caste Determination in Advanced Eusocial Insects. In the honey bee, behavioral dominance by the queen has been replaced by secretion of a pheromone (queen substance) that prevents development of the ovaries in the workers. Queens are reared in especially large cells and receive not only more food but food enriched by the contents of hypopharyngeal glands in the heads of the workers ("royal jelly") (see Fig. 5–1). In most of the more advanced social insects, the role of queen is sustained by pheromones. In colonies containing thousands of individuals, control by behavioral interactions between queen and workers would scarcely be practicable. (See also Chapter 11, pp. 259–260.)

By and large, polymorphism in social Hymenoptera has a **trophogenic** basis; that is, it is the result of differential feeding of the larvae or of different-sized eggs. Larvae receiving a minimal amount of food develop into small, sterile workers, a phenomenon sometimes called "nutritional castration." Better-fed larvae tend to produce larger individuals. In the ants these may have disproportionately large heads and mandibles, the result of **allometric growth,** and may serve as soldiers. Allometric growth is a genetically determined tendency for certain body parts to grow at a more rapid rate than other parts and thus to be larger in the adult stage (see Fig. 8–4). Reproductives—the queens of the next generation and their mates—are produced from larvae that are abundantly fed and sometimes provided with food of different quality. Often they are reared in larger cells, the combination of large cells and a high worker:larva ratio ensuring a crop of reproductives.

Hymenoptera appear especially well suited for the development of social behavior, as reflected in the fact that eusociality has evolved several times independently—in the ants, in the wasps, and several times in the bees; but only one other group of insects, the termites, has evolved eusociality. The Hymenoptera appear to have several features that together render them likely to become eusocial. Many solitary species build *nests*, such as the bee-wolf we discussed in the previous chapter (Fig. 7–10). These insects must have evolved fail-safe homing abilities as well as the ability to protect the larvae from predation, desiccation, and so forth. Limitation in sites suitable for nesting may cause them to aggregate in certain places and to produce more permanent, many-celled nests. Aggregated nests may improve the opportunities for parasites and predators, who do not have to look far for their hosts—thus the need for the development of defense mechanisms such as the presence of guards in the nest entrance. Another potent defense, the sting, evolved from an ovipositor that had first evolved into a mechanism for paralyzing prey (in solitary wasps) and then, in the eusocial Hymenoptera, came to serve primarily as a defensive weapon.

One additional factor may have been important in setting the stage for social behavior in Hymenoptera. Male Hymenoptera are haploid; that is, they have half the usual number of chromosomes and are produced from unfertilized eggs. Females have the capacity to lay either fertilized (diploid, female-producing) eggs or unfertilized (haploid, male-producing) eggs by controlling the flow of sperm from the spermathecae when the eggs are laid. Since the female sex is the one involved in nest making, feeding the young, and so forth, even in solitary forms, it proved advantageous for queens of social species to produce a large number of females capable of assisting in the work of the colony, delaying male production until time of production of a new brood of potential queens. The ability to produce progeny of the desired sex, as needed, is one that we humans may envy.

Termites.   Most of these remarks do not apply to termites, which are believed to have evolved from the ancestors of cockroaches, which do not make nests and are diploid. Hence they seem to lack the preadaptations of Hymenoptera. Nevertheless many do aggregate. The group of cockroachlike insects that gave rise to the termites probably lived gregariously in logs and digested cellulose with the aid of intestinal microorganisms, as some species of cockroaches still do. Since these symbionts are cast off with each molt and must be reacquired from others, and newly emerged individuals must obtain them from their elders, isolated individuals would soon starve. Both wood-feeding cockroaches and termites frequently feed at the anus of other individuals, thereby obtaining intestinal symbionts needed for digesting their food. It is apparently this factor that enhanced the development of eusocial behavior in the Dictyoptera.

Caste determination in termites differs considerably from that in Hymenoptera. Since termites have gradual metamorphosis, the young are not helpless, and in fact by the third instar members of both sexes assist in work of the colony—an example of "child labor" unique in the insects. These immatures molt further and may in some cases develop into soldiers or reproductives, depending on a complex system of pheromones circulating in the colony. Soldier termites are typically wingless adults that differ greatly from the workers and reproductives (Fig. 8–6). Termite colonies also con-

Figure 8–6
The castes of termites. In these insects, in contrast to Hymenoptera, the "king" (not shown here) is a permanent member of the colony. The queen (c) may ultimately become very large. The termites in (d) and (e) are supplementary queens; the one in (a) is a worker; in (b), a soldier. The worker is about 5 mm long in the species figured. (Drawings by Sarah Landry; reprinted by permission of the publishers of *The Insect Societies*, by Edward O. Wilson, Cambridge, Mass.: The Belknap Press of Harvard University Press, copyright 1971 by the President and Fellows of Harvard College.)

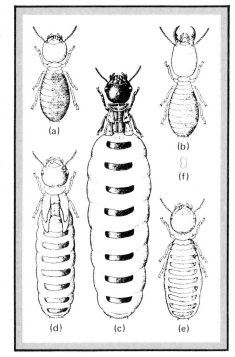

Figure 8–7
The nest of an Australian species of termite. A nest of this size may have been occupied by many successive generations of termites over many decades. Such nests provide a major tourist attraction in northern Australia. (Photograph by Robert W. Matthews, University of Georgia.)

tain "supplementary reproductives," which have wing pads and the capacity to become winged reproductives if the queen or king dies. Development of these various castes is mediated by the interaction of pheromones and hormones in ways that are far from fully understood. Presence of supplementary reproductives means that colonies are potentially immortal, and indeed the nests of termites sometimes persist for many years and result in structures of incredible size, considering the size of the builders (Fig. 8–7). The nests of each species of termite are distinctively different, so that taxonomists sometimes find it easier to identify the species by the nest rather than by the termites themselves.

Social Homeostasis

Maintenance of a functional steady state in an organism or in a colony of organisms is termed **homeostasis.** The colonies of social insects have sometimes been compared to organisms; the individuals, to cells, some of which are specialized for reproduction, others for nutrition, others for protection. Colonies have a birth, a growth cycle, and an eventual death. As we have said, the nests are species specific, just as individuals can be recognized as members of different species. Thus the term *superorganism* is sometimes applied to these colonies, though a little reflection will show that analogies between colonies and organisms are at best very rough and not overly instructive.

Nevertheless the superorganism concept does help one to visualize the vast amount of coordination that must occur within the colony—coordination that is achieved by complex systems of communication operating with respect to individuals that are diverse but limited in their responses and in their abilities to control other colony members. Integrity of the colony is maintained via nest structures that cannot easily be breached and via workers and soldiers prepared to respond to intruders with appropriate physical and chemical attacks.

In temperate climates, colonies of most species of bees and wasps are annual affairs, begun anew each spring by overwintered, inseminated females. Thus a nest of yellow jackets may contain only a few hundred or at most a few thousand individuals. Ant and termite colonies are, however, perennial, and in tropical climates colonies of some species reach enormous size. A single colony of army ants may sometimes contain over a million workers. Queens of certain termites are reported to lay as many as 30,000 eggs per day, or 10 million per year. Homeostatic mechanisms in colonies this size must be complex beyond belief.

Most social insects are able to control temperature and humidity of the nest at least to some extent. In temperate climates, many ants start to breed under stones in early spring, taking advantage of the capacity of stones to absorb heat, then they move deeper into the soil in summer. Aerial nests of wasps are built in protected places, and those of hornets and yellow jackets are enclosed in protective paper sheaths. Fanning at the nest entrance and the carrying of water to the nest play a role in cooling the nest in hot weather for many species. Honey bees survive the winter by clustering in the hive. Clustering bees consume honey and move about within the cluster, generating enough heat so that the temperature of the cluster does not fall below about 20°C, even though the outside temperature may frequently be below 0°C. Similarly honey bees are able to maintain a hive temperature of about 35°C even on the hottest summer days, so long as water is available to them.

Some of the most remarkable instances of "air conditioning" occur among some of the fungus-growing termites of Africa (Fig. 8–8). Warm air generated in the core of the nest rises and passes into a series of small chambers in the nest walls, where it is cooled and where fresh oxygen is obtained; it then flows back into the core of the nest at the bottom, and a slow but continuous circulation is maintained.

Slavery in Ants

Before leaving the social insects, mention should be made of another unique phenomenon: slavery among ants. This behavior is justly termed unique, since it differs from slavery among humans in usually occurring between species; that is, members of one species enslave those of another, related species. It also differs in being a biological rather than a cultural phenomenon. Thus one could argue that it should not be called slavery at all, though the superficial resemblances to slavery in humans are striking.

Slave-making species of ants are widely distributed, and most of the reported "battles" between ant colonies are actually slave raids. Certain reddish species of the genus *Formica* undertake frequent raids on colonies of some of the common and rather docile black species of the same genus. Columns of workers approach and surround the nest to be plundered, causing alarm, attempts to escape, and often some combat among the workers. Eventually the more aggressive slave makers enter the nest, seize larvae and pupae, and carry them back to their own nest. When these develop into adult workers, they acquire the colony odor and perform their normal behavior in the alien nest, effectively supplementing the worker force of the slave-making species.

Recent research by E. O. Wilson and his colleagues at Harvard University has shown that slave makers of at least some species have enlarged Dufour's glands in their abdomens (Fig. 8–9). When they approach a nest of the slave species, they spray the contents of these glands onto the workers. The secretion consists of a mixture of acetates that causes the workers of the slave species to become alarmed and disorganized. Curiously the effect of the acetates on the slave makers is quite the reverse: They produce attraction and excitement—as Wilson says, "exactly the responses needed to conduct successful slave raids."

By and large, colonies of most ant species defend themselves well against incursion by alien colonies, sometimes resorting to physical combat to maintain territorial boundaries. In honeypot ants of the deserts of the southwestern United States, territories are defended by ritualized displays in which workers of opposing colonies "stilt walk," at first facing one another (Fig. 8–10), then beginning to circle and trying to push one another sideways. Such "tournaments," recently described by Bert Hölldobler, of Harvard University, may go on for days. If one of the colonies is unable to recruit enough workers to the tournament area, it may be overrun by the stronger colony and the brood and other nest contents carried off. The surviving workers then become incorporated into the raider's nest. This is the only known example of slavery occurring within one species of ant.

Symbionts in the Nests of Social Insects

In spite of the various behavioral and chemical defenses of the colonies of social insects, quite a number of arthropods have evolved ways of breaching these defenses and taking advantage of the security of the nest and the abundance of food provided. They do this by tapping into the communication codes—that is, by acquiring the ability to "speak the language" of their hosts. Many of them have glands that secrete substances attractive to their hosts, who "adopt" them as part of the colony. Altogether, accord-

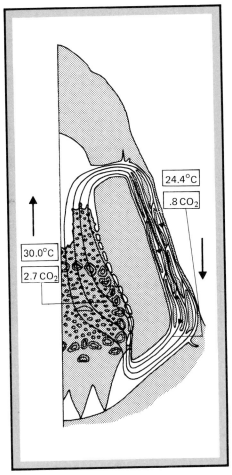

Figure 8–8
Temperature regulation in the nest of an African termite. Only half of the nest is shown here. (From "Air-conditioned termite nests," by M. Lüscher. Copyright © 1961 by Scientific American, Inc. All rights reserved.)

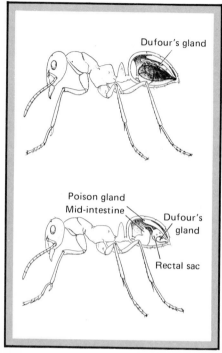

Figure 8–9
The enlarged Dufour's gland of a slave-making ant (*Formica subintegra*) (above) as compared with that of a member of a slave species (*Formica subsericea*) (below). (From *Slavery in Ants*, by Edward O. Wilson; copyright 1975 by Scientific American, Inc. All rights reserved.)

Figure 8–10
"Stilt-walking" display by territorial worker desert honeypot ants from adjacent nests. (B. Hölldobler, from *Science*, vol. 192, pp. 912–14. Copyright 1976 by the American Association for the Advancement of Science.)

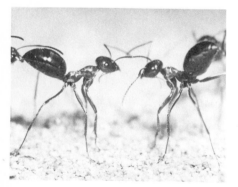

ing to E. O. Wilson, insects of at least 120 different families and 17 different orders have become "guests" of social insects, along with a variety of mites, millipedes, and other arthropods. As the noted student of social insects W. M. Wheeler put it many years ago:

> Were we to behave in an analogous manner we should live in a truly Alice-in-Wonderland society. We should delight in keeping porcupines, alligators, lobsters, etc., in our homes, insist on their sitting down to table with us and feed them so solicitously with spoon victuals that our children would either perish of neglect or grow up as hopeless rachitics.

Symbionts associated with ants are termed **myrmecophiles** (literally, "ant lovers"), while those found with termites are called **termitophiles.** Relatively few occur in the more open societies of the social bees and wasps.

One of the simplest associations is shown by the "highwayman beetle" (Fig. 8–11), which intercepts black wood ants on their trails and solicits food from them by tapping their mouthparts. Frequently the ants soon realize they are being "tricked" and begin to attack the beetle. The beetle responds by retracting its legs and flattening itself against the ground in such a way that it cannot be readily grasped by the ants.

In most myrmecophiles and termitophiles the relationship with the hosts is much more elaborate and intimate. Rove beetles of the genus *Atemeles* also approach ants, in this case brown ants of the genus *Myrmica*, but these beetles possess glands that the ants find attractive (Chapter 5, p. 137; Fig. 5–25). The ants carry the beetles into their nests, where they spend the winter among ample food. In the spring the *Atemeles* beetles migrate to colonies of the mound-building wood ant, a species of *Formica*, which provide a richer source of food in the summer. Here they lay their eggs, and the larvae, also provided with special glands, solicit food from their hosts and also consume larvae of the host ants.

Not all symbionts have such elaborate adaptations, but all have some means of escaping, appeasing, or otherwise "duping" their hosts. Termitophiles often have grossly swollen abdomens that exude substances that are licked by the termites (Fig. 8–12). In many cases we do not know precisely how the symbionts "break the code" of their hosts or what they feed on in the nest. Although the large colonies of army ants and some of the tropical termites contain hundreds of "guests" of great variety, many of the symbionts appear quite rare. Evidently this is a precarious existence; as in espionage among humans, a slip may be fatal.

The Advantages and Disadvantages of Group Living

A great many insect species live essentially solitary lives, contacting other individuals chiefly at time of mating—and a few even dispense with this "inconvenience," reproducing parthenogenetically. Clearly a solitary way of life has advantages: Such individuals can more effectively hide from predators and avoid competition with others with similar lifestyles; they can also more effectively live in small spaces and exploit dispersed food sources. Group living also has obvious disadvantages: There are opportunities for the spread of disease and for the incursion of unwanted "guests" into large societies. Why then live in groups? One advantage might be that escape from predators is actually enhanced if, via communication resulting in alarm and defense behavior of the group, a large predator is driven off. Another might be that group living permits better exploitation of food sources that are strongly localized or so large and tough that one individual could not well exploit them, as seen in first-instar sawfly larvae feeding on pine needles or ants attacking a beetle. Still another factor might concern limitations in living space—that is, cockroaches may aggregate because only limited areas provide suitable conditions of warmth and moisture; or ground-nesting bees or wasps, because soil of suitable consistency is patchily distributed.

Individuals of a species commonly compete severely for food and living space (Chapter 17). For group living to succeed, individuals must survive and reproduce more successfully within the group than apart from it. This means that they must evolve a degree of tolerance, if not actual cooperative behavior, with other members of that species. Clearly it will not pay an individual within a group to devote energies to competitive interactions; but it may very well pay to produce or respond to an alarm signal. Wherever the advantages of group living outweigh its disadvantages, in terms of individual reproductive success, group living may evolve.

Social groups among insects (and among animals generally) usually consist of related individuals—parents and offspring, siblings, or at least members of a local population having many genes in common. Often there are mechanisms for occasional outbreeding, such as dispersal at time of mating. Webs of a tent caterpillar commonly contain individuals from a single egg mass; lady beetles clustering for the winter are commonly the descendants of those that clustered there the previous winter. As we have seen, the colonies of eusocial insects are essentially "extended families;" even when the queen has mated more than once or when there is more than one queen, there is genetic relatedness among colony members. Individual **Darwinian fitness** (that is, reproductive success) may in such cases be replaced by **inclusive fitness** (defined as net genetic representation in succeeding generations, including other relatives in addition to offspring). That is to say, an individual may devote energies to cooperative behavior to the detriment of its own reproductive success to the extent that the individual it helps carries genes identical to its own.

The expenditure of energy in cooperative behavior that might otherwise be spent in individual effort is often spoken of as **altruism,** and natural selection that involves inclusive fitness is often spoken of as **kin selection.** Whether or not unselfish behavior among humans is a product of kin selection is a controversial topic we need not consider here; but in any case the word *altruism* as applied to insects does have this connotation and does not imply purposefulness on the part of the performers. Even with low degrees of relatedness among members of a colony, kin selection favors the evolution and maintenance of sterile castes if the benefit:cost ratio is high enough. In this context it is much easier to understand the behavior of a worker honey bee, for example, who leaves her sting in the body of a predator and dies thereafter: She is ensuring the survival of many thousands of individuals of very similar genetic constitution (Fig. 8–13).

All of this is helpful in explaining the unique properties of eusociality. Workers and soldiers have lost all individual fitness (or most of it, since workers of Hymenoptera do in some cases lay viable, male-producing eggs). But the colony, as an extended family, may be enormously successful in survival and reproduction and may exploit and even modify the environment in ways no individual could do. A soldier termite guards a colony of thousands of individuals, and at much risk to its own life. But ultimately its genes (which are also those of the reproductive caste) will persist in succeeding generations via inclusive fitness. We have suggested reasons why eusocial behavior may have evolved in termites and in Hymenoptera but not elsewhere. In Hymenoptera, eusociality has in fact evolved several times independently. In addition to permitting females to control the sex of their offspring, male haploidy, in this order, may result in an unusually close relationship among siblings. Since each male is the result

Figure 8–11
A "highwayman beetle" of the genus *Amphotis,* which has intercepted a black wood ant and is soliciting food from it. (Photograph by Bert Hölldobler, Harvard University.)

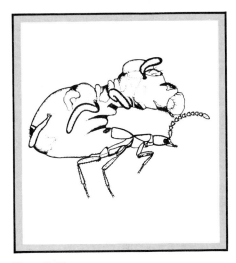

Figure 8–12
A rove beetle with the formidable name *Coatonachthodes ovambolandicus,* a resident of the nests of certain African termites. The swollen abdomen is believed to copy the body of the termites; the slender appendages, the termites' legs. (David Kistner, California State University, Chico.)

Figure 8–13
The ultimate act of altruism by a eusocial insect is here performed by a worker honey bee, who leaves her sting in the wound, where the muscles of the venom sac continue to pump fluid into the wound. This worker will shortly die. (Photograph by Bernd Heinrich, University of Vermont.)

of "virgin birth," every sperm he produces has all his genes, while each egg produced by a female has only half of hers (as in diploid species). Because any daughter of a male has a full set of his genes, sisters related through both parents are unusually closely related (by 3:4). On the other hand, daughters are related to their mother only 1:2. The assumption is that females are therefore more likely to evolve a tendency to cooperate with their sisters to rear more sisters than to start their own nest and produce daughters. How important this factor is in enhancing altruism in colonies of ants, bees, and wasps is a moot point, but one that must be kept in mind.

Summary

Social insects occur in colonies in which there is a division of labor involving sterility of many colony members. The term *eusocial* is often used to distinguish this specialized condition from broader usages of the word *social*. Entomologists tend to apply the term *presocial* to aggregative behavior and brood care when these do not involve a reproductive division of labor.

The behavior of the many individuals in social insect colonies is integrated through complex communication systems involving the broadcasting of messages throughout the colony. Social signals are occasionally visual, more often tactile or acoustic, most commonly olfactory. Alarm signals are usually volatile pheromones that fade quickly; but they may be acoustic. Attractants tend to be more complex, less volatile substances, such as the queen substance of the honey bee. Nest odors and recognition signals are complex mixtures of substances that identify nests and nest members.

Recruitment to food sources or new nest sites also involves chemical cues, and many species lay odor trails from specialized glands. Odor trails on the ground are characteristic of many ants and termites, while flying insects (certain wasps and bees) may form odor trails on twigs and leaves. Recruitment in the honey bee is most unusual and consists of the "round dance" when the food source is close and the "waggle dance" when it is more distant, the latter of these dances conveying information on distance and direction.

Division of labor in the colony is enhanced by the presence of several castes: queen, workers, and soldiers. The worker caste may be polymorphic, and different duties may be performed by the different morphs or according to the age of the worker. In Hymenoptera, polymorphism is largely a result of different quantities of food fed to the larvae and sometimes a difference in quality, workers and soldiers being females that are sterile as a result of "nutritional castration."

Termites differ from Hymenoptera in that workers and soldiers may be either male or female, and immatures often act as workers. In these insects the "king" is long lived, and the colony contains "supplementary reproductives" capable of replacing the reproductives. Thus colonies are potentially immortal and capable of constructing nests of great size.

Homeostasis is maintained through complex communication systems and through nest structures and behavior patterns functioning to maintain more or less constant conditions within the nest. Honey bees, for example, are able to cool the hive in summer by fanning and water carrying, and in the winter they form a cluster in which the temperature is maintained not lower than about 20°C.

Slavery is one of the more remarkable phenomena occurring among social insects. It is restricted to a relatively few species of ants that raid colonies of related species and take larvae and pupae back to their own nests, where the emerging adults are added to the work force. In desert honeypot ants, stronger colonies are known to invade and to enslave the workers of weaker colonies of the same species.

Colonies of ants and termites often contain diverse symbionts or "guests," called myrmecophiles and termitophiles, respectively. These symbionts have diverse adaptations for gaining entry into the colony and for being tolerated there; commonly they "break the communication code" of their hosts, sometimes using glands that produce substances that mimic the pheromones of their hosts.

Social groups among animals generally have arisen whenever the advantages (in terms of survival and reproduction) exceed the disadvantages. Solitary individuals may do best at hiding from predators and at living in small spaces and exploiting dispersed food sources. Group living may be advantageous for avoiding large predators (via group defense) or for exploiting localized or large and difficult food sources; and aggregations may be favored when optimal living space is patchily distributed.

Social groups commonly consist of related individuals. In such groups inclusive fitness may be important; that is, an individual may devote energies to cooperative behavior to the detriment of its own reproductive success to the extent that the individual it helps carries genes identical to its own. Among eusocial insects, workers and soldiers have essentially lost the ability to reproduce themselves directly, but they do so via inclusive fitness, since by their activities they enhance survival and reproduction of the colony, all members of which carry similar genes.

Selected Readings

Breed, M. D.; C. D. Michener; and H. E. Evans, eds. 1982. *The Biology of Social Insects.* Boulder, Colorado: Westview Press. 420 pp.

Evans, H. E., and K. M. O'Neill. 1981. "Insect societies: an independent experiment in group living," In *Group Cohesion* (H. Kellerman, ed.), pp. 191–204. New York: Grune and Stratton.

Evans, H. E., and M. J. West Eberhard. 1970. *The Wasps.* Ann Arbor: University of Michigan Press. 265 pp.

Frisch, K. von. 1967. *The Dance Language and Orientation of Bees.* Cambridge, Mass.: Harvard University Press. 566 pp.

Hermann, H. R., ed. 1979–1982. *Social Insects.* Volumes 1–4. New York: Academic Press.

Hölldobler, B. 1971. "Communication between ants and their guests." *Scientific American,* vol. 224, no. 6, pp. 86–93.

Jeanne, R. L. 1980. "Evolution of social behavior in the Vespidae." *Annual Review of Entomology,* vol. 25, pp. 371–96.

Lin, N., and C. D. Michener. 1972. "Evolution of sociality in insects." *Quarterly Review of Biology,* vol. 47, pp. 131–59.

Michener, C. D. 1974. *The Social Behavior of Bees.* Cambridge, Mass.: Harvard University Press. 404 pp.

West Eberhard, M. J. 1974. "The evolution of social behavior by kin selection." *Quarterly Review of Biology,* vol. 50, pp. 1–33.

Wheeler, W. M. 1910 (reprinted 1926, 1960). *Ants: Their Structure, Development, and Behavior.* New York: Columbia University Press. 663 pp.

Wilson, E. O. 1971. *The Insect Societies.* Cambridge, Mass.: Harvard University Press. 548 pp.

———. 1975. *Sociobiology: The New Synthesis.* Cambridge, Mass.: Harvard University Press. 697 pp.

———. 1975. "Slavery in ants." *Scientific American,* vol. 232, no. 6, pp. 32–36.

It has been estimated that roughly half the known species of insects feed on plants or plant products, such as seeds or pollen. But in a sense all insects (and all animals) are dependent on plants, since it is plants that capture the sun's energy and use it to make organic molecules from inorganic ones. In the three chapters that follow we shall be concerned with the more direct relationships between plants and insects, reserving to Part V the more indirect relationships of insects as secondary or tertiary consumers. Many plants depend on insects for pollination, but for the most part plants do not profit by being fed on. Hence they have often evolved structures and chemicals that deter the feeding of herbivores. Many insects have in turn

The Relationships of Plants and Insects

P A R T

evolved ways of overcoming the defenses of plants and in some cases of using the repellent or poisonous substances obtained from plants in their own defense (as we saw in Chapter 5) or as feeding or oviposition cues. Study of plant defenses can be of much practical value; indeed, certain plant substances, such as nicotine, have long been used as insecticides.

When plant damage is slight or the plants are of no immediate significance, we seldom notice the presence of insects. Often the effects of herbivorous insects are in the long run beneficial, since the insects are playing subtle but essential roles in maintaining natural ecosystems. It is usually the mass attacks of insects on our crops or shade and forest trees that attract

F O U R

our attention. Locust swarms have plagued humans since ancient times; the Colorado potato beetle exploded from its native home in the Rockies in the mid-nineteenth century and devastated a major food crop throughout North America and later Europe; the gypsy moth causes massive defoliation in the northeastern United States every year; and the Mediterranean fruit fly, though several times eradicated from our shores, continues to haunt orchardists (and even politicians) in California and Florida. Of the vast array of insects, it is the phytophagous species that most often impinge on human economy and lifestyles. Plant feeders not only injure plants directly but may weaken them so that they are susceptible to attack by other

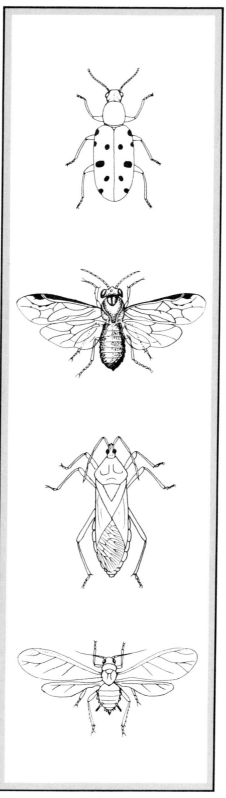

insects or by pathogens. They may also serve as vectors of such plant diseases as the all-too-familiar Dutch elm disease.

In Chapter 9 we shall discuss phytophagous insects in general terms, with emphasis on the ways insects attack plants, on their nutritional requirements, and on their role in transmitting plant diseases. In Chapter 10 we shall examine plant defenses against insect attack as well as those plants that include insects in their diet. Chapter 11 is devoted to a survey of the roles insects play in pollination. As we shall see, the relationships of plants and insects are diverse and often complex. In this field basic knowledge from several fields must be called on in our effort to manage insect populations effectively.

Phytophagous Insects

Phytophagy, or herbivory, implies the ability to macerate plant tissue or to imbibe plant fluids as well as the ability to digest these tissues and to utilize them for energy, growth, and reproduction. Oddly, insects have limited abilities to digest cellulose, the major substance of which plants are constructed (although some have symbiotic microorganisms that assist in its breakdown). Plant tissue is a source of sugars, proteins, fats, salts, water, and vitamins. Most plant tissue provides adequate nourishment for insects, although different species do have different nutritional requirements. That insects are to varying degrees discriminating feeders reflects a long and fascinating history of plant–insect coevolution, which we shall explore briefly in this chapter and the following two.

Insects feed on leaves, buds, stems, roots, fruits, and seeds, as well as on plant tissue in various stages of decay (as in Fig. 13–14). They feed externally or internally, as borers or leaf miners; they produce galls and other distortions; sucking insects imbibe plant juices and may cause weakening and yellowing. Many insects, even nonphytophagous species, also use plants as shelter, and some make still other uses of plants that are outside the scope of this chapter. Leaf-cutter bees, for example, use pieces of leaves to line their nest cells (Chapter 11), and leaf-cutter ants harvest bits of leaves that they use as a substrate for growing fungi in their nests.

Returning to more strictly phytophagous insects, we find scarcely any plant immune to attack. Clover is fed on by some 200 insect species; corn, by over 300. Over 200 species attack citrus trees; 400 or more, apples. Among forest trees, elms harbor at least 600 species, while oaks hold the record with nearly 1500 insect enemies (many of them gall formers). Out of its normal habitat, a plant may sometimes be relatively free of insect pests. Eucalyptus grown in Mexico or California, for example, shows little evidence of insect attack; yet in its native home in Australia, eucalyptus harbors a great number of insects. Obviously its foliage is perfectly edible for adapted insects; but insects elsewhere have not evolved mechanisms for overcoming the repellency of eucalyptus oils. Ginkgo trees are seldom if ever subject to insect attack wherever they are grown (they no longer exist in the wild). These trees are relics of a very ancient group of plants that have evidently survived as a result of repellent substances in their tissues that deter both insects and disease. The degree to which a plant species is immune to insect attack is, in general, a reflection of the defenses it has evolved and the evolved abilities of insects to overcome these defenses. These are matters to be explored in the next chapter.

Among the insects inhabiting a particular plant, some may restrict their feeding to that plant species alone and are said to be **monophagous.** Others may be general feeders and include a diversity of plants in their diet; such insects are said to be **polyphagous.** White oak, for example, plays host to several species of gall wasps that not only form galls on white oak alone but even form a distinctive type of gall on only one part of the tree. White oaks may also be attacked by gypsy moth caterpillars, which are decidedly polyphagous insects, attacking a wide variety of broad-leaved trees and even, at times, conifers. Under artificial conditions these caterpillars have been reared on over 400 plant species; but they reject certain plants (such as larkspur) and in nature will feed preferentially on a limited group of shade and forest trees. Some of the Orthoptera are strongly polyphagous: Hungry migratory locusts will gnaw on wooden fence posts, and crickets will consume clothing. Indeed crickets, many ants, and some other insects are essentially **omnivorous,** including both plant and animal tissue and sometimes detritus in their diet.

Much more commonly, insects may be described as moderately discriminating in their tastes. Such insects are said to be **oligophagous.** The Colorado potato beetle (Fig. 9–1) feeds on plants of the genus *Solanum;* the imported cabbageworm on various Brassicaceae; the monarch butterfly, on various kinds of milkweeds. We shall consider the chemical basis of host selection in a later section of this chapter. In behavioral terms, a phytophagous insect may reach its host in one of three ways:

1. The host may be selected by trial and error—that is, by moving about and tasting several plants before settling to feed. Many grasshoppers fall in this category.
2. The host may be selected by the mother, who lays her eggs in response to some particular cue, often chemical. Most Lepidoptera find their host in this way.
3. The insect may live in an aggregation that extends through several generations, so that emerging young find themselves already settled on their host plant. Aphids are an example. However, aphids also have a type 1 phase in their life cycle (Chapter 16 and Fig. 16–1). Many aphids (and some other insects) have alternate hosts that are occupied seasonally; often the winter is spent on a woody, perennial plant; the summer, on an herbaceous annual.

Figure 9–1
A pair of Colorado potato beetles mating on their host plant. (Photograph by O. Wilford Olsen, Colorado State University.)

Types of Plant Feeding

Chewing Insects.   Consumers of plant tissue commonly have mouthparts of generalized biting type, with stout, strongly musculated mandibles. Caterpillars may consume many times their own weight in plant tissue in the course of their development. Much fibrous tissue passes through the gut undigested and forms a major part of the large fecal pellets.

Many insects begin feeding at the margin of the leaf, while others feed on either the upper or, more commonly, the lower surface. The leaf may be eaten all the way through, or the insect may scrape off the epidermis and parenchyma, leaving one layer of epidermis and its supporting veins intact. Such insects are said to be **leaf skeletonizers.** Others mine the interior of the leaf, feeding on the parenchyma and leaving both the upper and the lower epidermis intact (Fig. 9–2). Such **leaf miners** tend to be small larvae that are pale in color and flattened, and whose mouthparts project forward (prognathous) rather than downward (hypognathous); most of them are legless or nearly so. The patterns they form are often intricate and diagnostic of the species. There

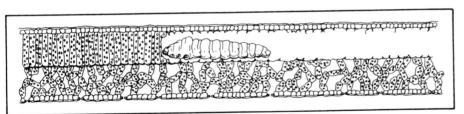

Figure 9–2
Cross section of a leaf being mined by a beetle larva. (Drawing by Peter Eades.)

Figure 9–3
A leaf of trembling aspen that has been mined by several lepidopterous leaf miners. The dark lines represent a trail of feces left behind by the larvae. (Photograph by Howard E. Evans.)

are two general types. Some leaf miners chew out a broad patch, forming a **blotch mine,** while others move along a slender path, forming a **linear mine,** which is often quite tortuous (Fig. 9–3). Linear mines tend to be very slender when the larva is small, but they broaden gradually, eventually ending in a small blotch. In North America there are more than 1000 species of leaf-mining insects, belonging to four orders of Endopterygota (Lepidoptera, Coleoptera, Diptera, and Hymenoptera).

Leaf tiers, leaf rollers, and **leaf folders** belong primarily to the order Lepidoptera. These insects use silk to hold leaves together in various ways to provide a retreat in which they molt or spend other inactive periods. Sometimes several such larvae live together, forming a large, often unsightly mass of leaves and silk; one such species is appropriately called the uglynest caterpillar. Tent caterpillars build communal retreats primarily of silk, and they build silken trails that they follow to and from the nest during foraging trips.

Borers in stems, trunks, and roots have some features in common with leaf miners: They tend to have larvae that are pale in color, prognathous, and more or less legless. Borers in trees have particularly powerful mandibles as well as a proventriculus capable of grinding the hard particles to a usable size. Even so, much undigested material is passed off in the dry feces. Many borers have intestinal symbionts that assist in digestion, while others feed not on wood but on fungi that grow in the galleries.

Larval insects of several orders live in fleshy fruits; others live in nuts and seeds, which like fruits provide a rich source of food, but unlike fruits provide a very dry environment. Thus the many species of insects that infest seeds, grains, flour, and the like have mechanisms for conserving the limited water available in their food. The feces of such insects are exceedingly dry as a result of water extraction by the rectum, and much of the required water is obtained through the metabolism of starches.

Sucking Insects.   The mouthparts of Hemiptera, as we saw in Chapter 1, Fig. 1–5c, are admirably adapted for piercing tissues and extracting fluids. Through muscular action, the four very delicate stylets are able to penetrate leaves, stems, and even the bark of trees. They are sufficiently flexible to pass between fibrous elements and, after considerable probing, reach the phloem or vascular bundles. Here the pressure of sap within the plant may induce a flow up the food channel between the stylets; or the flow may be assisted by the pharyngeal pump. At the same time saliva is being injected via the salivary channel in the stylets. Many sucking insects produce two kinds of saliva, from different parts of the salivary glands. One is a thin fluid that mixes with the sap and initiates digestion; the other, a more viscous substance that combines with fluids from accessory glands to form a sheath around the stylets. This sheath is formed of lipoproteins that gel on contact with air as a result of the formation of hydrogen and disulphide bonds. Evidently the function of the sheath is to prevent the loss of sap and saliva.

Sucking insects must imbibe much fluid in order to obtain their requirements of protein, minerals, and vitamins. Much of this fluid—chiefly water, carbohydrates, and some amino acids—is passed from the anus unmodified, and in fact in some Homoptera it bypasses the midgut by means of a filter chamber (Fig. 1–17). The sweet, watery excrement of aphids, leafhoppers, and scale insects forms a sticky deposit on foliage, on the ground, and on the tops of automobiles parked under infested trees. This **honeydew** is fed on by bees, wasps, ants, and other insects, and it is also the medium on which a sooty fungus grows, often causing disfigurement of ornamentals. The honeydew of certain scale insects of the Near East is sometimes so abundant that it has been used as food by humans—the "manna from heaven" of the Israelites.

Gall Insects.   Galls are abnormal growths on the buds, leaves, stems, or roots of plants. They result from the action of bacteria, fungi, nematodes, mites, or insects of several groups. Gall midges (Diptera, Cecidomyidae) and gall wasps (Hymenoptera, Cynipidae) are most frequently involved, but there are also gall makers in the orders Coleoptera, Hemiptera, Diptera, and Lepidoptera. Of the approximately 2000 species of gall makers in North America, about 1500 are gall wasps or midges. Although gall midges attack plants of more than 50 families, gall wasps largely restrict their attacks to two groups: Rosaceae and the genus *Quercus* (oaks) in the family Fagaceae. Some 800 species of gall wasps in North America form galls on oaks. These occur on roots, buds, twigs, leaves, flowers, and nuts, and are so diagnostic of species that identification based on the galls is often simpler than that based on the wasps themselves.

The simplest galls are mere swellings that involve no major distortion or discoloration; these are said to be **indeterminate galls.** In some cases these are expanded to form **pouch galls,** which are often open to the outside and are especially characteristic of aphids and related Homoptera. The majority of gall midges and wasps make **determinate galls,** which have a form and color quite different from that of the host plant; good examples are provided by the willow cone gall (Fig. 9–4) and the oak apple gall (Fig. 9–5). These exist in such variety that many different terms have been employed to describe them, and guides have been written for their identification. Yet these elaborate structures consist entirely of tissue supplied by the host plant under stimulation by invading insects. Often they resemble abnormal fruits (cones on willows!), but rather than containing plant embryos, they provide insects with a rich source of food as well as protection from predators and from the elements. Galls may continue to grow as long as the stimulation persists, even though normal leaf or stem growth may have ceased for that season.

Galls begin as small swellings at the point of oviposition by the female. In a sawfly gall of willow, studied by William Hovanitz, of the California Institute of Technology, the egg hatches in about five days, during which time the gall grows slowly. When the egg hatches and the larva begins to feed, growth of the gall continues. Dr. Hovanitz found that a fluid injected at the time of egg laying initiates gall formation. However, after about eight days the presence of the larva is required for continued growth. When the larva is removed from a gall, growth stops in about two days. Evidently both the mother, at the time of oviposition, and the larva secrete a growth-promoting substance, though whether the two substances are the same has not been determined.

It has often been proposed that plant growth hormones are involved in gall formation, since abnormal growths produced by these hormones are similar to the cells and tissues in natural galls. Studies of needle galls of pinyon pine (*Pinus edulis*) by J. Wayne Brewer and his associates at Colorado State University have revealed levels of auxin and gibberellin in these galls many times higher than in normal needles of the same age (Fig. 9–6). Extracts of the midge larvae that form the gall did not contain auxin and showed only traces of gibberellin, and it seems likely that the plants themselves produce these abnormal amounts of hormones under stimulation by the larvae. The substances secreted by the insects have so far defied full analysis. They must surely contain some components that differ from species to species, since the resulting galls are so different, even when related species make galls on leaves of the same plant. It is possible that a fuller understanding of gall formation will shed light on processes of cell differentiation in all living things. What little we know of the gall-producing secretions of insects suggests that they contain adenine and other amino acids as well as nucleic acids. It is even possible that they may shed light on the origin of cancer in humans.

Figure 9–4
A cluster of willow cone galls, formed by gall midges. (Photograph by Howard E. Evans.)

Figure 9–5
The oak apple gall, formed on a leaf of red oak by a gall wasp. (Photograph by Howard E. Evans.)

Figure 9–6
Auxin (a) and gibberellin (b) levels in normal needles of pinyon pine as compared with those in needles having galls. Auxin was measured in terms of its equivalent in indole-3 acetic acid (IAA); gibberellin, in terms of its equivalent in gibberellic acid (GA). (After J. A. Byers, J. W. Brewer, and D. W. Denna, 1976, "Plant growth hormones in pinyon insect galls," *Marcellia*, vol. 39, pp. 125–134.)

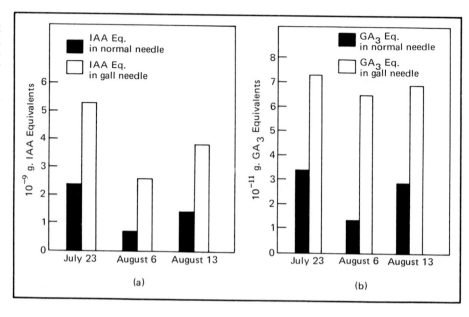

Host Plant Searching and Acceptance

How does an aphid or a butterfly locate an acceptable host plant for settling or for oviposition? How does a grasshopper "decide" which plants in the environment provide the most suitable food? These are questions of considerable importance, but only partial answers are available. The reaction chains may involve several sensory modalities and highly specialized receptors. It is useful to recognize four stages in host finding:

1. Search for a suitable habitat;
2. Settling on a plant and the most suitable part of a plant (either for oviposition or prior to feeding);
3. Initiation of feeding ("tasting");
4. Feeding to satiation.

The insect may fail to receive appropriate stimuli or may actually be repelled, at any stage, necessitating a return to an earlier step in the sequence. Starvation or failure to lay eggs on a suitable substrate may result if the insect is unable to complete the series.

Search for a Suitable Habitat.
As we have seen, apple maggot flies are attracted to yellow surfaces. Similarly, winged aphids can be collected in yellow-painted pans filled with water. In both cases yellow is probably not distinguishable from the green of the host plant, but yellow has higher reflectance properties than green and is thus a supernormal stimulus. Form as well as color may be involved. Immature desert locusts are attracted to patterns of vertical stripes, evidently because these simulate the grassy habitats they prefer. During dispersal, insects may become increasingly hungry and increasingly responsive to specific cues likely to guide them to a desirable habitat. These cues are evidently largely visual, but olfactory cues may also play a part, at least at short range.

Settling.   It is probable that visual cues often continue to play a role as the insect settles on a plant within a selected habitat. Female pipevine swallowtail butterflies (*Battus philenor*) alight on any leaves that resemble in shape those of their host plants, *Aristolochia*, but they lay their eggs only when, after drumming the leaves with their fore tarsi, they detect specific odor cues. According to Lawrence Gilbert, of the University of Texas, *Heliconius* butterflies (similar to that in Fig. 14–8) locate their passion fruit vine hosts in tropical forests partly in response to their characteristic leaf shapes. These relatively long-lived butterflies return to the same roosting site each night and learn the location of the widely dispersed plants that provide them with nectar and with oviposition sites, returning to these regularly.

Learning probably plays little if any role in host finding in most insects. Most evidence suggests that olfaction is involved and that the insects respond innately to sign stimuli in the form of odors of plant essential oils. In most cases these odors are attractants only within a range of a few centimeters to a few meters. They are difficult to demonstrate experimentally, since insects often are responsive only for a short period following dispersal. However, there is a growing body of evidence as to the reality of such **sign** (or "token") **stimuli.** Sweet clover weevils are attracted to coumarin, an odorous constituent of *Melilotus*, their normal host plant. Adult females of the imported cabbageworm lay their eggs on cabbage, broccoli, mustard, and other members of the family Brassicaceae, all of which produce odorous mustard oils. Instances such as this have given rise to the statement that insects are often "good botanists," capable of selecting plants that are taxonomically related. However, similar essential oils sometimes occur in quite different groups of plants. For example, methyl chavicol, anethole, and anisic aldehyde occur both in citrus and in members of the parsley family, and members of the black swallowtail butterfly group will oviposit and the larvae will feed on members of either group of plants—which may in fact be more closely related than is usually appreciated (Fig. 9–7).

Chemical substances that attract and stimulate attack are often called **kairomones,** defined as interspecific messages that benefit the receiver rather than the sender. Thus they stand in contrast to allomones, which benefit the sender (Chapters 5 and 10) and with pheromones, which are *intra*specific messages.

As we have seen, the olfactory receptors of insects reside mainly in the antennae, and there is evidence that insects are frequently able to distinguish between the odors of many plant species, even when the volatile substances are closely related chemically. Vincent Dethier, now at the University of Massachusetts, has shown that black swallowtail larvae do not confuse six essential oils, each of which is characteristic of one of the food plants of the parsley family. When specific odors reach the antennae of caterpillars, the 16 olfactory receptors respond differentially: Nerve impulses in some increase and in others decrease, producing a pattern characteristic of each substance. These are "generalist receptors," differing from the specialist receptors for pheromones (Chapter 6, p. 151). Response to a particular set of molecules is programmed in the central nervous system, and the behavior released is best described as a chemokinesis or chemotaxis leading to reduced locomotion when the source is reached.

While these remarks apply to many monophagous and oligophagous species, it is probable that many polyphagous species respond to more generalized cues and accept plants on the basis of the presence of phagostimulants (see below) or the absence of feeding deterrents or toxins (see next chapter). Polyphagous species feeding below ground, such as white grubs and wireworms, may respond to respiration products of plants, chiefly CO_2; but selective feeders such as cabbage and onion maggots may respond both to CO_2 and to volatile odors of the host.

Figure 9–7
Plants of the citrus family (a) and the parsnip family (b) produce the same three essential oils (c). These substances are phagostimulants for larvae of black swallowtail butterflies, which will feed on members of either plant family and will attempt to feed on filter paper soaked with these substances. (From *Butterflies and Plants,* by P. R. Ehrlich and P. H. Raven, copyright © 1967 by Scientific American, Inc. All rights reserved.)

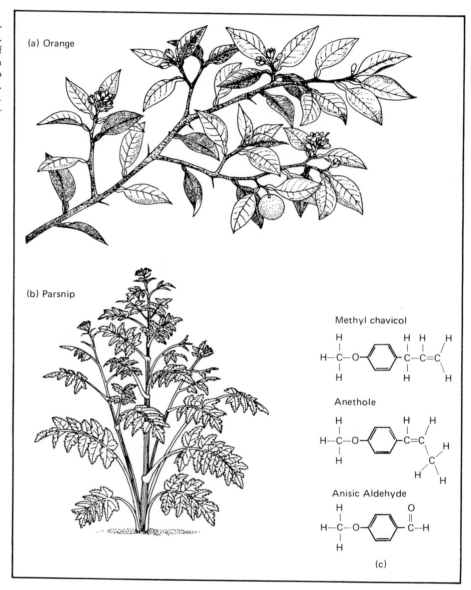

The identification of attractants may have great practical value. The oriental fruit fly, a major pest of many tropical fruits, is known to be strongly attracted to plants containing methyl eugenol. Oddly, it is mainly the males that are attracted to this host plant odor, which is believed to serve as a "rendezvous stimulant," bringing the sexes together for mating. Traps baited with methyl eugenol have been used for many years to monitor populations of oriental fruit flies. On the island of Rota, the fly has been eradicated by dropping fiberboard squares impregnated with the kairomone mixed with insecticide.

Food Plant Acceptance.   A food plant is finally accepted or rejected as a feeding or oviposition site through stimuli received on actual contact with the plant. These may be visual, relating to the shape or color of the substrate, but are more often tactile, olfactory, or gustatory. A. J. Thorsteinson, of the University of Manitoba, Winnipeg, showed that female diamondback moths lay eggs more readily on rough than on smooth surfaces. When the rough surfaces are coated with mustard leaf juices, oviposition is further increased. In nature the moths lay eggs on various members of the mustard family, but to varying degrees, depending on the combined effects of tactile and olfactory cues.

The initiation of feeding is commonly mediated by gustatory stimuli. In caterpillars, taste receptors are located primarily on the maxillary palpi. Removal of the palpi (along with the antennae) often causes these insects to accept plants they would normally refuse. Tobacco hornworms feed only on solanaceous plants in nature, but when the palpi and antennae are removed, they will accept such plants as dandelions and plantains.

Any substance that induces feeding is said to be a **phagostimulant.** Experimental evidence of the importance of phagostimulation has involved placing the substance to be tested on an abnormal plant or on agar or filter paper and recording the amount of feeding that occurs. As long ago as 1910, the Dutch entomologist E. Verschaffelt showed that imported cabbageworm larvae (*Pieris rapae*) would eat nonhost plants when these were smeared with the sap of cabbage plants or with sinigrin, a glycoside that is characteristic of members of the cabbage family. These substances serve both as attractants for ovipositing females and as stimulants for larval feeding.

Eastern tent caterpillars restrict their feeding mostly to Rosaceae, especially to wild cherry. According to Vincent Dethier, the leaves of wild cherry contain hydrocyanic acid (HCN) in sufficient quantity to poison cattle. Yet HCN, in combination with benzaldehyde, constitutes an odorous substance often called oil of bitter almonds, which is a mild attractant and a feeding stimulant for tent caterpillars. When the juice of wild cherry leaves is sprayed on filter paper, the larvae readily eat the paper; they also respond to emulsions of equal parts of HCN and benzaldehyde.

Chemical sign stimuli with similar effects have been identified with respect to many insects. Feeding by certain leaf beetles of the genus *Chrysolina* (Chrysomelidae) occurs only in the presence of hypericin, a substance present in the leaves of Klamath weed. As a result, the beetles have been used effectively in the biological control of these noxious weeds with little danger of them attacking desirable plants (see "Biological Control of Weeds," Chapter 18). The examples we have cited so far have involved **secondary plant substances,** which play no role in the basic metabolism of plants and have no apparent nutritive value for insects. In some cases these have evidently evolved as feeding deterrents. Mustard oils, for example, deter feeding by many polyphagous insects, although *Pieris* butterflies, diamondback moths, and some other oligophagous insects have evolved mechanisms not only for accepting these substances but also for using them as cues for host finding and feeding.

Phagostimulants are, however, by no means always secondary plant substances. Nutrients, including minerals, amino acids, and especially sugars, often elicit feeding behavior. European corn borers, for examples, show a preference for substances containing certain levels of sucrose. Most insects that have been studied have sucrose receptors but are capable of rejecting high concentrations of sucrose, which may be toxic. Often other plant substances are synergistic with sucrose. In diamondback moth caterpillars, for example, sinigrin acts synergistically with sucrose, and in *Pieris* ascorbic acid enhances the effects of sucrose.

The silkworm of commerce, *Bombyx mori,* is one of the best studied of insects, and many experiments with artificial diets have been performed with these insects by Japanese workers. In this instance, substances such as morin and inositol synergistically increase the response to sucrose, although by themselves eliciting no feeding response at all. Yasuji Hamamura, of Konan University, Japan, has identified substances involved as attractants and in the biting and swallowing responses of silkworms. Some of the results of his experiments demonstrating feeding stimulants are shown in Table 9–1, the number of fecal pellets being used as a measure of feeding. When all necessary substances know to elicit attraction, biting, and swallowing are added to agar, along with known necessary growth substances, the larvae can be reared successfully in the total absence of the normal host, mulberry leaves. However, the resulting cocoon shells do not have the normal weight of silk, and development is retarded. It is evidently the diet of the first-instar larvae that is critical (Fig. 9–8). Thus silk producers do best to add mulberry leaves to the diet or to use them exclusively, at least until all dietary elements have been identified.

It is obvious from these and similar studies that even though plant tissue generally provides a suitable diet for many insects, many species require certain specific substances for normal development. Much progress has been made in the study of insect nutrition in the past two or three decades, and it will pay us to look at this important field of study at least briefly.

Nutritional Factors.   Like insects, many vertebrate animals also subsist largely or entirely on plant tissue, and it is interesting to compare the two groups. As we have said, insects have limited ability to digest cellulose, although some have intestinal microorganisms that break down cellulose so that it is available to them (see below). Grazing mammals, on the other hand, have complicated digestive processes that take advantage of bacterial fermentation to break down cellulose into digestible fatty acids. If we compare silkworms with cattle, for example, we find that both are about equally efficient at utilizing the protein in their diet; but cattle (with the aid of symbionts in their rumen) digest over 70% of the crude fiber in grass, while silkworms pass nearly all of the fiber of mulberry leaves undigested in their feces.

Table 9–1. Feeding activity of silkworms on certain diets	
Diet	**No. of feces**
Basic diet (BD) only[a]	0
BD + Sitosterol + R_1[b]	89
BD + Sitosterol + Inositol + R_1	200
BD + Sitosterol + Inositol + Morin + R_1	254
BD without sucrose + Sitosterol + Inositol + Morin + R_1	60

Source: After Y. Hamamura, 1970, "The substances that control the feeding behavior and the growth of the silkworm *Bombyx mori*," in *Control of Insect Behavior by Natural Products,* D. L. Wood, R. M. Silverstein, and M. Nakajima, eds., pp. 55–80. New York: Academic Press.
[a]Basic diet includes all essential nutrients (see p. 227).
[b]R_1 is a group of substances that induce swallowing.

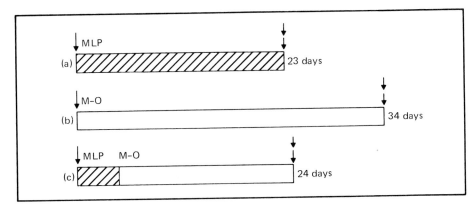

Figure 9–8
Length of larval period of silkmoths from hatching (arrow) to cocoon spinning (double arrow): (a) when reared throughout on artificial diet enriched with mulberry leaf powder (MLP); (b) when reared on a similar diet without mulberry leaf powder (M-O); (c) when reared on MLP during the first instar only. (After M. Kato, 1978, "Phenols as indispensible components of the synthetic diet of the silkworm, *Bombyx mori*," *Entomologia Experimentalis et Applicata*, vol. 24, pp. 285–290.)

Insects also differ from vertebrates in being unable to synthesize cholesterol and therefore requiring this or a similar sterol in their diet. Also, they do not require vitamins D and K, which vertebrates need for bone development. They do require several vitamins of the B group. They also require vitamin C. This is normally available in fresh plant tissue, but some insects are able to synthesize vitamin C. Vitamin A deficiency has been shown to result in reduced visual function in moths; apparently this vitamin is essential for the full development of visual pigments, as it is in vertebrates.

Insects also require water, nitrogen, carbohydrates, amino acids, lipids, and minerals such as iron, phosphorus, zinc, magnesium, and sometimes sodium and others. In a recent review by J. M. Scriber, of the University of Wisconsin, and F. Slansky, Jr., of the University of Florida, it was pointed out that larvae of Lepidoptera show superior relative growth rates on leaves with a water content of from 60% to 90% and levels of nitrogen between 2% and 6%. However, species adapted for feeding on tree leaves subsist on somewhat lower levels of water and nitrogen than do those feeding on forbs. Leaves show seasonal trends in the percentages of water and nitrogen (Fig. 9–9), which may be accompanied by other changes in nutrients and in defensive mechanisms.

Qualitative variations in nitrogen may be as important as variations in quantity. In young, moisture-rich tissues, nitrogen is available in amino acids and soluble proteins, and in nitrates, vitamins, and other substances. However, in older tissues it is largely in the form of insoluble proteins. Much work has been done on the amino acid requirements of insects. Apparently most insects require the ten "essential" amino acids also required by vertebrates. Various amino acids regarded as "nonessential" in the diet are commonly synthesized by insects. However, proline is required by silkworms, cystine by pale western cutworms, and others not among the ten essential amino acids by still other species.

Balance of nutrients is of overriding importance in any diet. Dietary deficiencies may restrict the laying down of sufficient food reserves, in the form of fat body, to carry the insect through the pupal and adult stages; or they may cause reduction in the size and functioning of the endocrine glands, resulting in abnormal development or the suppression of ovarian function. In extreme cases, they may of course cause starvation and death.

Even accepted food plants may not always provide a perfect diet if foliage is old, for example, or deficient in water, or if the plant is growing in soil deficient in certain minerals. Rates of food intake by aphids are known to be affected by the levels of nutri-

Figure 9–9
Relative growth rates of phytophagous insects as related to seasonal trends in leaves as food. Species feeding on tree leaves (oak and cherry) fall to the left and subsist on leaves of lower nitrogen and water content, while those on the right, feeding on annuals (corn and alfalfa), subsist on leaves of higher water and nitrogen content. However, all leaves show decreasing content of N and H_2O as the season progresses. (From J. M. Scriber and F. Slansky, Jr. Reproduced, with permission, from *Annual Review of Entomology,* Volume 26. © 1981 by Annual Reviews Inc.)

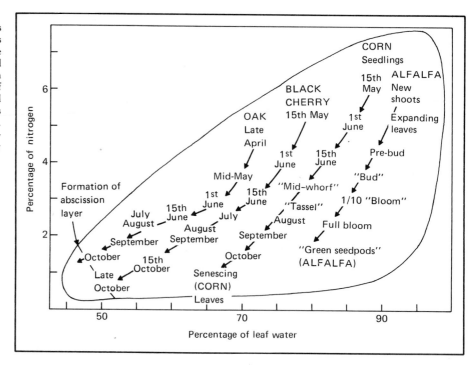

ents (sucrose, amino acids) in the plants. Varieties of peas more resistant to pea aphids, *Acyrthosiphon pisum,* are those containing lower concentrations of amino acids at stages of growth corresponding with the period of aphid attack in the field. Clearly there are advantages to plant breeders in knowing what factors retard the feeding and development of insect pests.

Adults of many insects, such as butterflies and moths, require carbohydrates in the form of nectar; but many also obtain amino acids from nectar. Adults of some insects do not feed at all (giant silkworm moths, for example), subsisting for their short lives entirely on fat stored in the larval stage. Sucking insects, since they do not imbibe the fibrous parts of plants, are able to digest most of their intake; but as we have seen, they frequently discharge a large part of the water and sugars as liquid feces. Nutritional changes in plants, such as those occurring in senescence, may result in the production of winged morphs of aphids that disperse to other hosts (Chapter 16).

Diverse insects have symbiotic microorganisms, either in the gut or in special organs called **mycetomes,** which play various roles in nutrition. In the case of aphids and similar sucking insects, symbiotic bacteria are believed to supply nitrogen, which is not obtainable in adequate quantities in plant sap. In other instances symbionts are known to synthesize B-group vitamins, which cannot be obtained from foods such as dried grains. In termites and some other wood-eating insects, intestinal protozoa and bacteria play important roles in the breakdown of cellulose.

Artificial Diets.   In the past 30 years much progress has been made in developing artificial diets for insects. These provide a tool for investigating the precise dietary requirements of insects. By omitting substances or altering the quantities of substances and determining the effect on development, number of eggs laid by the resulting females, and the like, one may determine the balance of nutrients required in nature. For precise nutritional studies, one must provide a diet in which the chemical nature of all ingredients is known. A basic diet must contain the following: water, carbohydrates, fatty acids, the 10 essential amino acids, cholesterol, choline, pantothenic acid, nicotinamide, thiamin, riboflavin, folic acid, pyridoxine, biotin, vitamin B_{12}, vitamin A, vitamin C, and several minerals. For chewing insects, it must also be made of the right consistency, usually by adding agar or cellulose (both nutritionally inert). Sucking insects require a liquid diet, which must be imbibed through a membrane. In every case the artificial diet must be sterilized and/or supplied with a substance that inhibits microbial growth.

Quantities of these ingredients must be adjusted appropriately, and phagostimulants and additional nutrients added as required—the latter may depend on the ability of the species or its symbionts to synthesize the substance. Research on precise dietary requirements is a very active field at the present time. It was found, for example, that a diet satisfactory for pea aphids was not adequate for rearing green peach aphids. R. H. Dadd and T. E. Mittler, of the University of California at Berkeley, found that the addition to the diet of small amounts of iron, zinc, and manganese, as well as greater care to prevent loss of vitamin C, permitted rearing green peach aphids through many generations. Discoveries of differences between species in their requirements for minerals, amino acids, vitamins, and other substances, combined with knowledge of phagostimulants and the ability to resist or detoxify plant defensive substances, provide the basis for a biochemical definition of an insect's food niche, a matter of much potential importance in the effort to control insects without resort to insecticides.

The development of artificial diets has also been a boon to mass rearing of insects for experimental studies of many kinds, as well as for the rearing of predators and parasitoids of phytophagous insects for biological control (Chapters 12, 18). For mass rearing, it is cheaper and more convenient to use diets of unrefined substances such as wheat germ or soybean meal, which contain protein, fatty acids, minerals, and other essentials. To these water, sugars, and other more specific requirements must be added to provide a balanced diet. Artificial diets for many phytophagous insects are now available commercially, and "cookbooks" for diverse insects are now available (for example, Singh's *Artificial Diets for Insects, Mites, and Spiders*, 1977).

The Advantages of Polyphagy and Monophagy. Polyphagous insects are able to take advantage of the nutrients in a variety of plant species, since they are not cued into specific attractants and phagostimulants. Their food is available with limited searching, and they are in no danger of food shortage if a particular plant is decimated. (We live in a time when many plant species are regarded as endangered, and some are already extinct, carrying with them any insects that are closely tied to them.) A polyphagous species may be able to extend its range widely, even to other continents if opportunities are provided. Some of our worst imported pests are polyphagous—the Japanese beetle and the gypsy moth, for example. So why be monophagous?

Monophagy involves close adaptation to a single host species. The plant is easier to locate, since the sense organs and nervous system of the insect are programmed to respond to specific cues. Mate finding may also be enhanced, since many phytophagous insects mate on the host plant. On its host, the insect is able to compete successfully with other, polyphagous species, since it is able to resist or detoxify the substances the plant has evolved to deter its enemies. Toxins in the plant may even be used by the insect to deter its own enemies, as has happened in the case of the monarch butterfly (Chapter 5) and other insects. Many of the plants utilized by monophagous or narrowly oligophagous insects are, in fact, avoided by generalist feeders because of their toxic or repellent properties. A good example is provided by ferns, which are generally avoided by grasshoppers and other generalists but exploited by a small group of specialists belonging to several orders of insects. There is, incidentally, little evidence that specialized feeders use their host plants with greater physiological efficiency than generalists. For the most part, the advantages they gain are ecological rather than physiological.

Despite the advantages of close adaptation to a specific host plant, it is probable that insects so adapted have limited capacity to reverse the process or to broaden their host acceptance. Their receptors may have evolved so as to perceive only certain molecules, and they may lack the capacity or the necessary symbionts to digest a different plant tissue. They may be able to detoxify (at some metabolic expense) a plant substance that would deter another insect, but that ability would not serve them well on another plant. Some aspects of the relationships of phytophagous insects with their hosts are diagrammed in Fig. 9–10. It should be added that monophagous species can sometimes be put to use in the biological control of weeds; in this instance one must be sure that the insects lack the capacity to switch readily to another host.

Clearly it is not correct to approach these questions from a purely entomological point of view. Plants undergo their own evolution, which is directed by many environmental influences. The development of secondary plant substances, of particular types of foliage or bark, of repellents or toxins—all of these require specific genetic events and many of them are energetically expensive, requiring physiological commitments

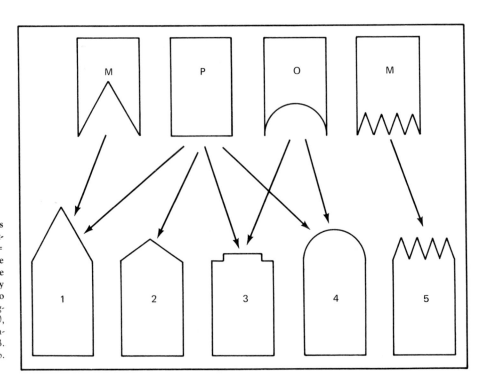

Figure 9–10
Diagrammatic representation of the relationships between insects (top) and their host plants (bottom). M = monophagous; P = polyphagous; O = oligophagous. Plants 1–5 are assumed to produce chemical signals that impinge on receptors in the insects. Plant 5 produces substances to which only one monophagous species responds positively; to the remainder the substances are repellent. (Suggested by a diagram by Vincent Dethier, 1970, "Chemical interactions between plants and insects," in *Chemical Ecology*, E. Sondheimer and J. B. Simeone, eds., New York: Academic Press, pp. 83–102.)

that influence other events (for example, seed production). Under certain conditions it may be advantageous for a plant to protect itself against insect attack or against disease at a certain cost in seed output. When this occurs, there are likely to be groups of insects that themselves find it advantageous (in terms of natural selection) to develop mechanisms for overcoming the plant's defenses, even though it may have disadvantages from other points of view.

In the many tens of millions of years in which insects and plants have been on earth, associations of many kinds have evolved. As a general rule, most plants are attacked by a somewhat limited number of pests, and most insects attack a somewhat limited number of hosts—that is, most are narrowly to broadly oligophagous. Natural selection, like politics, is often a matter of compromises that compound themselves to a degree of complexity that is often hard to fathom. Since polyphagy is prevalent among groups of insects that are usually considered more primitive—bristletails, cockroaches, crickets, grasshoppers, earwigs, and the like—and most of the more advanced groups of insects show varying degrees of food specialization, it does appear that in the course of time the trend has been toward more intimate associations with specific food sources.

The Evolution of Host Specificity.   In the course of evolution, insects have obviously shown the capacity to change hosts and in particular to become closely adapted to one host to the exclusion of others. Some suggestions as to the manner in which this may come about may be gleaned from experiments by Vincent Dethier and his colleagues, then at the University of Pennsylvania. They were concerned with broadly oligophagous species, such as the corn earworm, also called the tomato fruitworm and the cotton bollworm, an indication of the variety of plants attacked by this species. Larvae were fed on artificial diet until the molt to the fifth instar, then transferred to leaves of geranium, dandelion, or cauliflower, all within the normal host range. Here they fed through the fifth instar. After the next molt they were fed on artificial diet for a period, to clear the gut of leaf material, then given a choice of the three kinds of food. The majority chose the leaves they had previously fed on, even after undergoing a molt. Such **induction** of food preferences suggests a mechanism whereby a population may come to select a particular host plant within an initially broad host range.

It should be noted that in early instars, phytophagous species are often fastidious feeders, but in later instars they may be less so. The case of the silkworm has already been mentioned (Fig. 9–8). Eastern tent caterpillars begin feeding on the plant on which the eggs have been laid (usually cherry), but as they approach maturity, they will accept leaves of oaks and other trees on which the females do not normally lay their eggs. There are numerous records of the laboratory production of stocks of phytophagous species that will survive on plants outside the normal host range. Often much mortality occurs on the abnormal host, but after a few generations selection may produce a stock that thrives on the new host.

Insects sometimes adapt to new host plants quite rapidly. In the past few decades, in California, the apple-feeding codling moth has evolved populations that show a preference for plums and for walnuts. Such a switch to a new host may involve a genetic change only in the behavioral response to certain physical or chemical cues; later this may be reinforced by behavioral and physiological "fine tuning." Many species of insects are known to have two or more **biotypes,** or "host races," that is, subpopulations on several related host plants, the insects differing in such matters as time of emergence and responsiveness to specific token stimuli. There is evidence that monophagous species sometimes arise through continued isolation of such host races. This topic is developed further in Chapter 15.

Insect Production and Transmission of Plant Diseases

The effect of insects on their host plants often goes well beyond the actual consumption of plant tissue. This is especially true of sucking insects. As the stylets reach the vascular tissue, salivary fluids are released and may have localized toxic effects or toxic effects that are carried throughout the plant. Insects may also pick up pathogenic microorganisms from one plant and carry them to another. An insect whose feeding produces symptoms of disease is said to be **toxicogenic,** and the condition is spoken of as a **phytotoxemia.** An insect that transmits disease organisms is called a **vector.** A variety of insects, chiefly those with sucking mouthparts, may serve as vectors of plant diseases. This is a very large subject we can pursue only briefly here.

Phytotoxemias. Little is known of the chemistry of the salivary secretions of sucking insects; they are believed to contain enzymes such as amylase but may also contain inhibitors of plant growth substances. Plugging or localized destruction of vascular tissues may also produce disease conditions. The reasons why certain insect–plant associations result in disease and others do not remain obscure. Several kinds of phytotoxemias are recognized, and we shall consider each briefly in the following paragraphs:

Figure 9–11
Leaf curl on plum produced by the feeding of large numbers of green peach aphids. (Drawing by Miriam Palmer, Colorado State University.)

1. Localized lesions at the feeding site, resulting in spotting or stippling. Leafhoppers and mealybugs have been indicted as causative agents of leaf spotting on citrus, pineapples, sunflowers, and other plants. Spots may be paler or darker in color than the surrounding tissue, or in some cases reddish in color. Size of the spot may depend on the time spent feeding at that site, since there is little diffusion of toxin from the point of insertion of the stylets and only localized damage to tissue.

2. Localized lesions with development of more general symptoms. Two-lined spittle bugs (Cercopidae) may cause initial spotting followed by streaking and browning of leaf blades; tarnished plant bugs and other members of the family Miridae often produce disfiguring blotches on leaves or fruits. In these instances vascular tissues carry the toxins some distance from the point of feeding.

3. Malformations of plants, including leaf curling (Fig. 9–11), production of witches' brooms, shortening of internodes, and other distortions. One of the most common examples of this is a browning and curling of leaf edges produced by the feeding of leafhoppers (Cicadellidae) and often called "hopperburn." Several crop plants are subject to hopperburn, such as potatoes, melons, and lettuce. Gross malformations of plants are sometimes caused by sucking insects as well as by mites; they are sometimes difficult to distinguish from certain microbial diseases or from true galls.

4. Systemic conditions, including yellowing, wilting, reduction in growth, or killing of part or all of the plant. These conditions result from translocation throughout the plant of toxins produced by sucking insects. Aphids, leafhoppers, mealybugs, and other Homoptera have been indicted, and crops attacked include celery, sugar beets, corn, and others. Psyllid yellows of potatoes is one of the best known of systemic toxemias. At least 15 or 20 immature potato psyllids per plant are necessary

to produce symptoms. If the psyllids are removed after a week or two of feeding, the plants improve and may recover fully, demonstrating that toxins produced by the insects are responsible rather than introduced disease organisms.

Since phytotoxemias, especially when systemic, often resemble diseases caused by viruses, it is useful to recall that there are several fundamental differences:

Toxemias	Virus diseases
1. Toxin does not reproduce in plant.	1. Virus reproduces in plant.
2. Symptoms subside when insects are removed.	2. Symptoms persist when insects are removed.
3. Recovery common.	3. Recovery uncommon.
4. Degree of injury related to number of insects and length of time they feed.	4. Degree of injury not necessarily related to number of insects or length of time they feed.
5. Disease not perpetuated by vegetative propagation or transmitted by grafting.	5. Disease can be perpetuated by vegetative propagation and transmitted by grafting.

Insects as Vectors of Plant Diseases.   Insects are involved in the transmission of bacterial, fungal, viral, and mycoplasmal diseases of plants. Some of these are all too well known: Dutch elm disease, a fungus disease transmitted by bark beetles; fire blight, a bacterial disease of fruit trees, transmitted by flies and other insects; and potato leaf roll virus, transmitted by aphids. Note that vectors need not be sucking insects, although these insects are especially well suited for inoculating plants, especially with viruses. Insect transmission of plant diseases is a large subject we can treat only briefly here. More extensive treatments are provided in books by J. G. Leach and by Walter Carter, cited at the end of this chapter.

Disease organisms may be spread in one of two ways:

1. Transmission may be **mechanical;** that is, the organisms are borne on the surface of the insect, usually the mouthparts, and in this way carried from plant to plant. When sucking insects are involved, this type of transmission is spoken of as **stylet borne.** It is also spoken of as **nonpersistent,** since the microorganisms survive only a short time unless deposited on or in a plant.
2. Transmission may be **circulatory;** that is, the organisms are ingested by an insect, circulate in the body, and are later discharged in salivary fluids. In some cases the microorganisms multiply in the insect. This type of transmission is **persistent,** in the sense that it may occur over a considerable period of time.

Transmission may be by **inoculation,** in the case of sucking insects, or by **surface deposit** of pathogens, which must then invade the tissue, often through a wound made by the insect. In some cases the relationship between insects and pathogens is **obligate,** which means that the pathogens are transmitted only in this way. In other cases the relationship is **facultative,** since the pathogens can be transmitted in other ways, for example, by wind, rain, or other organisms.

We present here a single example of a disease caused by each major type of pathogen. Several others are listed in Table 9–2. Throughout the world several hundred plant diseases have been reported to be transmitted by insects and mites, and several thousand insect species have been incriminated as vectors. The brief overview given here will serve only as the briefest of introductions to this field.

Table 9–2. Selected examples of plant diseases transmitted by insects

Disease	Vectors	Disease organism	Host plants
Cotton boll rot	Various Hemiptera, such as stink bugs	*Bacillus gossypina* (bacteria)	Cotton
Fire blight	Various beetles, flies, other insects	*Erwinia amylovora* (bacteria)	Apples and other fruits
Soft rot	Beetles, flies, other insects	*Erwinia carotovora* (bacteria)	Cabbage, carrots, onions, other vegetables
Stewart's wilt	Beetles, bugs, other insects	*Erwinia stewartii* (bacteria)	Corn
Ergot	Various flies, beetles	*Claviceps purpurea* (fungus)	Grasses, cereals
Brown rot	Fruit-feeding Coleoptera and Lepidoptera	*Monilinia* species (fungus)	Apples, peaches, other fruits
Dutch elm disease	Bark beetles of two species	*Ceratostomella ulmi* (fungus)	Elms
Oak wilt	Beetles of several kinds	*Ceratocystis fagacearum* (fungus)	Oaks
Corn stunt	Leafhoppers of several kinds	Mycoplasma	Corn
Cucumber mosaic	Aphids of many species	Virus	Cucumbers, various weeds, and other plants
Potato leaf roll	*Myzus persicae* (green peach aphid)	Virus	Potatoes and other Solanaceae
Sugarcane mosaic	Aphids of several species	Virus	Sugarcane, sorghum, millet, corn
Spotted wilt	Thrips	Virus	Tomatoes, other crops, weeds

Bacterial Diseases. Plant pathogenic bacteria are rod-shaped bacilli, usually non–spore formers, which are able to enter plant tissue only through wounds or by inoculation. They have little capacity to live outside their plant host, and transmission is usually mechanical and nonpersistent. Obligatory relationships between the bacteria and a particular insect host are unusual; more commonly a number of different insects and sometimes other agents are involved in transmission. However, the example we select does involve specific vectors.

Cucurbit wilt is caused by *Erwinia tracheiphila*, a bacillus that invades and blocks the vascular bundles of cucumbers and to a lesser extent those of cantaloupes, squash, and pumpkins. The first symptoms are localized wilting; when stems are cut a milky ooze emerges. Eventually the entire plant may wilt and even die, a condition resembling dehydration resulting from drought. This disease occurs in many parts of the world and may be devastating if not controlled. Experiments have shown that when plants are protected from the attacks of two species of leaf beetles, the striped and the spotted cucumber beetle, the plants are also protected from cucurbit wilt. Apparently these beetles carry the bacteria from infected to noninfected plants on the mouthparts or in their fecal pellets, the bacteria gaining entry via feeding wounds.

Fungus Diseases. Relationships between insects and fungi are common and often complex. The relationship may be intimate and mutually beneficial, as in the case of fungi cultivated by certain ants, termites, and ambrosia beetles, or it may involve the casual transmission of spores from one plant to another. A great variety of rots, wilts, cankers, and root infections are produced by fungi that may be transmitted by insects. Unlike bacteria, fungi are often able to penetrate plant tissues directly, without requiring a wound. Transmission is usually mechanical, and sometimes several kinds of insects, or even physical agents such as wind or water, may serve as vectors. The example we select is, however, one involving a specific vector and a mutualistic relationship between insect and fungus.

Blue stain of conifers is caused by fungi of the genus *Ceratostomella*. These fungi cause discoloration of felled timber, reducing its value, and they also invade living trees via holes through the bark made by invading bark beetles. The relationship is beneficial to both beetles and fungi. The latter proliferate in the new host, weakening and eventually killing the tree. This renders the tree more suitable for development of the beetle larvae. When these complete their development, they fly to another tree while carrying the spores. The result is an expanding group of dead and dying trees. The mountain pine beetle of the Rockies and the southern pine beetle of the Gulf states are especially notorious vectors. Both are members of the genus *Dendroctonus*. As we saw in Chapter 5, bark beetles employ a complex chemical signaling system that enables them to attack trees en masse. The trees, in turn, have evolved chemical defenses, which we shall review in Chapter 10.

Virus Diseases. A great number of plant diseases are produced by viruses; in Walter Carter's book *Insects in Relation to Plant Disease*, a partial listing of these occupies 38 pages, while a list of the insect vectors occupies 41 pages. Symptoms of virus disease are diverse; they include blotching and mottling of leaves (termed **mosaic**), leaf curl, tumors (Fig. 9–12), rosettes, distortions of flowers and fruits, yellowing, and necrosis. Symptoms are produced by destruction of tissue cells, often accompanied by abnormal growth of other cells, resulting in disturbances in respiration and photosynthesis. Various inclusions are sometimes visible in prepared sections; these represent accumulations of virus particles (Fig. 9–13). Like all viruses, they are unable to survive apart from host tissue. Transmission may, however, sometimes be mechanical and nonpersistent, in which case either chewing or sucking insects may be involved. The more usual vectors are sucking insects such as aphids, leafhoppers, and thrips. Transmission is frequently circulative and may involve relationship with a specific vector and multiplication within the vector. In some cases the virus is transmitted from one generation to the next transovarially.

Curly top of sugar beets is transmitted by the beet leafhopper, an insect occurring in the southwestern United States but undertaking seasonal migrations into the beet-growing regions of the Great Basin and the western Great Plains (Fig. 16–5). Symptoms include leaf curl, stunting, and distortions of the roots. The only important mode of transmission is by the feeding of beet leafhoppers. The virus is retained within the blood of the leafhoppers through molts, but there is no multiplication within the body. The virus overwinters in various wild plants and in beets that have been left in the field. It is picked up by leafhoppers feeding in the spring and transmitted to seedlings.

Mycoplasmal Diseases. Before 1967, mycoplasmas were classed as viruses, since like viruses they pass through filters capable of retaining bacteria. However, they are now regarded as quite a different group of organisms having some features in common

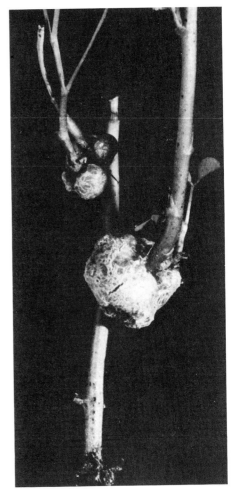

Figure 9–12
Stem tumors on sweet clover plants infected with wound tumor virus. This virus multiplies in susceptible plants and also infects and multiplies in several species of leafhoppers, which serve as vectors and alternate hosts. (K. Maramorosch, Waksmann Institute of Microbiology, Rutgers University.)

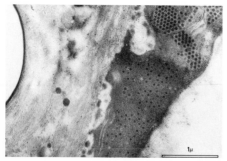

Figure 9–13
Microcrystal formation (top right) of wound tumor virus in the salivary gland of a leafhopper. Below the crystalline formations is a site of virus multiplication. Bar represents 1 micron. (From *Virology* 32 (1967): 363–377, Shikita and Maramorosch. By courtesy of the authors and Academic Press.)

with bacteria: They can be cultured on agar media apart from the host, and they are susceptible to certain antibiotics, especially tetracycline. Mycoplasmas are pleomorphic, the cells undergoing changes in form through their life cycle. The most common disease symptoms are yellowing, stunting, and the development of "witches' brooms." At least 15 plant diseases are now attributed to mycoplasmas, all of them transmitted by leafhoppers.

Aster yellows occurs in many parts of the world and infects not only asters but also plants of at least 40 families, including vegetable crops such as potatoes, carrots, lettuce, and spinach. In addition to yellowing, plants may show dwarfing or various malformations. The common vector in North America is the aster leafhopper, *Macrosteles fascifrons*. Transmission is circulative, with multiplication occurring in the body of the vector. Leafhoppers are not infective until at least nine days after the mycoplasma has been ingested.

Summary

Phytophagous (or herbivorous) insects consume leaves, buds, stems, roots, fruits, and seeds. They feed externally or internally, either by chewing or by sucking plant juices. Scarcely any plant species is immune to attack, and some are attacked by hundreds of insect species. Some insects restrict their attacks to a single plant species and are said to be monophagous; others attack a few, often related species and are said to be oligophagous; while some polyphagous feeders are rather general, unselective feeders. Phytophagous insects may reach their host by moving about and tasting several plants before feeding; or the host may be selected by the mother, who lays her eggs in response to some cue, often chemical; or the insect (such as an aphid) may be born into a feeding aggregation.

Chewing insects may be whole leaf feeders, skeletonizers, miners, tiers, rollers, or folders, or they may bore in stems, trunks, or roots. Larvae may live in fleshy fruits or in nuts or seeds. Sucking insects draw plant sap through their stylets, generally obtaining an excess of water and carbohydrates, which they excrete as honeydew. Gall formers provide a special category of either chewing or sucking insects that, through secretions, stimulate the plant to form an abnormal growth that provides them with food and shelter.

Insects seek a suitable habitat by the use of visual and chemical cues. Within the habitat, they may respond to visual and tactile cues such as leaf shape or roughness. More commonly they are attracted to the host plant through chemical stimuli, often in the form of volatile essential oils, such as mustard oil. Acceptance and feeding are mediated by the presence or absence of gustatory cues termed phagostimulants. Feeding specialists will often respond to phagostimulants in the absence of the host plant itself, for example, if they are smeared on an abnormal host or on filter paper. Phagostimulants are often secondary plant substances, sometimes those with repellency to other insects, but they may also be nutrients such as sucrose.

In terms of nutrition, leaf-feeding insects differ from herbivorous vertebrates in requiring cholesterol or a similar sterol and in not requiring vitamins D and K. They do require other vitamins as well as water, nitrogen, carbohydrates, amino acids, lipids, and certain minerals. Diverse insects have symbiotic microorganisms in their gut or in special organs called mycetomes, which play various roles in nutrition.

In recent years much progress has been made in developing artificial diets for insects. Such diets are useful in the mass rearing of insects for experimental studies as well as for rearing predators and parasitoids for biological control.

Polyphagous insects have the advantage of being able to find abundant food with limited searching and to be able to extend their range widely. However, on a specific host a monophagous species may be a superior competitor, since it is specifically adapted to the signals emanating from that host and to its nutrients. In the course of time there has been a trend toward monophagy, best understood in the light of experiments on induction of food preferences and of observations of host races in nature.

Insects also cause adverse effects in plants by producing localized or more generalized toxic effects via salivary fluids injected while feeding (called phytotoxemias) or by serving as vectors of plant diseases. Phytotoxemias often bear much resemblance to virus diseases, but they do not persist when the insects are removed, and the degree of injury is directly related to the amount of feeding that has occurred. A great many plant diseases are transmitted by insects. These include, for example, fire blight, a bacterial disease of fruit trees transmitted by flies and other insects; Dutch elm disease, a fungus disease transmitted by bark beetles; potato leaf roll virus, transmitted by aphids; and aster yellows, a mycoplasmal disease transmitted by leafhoppers. Transmission may be purely mechanical, or the pathogens may circulate or even reproduce within the insect.

Selected Readings

Carter, W. 1973. *Insects in Relation to Plant Disease*. Second edition. New York: Wiley Interscience. 705 pp.

Dadd, R. H. 1973. "Insect nutrition: Current developments and metabolic implications." *Annual Review of Entomology*, vol. 18, pp. 381–420.

Dethier, V. G. 1980. *The World of the Tent-makers: A Natural History of the Eastern Tent Caterpillar*. Amherst: University of Massachusetts Press. 148 pp.

Ehrlich, P. R. and P. H. Raven. 1967. "Butterflies and plants." *Scientific American*, vol. 216, no. 6, pp. 104–13.

Felt, E. P. 1940. *Plant Galls and Gall Makers*. Ithaca, N.Y.: Comstock. 364 pp.

Leach, J. G. 1940. *Insect Transmission of Plant Diseases*. New York: McGraw-Hill. 615 pp.

Maramorosch, K. 1963. "Arthropod transmission of plant viruses." *Annual Review of Entomology*, vol. 8, pp. 369–414.

Metcalf, R. L. 1979. "Plants, chemicals, and insects: some aspects of coevolution." *Bulletin of the Entomological Society of America*, vol. 25, pp. 30–35.

Singh, P. 1977. *Artificial Diets for Insects, Mites, and Spiders*. New York: Plenum. 594 pp.

Thorsteinson, A. J. 1960. "Host selection of phytophagous insects." *Annual Review of Entomology*, vol. 5, pp. 193–218.

Vanderzant, E. S. 1974. "Development, significance, and application of artificial diets for insects." *Annual Review of Entomology*, vol. 19, pp. 139–60.

Wood, D. L.; R. M. Silverstein; and M. Nakajima. 1970. *Control of Insect Behavior by Natural Products*. New York: Academic Press. 343 pp.

The Defenses of Plants against Insects

Despite the great numbers of phytophagous insect species and the destruction they sometimes cause, green plants continue to dominate the landscape. True, insects are to some extent kept in check by their own natural enemies; but the fact that most plants escape or survive the attacks of insects is often the result of defenses the plants themselves have evolved. Plants and plant-feeding insects have evolved together for many millions of years. Many features of plants, such as particular life histories, leaf forms, and secondary plant substances, have evolved (at least in part) in response to attacks by insects. Certain insects have, in turn, developed mechanisms for overcoming these defenses, even to the point of using "repellent" substances characteristic of certain plants (such as mustard oil) as oviposition cues or phagostimulants, as we saw in the preceding chapter. Plants have also acquired toxins that deter feeding by herbivorous mammals, such as hydrogen cyanide in cherry leaves and cardiac glycosides in milkweeds; but these too may serve as phagostimulants or sometimes may even be used by insects in their own defense, as we saw in the case of the monarch butterfly (Chapter 5, p. 131). When an insect species breaches the defenses of a plant, natural selection may then result in a strengthening of those defenses.

The coevolution of plants and animals is a subject of much current interest. Coevolution has been defined by Daniel Janzen, of the University of Pennsylvania, as "an evolutionary change in a trait of the individuals in one population in response to a trait of the individuals of a second population, followed by an evolutionary response by the second population to the change in the first." Paul Feeny, of Cornell University, describes this aspect of plant–insect relationships as "an evolutionary arms race in which the plants, for survival, must deploy a fraction of their metabolic budgets on defense (physical as well as chemical) and the insects must devote a portion of their assimilated energy and nutrients on various devices for host location and attack."

This chapter considers first some of the physical features of plants that deter insect feeding. Then it reviews the role of plant-produced chemical substances in repelling insects, deterring their feeding, or poisoning them. The effect of nutritional factors on insect feeding will also be considered, as well as the recruitment of natural enemies to assist in the plant's defense. Finally, this chapter discusses briefly a rather special group of plants: those that include insects in their diet—the insectivorous plants.

Morphological Resistance

Morphological, or physical, resistance against insects involves plant structures that interfere physically with the insect's locomotive, feeding, or reproductive functions. These functions may include host selection, feeding, ingestion, digestion, mating, or oviposition. The plant morphological features that act to disrupt these functions may involve plant color and shape or more specialized defensive adaptations such as trichomes, tough tissues, surface waxes, or cell silication. As with chemical factors, physical defenses may act at **close range,** as when plant trichomes prevent feeding or oviposition. They may also act at greater distances, as when host color determines whether or not an insect alights on a plant; in these cases we refer to them as **remote factors.**

Remote Factors.   A number of host cues are apparently used by insects in determining whether or not they will visit a plant. These include color, plant size, shape, and density.

Color.   Most of the research on the use of plant color as a cue in insect visitation has been done with aphids and their relatives. These insects seem to be most attracted to leaves that are yellow-green—that is, those that reflect light with wavelengths within the 500-to-600-nm range. Alate aphids are attracted to leaves reflecting about 500 nm regardless of the plant species. Thus in many cases healthy, dark green plants are less attractive to these insects than yellowing plants under stress.

Generally it is difficult to alter the natural color of a plant as a means of increasing resistance to insect attack. Specific color-related resistance does, however, exist. For example, S. G. Stephens, at North Carolina State University, suggested some years ago that red cotton plants were less attractive to the cotton boll weevil than green plants when both types were grown together. Later, Edward Radcliffe and Keith Chapman, working at the University of Wisconsin, clearly demonstrated that the red cabbage varieties were less susceptible to oviposition of the imported cabbageworm, *Pieris rapae*, than any of the green varieties. Their work also showed that these preferences were not related to the value of the plant as an insect food source, since larval survival was favored on the red varieties. Other workers have reported similar findings. For example, oat cultivars with red tiller bases were less susceptible than others to attack by the frit fly, *Oscinella frit*. This resistance in oats is likely the combined effect of color and pubescence, since the latter has also been reported to inhibit oviposition, as we shall discuss later in this chapter.

As we saw in Chapter 7, Ronald Prokopy has demonstrated that apple maggot flies, *Rhagoletis pomonella*, respond visually to colors, shapes, and sizes. The response evoked by various stimuli varies with age and reproductive condition. His studies revealed that early in life adult flies are attracted to large yellow objects, probably because they are seeking feeding sites (usually honeydew on foliage). Yellow and green may in fact be indistinguishable to the flies, but yellow is more intensely reflective. When the females are sexually mature, they are attracted to dark, apple-sized objects where they normally find mates and oviposition sites. Prokopy has demonstrated that this information can be used in conjunction with sticky traps to monitor the number of flies in an orchard and to reduce populations.

Shape.   As mentioned in the preceding chapter, tropical *Heliconius* butterflies locate their passion vine hosts (*Passiflora* species) at least in part by visual cues relating to leaf shape. Passion vines have a range of defensive chemicals and also have extrafloral nectaries that attract ants and parasitoid wasps, which add to the defense of the plants. *Heliconius* butterflies are among the few insects that exploit *Passiflora* species successfully. Passion vines show unusual variation in leaf shape, both within and between species. Lawrence E. Gilbert, of the University of Texas, believes that diversity in leaf shape has evolved so as to render the vines more difficult for *Heliconius* butterflies to locate. Furthermore, the leaves of some *Passiflora* species closely resemble those of other tropical plants that *Heliconius* caterpillars find inedible, an example of plant mimicry similar to that employed by many insects (Chapter 14).

Close-Range Factors.   Most of the known plant physical-defense factors function only at close range. In most cases these factors make it difficult for the attacking insect to feed or oviposit on the plant. These include such characteristics as thickened cell walls, increased toughness of tissues, proliferation of wounded tissues, solid stems, trichomes, surface waxes, silica in the cell wall, and other protective mechanisms.

Thickened Cell Walls.   Cell walls that are thicker than normal, usually owing to deposition of additional cellulose and lignin, are more resistant to the tearing action of insect mandibles or penetration of the stylets or ovipositor. M. T. Tanton, of the Imperial College Field Station, England, working with the mustard beetle, *Phaedon cochleariae*, found that leaf toughness affected the amount of food eaten in a given time. He also demonstrated that there was increased larval and pupal mortality on leaves of greater toughness and that growth rates were reduced, especially for the early instars, when insects were reared on leaves with higher toughness ratings. Tanton speculated that growth rates of later instars were not affected as much because of the increased power of their mandibles. Researchers have also observed that the cowpea curculio, *Chalcodermus aeneus*, has difficulty penetrating cowpea pods that are thicker than normal, and thus these suffer less damage than pods with thinner walls. Thick hypodermal layers have also been considered a factor in resistance in rice to the rice stem borer.

In some cases thick cell walls inhibit not only feeding but also digestion. For example, Hal Caswell and Frank Reed, working at Michigan State University, found that the grasshopper *Melanoplus confusus* was not capable of breaking down the cells of three grass species that were characterized by thick-walled bundle sheath cells. The cells of these plants passed through the insect's digestive tract largely unbroken and with the cellular contents intact. These workers suggest that the inability of the insects to digest these grasses properly accounts for the lower survival rates on those plants reported by other investigators.

Wound Response.   In some plants the response to wounding by insect feeding results in a proliferation of cells, or other plant products, that may act as a defense mechanism. For example, larvae of the pink bollworm, *Pectinophora gossypiella*, may be crushed or drowned by proliferating cells of injured tissues in certain varieties of cotton.

Another example is evident in trees of the family Pinaceae, which are characterized by the presence of oleoresin systems. Many persons consider oleoresin systems primarily physical mechanisms that "pitch out" adult bark beetles by trapping them in the resin and forcing them out of the tree. However, individual trees within and among species differ in their ability to "pitch out" bark beetles. Healthy, vigorous trees are apparently able to withstand repeated attacks, while unhealthy or injured trees quickly succumb. It is believed that 20% to 30% of the ponderosa pines in the western United States are highly susceptible to attacks of bark beetles, such as the mountain pine beetle, because of a lack of a sufficient resin flow system or oleoresin pressure to counter the attacks.

Stem Characteristics.   Stem-inhabiting insects are sometimes seriously affected by differences in stem characteristics, and in many cases resistance to stem borers is related to the nature of the stem tissues. For example, solid stems are much more resistant to wheat stem sawfly, *Cephus cinctus*, than are hollow stem varieties. Also, the hard, woody stems of some species of *Cucurbita* with closely packed, tough vascular bundles are reported to be the main resistance factors against the squash vine borer, *Melittia satyriniformis*. Apparently, penetration of the stems and subsequent feeding by

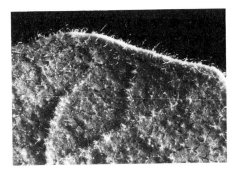

(a)

larvae are inhibited by the structural characteristics of the plant. This seems to be the case, too, with the wild tomato *Lycopersicon hirsutum*, where the thick cortex in the stem prevents the potato aphid, *Macrosiphum euphorbiae*, from reaching the vascular tissue.

Trichomes.   Trichomes are cellular, hairlike outgrowths of the plant epidermis, which may occur on leaves, shoots, or roots. Trichomes are important for various physiological reasons but are of particular value in water conservation and are probably the plant's most important morphological defense against insect attack (Fig. 10–1). Insect species vary greatly in their response to the presence of plant trichomes. These structures may interfere with insect oviposition, attachment of the insect to the plant, feeding, or ingestion. It is generally believed that the mechanical effects of trichomes depend on four main characteristics: density, erectness, length, and shape. Some trichomes possess glands that exude secondary plant products. If these are defensive chemicals, the plant may combine physical and chemical defense in one structure. In other cases, the exudate is a sticky material that acts physically to glue the insect's legs or other body parts together and thus reduce locomotion.

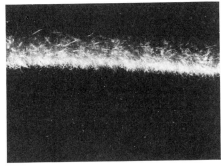

(b)

Small insects or other arthropods with piercing-sucking mouthparts are generally thought to be unable to feed on plants that are pubescent (those with high densities of trichomes) because the "hairs" prevent them from reaching the conductive tissues with their stylets. However, the effect of such plant "hairs" is determined in part by the tissues that the insects feed on. For example, phloem or xylem feeders must insert their stylets deeper into plant tissue to reach their food source than mesophyll feeders. Thus short trichomes may be an effective barrier against the former but not the latter. Obviously the length of the trichome in comparison with that of the proboscis of insect pests is important in determining the effect of this defense mechanism. For example, the trichomes on soybean are commonly about 1 mm long and cover the plant at a density of about 8 trichomes per mm. Because the proboscis of the potato leafhopper, *Empoasca fabae*, a common pest, varies from 0.2 to 0.4 mm in length, it is possible that the trichomes present a barrier to feeding by this insect. However, insects with longer stylets might be less deterred by this defense mechanism.

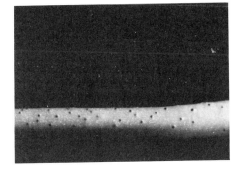

(c)

The presence of trichomes may also be detrimental to insects with chewing mouthparts. M. A. Schillinger and R. L. Gallun, of the U.S. Department of Agriculture, showed that leaf pubescence in certain wheat varieties adversely affected oviposition behavior of adults of the cereal leaf beetle *Oulema melanopus*. Egg viability and survival and growth rates of young larvae were also reduced on varieties having densely pubescent leaves. Females laid fewer eggs on such leaves, and less than 10% of those laid hatched, apparently owing to desiccation. Young larvae restricted to densely pubescent leaves died quickly, and those that did survive grew more slowly than their counterparts on less pubescent leaves. Schillinger and Gallun explained the high mortality by noting that the young larvae had to eat trichomes to reach the epidermis, where they would normally feed. In doing so, they ingested large amounts of cellulose and lignin, the basic constituents of the trichomes, and death resulted from this inadequate diet. Stanley Wellso, also of the U.S. Department of Agriculture, later demon-

Figure 10–1
Movements of first instar larvae of pink bollworm, *Pectinophora gossypiella*, are greatly retarded by pubescence on leaves and petioles. Thus cotton varieties such as TM-1(H₂) (a and b) would be expected to be more resistant to this pest than smooth varieties such as TM-1(sm) (c and d). (R. L. Smith, R. L. Wilson, and F. D. Wilson, from *Journal of Economic Entomology*, copyright 1975 by the Entomological Society of America.)

(d)

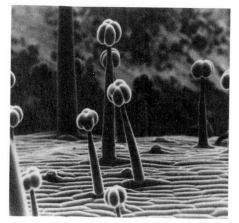

(a)

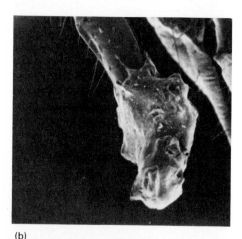

(b)

(c)

strated that cereal leaf beetle larvae that fed on densely pubescent wheat leaves were filled with undigested trichomes and that some of these sharp "hairs" pierced the gut wall. It appears, then, that the trichomes of wheat leaves may act as defense mechanisms in several ways. First, they deter the females from ovipositing on the leaf surface. Second, the eggs that are deposited tend to desiccate, perhaps because they are raised above the leaf epidermis where air currents circulate more freely. Finally, many of the surviving larvae starve because of the nutritive-inadequacy diet of trichomes, and some of those that might survive such a diet may have their gut walls pierced by the spike-like trichomes.

The presence of trichomes is not always detrimental to oviposition. For example, glabrous (smooth) leafed cotton strains are less favored sites for oviposition by *Heliothis zea* and *H. virescens* than are pubescent strains. It has been suggested that the pubescent leaves provide a better foothold for the females and thus facilitate oviposition.

Trichomes may also act as purely mechanical barriers to insect activity. In some instances insects and other arthropods are trapped or impaled by glandular or nonglandular trichomes. Certain varieties of bean plants possess hooked trichomes, and entomologists observed long ago that soft-bodied insects such as aphids or leafhoppers are sometimes impaled on these hooks. More recently E. A. Pillemer and W. M. Tingey, of Cornell University, demonstrated the efficiency of this defense through their studies on potato leafhoppers, *Empoasca fabae*, attacking field beans. These workers noticed leafhopper nymphs "clinging" to the leaves of the plants of some of the varieties. Scanning electron microscope examination revealed that some of the leafhoppers were impaled through the intersegmental membranes of the abdomen; others were captured by trichomes impaled in the tarsal segments of the leg or entangled in the tarsal claws. These workers demonstrated that the frequency of capture and mortality were highly correlated with trichome density of the leaves. They noted that the lower leaf surfaces had higher densities of trichomes than the upper surfaces. Since the leafhoppers feed almost exclusively on the underside of the leaves, the trichomes are ideally suited for defensive purposes.

Glandular trichomes may produce exudates that act as mechanical traps for small arthropods. For example, trichomes on wild potato, *Solanum polyademium*, produce an oozing liquid when contacted by aphids, leafhoppers, or other small arthropods. Upon contact with the air, the liquid hardens around the legs of aphids, and the immobilized insects quickly die (Fig. 10–2). In other cases the glandular materials produced by trichomes cause toxic effects on the insect that encounters them. The trichomes of several species of *Nicotiana* (tobacco), for example, exude materials that produce symptoms in aphids similar to those caused by nicotine poisoning, that is, leg paralysis, loss of equilibrium, and death.

Silica. Plant species in a number of families have silica incorporated into the epidermal walls, and this seems to be an effective defense mechanism against attack by some insects. Rice plants, in particular, seem able to incorporate silica from the soil into the epidermal walls. It has been noted that mandibles of rice stem borers feeding on varieties with high levels of silica were markedly worn. Given a choice between rice varie-

Figure 10–2

(a) Glandular trichomes on petiole of wild potato, *Solanum polyadenium*. (b) Trichome exudate that has encased the labium of a potato leafhopper. (c) Trichome exudate covering the tarsus of a potato leafhopper. (Scanning electron micrographs by W. M. Tingey and R. W. Gibson, from *Journal of Economic Entomology*, copyright 1978 by the Entomological Society of America.)

ties with high and low levels of silica, the stem borers select the latter. Agricultural workers have demonstrated that borer infestations can be reduced by adding silicon to low-silicon soils.

Surface Waxes.

The cuticle of most vascular plants is covered by a thin layer of "waxy" material, mostly of hydrophobic constituents. This layer functions primarily in maintaining the water balance of plants, but it also contains substances that reduce attacks by pathogens and some insects. For example, the normal waxy leaves of broccoli, *Brassica oleracea*, are more resistant to attack by the cabbage flea beetle, *Phyllotreta albionica*, than a glossy-leafed strain. In some cases, however, the waxy surface seems to promote success of insect infestation. The cabbage aphid, *Brevicoryne brassicae*, and the whitefly, *Aleyrodes brassicae*, developed large colonies on the waxy leaves of broccoli but tended not to colonize the nonwaxy strain. It appears, however, that the wax affected only the initial infestation level. Once infestation was established on either plant type, the aphids did not differ significantly in their reproductive success.

Chemical Defenses

As long ago as 1888 a German worker, E. Stahl, suggested that plant chemical systems may be important in the defense of plants against insects. We now recognize that they are indeed the most important deterrents to insect predation. Chemicals that benefit the producing plant may be called **allomones,** in much the same way that the term is used with respect to insects (Chapter 5, p. 114). Allomones may act as repellents, feeding deterrents, toxins, or growth regulators and may also affect the ability of insects to extract nutrition from ingested plant tissue. While there are thousands of chemical compounds that act in one or more of these ways, most can be grouped into five major categories: nitrogen compounds (primarily alkaloids), terpenoids, phenolics, proteinase inhibitors, and growth regulators related to insect hormones (Fig. 10–3).

Nitrogen Compounds.

Some nitrogen compounds, such as nonprotein amino acids, act as antimetabolites. The insect may mistakenly incorporate a nonprotein amino acid and produce an unnatural protein. More commonly, however, nonprotein amino acids act as feeding deterrents. They are especially common in seeds, which are normally rich sources of nutrients for herbivores.

Alkaloids are complex nitrogenous bases of diverse molecular structure occurring in many plants. Their toxic properties have long been appreciated; Socrates was put to death with an alkaloid from hemlock. Alkaloids are among the best known of the toxins that serve as defenses against insects. One of them, nicotine, has a long history of use as an insecticide. Another, tomatine, is a major alkaloid in tomato and many other *Solanum* species. Colorado potato beetles, *Leptinotarsa decemlineata*, are deterred from feeding on tomatine-containing tissue, and if feeding does ensue, beetle mortality may result. Hence, potato beetles are not normally damaging to tomato, while potato (which is closely related but lacks tomatine) is very susceptible to injury. When tomatine solution is infiltrated into potato leaves, resistance results, suggesting that breeding programs to transfer tomatine to potato might be productive in reducing insect damage.

Ragwort, *Senecio jacobaea*, contains toxic pyrrolizidine alkaloids that provide protection from a wide variety of herbivores. Cattle poisoning is commonly associated with *Senecio*. However, some insects such as the cinnabar moth, *Tyria jacobaeae*, are not affected by the alkaloid and feed with impunity on *Senecio*. Moth larvae sequester the toxic principles in their bodies, imparting protection against birds and other predators.

Figure 10–3
Some of the diversity in molecular structure among plant defensive chemicals.

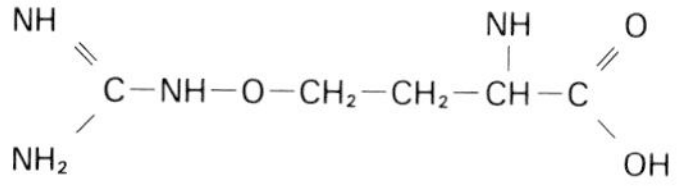

Canavanine: a toxic nonprotein amino acid

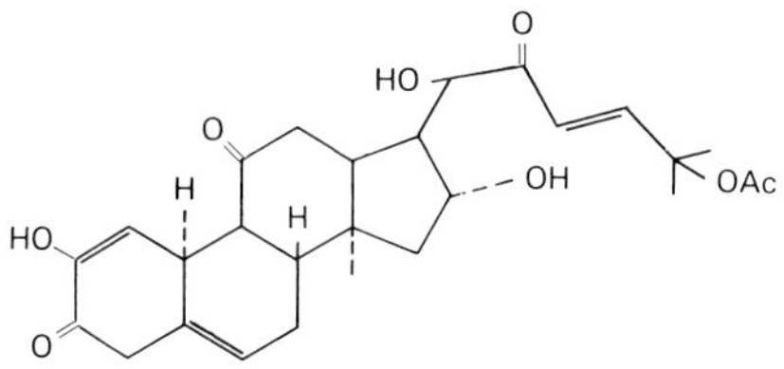

Cucurbitacin E: a bitter, toxic triterpenoid

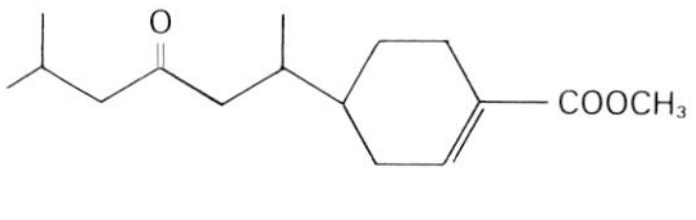

Juvabione: a juvenile hormone mimic

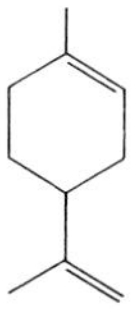

Limonene: a toxic monoterpene

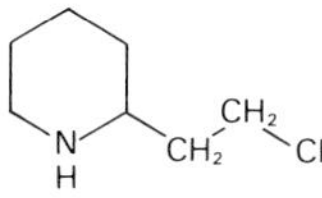

Coniine: a toxic alkaloid

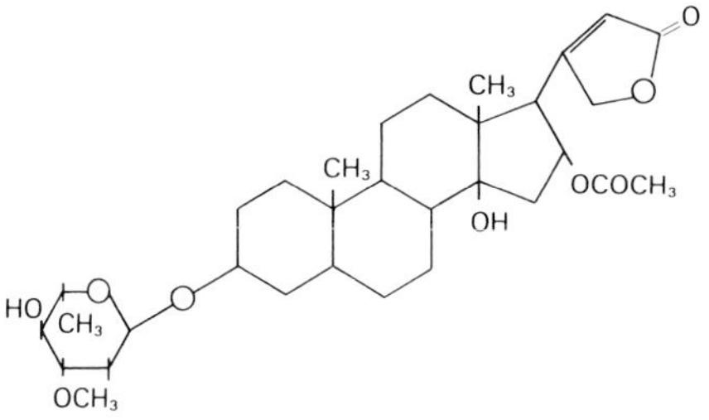

Oleadrin: a toxic cardiac glycoside

Terpenoids.   Terpenoids are among the largest and biologically most important classes of natural plant products. They are widely distributed and extremely diverse, both structurally and functionally. They are nonnitrogenous and use a 5-carbon, branch-chained hydrocarbon as a building unit. Terpenoids function as attractants for pollinators, as feeding deterrents, and as toxins.

Pyrethroids are toxic monoterpenes from *Chrysanthemum*. Dried flowers and various extracts of *Chrysanthemum* have been used for centuries because they possess insecticidal properties. Synthetic formulations are widely used in agriculture and have largely replaced natural extracts. Pyrethroids exhibit low mammalian toxicity but rapid knockdown of insects. Other monoterpenes provide the major defense of some conifers against attacking bark beetles. When trees are attacked, the concentration of toxic or repellent monoterpenes found in resin increases.

The sesquiterpenoid gossypol provides a major defense of cotton against herbivory. Resistance against *Heliothis* species (bollworm and tobacco budworm) is directly correlated with gossypol content. Cotton leafhoppers show more than 50% higher survival on susceptible varieties of cotton than on resistant varieties, and development is more rapid on susceptible varieties. Although low-gossypol varieties have been developed to obtain cottonseed that is suitable as food, where insect damage is a problem high gossypol varieties are recommended.

Cucurbitacins are triterpenoids that impart a bitter taste to plant materials; they are found throughout the Cucurbitaceae (squash family). Cucurbitacins are potent feeding deterrents for a wide variety of herbivores. However, they serve as attractants to cucumber beetles.

One of the most promising natural feeding deterrents to insects is azadirachtin, a triterpenoid isolated from the neem tree, *Azadirachta indica*. It has long been known that migratory locusts avoid feeding on these trees but consume almost all other foliage. Azadirachtin is effective against many species and may be an effective deterrent at concentrations as low as 0.04 ppm.

Phenolics.   Phenolics are nonnitrogenous compounds that contain one or more hydroxyl groups attached to benzene rings. Among the more important phenolics are the flavonoids. One isoflavonoid, rotenone, is used commercially as an insecticide. Others are effective feeding deterrents because of their bitter taste. Tannins are polymeric phenolic compounds that have strong protein-adsorbing properties. Proanthocyanidins (condensed tannins) are feeding inhibitors and also reduce the digestibility of ingested foliage. Hydrolyzable tannins elicit variable responses from herbivores, depending on concentration and herbivore. Anthocyanin provides color to flowers, promoting attraction of pollinators, and is also important in fruit coloration, thus aiding in attraction of fruit and seed dispersers.

The effects of condensed tannins in oak leaves on feeding behavior of winter moth larvae, *Operophtera brumata*, dramatically illustrate the importance of phenolics. Moth larvae feed readily on oak foliage during the spring months, but switch abruptly to other species in mid-June. The switch is correlated with a sharp increase in condensed tannin levels of leaves. Oak trees that are defoliated also produce new foliage containing increased tannin levels. This presumably helps prevent additional defoliation.

Proteinase Inhibitors.   Proteinase inhibitors are proteins or polypeptides that bind to the enzymes that split peptide bonds of proteins. This inhibits the proteolytic activity of the enzymes. Proteinase inhibitors in plants are found in large quantities in seeds and tubers but also occur in foliage. The inhibitory activity of proteinase inhibitors is rather specific for digestive proteinases and thus differs considerably from the generalized protein complexing compounds, such as condensed tannins. Presumably they provide protection against insects, other animals, and microorganisms. The level of proteinase inhibitors in potato plants increases when the plants are attacked by insects; even leaves distant from the site of attack respond. Foliage or other plant parts with elevated proteinase inhibitor levels should be less digestible to herbivores. Some plant species produce a variety of proteinase inhibitors, each having different specificities. Thus these plants possess defenses against a wide variety of herbivores.

Insect Growth Regulators.   **Phytoecdysones** are ecdysone relatives found in plants. Phytoecdysones were discovered in primitive gymnosperms by K. Nakanishi and his co-workers at Columbia University and the University of Tokyo while they were searching for anticancer drugs. The phytoecdysone content in some plants is astonishingly high. One gram of the rhizomes of an oriental fern contain ecdysone activity equivalent to 200,000 grams of silkworm pupae! Several dozen phytoecdysones have been isolated and identified from over 80 plant families.

One insect growth regulator from plants was discovered because of a careful investigation of a biological puzzle. When Karel Sláma brought his insects (*Pyrrhocoris apterus*, Hemiptera) from Czechoslovakia to Harvard, the cultures would not produce healthy adults in Massachusetts. In a beautiful series of experiments, Sláma and Carroll Williams demonstrated that the lack of maturity in these bugs was related to the paper lining the rearing cages. Lining paper made in the United States and Canada was juvenilizing to these bugs, but that from Europe allowed *Pyrrhocoris* to reach adulthood. The papers in America come from different plant sources from those in Europe. Sláma and Williams went across the courtyard to the forestry collections and tested samples of wood from pulp trees for juvenilizing activity. The culprit was balsam fir. North American paper is partly derived from the pulp of balsam fir, and William Bowers, of Cornell University, found that within that pulp is a juvenile hormone mimic (juvabione). When cages were lined with paper from the *New York Times* (which contains juvabione), the *Pyrrhocoris* failed to metamorphose. Paper from the *London Times* (which is not derived from balsam fir and so lacks juvabione) did not inhibit metamorphosis or reproduction. It was later found that juvabionelike compounds are produced by relatives of balsam fir when those trees are infested by aphids. It may be that some of the compounds are produced only in response to herbivore attack.

In a systematic search for insect growth regulators from plants, Bowers later isolated two interesting substances, which he called **precocenes,** from the common bedding plant *Ageratum houstonianum*. When precocenes are applied to the cuticle of some insect species, the cells of the corpora allata are killed. By destroying the source of juvenile hormone, precocenes accelerate metamorphosis to yield a precocious (and sterile) adult. But many species are insensitive to precocenes. Resistant insects can degrade precocenes and thus nullify the effect of the poison.

The Evolution of Chemical Defenses

Many thousands of defensive substances from a wide variety of plants are now known. These are highly diverse in their molecular structure (Fig. 10–3), in their mode of action, and in their effects on specific insect predators. Fortunately a general theory of the evolution of plant defenses has been recently proposed, permitting us to see important patterns in this diversity. This theory was proposed simultaneously and independently in 1976 in two different laboratories that were involved in various aspects of research in the area of plant–herbivore interactions. Paul Feeny, at Cornell University, studied the tannin defenses of oaks in England as well as the mustard oils that occur in various Brassicaceae. David Rhoades, of the University of Washington, and Rex G. Cates, of the University of New Mexico, working together, were investigating grasshoppers, caterpillars, and phytophagous beetles in the deserts of Argentina and Arizona.

Both of these groups converged on the ideas that (1) the defenses of plants are primarily a result of life history characteristics of the plants along with their predictability and availability as food resources to herbivores, and (2) based on the mode of action of natural plant products, there are two major types of chemical defenses that have been evolved by plants. *Predictable* resources are those that are abundant and persistent so they are subject to attack over longer periods of time, such as the stems and mature leaves of trees and other woody plants. In contrast, *unpredictable* resources are more ephemeral, such as plants and plant tissues that grow and disappear rapidly, including annual plants, flowers, buds, and new foliage of trees. One should note that woody plants will contain both ephemeral, unpredictable tissues (buds, new foliage) and predictable tissues (branches, mature leaves).*

The two major types of chemical defenses associated with these are spoken of as **quantitative** and **qualitative defenses,** respectively. Predictable resources that are exposed to phytophagous insects for long periods of time are chiefly characterized by protein-complexing and digestibility-reducing substances (resins, tannins, and the like) (Table 10–1). These are usually produced in high concentrations, hence the term *quantitative defenses*. Mature leaves, for example, remain on the plant for some time and may encounter high levels of herbivory. Consequently selection may favor a large commitment to defense. Autotoxicity problems are reduced in these tissues by storing these compounds in vacuoles or specialized structures, but at considerable energy expense.

In contrast, in ephemeral tissues most of the energy produced is devoted to rapid growth, flowering, and the production of large numbers of seeds that are readily dispersed. Thus natural selection has favored a lower commitment to defense, chiefly to low-molecular-weight toxins. Toxins are active at low concentrations, often no more than 1% or 2% of the dry weight of the plant tissue; hence it is the quality of the defense and not its quantity that is critical. Toxins include alkaloids, mustard oils, hydrogen cyanide, and others (Table 10–1).

In young leaves, for example, selection has favored rapid growth and maturation. During this time little energy is used to form storage structures, and the cells are not well vacuolated. Rapidly growing tissues need a defense system that will not require large amounts of energy for synthesis or storage. Under these constraints, qualitative systems are likely to evolve, since in general toxins are physiologically cheaper to synthesize in the amounts required for defense. These substances are either not toxic to the plants or they can be rendered innocuous to the plants with little energetic effort.

Host Specificity and Plant Defensive Systems.   It has been observed that monophagous and oligophagous herbivores often show strong preferences for the more nutritious, younger leaf tissues that are high in toxins, whereas polyphagous herbivores demonstrate a strong preference for the less nutritious, mature leaf tissues. Investiga-

---

*The terms *predictable* and *unpredictable* are those of Rhoades and Cates (see Table 10–1). Feeny (1976, *Recent Advances in Phytochemistry*, vol. 10, pp. 1–40) used the alternative terms *apparent* and *unapparent*, stressing the detectability of the plants to herbivores. Hence this concept is sometimes referred to as the "apparency hypothesis."

Table 10–1. Essential Components in the Theory of the Evolution of Plant Defenses

I. Predictability of Resources

Unpredictable resources		Predictable resources
1. Annuals, biennials, herbaceous perennials, deciduous trees and shrubs (ephemeral leaf and other tissues)	vs.	Evergreen, long-lived woody perennials, mature leaf tissues, heartwood
2. Short generation times, rapid growth rates, numerous seeds	vs.	Long generation times, slow growth rates, fewer large seeds
3. Available as food resources for a short time period	vs.	Reasonably stable food resource over longer periods of time
4. Unpredictable in space and time	vs.	Predictable in space and time
5. Lower commitment to defensive chemistry (qualitative)	vs.	Complicated, well-developed defensive chemistry (quantitative)

II. Expected Defensive Chemical Types Based on Function

Toxins		Digestibility-reducing substances
1. Alkaloids, hydrogen cyanide, cardiac glycosides, mustard oils, phytoecdysones, etc.	vs.	Tannins, enzyme inhibitors, resins, silica
2. Highly active in low concentrations against physiological systems of animals	vs.	Form complexes with nutrients needed by herbivores
3. Low concentrations (1%–3%)	vs.	Much higher concentrations (5%–50%)
4. Stored in "bound" form in plant tissues or in an inactive form	vs.	Usually compartmentalized in cells or in specialized structures to reduce autotoxicity problems

Source: Modified from Rhoades and Cates, 1976, "A general theory of plant antiherbivore defenses," *Recent Advances in Phytochemistry*, vol. 10, pp. 168–213.

tions suggest that specialized herbivores become well adapted to their hosts and are able to detoxify the chemicals, thereby being able to utilize a much more nutritious tissue. Polyphagous herbivores, however, feed on several plant species, all of which may contain several different toxin systems. It is thought that these herbivores cannot detoxify all of the numerous toxins in their various host plant tissues and therefore are forced to use a less nutritious tissue. The less toxic tissues may be heavily defended by quantitative defenses, but are usually much lower in lethal or growth-retarding chemicals. There are exceptions to these generalizations, but strong evidence does suggest that toxins are most effective against generalized herbivores.

Students of plant-insect systems are often impressed by the intricate detoxification systems that have evolved in some insects, chiefly specialized feeders. Degradation of toxins is accomplished by a variety of enzymes known as mixed function oxidases (MFOs) located in the fat body and the midgut, which is the major area for nutrient and toxin uptake of insects. Mixed function oxidases catalyze numerous oxidative reactions. They are remarkably nonspecific in many cases and are often rapidly induced. Numerous other enzymes are capable of transforming extremely toxic compounds into innocuous chemicals that can be rapidly excreted by insects before poisoning occurs. David Rhoades, of the University of Washington, suggests that the MFO activity in the fat body may act as a secondary system for the detoxification of toxins that have survived the MFO activity in the midgut.

Chemical Defenses in Relation to Stress. Concentration and composition of defensive natural plant products are greatly modified by abnormal stresses that a plant may encounter during its lifetime. Changes in natural product chemistry due to drought, nutrients, shade, pollution, and other stresses may contribute significantly to population fluctuations of herbivorous animals. For example, changes in monoterpene, tannin, polyphenol, and nitrogen chemistry of Douglas fir foliage result in a 35% increase in the dry weight of female western spruce budworms. This increase is directly correlated with a highly significant increase in fecundity and larval survival.

Geographic Variation in Plant Chemistry. Wide-ranging plant species often show variation in the chemistry of their tissues. Ponderosa pines, for example, range throughout most of the western United States and have been shown to be highly polymorphic with respect to monoterpenes. Bark beetles of the genus *Dendroctonus* are serious pests of these pines. Certain of these monoterpenes, such as myrcene, play roles in aggregation and attack by the beetles (see "Aggregation Pheromones," Chapter 5), while another, limonene, is inhibitory or toxic. Kareen Sturgeon, of the University of Colorado, has shown that populations of trees having a past history of attack by bark beetles have a high frequency of trees with high concentrations of limonene, evidently as a result of natural selection for resistance.

Nutritional Effects on Insect Feeding

In addition to variation in concentrations of defensive substances, there are natural and induced changes in nutrient availability that affect herbivores. The mineral elements nitrogen, potassium, and phosphorus are among those that have long been recognized as important. Nitrogen is a key component of protein, and when plant protein is made unavailable to insects because of proteinase inhibitors or condensed tannins, insects do not have access to the nitrogen necessary to create body protein. Potassium and phosphorus are also important for production of organic molecules.

Natural variation in nitrogen, phosphorus, and potassium levels in plants is commonplace. New, rapidly growing plant tissue often has higher concentrations of minerals, and reproductive tissue, such as seeds, is a rich source of nutrients. The attractiveness of these tissues may be reduced, however, by plant defensive chemistry.

Herbivores, in general, tend to select the more nutritious tissues, and thus natural selection favors lowered nutritional quality in plants. Tissues of poor nutritive quality will tend to lower the growth rate of insects feeding on that tissue, increasing their likelihood of succumbing to predators or disease. However, insects feeding on leaves of low nutritional quality may consume many more leaves. Thus in some cases there may be counterselection for the production of increased palatability and nutritional value.

When plants experience stresses, nutrient availability is sometimes enhanced. T. C. R. White, an Australian entomologist, has done much to increase our appreciation of weather-related change in plant nutrient concentration and its effect on insect population ecology. White argues that water stress increases the availability of nitrogen in plants, which results in a more nutritious food and enhanced survival of herbivorous insects. Outbreaks of a variety of insects are correlated with abnormal weather patterns, presumably resulting in plant stress that increases nutrient availability (Chapter 17 and Fig. 17–7).

Agricultural practices also result in modification of plant nutrient content, primarily through fertilization. Application of nitrogen frequently increases host plant suitability, and herbivores such as aphids, scales, mites, leafhoppers, and caterpillars exhibit significant population increases following heavy nitrogen applications. On the other hand, chinch bugs, *Blissus leucopterus* (Lygaeidae), are usually associated with nutrient-poor crops. This may be due to microclimate effects as well as nutrient effects. Fungal epizootics are more likely to occur in dense, well-fertilized crops, as the humidity is maintained at a higher level. Potassium applications often are deleterious to insect populations, while phosphorus applications produce variable results.

Escape in Time and Space

Resource availability, as well as suitability, is important in plant defense against herbivory. Plants or plant tissues that are short lived escape from insect attack in time or space. New, succulent tissue, which is highly favored by aphids, is an example of plant material that occurs for a relatively brief time. Aphid populations often dissipate as plant tissue senesces. Genetic variability in plant development may be a form of "bet hedging," whereby at least some susceptible plants or plant products occur outside the range of normal temporal occurrence of the herbivore.

Annual plants produce seeds and die after a single season. If the seeds do not germinate in the following season, any herbivores that had fed on the plants in the previous season would be forced to find a new food supply. This could pose a significant problem for specialist herbivores, as there often are dramatic differences in abundance of annuals from season to season. Plant seeds frequently can retain their germination ability for many years, while insects usually cannot persist for more than a few months without food. Agriculturalists assist this "natural" process by rotating among susceptible and nonsusceptible crops and by timing planting and harvesting procedures to minimize exposure of susceptible stages to insect pests.

Early-successional plant species usually are short lived, and their distribution is often patchy. Insect herbivores therefore must be very mobile if they are to locate their specific host plant in a mosaic of vegetation types. Late-successional plants, however, tend to be long lived, and many generations of insects could develop on the same plant. Although a late-successional plant, such as a tree, cannot escape readily either in space or time, it nevertheless can reduce its susceptibility. A tree can escape by having susceptible or nutritious tissues available for only a brief period or by presenting a variety of defenses.

Attraction of Natural Enemies

Plants sometimes are attractive to the natural enemies of insects. The presence of natural enemies reduces the susceptibility of the plant to herbivory. The most common method of attracting beneficial insects is to provide them with food, such as nectar from extrafloral nectaries. Extrafloral nectaries are glands located outside the flower that produce water and sugar secretions (Fig. 10–4). Wasps, ants, flies, and moths are

Figure 10–4
Extrafloral nectaries on the petiole of passion fruit (a) and on *Euphorbia* (b). (T. Nishida, from Hawaiian Entomological Society, *Ex.-Proc.* 16:381, 382 [1958].)

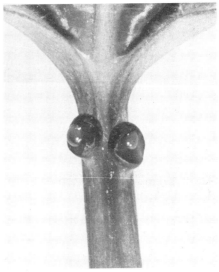

(a)

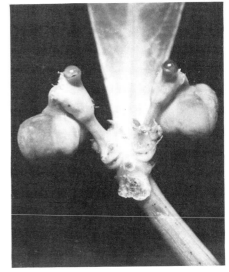

(b)

Figure 10–5
Melon flies (*Dacus cucurbitae*) feeding on extrafloral nectaries of castor bean. (T. Nishida, from
Hawaiian Entomological Society, *Ex.-Proc.* 16:381,
382 [1958].)

attracted to this food (Fig. 10–5). While moths are generally considered to be pests, the presence of predators and parasites may offset an increase in caterpillars. However, in predator- or parasite-deficient systems, nectaries could definitely be detrimental. Nectariless cotton varieties, for example, tend to have fewer pink bollworms than do cotton varieties with nectaries.

Certain species of *Acacia* not only have extrafloral nectaries but also proteinaceous bodies that ants use as food; in addition, the hollow thorns of these *Acacias* provide nesting sites for the ants. The ants aggressively attack other insects, and when these are removed experimentally the plants suffer increased herbivory. Species of *Acacia* that do not harbor ants are commonly defended by cyanogenic glycosides, but these substances are lacking in *Acacias* defended by ants—a complementarity that serves to support the defensive role of both the ants and the glycosides.

It is possible to apply nectar or nectar equivalents to crops that lack nectaries. Honey, molasses, and yeast products have been applied experimentally to potatoes to increase the supply of food available to beneficial insects, with observed reductions in pest numbers.

Entomophagous Plants

Charles Darwin spent much time studying plants that use insects to supplement their nutrition, and he declared that he was "frightened and astounded" by his results. His book *Insectivorous Plants* was published in 1875. We now know that there are more than 500 species of photosynthetic plants that make use of organic compounds from insects they trap and digest. These plants can survive without their prey, but when they are successful "hunters," the nutrients they obtain stimulate more rapid growth. Apparently it is nitrogenous compounds that are of most benefit to the plant. Entomophagous plants are usually found in nitrogen-poor soils, particularly in acid bogs and heavy volcanic clays, and their root systems are not extensive. The acidic sphagnum bogs of the eastern United States frequently have several kinds of these insect-eating plants.

Entomophagous plants trap insects using one, or a combination, of three basic mechanisms. First, a number of relatively normal-looking plants are covered with a sticky exudate that entangles the insects. Second, some plants have developed structural modifications that entrap insects, but the plants do not move. Third, a group of plants has developed mechanisms that move modified leaves to entrap insects. Regardless of the trapping mechanism used, the entomophagous plant must have three essential parts. First, it must have an attractant to entice the insects close enough to capture; second, a system to entrap or entangle the insect; and third, a mechanism to digest the prey.

The fly catcher, *Drosophyllum*, is an example of the plant group that entraps insects with sticky exudates. The stem of this plant is covered with special glands that secrete drops of a very sticky, nectarlike liquid that attracts insects. When an insect lands on the stem, it becomes entangled in the sticky fluid and cannot escape. After an insect has been trapped, a second type of gland, also on the stem, secretes a fluid rich in enzymes that dissolves and digests the insect, except for the exoskeleton. The digestive fluid, along with the dissolved insect material, is then reabsorbed, and the exoskeletal remains drop from the plant. This plant is found in arid regions of Portugal and Morocco.

The pitcher plants, *Sarracenia*, *Darlingtonia*, and *Nepenthes*, capture insects by special structural modifications similar to a pitfall trap. The pitcher consists of the petiole of a leaf modified into a vaselike structure partially filled with water. The leaf itself is usually small, sometimes existing as a caplike structure that prevents excess rain from entering the pitcher. In pitcher plants the attractant is usually a bright spot of color, often purple, yellow, or white, near the opening. This seems to attract insects much as colored flowers attract pollinators. Drops of an attractant nectarlike fluid are sometimes secreted along the rim of the pitcher. Insects that enter the pitcher are prevented from climbing out by numerous stiff, downward-pointing hairs on the inside rim. The proteins of the trapped insect are digested by enzymes secreted into the water, and the products of digestion are absorbed by the inner surface of the pitcher. These plants are common in the sphagnum bogs of eastern and midwestern North America.

Sundew plants of the genus *Drosera* (Fig. 10–6) combine a sticky exudate trap with plant movement into a very effective insect-trapping mechanism. The plant grows in a rosette pattern close to the ground. Each leaf of the rosette is broad and spatulate at the tip. The upper surface is covered with many tentaclelike filaments, each tipped with a drop of glistening, sticky liquid. Insects are attracted to the glistening droplets, land on the leaf, and become entangled in the fluid. When an insect is caught, the other tentacles bend inward, and within minutes all the sticky tips touch the insect. The tentacles then secrete digestive enzymes that break down the insect body parts, and the amino acids are eventually absorbed. The sundew plant responds only to nitrogenous substances and not to objects such as sand, rain, or other nonnitrogenous materials. When digestion has been completed, the tentacles reassume their original position.

The Venus flytrap, *Dionaea muscipula,* is the most striking of the entomophagous plants. Like the sundew, the flytrap grows in the form of a rosette on the ground of bogs in the eastern part of the United States. The leaf of the flytrap has an expanded blade with a hinge down the middle. There is a row of long, stiff spines along the margin of each half of the blade. The leaves have a purple area on the upper surface and also secrete a nectarlike substance. Apparently both act as attractants to insects. In the center of the upper surface of each leaf are three short hairs that act as a triggering mechanism. When an insect touches these sensitive hairs, the leaf quickly folds at the hinge and the two halves come together with the long marginal spines interlocked. The trapped insect is then digested by enzymes secreted from glands on the leaf surface, and the resulting amino acids are absorbed by the plant. For many years the rapid movement of the leaves of the flytrap was thought to involve changes in turgor pressure of the cells near the hinge in the center of the leaf. Recently, however, Stephen Williams, of Lebanon Valley College, and Alan Bennett, of Cornell University, demonstrated that this is not the case. Rather the rapid closure movement involves irreversible acid-stimulated cell enlargement in the midportion of the leaf blade. These workers suggest that such enlargement results from pH changes inside the cells caused by a rapid hydrogen ion pump. When leaves are infiltrated with neutral buffers that keep the pH above 4.5, the leaves do not close in response to stimulation of the trigger hairs.

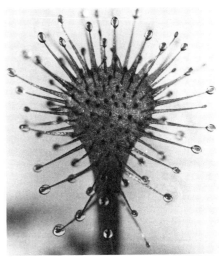

Figure 10–6
Leaf of a sundew plant (*Drosera*). Each hair is tipped with a viscid droplet of secretion in which insects become ensnared, later to be digested by the plant. (Photograph by Thomas Eisner, Cornell University.)

Summary

The fact that most plants escape or survive the attacks of insects is often the result of morphological or chemical defenses that plants have evolved in the course of a long period of coevolution with insects. Morphological (or physical) defenses include such features as color and leaf shape, which may determine whether or not an insect alights on the plant. Or, more often, they involve close-range factors such as thickened cell walls, increased toughness of tissues, proliferation of wounded tissues, solid stems, trichomes, surface waxes, and silica in the cell walls. Research on the effects of trichomes suggests that they may either interfere with oviposition or deter attachment to the plant or the insects' ability to feed.

Chemical defenses, called allomones, include thousands of compounds that can be grouped into five major categories: nitrogen compounds (primarily alkaloids), terpenoids, phenolics, proteinase inhibitors, and insect growth regulators. Alkaloids function as toxins and include such substances as tomatine, which occurs in tomato and some other *Solanum* species and which deters feeding by Colorado potato beetles. Terpenoids are toxins and feeding deterrents and include pyrethroids produced by *Chrysanthemum*, which have low toxicity to mammals but serve as effective insecticides. Phenolics include rotenone, also used as an insecticide, and condensed tannins, which deter feeding and inhibit digestion. Proteinase inhibitors are proteins or polypeptides that bind to enzymes that split peptide bonds of proteins, inhibiting the proteolytic activity of the enzyme. Insect growth regulators include phytoecdysones and juvabione, which influence development through their resemblance to molting and juvenile hormones in insects, respectively.

It has recently been proposed that chemical defenses are of two major types. Qualitative defenses are those substances (chiefly toxins) that are produced in small quantities, mainly by plants that are unpredictable resources for insects, including ephemeral growths such as buds and new foliage of trees, annual plants, and flowers. Quantitative defenses are produced in high concentrations and include protein-complexing and digestibility-reducing substances (resins, tannins, and the like). These are chiefly produced by plants that provide predictable resources, such as perennials, trees, and mature leaf tissues. These are exposed to insect attack for longer periods, and selection has favored a higher commitment to defense.

By and large, monophagous and oligophagous herbivores show a preference for unpredictable resources, such as younger leaf tissues, that are often protected by toxins; such insects often have intricate detoxification systems that can cope with these specific defenses. On the other hand, polyphagous species feed on a variety of plant species that may contain a diversity of defenses. Lacking specific detoxification mechanisms, they are forced to use less nutritious tissues that may be heavily defended by quantitative defenses.

In addition to variation in concentrations of defensive chemicals, there are natural and induced changes in nutrient availability that affect herbivores. Levels of nitrogen, potassium, and phosphorus are especially important. Stresses resulting from unusual weather patterns may thus be associated with insect outbreaks.

Plants may sometimes attract the natural enemies of herbivores through the development of extrafloral nectaries or other mechanisms, thus reducing their susceptibility to herbivory. Certain plants, such as some acacias, have hollow cavities that are colonized by ants that serve to reduce herbivory caused by both insects and browsing mammals.

More than 500 species of plants supplement their diet with organic compounds from insects they trap and digest. These entomophagous plants use one or a combination of three basic mechanisms: They may be covered with a sticky exudate; they may have structural modifications that entrap insects; or they may exhibit leaf movements that enclose insects. Examples include pitcher plants, sundews, and Venus flytraps.

Selected Readings

Bell, E., and B. Charlwood, eds. 1980. *Secondary Plant Products.* New York: Springer-Verlag. 664 pp.

Cates, R. G., and D. F. Rhoades. 1977. "Patterns in the production of antiherbivore chemical defenses in plant communities." *Biochemical Systematics and Evolution,* vol. 5, pp. 185–93.

Darwin, Charles. 1889. *Insectivorous Plants.* New York: D. Appleton-Century. 462 pp. (First edition published in 1875).

Eisner, T. 1967. "Life on the sticky sundew." *Natural History,* vol. 76, no. 6, pp. 32–35.

Futuyma, D. J. 1983. "Evolutionary interactions among herbivorous insects and plants." In *Coevolution,* D. J. Futuyma and M.

Slatkin, eds. Sunderland, Mass.: Sinauer. pp. 207–231.

Gilbert, L. E., and P. Raven, eds. 1975. *Coevolution of Animals and Plants.* Austin: University of Texas Press. 246 pp.

Harborne, J. B., ed. 1978. *Biochemical Aspects of Plant and Animal Coevolution.* New York: Academic Press. 435 pp.

Harborne, J. B. 1982. *Introduction to Ecological Biochemistry.* Second edition. New York: Academic Press. 278 pp.

Heslop-Harrison, Y. 1978. "Carnivorous plants." *Scientific American,* vol. 238, pp. 104–14.

Jermy, T., ed. 1976. *The Host-Plant in Relation to Insect Behavior and Reproduction.* New York: Plenum Press. 322 pp.

Levin, D. A. 1976. "The chemical defenses of plants to pathogens and herbivores." *Annual Review of Ecology and Systematics,* vol. 7, pp. 121–59.

Rosenthal, G., and D. Janzen, eds. 1979. *Herbivores: Their Interaction with Secondary Plant Metabolites.* New York: Academic Press. 718 pp.

Stack, A. 1980. *Carnivorous Plants.* Cambridge, Mass.: The MIT Press. 240 pp.

Wallace, R., and R. Mansell, eds. 1976. *Biochemical Interaction Between Plants and Insects.* New York: Plenum Press. 425 pp.

Williams, S. E., and A. B. Bennett. 1982. "Leaf closure in the Venus fly trap: an acid growth response." *Science,* vol. 218, pp. 1120–22.

Insect Pollination of Plants

O ne of the most highly developed and fascinating associations in biology is that between insect pollinators and flowering plants. The association has been instrumental in the evolution of many plant and insect species and has resulted in the development of numerous intimate and specialized relationships between members of the two groups. In many cases the insects and plants involved have developed elaborate structural or behavioral mechanisms that ensure pollination. While these mechanisms are interesting from a scientific standpoint, their practical value cannot be overstated. Without these plants and the insects that pollinate them, the basic quality of our environment and lives would be greatly altered. In this chapter we shall first discuss the coevolution of this association and the advantages it conveys on the partners involved; then we shall consider the mechanism, results, and importance of pollination; lastly we shall describe the biology and attributes of the basic pollinator groups.

Coevolution of Insects and Flowering Plants

Plant pollination by insects is a coevolutionary process that has been ongoing for over 200 million years. Fossil records show that winged insects were abundant in the Carboniferous period, long before any flowerlike structures were present. The earliest flowering plants were probably fertilized by lightweight, wind-borne pollen. Such pollen would have been difficult to collect by insects and may not have been an important food source for them at that time. Early insect pollination was undoubtedly accidental, with plant-feeding insects bumbling into anthers, becoming contaminated with pollen, and transporting a few grains to the next plant visited. Since these plant-directed vectors were much more efficient than the randomly directed wind, there must have been enormous selective pressure on plants toward the development of new and more effective pollinating mechanisms. An early step in this direction was probably the development of sticky pollen grains that adhered to the insects and thus would have been carried more readily on the insects' bodies to other flowers where some would have been transferred to the sticky or feathery stigmas of those flowers. Insects would also find this sticky pollen more accessible as a food source. The flowers eventually began to secrete small amounts of sweet fluid (nectar), and thus floral visitation by insects was more encouraged. Later, or perhaps concurrently, the flowers also developed attractive odors that increased insect visitation, and it was probably then that insects become important pollinators. With time the flowers acquired colors that made them stand out from the green plant and allowed them to be more easily seen by insects. Paralleling these changes, of course, was an evolution in the senses and behaviors of associated insects, especially with regard to the perception of colors and odors and in the ability to associate these characteristics with food.

In general, insects' perception of color differs from that of humans in that their visual range is shifted toward the shorter wavelengths of the electromagnetic spectrum. Although they cannot perceive long red wavelengths, they do see shorter ultraviolet ones quite well. Honey bees, and presumably many other insects, distinguish fewer colors than humans do. They normally respond to only four areas of the spectrum: yellow-green (650–500 mμ), blue-green (500–480 mμ), blue-violet (480–400 mμ), and ultraviolet (400–310 mμ). Since honey bees and probably other bees do not perceive pure spectral red, it is not surprising that few pure red insect-pollinated flowers are found in Europe and Asia, where bees are the major pollinators. The native flowers of that area are mostly yellow, blue, or purple-red—colors readily perceived by bees; deep red flowers, like the field poppy, are visible to bees only because they reflect large amounts of ultraviolet. On the other hand, most pure red flowers are pollinated by birds, especially hummingbirds, which are confined to the Americas. Birds are known to have vision especially sensitive to the red wavelengths. An exception is a group of red flowers pollinated exclusively by butterflies, which are among the few insects tested that have been shown to perceive this color.

Ultraviolet colors appear to be an important characteristic of insect-pollinated plants. Recent investigators of pollination biology have discovered that there is a whole world of ultraviolet in the plant kingdom, invisible to humans but clearly of great importance to plant-visiting insects. When plain flowers are photographed with film sensitive to ultraviolet light, many are seen to have striking colors and/or patterns that probably act to attract bees (Fig. 11–1). Specialized floral patterns called **nectar guides,** which appear to be used by insects in locating the nectar source of a particular flower, have been recognized for many years. However, the use of the photographic technique described above has revealed that ultraviolet nectar guides are quite common and are probably more important in encouraging floral visitation by insects than those we can see. Obviously, floral colors and other characteristics have evolved in conjunction with insect senses, and we shall not completely understand this complex association until floral characters can be considered from the standpoint of what insects perceive.

Continued coevolution of insects and plants has resulted in a broad range of relationships between the two groups, from accidental visitation to very intimate associations between some members. In extreme examples the relationship becomes a form of obligate symbiosis, in which both the plant and the insect are completely dependent on each other for survival.

The Yucca Moth.

A famous example of the latter is the obligate association between the Spanish dagger, or yucca plant, and the yucca moth. All species of yucca are American, and those east of the Rocky Mountains are pollinated exclusively by a single species of moth, *Tegeticula (Pronuba) yuccasella*. The only food of the larva of this moth is provided by the yucca ovules, which grow abnormally large in the neighborhood of the moth's egg. Because an unpollinated yucca flower soon dies, pollination of the flowers by the moth is necessary for survival of the larvae. Pollination by casual insect visitors does not occur, because the plant's vaselike style requires that pollen be carefully placed for fertilization to occur. However, the yucca moth has developed a behavior pattern that results in pollination of the yucca flowers and survival of her offspring. After these small, undistinguished nocturnal moths mate, the female visits the large, creamy white, pendulant yucca flowers to collect pollen. Through evolution the mouthparts of the moth have been greatly modified into curved tentacles that cannot

(a)

(b)

Figure 11–1
(a) Blossom of a marsh marigold photographed with a normal camera and as we would see it. (b) The same blossom photographed with a television camera sensitive to ultraviolet light, as an insect might see it. (T. Eisner, from *Science*, vol. 166, p. 1172, copyright 1968 by the American Association for the Advancement of Science.)

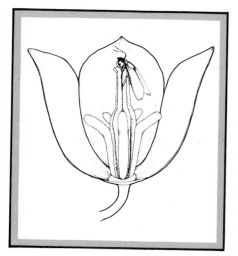

Figure 11–2
Cross section of a yucca blossom, showing a female pronuba moth stuffing a mass of pollen (gathered from another blossom) down the style. (Drawing by Peter Eades.)

be used to feed on either the plant's pollen or its nectar. Apparently the adults do not feed at all and live only a short time. The curved tentacles of the yucca moth are used instead to collect the sticky pollen from the open anthers of a yucca flower and to form it into a ball. The female moth holds this under her head with her tentacles and front legs, and generally carries the pollen ball to a second flower. There she examines the plant's ovary, and if she determines it is suitable, bores into it with her ovipositor and lays an egg. She then climbs up the vaselike style and thrusts some pollen down the tube, thus pollinating the flower (Fig. 11–2). This process is repeated several times per plant. Since the moth usually moves to a second plant before applying the pollen, greater genetic diversity is ensured for the yucca through cross-pollination than if the moth had pollinated flowers of the same plant. An individual female will repeat this process of egg laying and pollination on a number of flowers, apparently laying only a few eggs in each, with the result that there are normally more developing seeds than the moth larvae can consume. When full grown, the larvae leave the seed pod and enter the soil, where they overwinter as pupae, and the "excess" yucca seeds ripen and disperse. The emergence of adults from one season's moth brood occurs over a three-year period after pupation, thus ensuring survival of some individuals even if the yuccas fail to flower, as happens some years. In this relationship the moth pollinates the plant and ensures seed production, while the plant provides food and shelter for the insect larva.

Fig Insects.   A similar obligate symbiotic association is found between plants of the genus *Ficus* (fig trees) and their exclusive chalcidoid wasp pollinators of the genus *Blastophaga* (Fig. 11–3). The fig "flower" actually consists of numerous tiny imperfect flowers arranged inside a pear-shaped receptacle, the small opening of which is frequently closed with flexible scales. In the wild fig there are three types of receptacles, each associated with the reproductive cycle of the plant's pollinators, which can only develop in the fig flowers. The first type of receptacle is formed in winter. It contains many neuter (modified female) flowers and a few male flowers, the latter located at the entrance of the receptacle. Females of the fig wasp enter the receptacle, lay eggs in the neuter flowers, and die. The larvae develop in the flowers' ovaries, maturing to adults in the spring. The wingless, nearly blind males emerge first, and individuals crawl about until they locate the pupal cocoons of females. The male chews a hole in the cocoon and inseminates the female before she emerges. He dies in about a day, never leaving the flower. The female emerges shortly after copulation and leaves the receptacle, acquiring pollen from the male flowers at its entrance. A second type of receptacle has now appeared on the fig, containing either a mixture of neuter and female flowers or flowers of the female type only. Wasps that enter the fig lay eggs on both types of flowers, but only those in the neuter flowers develop. Oviposition activities also result in the accidental pollination of the female flowers, and these develop seed. Wasp development in the neuter flowers occurs as before, but this time with the inseminated females emerging in the fall and going to the third type of receptacle. This small receptacle contains only neuter flowers, in which the females lay their eggs. The larvae develop successfully in these flowers, emerging in the winter, and the cycle continues.

The fig tree has developed a number of unusual characteristics that promote the survival of its wasp pollinators. For example, the flexible scales located at the entrance to the receptacle probably discourage entry by predators and parasitoids and thus increase the survival of the wasps. The plant has also evolved special flowers (the neuters) in which the developing larvae feed, with the winter receptacles entirely devoted to this use. Generally each species of fig is pollinated by a single species of fig wasp. This species-to-species relationship offers strong evidence for the coevolution of these plants and their insect associates.

Flower Constancy. Another effect of coevolution is the flower constancy exhibited by some pollinators, especially bees. Where flower constancy is high, the pollinator restricts foraging to one plant species during single trips or for longer periods of time. A major result is to increase the probability of fertilization, in that pollen deposited in a flower by a visiting bee is likely to be from the same species rather than from an "alien" species of plant. High levels of flower constancy may be seen in the behavior of individuals and colonies of honey bees, although other bee species exhibit similar behavioral patterns. The constancy of a bee is best estimated by examining the types of pollen on the bee's body after a foraging trip, and such studies reveal an amazing fidelity for several bee groups (Table 11–1). Not only do individual bees work the same flower species on separate trips, as indicated by their pollen loads, but entire honey bee colonies, or groups of colonies, may work one flower species for 10 to 11 days or until the source is depleted. Such constancy must influence the plant species composition in an area, since flowers visited selectively by the bees would be more productive and thus eventually out-compete less favored plants. Plant communities and bee populations have become established in patterns similar to the coevolution of individual flower species and their insect pollinators.

Flower constancy is of value to both the plant and the insect, and there is likely to be a coevolutionary trend in that direction. The general pattern seems to have been the evolution of floral structures that favor certain types of efficient pollinators and discourage less efficient ones. Natural selection has also favored the pollinators with more efficient food-gathering behavior. A bee's fidelity to a plant species whose whole local population is synchronously providing nectar and pollen is more efficient than random flower visitation. Bees can readily learn to go to the food source without wasteful efforts and are thus able to visit more flowers per unit time. In extreme cases of constancy, developing larvae are thought to have specialized in metabolizing floral products of individual plant species and are able to utilize the resources more efficiently and develop more quickly.

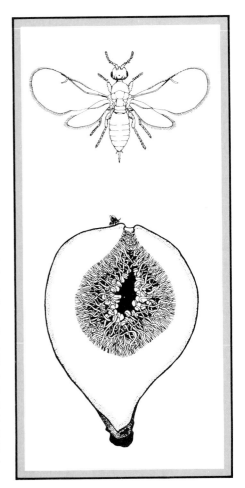

Figure 11–3
Schematic drawing of a fig inflorescence, with a female fig wasp at the entrance. An enlarged drawing of the female wasp is shown at the top. (Drawing by Peter Eades.)

Table 11–1. Flower constancy as indicated by the percentage of individuals in a sample containing pollen from one plant species

Genus of bee	Percentage pure pollen loads
Andrena (a solitary bee)	54.0
Apis (honey bee)	81.0
Anthophora (a solitary bee)	20.0
Bombus (bumble bee)	55.0
Halictus (a solitary bee)	80.5
Megachile (leafcutter bee)	65.0

Calculated from data presented by V. Grant, 1950, "The flower constancy of bees," *Botanical Review*, Vol. 16, pp. 379–398.

Structural plant adaptations that facilitate pollinator constancy are selected for in that nectar and pollen production can be more efficient, since fewer insect groups must be "fed" to achieve pollen transfer. Nectar and pollen sources are usually partially protected by floral parts, and this conserves the food resources for the specialized pollinators of the plant species. An early development in flower constancy is thought to be the fusing of petals, or other floral parts, into a tube around the stamens. This reduces the number of general insect visitors and lowers competition in favor of those that have special adaptations to enter the blossom, thus reinforcing flower constancy.

Insects that are genetically programmed to exploit only certain species of flowers are said to be **oligolectic** (as opposed to **polylectic** species, which exploit flowers of many kinds). Such insects tend to have emergence times synchronized with the flowering periods of their hosts, and there are often intricate behavioral and morphological adaptations for gathering pollen from the host plants. Oligolectic bees tend to be plentiful in arid regions, when many plant species flower simultaneously following rains.

Pollination

The sexual organs of plants are found in the flowers. Flowers that carry both the female pistil and the male anthers are said to be **perfect. Imperfect** flowers, on the other hand, may possess either the pistil or the anthers, but not both. Plants that possess imperfect flowers of only one sex, either male or female, are called **dioecious.** Those having imperfect flowers of both sexes on a single plant are called **monoecious.** As one would expect, insects and other pollen vectors are generally most important to plants that have imperfect flowers, particularly those that are dioecious. However, many perfect flowers also profit from insect pollination and have evolved mechanisms to ensure that it occurs.

Pollination refers only to the transfer of pollen from anther to stigma and technically does not include fertilization, since the latter indicates the successful union of two gametes to form a zygote. Pollination can occur, and frequently does, without fertilization. The term *pollinator* refers to the *vector*—a bee, for example—that carries the pollen from flower to flower. *Pollinizer,* on the other hand, refers to the *plant* that produces the pollen.

There are two basic types of pollination in nature. In **self-pollination,** pollen is transferred from the anther to the stigma of the same flower, to other flowers on the same plant, or to flowers of another plant having the same genetic makeup. The latter situation frequently occurs in the fruit industry, where an entire orchard may theoretically be composed of plants reproduced vegetatively from a single parent and that are thus genetically identical. Self-pollination may occur without any external agent involved, or as a result of some vector.

Cross-pollination involves the transfer of pollen from the anther of one plant to the stigma of another having a different genetic makeup, and it always involves an external agent, usually wind or insects. Cross-pollination may take place between varieties of the same species, as frequently occurs among the several closely related varieties of blueberries, or between less closely related plant varieties, such as cauliflower and cabbage. However, if pollination occurs between species with greatly different genetic compositions, fertilization does not occur. Self-pollination is generally thought to result in less genetic diversity and plant vigor and thus in many instances is less desirable than cross-pollination.

When pollen with the proper genetic components is placed on a stigma, the pollen grain produces a **germ tube** that grows down through the style to the **embryo sac** in the flower's ovary. There it discharges two male cells, one of which fertilizes the female egg cell, or **ovule.** The second male cell unites with sterile female cells, called **polar bodies,** to form the **endosperm** nucleus. The result of this second union is the formation of the seed endosperm, which is the nutritive tissue used by the fertilized egg during seed germination.

Fruit formation is stimulated by fertilization and is a result of secondary growth in the ovary and nearby floral structures. In most cases pollination and fertilization are necessary for fruit formation to occur, although plant growth hormones and other artificial materials have been used successfully in some plants to stimulate fruit development without seed formation. In some species normal fruit formation is the result of the combined development of several, or even many, ovules, and fertilization of all ovules is necessary. Fertilization of only a portion of the ovules results in lopsided, deformed, or stunted fruits. In these cases adequate pollination is very important, and sometimes several visits by pollinators are necessary to ensure proper fruit development.

Importance of Insect Pollination

Plant succession, animal diversity, and soil composition are all interrelated facets of plant survival and reproduction. The role of insects in plant reproduction is far-reaching, involving not only plants and their insect associates but in the longer term the whole ecology of an area. Here we shall consider the value of insect pollination to the insects themselves, to plants, and to ourselves.

Importance to Insects.   A number of insects, but especially bees and adult Lepidoptera, have developed total dependence on floral products for food. **Nectar,** which is composed mainly of the sugars glucose, fructose, and sucrose, with traces of other materials such as proteins, salts, and acids, provides the sole energy source for many flower visitors. Some nectar-feeding insects obtain protein from plant foliage, meat, excrement, or blood, but bees are totally dependent on pollen for this cell-building component of their diet. This dependence has led to many specializations in structure and behavior, both of the bees and the flowers, as will be discussed below. These specializations have given rise to frequent speciation in the evolution of insect pollinators, and many of these insects could not survive without the plants with which they evolved.

Some insect-pollinated flowers provide advantages to their visitors in addition to food. Peter Kevan, now at the University of Guelph, Canada, demonstrated that flowers of the arctic plants *Dryas integrifolia* and *Papaver radicatum* act as solar reflectors, the corollas focusing heat on the plants' reproductive structures. These flowers track the sun during the day, following it at the rate of 15 degrees of arc per hour as it moves across the sky. Both flower species have open bowllike corollas in which several anthophilous insects commonly bask. Kevan's research showed that all basking insects developed considerably higher body temperatures than would be expected, and he suggested that the extra warmth is valuable for increased metabolism and greater mobility. The insects also obtained food in the form of nectar and sometimes pollen from the plants, and transported pollen from flower to flower during their foraging trips.

Importance to Plants.   Without insects as pollinators, many plant species in natural areas would eventually become extinct, ultimately reducing animal as well as plant diversity. It is known that insect pollination is extremely important to the continued survival of forbs of the grasslands, to shrubs and herbs of temperate forests, and to desert flowering plants. George Bohart, of the USDA Bee Research Laboratory at Logan, Utah, has pointed out that the most drastic effect of the elimination of pollinating insects would be in uncultivated areas, where many plants that act to enrich and hold the soil would soon die out. The value of this association is especially great in the arid western United States, where even minor plant damage can cause serious soil erosion.

Peter Kevan notes that although there is only meager information on the relationships between native pollinators and plant species, it is clear that disruption of the pollination mechanism would have far-reaching consequences to whole ecosystems. For example, without pollination seeds and fruits of some species would be eliminated, thus reducing food levels for animals dependent on these structures and, of course, preventing reproduction of many plant species. E. C. Martin, apiculturist of the U.S. Department of Agriculture, notes that hundreds of wild flowers, weeds, trees and other noncrop plants are apparently insect pollinated, with such complex interrelationships that a serious estimate of the value of insect pollination in native areas is almost impossible. Martin concludes that without such pollination we would live in a very different, less productive, and less interesting world.

Importance to Humans.   Many of the world's food crops, such as rice, wheat, and corn, are wind pollinated and do not benefit from insect visitation. However, most fruits, vegetables, and nuts cannot be produced commercially without insect pollination. In addition, many of the animal products that we consume, including beef, pork, poultry, and dairy products, are produced by using insect-pollinated legumes, such as alfalfa, clover, and trefoil, as major food sources for the animals. It has been estimated that when all sources are considered, about one-third of our total diet is directly or indirectly dependent on insect-pollinated plants. It is generally believed that modern agricultural practices in the United States, especially the widespread use of pesticides, have reduced native pollinator levels to the point that they are no longer adequate for proper pollination of these plants. As a result, fruit, vegetable, and seed producers commonly rent honey bee colonies to place in their fields during the blossom period to ensure adequate pollination. In many cases crop yields have been drastically increased by such practices. For example, researchers in Michigan demonstrated that blueberry yield can be increased as much as 200% by providing additional pollinators in the form of honey bees, as compared with yields obtained under normal conditions.

Insect pollination may provide advantages other than increasing crop yield. When pollinators are abundant, a greater proportion of early flowers set fruit, which results in earlier and more uniform ripening. With high pollinator levels approximately 90% of a blueberry crop may be harvested after two pickings, when under normal conditions the 90% level is not reached until the fifth harvest.

It is apparent from numerous studies that without insect pollination, many foods we take for granted would not be available, because the articles either could not be produced at all or could only be produced at prohibitive costs.

Pollinators

Hymenoptera: Bees.   The most important pollinating insects in temperate areas are the bees, which collect pollen and nectar throughout the summer to feed themselves and their young. This group, which is totally dependent on flowers for survival, has developed a number of structural and behavioral modifications to facilitate the collection of these plant products.

Most bees have abundant branched setae, or hairs, on their bodies that act to collect and hold pollen during floral visits. Often these occur in localized patches, forming a specialized pollen-collecting area, or **scopa,** with the location on the body varying according to the group. For example, solitary bees in the genera *Colletes* and *Andrena* have scopal hairs on the propodeum, hind trochanters, femora, and tibiae, while leafcutter bees (Megachilidae) have the scopa located on the underside of the abdomen. The most intricate specialization is the **corbicula,** or pollen basket, with its associated structures, found in bumble bees and honey bees (Fig. 11–4). The corbicula consists of the broad hind tibia, which has its smooth, slightly concave surface surrounded by a fringe of long, curved hairs. An associated "tool," the pollen comb, consists of a row of spines at the lower end of the tibia. Below the pollen comb, on the first tarsal segment, which is also broad and flat, is a dense "brush" of hairs on the inner surface. With this device pollen is brushed from various parts of the bee's body and mixed with a bit of nectar. It is then removed from the brush with the pollen comb on the opposite leg. The pollen-nectar mixture is next pressed upward into the corbicula by means of a long "spur" at the upper end of the broad tarsal segment. The process continues until the baskets are filled with the sticky pollen-nectar mixture when the bee returns to her nest.

Bees collect nectar with a long, tonguelike **proboscis** (Fig. 1–5b). The bee proboscis is not a permanent structure as in most sucking insects, but instead is a temporary organ created by bringing together the galeae, labial palpi, and glossa to form a tube for ingesting liquids. When the proboscis is not in use, it is withdrawn by pulling the basal parts up behind the head and folding the distal structures back against the prementum.

The activity of social bees centers around the collection of nectar and pollen. The abilities of bees, especially honey bees, to tell time, distinguish colors and shapes of flowers, communicate with their sisters, and in general find their way about evolved as means of increasing foraging efficiency. These and other characteristics make social bees the most important and effective pollinators of flowering plants. In the section below we shall discuss examples of social and solitary bees in some detail.

Figure 11–4
A bumble bee gathering pollen. The pollen has been collected in a mass in the corbicula, or pollen basket, on the hind tibia. (Photograph by Bernd Heinrich, University of Vermont.)

Honey Bees. Honey bees are probably the most important pollinators of commercial crops. They are maintained by humans because they produce and store honey in large quantities and, in recent years, because of their value as pollinators. Honey bees are especially valuable because the entire colony, except for the males, survives the winter, and thus high populations are available in early spring when pollination requirements are high.

During the summer a honey bee colony normally consists of 15,000 to 100,000 sterile female workers, a single fertile female queen, and a few hundred males, or drones. These bees live in a nest made of parallel wax combs of hexagonal cells that contain the larvae and stores of honey and pollen. In nature colonies occur in hollow trees, caves, or similar protected areas.

As we saw in Chapter 8, three different types of individuals are found in a colony of honey bees. **Males,** or drones, inseminate virgin queens but otherwise seem to serve no useful function for the colony; they are driven from the hive or killed by workers in the fall. They cannot even defend the hive since they have no sting, which, being a modified ovipositor, is found only in females. The **workers** develop from larvae that received **royal jelly** (often called "bee milk"), a high-quality food produced by hypopharyngeal glands of worker bees (Fig. 5–1), for the first two to three days of larval life;

thereafter they are fed a pollen-honey mixture called **bee bread.** Workers are reared in normal wax cells later used to store honey or pollen. Adult workers perform all the labor in the colony except for egg laying.

Worker bees seem to go through something of an "apprenticeship" in that their duties tend to change as the bees grow older and their ability to do certain tasks improves with practice. There is much variation, but usually during the first few days as an adult the worker's only task is janitorial, cleaning cells so that they can be reused. At about three days of age she becomes a nurse bee, first feeding bee bread to older larvae and later, as her hypopharyngeal glands develop, feeding one-to-three-day-old larvae royal jelly. When the wax-producing glands on the underside of her abdomen develop, she becomes a comb builder. Later still she becomes a receiver bee, taking nectar from successful forager bees and transferring it to cells or performing any of a number of warehouse-type duties associated with food collection. Just before becoming field bees, some workers act as guards at the hive entrances. Bees become field foragers at around 10 to 34 days of age and normally continue in that role for the rest of their lives, possibly only three to four weeks. Usually the specialized food glands in the head and wax glands in the abdomen atrophy and are no longer productive by the time a bee becomes a forager. However, bees can adjust their activities to some extent to meet the needs of the colony. If necessary, workers can continue to produce royal jelly for over 80 days when no young bees are available to take over this duty. Conversely, it is possible for hypopharyngeal glands of older workers to reenlarge after they have atrophied and for these bees to reassume nurse bee duties. Similarly the bees can redevelop wax glands and become comb builders to meet the needs of the colony, and bees as young as four days have been known to become field foragers. Apparently bees can change physiologically if necessary, although usually the physiology of the bee determines her role in the colony.

The **queen** is the largest bee in the colony and is responsible for all the egg laying. She can be distinguished from workers by her size and the absence of pollen baskets on her hind legs. She cannot perform any of the worker bee tasks and even must be fed by her workers, as she is unable to feed herself. The queen is genetically similar to the workers, and her different physical and behavioral characteristics are a result of her rearing environment and larval food. The queen is reared in a specially constructed queen cell, a long, tapered wax cylinder that usually hangs from the bottom of the comb. Queen larvae are fed royal jelly throughout their development and do not receive bee bread at all. A new queen leaves the hive a few days after emerging to mate high in the air with drones. She will often mate on several successive days, and perhaps with many drones, during this period but does not normally mate again after she begins egg laying. The sperm received from the drones are stored in her spermatheca and are released as needed to fertilize the eggs. A good queen may lay as many as 2500 eggs per day during the summer and often lives several years. When her egg-laying ability begins to decline with age, the workers usually replace her by rearing a new queen.

A honey bee colony reproduces by swarming. Under conditions of crowding or fast population growth, a new queen is reared by the workers. After she has matured, about half of the bees and the old queen will leave the colony and fly en masse to a nearby site, frequently a tree limb (Fig. 11–5). There the majority of bees will wait while scout bees seek an appropriate place for a new home. Eventually the scouts return, and the swarm flies off to establish a "new" colony. The "old" colony remains at the original site with the new queen and continues normal activities. During recruitment to a new nest site, worker bees employ much the same signals ("dances") as they do in recruitment to food sources (Chapter 8).

Figure 11–5
A swarm of honey bees clustered on a branch. A free swarm such as this will normally move to a new nest site that has been located by scout bees. (Photograph by Bernd Heinrich, University of Vermont.)

Bumble Bees.   Bumble bees are generally considered to be efficient pollinators, but populations are normally too low to pollinate large areas of agricultural crops. Also, the number of bees fluctuates greatly from year to year and area to area, and consequently commercial growers find bumble bees undependable general pollinators. However, bumble bees are regarded as one of the most efficient individual pollinators of many crops, especially tree fruits, where their large size improves the chances of pollination during nectar collection visits. Some researchers also believe that bumble bees are better than honey bees as cross-pollinators because they tend to work only a few flowers on a plant before moving to another, rather than continuing to work blossoms of one plant for long periods. Bumble bees are generally less valuable as pollinators of early blooming plants, however, because their colonies are annual, and thus early spring populations are low when pollination requirements of most commercial food crops are high.

The overwintering bumble bee queen emerges from hibernation in the spring and begins feeding on nectar and pollen. As a result, her ovaries develop and she begins to search for a nesting site. Some bumble bee species select sites below ground—an abandoned rodent nest, for example—while others choose concealed sites on the surface, such as under thick grass or in a shallow depression. After selecting the site, the queen digs a small cavity in the center, where she places a pollen mass. She constructs a wax cup on the pollen, in which she lays several eggs; then she seals the cup with more wax. The larvae feed on the pollen mass and on additional nectar and pollen introduced by the queen into the wax cup through temporary openings. The queen enlarges the cup as the larvae grow so that they remain enclosed. In about 10 days the larvae spin cocoons and pupate, and the queen removes and reuses the wax to build more egg cells on top of the pupae. New adults emerge after about 10 days. Total growth period from egg to adult requires about three weeks, although this is dependent on temperature and food availability. The old pupal cocoons become storage cells for pollen and nectar.

The first bees to emerge are all workers. At first they care for the larvae that are developing from a second egg group laid by the queen, but after a few days they begin to forage for pollen and nectar. When the foraging workers are collecting sufficient food, the queen "retires" and stays in the nest where her only duties are to lay eggs and care for the developing larvae. As the population of the colony grows, the number of eggs laid by the queen also increases and in general is adjusted to the number of workers available to care for the larvae. Males and new queens are produced near the end of summer. These mate, and the new queen develops large fat bodies that serve as her food reserve for the winter. In late summer she leaves the maternal nest, locates an overwintering site in a protected place, and goes into hibernation. There are usually several new queens produced by an individual colony that will survive, but the old queen, the males, and all the workers die with the onset of winter.

Solitary Bees.   Solitary bees are valuable pollinators of specific crops in many parts of the world, but like bumble bees their value is limited because of great fluctuations in population levels. However, certain species have proved so valuable that great efforts have been made to increase populations artificially, and with some success.

In studies on the relative efficiency of alfalfa pollinators, George Bohart found that honey bees visit an average of 7 to 17 flowers per minute, compared with 10 to 30 for bumble bees and 9 to 40 for the leafcutter bees (species of *Megachile*). Perhaps more important, researchers have found that honey bees generally visit alfalfa blossoms for nectar rather than for pollen and that they pollinate very few of the blossoms they

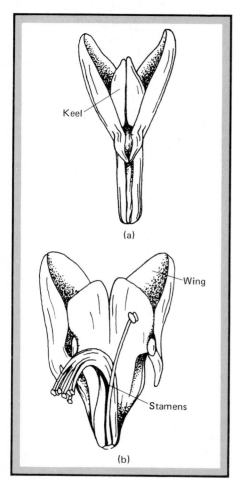

Figure 11–6
An alfalfa flower untripped (a) and after having been tripped by a bee (b). (Drawing by Peter Eades.)

visit. For example, leafcutter bees are reported to pollinate 95% to 100% of the flowers visited, compared with about 80% for certain bumble bee species; but only 2% or fewer of the blossoms visited by nectar-gathering honey bees are pollinated.

The inefficiency of honey bees as pollinators of alfalfa seems to be in part a learned response. The alfalfa flower is constructed such that the staminal column is held under pressure within the keel (Fig. 11–6). The keel makes a convenient landing pad, and when it is pressed down by the weight of a visiting bee, the stamens are released and snap forward, striking the bee on the underside of the head and depositing pollen there. This process, known as tripping the flower, ensures proper pollination and is generally necessary for seed set. After a nectar-gathering honey bee has been hit by the staminal column and/or trapped between the column and the flower petal several times, she learns to collect nectar from the side of the flower and thus to avoid tripping it. Nectar-gathering honey bees that do not learn this avoidance technique within a few days usually desert the crop for a more hospitable food source. Solitary bees, especially leafcutter and alkali bees, primarily collect pollen rather than nectar. These bees cannot normally collect pollen from an alfalfa flower without tripping it and are therefore efficient pollinators of this crop.

The life histories of solitary bees vary tremendously depending on the species involved. Here we shall present details on only two species important in alfalfa pollination: a cavity-nesting leafcutter bee *Megachile rotundata*, and the ground-nesting bee *Nomia melanderi*. *Megachile rotundata* was accidentally introduced into the eastern United States about 1930 and spread westward, reaching Oregon in 1958. On this continent *M. rotundata* occupies a variety of nesting sites but generally is found in hollow stems, beetle burrows, nail holes, or similar tunnels in wood. *M. rotundata* adults emerge in late May when alfalfa is in bloom (Fig. 11–7). After mating, the female selects a nesting site and makes a series of cells in it. She lines the walls of the cells with oblong leaf cuttings (hence the name) glued together with a salivary secretion. The cuttings are usually obtained from alfalfa, perhaps because it is often abundant, but other plants, including roses, are also used. (Perhaps you have seen the near-circular cuts made in rose leaves by related bees.) Each cell is filled with a nectar-pollen mixture on which a single egg is laid. The cell is then capped with 3 to 10 circular leaf cuttings, and another cell is started. The line of cells is closed below the top of the cavity by another cap of circular leaf cuttings. The larvae feed on the nectar-pollen mass in their individual cells, reaching maturity in late summer and then changing into the overwintering prepupal stage. In the spring the bees pupate, the adults emerge through the tunnel entrance, and the cycle continues. Interestingly, the bee that develops from the last egg laid in the tunnel, and therefore the youngest by as many as several days, is the first to emerge as an adult. Emergence continues on down the series of cells, and thus bees do not destroy the cells of their not-yet-emerged siblings to get out of the nest.

The alkali bee *Nomia melanderi* gets its common name from the fact that it nests in tunnels dug in alkali soil in the western United States (Fig. 11–8). Adults emerge from the overwintering state in the ground between late June and mid-July. The female mates and soon after begins digging a nesting tunnel in the soil, usually in the area where she originated. She completes the main burrow during the night and the next day digs a cell, provisioning it with a ball of nectar-pollen mixture. The following day she lays an egg on the ball and seals the cell. A completed nest usually consists of a main vertical tunnel with 3 to 4 branches and 15 to 20 cells. The larvae mature as they feed on the nectar-pollen mixture and overwinter in the prepupal stage. In the spring the bees pupate, the adults emerge, and the cycle continues.

Proper nesting sites of *N. melanderi* have narrow limits, especially regarding soil moisture. Under the right conditions nests may be highly concentrated, with over 500 entry tunnels per square meter in some natural areas. In order to be satisfactory, the nesting sites apparently require constant underground moisture extending up to the surface. It seems that alkali soil over an impermeable soil layer supplied with underground water provides the correct moisture levels, with the salt contained in the soil drawing moisture up at a relatively constant rate. These conditions occur infrequently and probably account for the heavy use of acceptable sites.

Artificially Increasing Bee Populations. The need for more efficient food production throughout the world has increased the need for pollinating insects. Also modern agricultural practices, especially the trend toward monoculture, have placed great demands on pollinator populations, and at the same time the use of pesticides has reduced the numbers of these valuable insects. In many areas researchers have attempted to protect existing pollinators and also to increase populations by artificially creating more favorable habitats and nesting sites. In the following section we present some of the approaches and methods used for a few pollinator species.

Increasing Honey Bee Populations.   The methods used to increase honey bee populations are well known and have changed little for several hundred years. Essentially beekeepers provide a nesting cavity, or **hive,** consisting of a rectangular wooden box containing a series of removable wooden frames (Fig. 11–9). These are often furnished with a wax foundation, to ensure that the bees will build their combs on them. The frames provide support for the fragile wax combs, allowing them to be removed without serious damage. The hive also has a removable top and bottom, providing easy access to the bees and honey. During the summer, as bees collect nectar and process it into honey, the beekeeper will add more boxes, or supers, to the top of the colony to provide additional storage space. Later the beekeeper will remove the "superfluous" honey for personal use, leaving enough for the bees to feed on over the winter. In addition to providing a nesting area, the beekeeper may also provide some protection from diseases and predators and, in more recent years, from pesticides. Generally honey bees survive well without assistance, however.

One major reason that honey bees are such important pollinators is that populations can be quickly increased in an area where they are needed. This can be done simply by moving hives of bees from one area to another or by buying "packages" of bees from southern bee producers to increase the number or strength of colonies.

Increasing Bumble Bee Populations.   Bumble bees have been especially damaged by the destruction of nesting and overwintering sites caused by intensified land use and the use of insecticides. In addition herbicides have destroyed many of the wild flowering plants on which bumble bees depend, especially in early spring. Growers can encourage bumble bees by planting small plots of nectar-producing flowers to help colonies survive periods of low food levels and by leaving areas of uncultivated land for nesting and overwintering sites. Reducing the level of pesticide applications is also important for protection of all bees. The use of artificial nest sites to increase bumble bee populations has been investigated by several researchers. These workers placed cans or similar objects containing nesting material in the soil to attract ground-nesting species, and wooden boxes, also containing nesting material, on the surface for above-ground nesters. A number of variations on both themes have been tested and some researchers have had as many as 50% of above-ground and 72% of below-ground sites accepted. Other approaches include the establishment of laboratory colonies for move-

Figure 11–7
An alfalfa leafcutter bee, *Megachile rotundata,* visiting an alfalfa blossom. (U.S. Department of Agriculture.)

Figure 11–8
The alkali bee, *Nomia melanderi:* **(a)** an adult bee at the nesting site; **(b)** cells dissected from the soil, showing eggs and larvae on pollen masses. (U. S. Department of Agriculture.)

(a)

(b)

ment into the field and artificial hibernation of queens to increase overwinter survival. Although these efforts have been reasonably successful, the prospects of producing bumble bees for use as pollinators on a commercial scale are not yet feasible. At the present time even the most successful methods involve too much effort relative to the number of bees produced to justify the cost.

Artificial Nests for Solitary Bees.   The use of artificial nesting sites to encourage populations of solitary bee species has been successful in some areas of the western United States. This is particularly true for *Megachile rotundata* and *Nomia melanderi.* Artificial nests for *M. rotundata* are most often made by drilling holes of the proper diameter in a large block of wood (Fig. 11–10). The bees occupy the tunnels as though they were hollow stems, and in areas where nesting sites are a limiting factor such nests have been used to increase populations substantially. Generally the nesting boards are mounted on posts provided with an overhanging "porch" for protection from the sun, wind, and rain and screened to prevent bird predation. Most boards are painted in a checkerboard pattern or otherwise marked to help the bees orient toward their individual tunnels. A great advantage of the nesting boards is that they can be moved to new locations where the bees are needed, as is done with hives of honey bees. The cost of increasing populations of *M. rotundata* is high, but the value of the bees for pollination is such that many alfalfa seed producers in the west have made use of this unusual approach to bee management.

Construction of artificial nesting sites for *Nomia melanderi* is considerably more difficult than for *M. rotundata* because of its ground-nesting habits and specific site requirements. However, research by William Stephen, of Oregon State University, George Bohart, and others has resulted in successful methods for construction of nesting sites for these bees. Their recommendations are basically attempts to duplicate conditions found at natural nesting areas. Essentially a pit is dug and an impermeable layer created by covering the bottom with waterproof plastic. This is covered with a shallow layer of gravel, then one of sand, and the remainder of the pit is filled with soil. Water pipes run down to the plastic from the surface to provide the necessary underground moisture. Salt is mixed with the top 5 cm of soil to draw water upward at the proper rate to meet the moisture requirements of the bees (Fig. 11–11).

Such beds have been quite successful. For example, an artificial site created by William Stephen, and the ample food supplied by an adjacent alfalfa field, resulted in more than 2600 nests per square meter. Stephen estimated that a well-populated artificial site 8 by 15 meters would provide sufficient bees to pollinate 16 hectares of alfalfa. Such a field would require from 100 to 200 hives of honey bees at present recommended levels.

Other Hymenoptera.   Hymenoptera other than bees are less intimately associated with flowers, but many wasps and ants feed on floral products during a part of their lives. Social and solitary wasps, for example, utilize insects or other meat for a protein source but usually visit flowers for nectar. Most wasps are smooth bodied and therefore are less efficient pollen collectors from a structural standpoint than are the hairy bees. In general wasps are not dependent on floral products for survival. An exception is an unusual wasp group, the Masarinae, most of which, like bees, provision their nests exclusively with pollen and nectar. Unlike bees, however, masarine wasps are not very hairy and have no pollen-collecting structures on the body. Instead the female carries the pollen-nectar mixture in her crop, regurgitating it at the nest. Many masarines forage on only one or a few plant genera, and many species have specialized

Figure 11–9
A beekeeper about to open a hive. He is fully clothed and wears a bee veil. He is puffing smoke into the entrance to calm the bees.   (J. W. Brewer and W. T. Wilson, 1981, from *Beekeeping in the Intermountain Region*, Colorado State University Experiment Station Bulletin 507A.)

Figure 11–10
An artificial nesting site for leafcutter bees. Many of the holes have been occupied by the bees, which have closed them off with plugs of leaf fragments.   (E. H. Erickson and W. T. Wilson, 1972, *Management of the Alfalfa Leafcutter Bee in Colorado*, Colorado State University Experiment Station Bulletin 552S.)

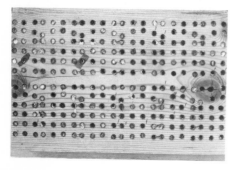

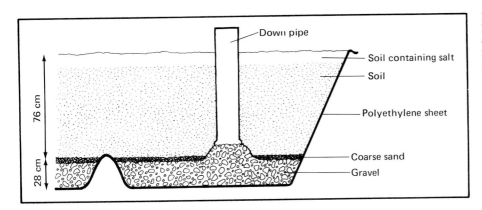

Figure 11–11
Diagrammatic cross section of an artificial bed for colonizing alkali bees. (Redrawn from W. P. Stephen, 1965, "Artificial beds for alkali bee propagation," *Oregon Agricultural Experiment Station Bulletin* 598, 20 pp.)

mouthparts with which they extract nectar from these plants. Most members of this group are not common insects, however, and it is apparent that the group has been less successful than bees at exploiting flowering plants as food sources. Regardless, these and other wasps are frequent nectar-directed floral visitors and undoubtedly account for a certain amount of pollination of native and commercial plants.

Flowers Pollinated by Hymenoptera. A number of plant groups have developed specialized flowers and associated structures that tend to favor pollination by Hymenoptera, especially bees. Most of these "bee flowers" produce nectar, and all advertise themselves with brightly colored, showy flowers, usually with a sweet odor. Most bee flowers are blue, yellow, or some related color, as one would expect from the range of colors bees perceive. Most bee flowers open during the day, closing, if they do, at night when bees are inactive. They often have a distinct floral pattern that may include nectar guides (Fig. 11–1). Nectar guides are especially valuable to bees since the nectar of bee flowers is frequently located at the base of a long floral tube and is not readily apparent. The deep floral tubes of many bee flowers and the long "tongues" of bees are another example of a coevolutionary development that favors a narrow group of pollinators.

Many bee flowers have a protruding lip that serves as a landing platform for the visitors. Some bee flowers also have pollen-dispensing mechanisms, frequently activated or tripped by the bee's weight on the platform, as in the alfalfa flowers previously discussed.

Some of the evolutionarily more advanced flowers have complex passageways or traps that force bees to follow a certain route to and/or from the nectar source. These mechanisms ensure that the bees encounter the anthers or stigma at a particular point and thus acquire or deposit pollen. In the case of the orchid *Coryanthes speciosa*, a bucketlike lip of the flower is filled with a fluid that seems to be somewhat inebriating. Visiting bees land on the upper part of the flower, lose their footing, fall into the bucket, and can get out only by crawling through a narrow opening. As they do so, they must brush against the anthers, and thus they acquire pollen that they eventually transport to other flowers. Male bees of *Euglossa cordata* are attracted by the odor of another orchid, *Gongora maculata*, and land on its flowers in search of secretions. Having penetrated the inner floral parts in its search, a bee suddenly loses its footing on the slippery surface of the curved column and gets a tobogganlike ride to the bottom. As it slides past the anthers, pollinia (specialized pollen masses) become attached to a specific spot on the bee's abdomen. These are later carried to another flower, where a similar ride leaves the pollinia on the stigma. A number of flower species use this "give-and-take" approach to pollination. It is important in these cases that the pollinia be placed in a specific location on the bee's body for it to be deposited on a stigma. Different flower species use different locations on the bee for depositing the pollen. The location is such that the normal activities of the bee, as influenced by the flower species, result in the proper part of the bee's body touching the reproductive structures of the flower and acquiring or leaving pollen.

A more elaborate adaptation is found in the orchids of the genus *Ophrys*. The labellum, or lip, of these flowers resembles female bees or wasps in both form and color. Perhaps more important, the odor of the flowers mimics the sexual attractant given off by females of particular bee or wasp species. Male bees or wasps are attracted to these impostors and attempt to copulate with them, inadvertently pollinating the flowers at the same time. This phenomenon, called **pseudocopulation,** is responsible for the pollination of a number of orchid species and, of course, involves a very intimate floral-insect association (Fig. 11–12).

Other Insect Pollinators.   As noted previously, Hymenoptera, especially bees, are the most valuable general plant pollinators, although other insect groups also visit flowers and collect pollen and nectar. In general, non–bee pollinators lack sufficient body hairs and the necessary behavior patterns to be important in plant fertilization. However, some groups may be important, or even essential, pollinators of certain plants. The orders Lepidoptera, Diptera, and Coleoptera contain members that are valuable pollinators, as discussed below.

Lepidoptera.   Adults of most species of Lepidoptera feed on nectar, and nearly all larvae are plant feeders. Most Lepidoptera have permanent tubelike sucking mouthparts formed from the galeae (Fig. 1–5e), well adapted for extracting nectar from plants. The proboscis tube is carried coiled beneath the head when not in use and is readily apparent only when the insect is actively collecting nectar. Tube length varies from 1 to 250 mm and in many groups apparently has coevolved with flowers having deep corollas. Butterflies generally frequent day-blooming flowers, and moths visit evening- or night-blooming varieties or those that remain constantly open. Since most past research on flower visitation has been done during the day, the value of moths as pollinators has likely been underestimated.

The most important groups of Lepidoptera from a pollination standpoint are probably the following: Papilionoidea, butterflies and skippers; Noctuidae, cutworm moths and owlet moths; Geometridae, loopers and inchworms; Pyralidae, snout moths; Arctiidae, tiger moths and woolly bears; Sphingidae, hawk moths.

Flowers pollinated by butterflies frequently are bright red or orange; at least some butterfly species can see red. Often they have long, narrow corollas with nectar at the bottom, accessible only to their specialized mouthparts. In most other respects flowers pollinated by butterflies and diurnal moths are similar to bee-pollinated flowers, since these insects are also guided to the flowers by sight and odor. In contrast most moth flowers are white with a heavy fragrance usually emitted only after sunset. The non-white moth flowers—the yellow evening primrose, for example—display colors that stand out against a dark background.

The hawk moths, perhaps the most interesting of the Lepidoptera pollinators, may be seen in the evening or at night darting rapidly from one blossom to another. Hawk moths often do not land on flowers but hover in the air while extending their long pro-

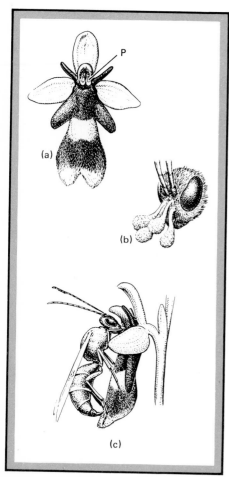

Figure 11–12
(a) Blossom of an orchid of the genus *Ophrys*, the pollinia labeled *P;* (b) the pollinia attached to the head of a male bee that has attempted "pseudocopulation" with the flower; (c) a male bee at a blossom. (From *Mimicry in Plants and Animals*, by W. Wickler, with permission of the publishers, McGraw Hill Book Co., copyright 1968 by Weidenfeld and Nicolson Archives.)

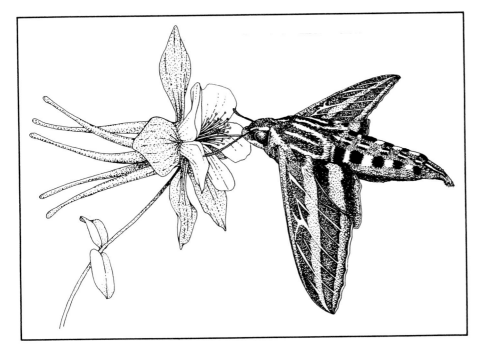

Figure 11–13
A hawk moth, *Hyles lineata*, at a blossom of Colorado columbine, *Aquilegia caerulea*. These flowers are specialized for pollination by hawk moths, and bees are unable to reach the nectar in the long spurs. Variation in color and spur length in columbines in various parts of the range is believed to be correlated with the presence of different species of hawk moths. (Drawing by Peter Eades, based on R. B. Miller, 1981, "Hawkmoths and the geographic patterns of floral variation in *Aquilegia caerulea*," *Evolution*, vol. 35, pp. 763–774.)

boscises into the blossoms in search of nectar (Fig. 11–13). Hence the flowers they frequent do not normally have the landing platforms or the complex passageways of bee flowers. Adult hawk moths do not feed on anything but nectar, but recent research by Robert Stockhouse, at Colorado State University, on *Oenothera* flowers has demonstrated that they may get nutrients other than the normal carbohydrates from the nectar. Stockhouse and others have shown that when pollen grains fall into the nectar, amino acids are leeched out and may be taken up by nectar-feeding insects. Thus it appears that some non–pollen-feeding insects may be able to obtain additional nutrients via this pathway.

Diptera. The best-known fly pollinators include the families Syrphidae, Bombyliidae, Tephritidae, Tachinidae, Calliphoridae, and Chironomidae.

Syrphidae. The syrphids, or hover flies, are common visitors to flowers, frequently hovering above the blossoms. Many resemble honey bees; others look like bumble bees; and others, like wasps. In some cases the mimicry is very striking. The syrphids may be the most important fly pollinators because, although they visit the same flower type as bees, they tend to continue their work under poor conditions when most bees are inactive. Also, like bees, they feed on pollen as well as nectar. Consequently these flies may be important pollinators in areas where plants frequently bloom during inclement weather. Syrphid larvae vary considerably in habits, but some species live in the nests of social Hymenoptera.

Bombyliidae. The bee flies are common insects especially in the arid areas of the southwestern United States. Most are large, stout-bodied, hairy flies with a long, slender proboscis. As the common name implies, these flies greatly resemble bees in appearance and behavior. Bombyliids are considered important pollinators but seem to be less active under poor weather conditions than syrphids. The bombyliid larvae are all parasitic, as far as is known, attacking Lepidoptera, Coleoptera, and Hymenoptera larvae and the eggs of grasshoppers.

Calliphoridae.   Blow flies are about the size of a house fly, and many species are metallic green or blue. Most blow flies are scavengers, with the larvae feeding in carrion, excrement, and similar materials, and have no specialized structure for collecting pollen or nectar. Even so blow flies of various species have been used for years in pollinating specific onion crops, especially for seed production, because they are attracted to the flowers and their activities result in high levels of fertilization.

Tephritidae, Tachinidae, Chironomidae, and Tabanidae.   Adults of these families are frequently seen on flowers, but detailed studies on their value as pollinators are generally lacking. It is assumed that they play only a supplemental role in pollination.

Flowers visited by flies in the families Syrphidae and Bombyliidae tend to be similar to those normally visited by bees. This is not unexpected, since both fly groups contain visual and behavioral mimics of bees and thus are presumed to have evolved in close association with those insects. The really distinctive "fly flowers" are visited by a diverse group of about 30 families of flies that have no particular specialization for feeding on flowers. Most, like blow flies, probably get the majority of their nourishment from other sources, such as carrion, excrement, plant sap, and blood. Unlike bees, butterflies, and the beelike flies, this group of diverse flies is attracted primarily by odor to the flowers they visit. Fly flowers, consequently, are usually dull colored and odorous, frequently having an objectionable smell that mimics the oviposition sites of the flies. The odor of some fly flowers resembles decaying flesh; another group smells like human excrement; and still another group, like fish oil. These are "deceit flowers," providing no actual reward for the flies.

Coleoptera.   Coleoptera are not as important as Diptera or Lepidoptera as pollinators, but there are numerous flower-visiting forms that may be of occasional significance. A few plant species are pollinated solely, or chiefly, by beetles. For example, flowers of magnolia, California poppy, and wild rose are generally pollinated by beetles. Deceit flowers often attract beetles as well as flies.

Some 16 beetle families are commonly seen in flowers, but the most commonly seen include the following:

- *Cantharidae:* Most species of soldier beetles are predaceous as larvae and some also as adults, but many adults also feed on pollen and thus are frequent flower visitors.
- *Meloidae:* Larvae of some species of blister beetles are parasitic in bee nests, others on grasshopper eggs, but most feed on nectar and pollen as adults.
- *Cleridae:* Larvae of the flower-visiting species of checkered beetles are mostly parasitic in the nests of bees and wasps. The adults, while sometimes predaceous, all feed on pollen.
- *Buprestidae:* Larvae of the flat-headed borers eat wood. The metallic adults are pollen feeders and are frequently seen on flowers.
- *Cerambycidae:* Like the Buprestidae, these long-horned beetles are wood borers as larvae but feed on pollen as adults.

An interesting beetle–flower association is found in the pollination of the giant water lily of the Amazon, *Victoria regina.* The flowers open in the evening and attract scarab beetles, which alight and begin to feed. The flowers soon close, and the beetles are trapped inside until the next evening, when the blossoms reopen. During the period they are trapped, the beetles feed on the interior parts of the flower and at the same time pollinate it. A number of species of angiosperms are pollinated mainly by beetles. Beetle flowers attract their pollinators primarily by odor rather than by sight and commonly have a sweet, spicy, or fruity smell. Like many flies, most beetles that pollinate flowers are general feeders and are not specially adapted for a floral diet. Most of their nourishment is probably obtained from other sources, such as fruit, leaves, excrement, and carrion.

Many beetles are quite destructive of the flowers they visit, and some devour the entire floral contents. Thus, although they probably pollinate the flower, their contribution is often of little consequence. As a result most flowers commonly pollinated by beetles have ovules well buried beneath the floral chamber, where they are safer from destruction by the feeding beetles. Beetles (and perhaps flies) have been suggested as the original insect pollinator of angiosperms, before bees and Lepidoptera evolved.

Other Animals as Pollinators.   Some birds are regular visitors of flowers, where they feed on nectar, floral parts, or flower-visiting insects. During these visits many birds also serve as pollinators. As noted previously, pure red flowers are not perceived by most insects unless they also reflect ultraviolet, and they are usually pollinated by birds. Birds have keen vision and see a range of colors similar to that of humans. Most bird-pollinated flowers are colorful, with reds and yellows predominating. On the other hand, birds have little sense of smell, and the flowers they pollinate are usually odorless, or nearly so. In the Americas the chief birds that serve as pollinators are hummingbirds.

Bat-pollinated flowers are found in the tropics of Africa and South America. Since bats are nocturnal, the flowers they visit are often dull and open only at night. Bats locate flowers largely through their sense of smell, and consequently bat-pollinated flowers are characterized by a strong odor. Bats that are close associates of flowers usually have slender, elongated faces and long, extensible tongues, sometimes with a brushlike tip. The front teeth of such bats may be reduced or completely missing. These bats will fly from tree to tree, often in groups, lapping nectar and eating pollen and other flower parts. Bat-pollinated flowers include the dowa-dowa tree (*Parkia clappertoniana*), agave (*Agave schottii*), and saguaro and organ pipe cactus, as well as others.

Wind Pollination.   Wind is the most important pollination agent for the gymnosperms, grasses, and some dicotyledonous plants. Successful wind pollination is purely a chance occurrence, but the chance is greatly increased by several floral characters. These plants usually produce enormous quantities of dry, lightweight pollen that is easily transported by the wind. Some plants have a mechanism for rapid, sometimes forceful dehiscence (release) of pollen, and most have anthers that are exposed to the wind. Wind-pollinated plants frequently have a feathery stigma that acts to trap wind-blown pollen. In some cases the flowers will appear before the leaves, reducing the possibility of the leaves' interfering with pollen transport. Clouds of yellow pollen from pine trees and corn plants may be frequently seen in the spring and summer as they are carried by the wind. Wheat, barley, oats, and many graminaceous weeds are wind pollinated, as are walnut and oak trees. Wind pollination is a highly evolved mechanism, with complex plant structures developed that act to ensure successful pollen transport.

Summary

Plant pollination by insects is a coevolutionary process often involving intimate and specialized relationships among members of the two groups. The earliest flowering plants were probably wind pollinated, as some still are. Early insect pollination may have been accidental but eventually led to the development of sticky pollen grains, attractive odors and colors, and structural and behavioral mechanisms in the associated insects. Insects, in general, perceive ultraviolet light, and ultraviolet colors are an important characteristic of insect-pollinated plants. For example, when flowers are photographed with film sensitive to ultraviolet light, many are seen to possess nectar guides that appear to be used by insects in locating sources of nectar.

Some of the more intimate associations between flowers and insects include the obligate association between yucca plants and the yucca moth, and between figs and certain chalcidoid wasp pollinators. Another effect of coevolution is the flower constancy exhibited by some pollinators, especially bees. Where flower constancy is high, the pollinator restricts foraging to one plant species during single trips or for longer periods. There is a trend toward the evolution of floral structures that favor certain types of efficient pollinators and discourage less efficient ones.

A number of insects have developed total dependence on floral products for food. Nectar, which is composed mainly of sugars with traces of other materials such as proteins, salts, and acids, provides the sole energy source for many insects. Bees and some other insects utilize pollen as a source of protein, and this has led to many specializations in structure and behavior. Plants, in turn, depend heavily on insects for pollination, and disruption of pollination can have far-reaching consequences in natural and in some artificial ecosystems. Most fruits, vegetables, and nuts cannot be produced commercially without insect pollination. In addition, products such as beef, pork, and dairy products depend on the consumption of insect-pollinated legumes such as alfalfa, clover, and trefoil.

Honey bees are probably the most important pollinators of commercial crops and have the advantage of surviving the winter as adults, thus being available in early spring when pollination requirements are high. Bumble bees are also efficient pollinators, but their numbers are not usually high enough to assist commercial growers consistently. Solitary bees are often efficient pollinators but are usually not present in consistently high numbers. Methods have been developed for increasing populations of leafcutter and alkali bees, which are effective pollinators of alfalfa. Alfalfa is only one of many flowers that have special devices for ensuring insect pollination. Many orchids, in particular, have elaborate lips, passages, and traplike devices; some even mimic insects in structure and have odors resembling the sex attractants of certain bees and wasps.

Pollinators other than Hymenoptera include butterflies and moths, flies, and some groups of beetles. In some cases the association of these insects with specific plants is quite intimate. Certain flowers are specialized for pollination by birds and by bats, while gymnosperms, grasses, and some dicotyledonous plants such as oak trees are wind pollinated.

Selected Readings

Bohart, G. E. 1972. "Management of wild bees for the pollination of crops." *Annual Review of Entomology*, vol. 17, pp. 287–312.

Eisner, T.; R. E. Silberglied; D. Aneshansley; J. E. Carrel; and H. C. Howland. 1969. "Ultraviolet video-viewing: the television camera as an insect eye." *Science*, vol. 166, pp. 1172–74.

Feinsinger, P. 1983. "Coevolution and pollination." In *Coevolution*, D. J. Futuyma and M. Slatkin, eds. Sunderland, Mass.: Sinauer. pp. 282–310.

Free, J. B. 1970. *Insect Pollination of Crops*. New York: Academic Press. 544 pp.

Gilbert, L. E., and P. H. Raven, eds. 1973. "Coevolution of animals and plants." *Symposium V/First International Congress of Systematic and Evolutionary Biology*. Austin and London: University of Texas Press. 246 pp.

Heinrich, B. 1979. *Bumblebee Economics*. Cambridge, Mass.: Harvard University Press. 245 pp.

Johansen, C. A. 1977. "Pesticides and pollinators." *Annual Review of Entomology*, vol. 22, pp. 177–92.

Kevan, P. G., and H. G. Baker. 1983. "Insects as flower visitors and pollinators." *Annual Review of Entomology*, vol. 28, pp. 407–53.

Martin, E. C. 1975. "The use of bees for crop pollination," in *The Hive and the Honeybee*, C. C. Dadant, ed. pp. 579–614. Hamilton, Ill.: Dadant and Sons.

———, and S. E. McGregor. 1973. "Changing trends in insect pollination of commercial crops." *Annual Review of Entomology*, vol. 18, pp. 207–26.

McGregor, S. E. 1976. *Insect Pollination of Crop Plants*. USDA Agricultural Research Handbook, no. 496. 411 pp.

Proctor, M., and P. Yeo. 1973. *The Pollination of Flowers*. New York: Taplinger. 418 pp.

Richards, A. J. 1978. *The Pollination of Flowers by Insects*. London: Academic Press. 213 pp.

Although we are so frequently impressed by the appetite of insects for plants, in fact a great many insects—perhaps as many as half the described species—feed on animal tissue of one kind or another. Many of them feed on other insects and thus are of importance in the regulation of populations of pest species. Others, such as lice and fleas, attack warm-blooded animals, and some act as vectors of serious diseases of humans and our domestic animals. Many predatory and parasitic insects are themselves attacked by other species, a phenomenon so prevalent that Jonathan Swift was moved to remark on it in verse (with considerable poetic license, since in fact fleas have no important predators):

So, naturalists observe, a flea
Has smaller fleas that on him
    prey;
And these have smaller still to
    bite 'em;
And so proceed *ad infinitum*.

One may, of course, look at this subject from the opposite point of view: that a great many animals subsist on insects. Not only are a great many insects **entomophagous** (insect eating), but many fish, birds, reptiles, and mammals depend on insects as a major source of food. In response; the insects have evolved a variety of mechanisms that reduce their chances of being eaten: con-

The Relationships of Insects with Animals

PART

FIVE

cealing coloration, repellent secretions, mimetic patterns, and still other features we shall consider in Chapter 14.

Major Lifestyles of Zoophagous Insects

Animals that feed on other animals may, in the broad sense, be said to be **zoophagous.** But so diverse are their behavior patterns and lifestyles that they defy classification. The following seven categories may conveniently be recognized, though not all cases will fit neatly into one or the other of them:

1. Predators capture and consume a succession of living individuals, or **prey.** Among birds, flycatchers provide an example; among insects, preying mantids and lady beetles. A special case is provided by the wasps, which take prey not for themselves but to feed their offspring; the adult wasps feed mainly on nectar and are thus properly phytophagous: an example of the difficulty of classifying insects according to feeding behavior.

2. Parasites live in close association with a single individual or **host,** usually without causing its death. Mammals, for example, are host to a variety of worms and protozoans that live internally **(endoparasites),** as well as to such **ectoparasites** as lice. With a few exceptions, parasitic insects attack warm-blooded animals.

3. Blood-feeders are predators in the sense that they may attack a series of hosts and that they live apart from the animals they attack; yet they inflict only minor damage to the host. Mosquito bites are all too familiar to everyone—and not a few have been felled by the disease organisms carried by these and other blood-sucking insects. Most blood-feeders attack warm-blooded animals, but a few attack other insects.

4. Parasitoids live in close association with a single host, which they consume slowly, feeding at first on nonvital organs and only causing the death of the host as they approach maturity. The resulting adults are free living, the females laying their eggs on or in a succession of new hosts. Although entomologists often speak of these as "parasites," in fact they are specialized predators. The term **protelean** (literally, "early ending") **parasites** has often been applied to these insects, since the close association with the host occurs only in the larval stage, the adults being free living and often nectar-feeders. Parasitoid insects attack other insects, spiders, ticks, and snails, but not higher animals.

5. Social parasites obtain their sustenance at the expense of a family or colony, which is not usually destroyed. The cowbird is an avian

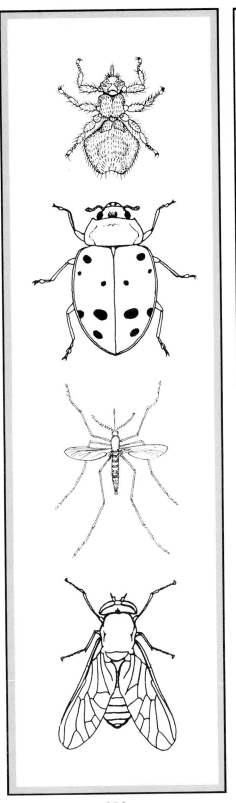

example of a social parasite; among the insects there are, for example, certain kinds of bumble bees that do not found their own colonies but invade colonies of other species, displace the queen, and exploit the workers to rear the invaders' young. The colonies of ants and termites contain many **commensals,** which feed on food or detritus in the nest, or sometimes even on the young of their hosts. Insects that feed primarily on food stored by the host species are often spoken of as **cleptoparasites** ("thief parasites").

6. Inquilines live inside the galls, burrows, or nests of other insects without interfering in any important way with their hosts.

7. Scavengers feed on the products of animal activity, for example, on wastes, droppings, carcasses, or animal secretions or structures. The wax moth provides an example of the last type, since it feeds on the cell walls of honey bee nests. There are many dung- and carrion-feeders among the insects, though in some cases the food consists of bacteria in the dung or decaying flesh, in which case the insects cannot truly be called zoophagous. Other difficulties in classification are provided by the fleas, which as adults are blood-feeders, but as larvae are scavengers in the nests of their hosts.

Entomophagous Predators and Parasitoids

Figure 12–1
A ground beetle (Carabidae) seizes a cutworm. These beetles and their larvae are effective predators in many farmlands. (Photograph by James E. Lloyd, University of Florida.)

The study of entomophagous insects is of growing importance in a world well aware of the effects of indiscriminate use of chemical pesticides. In this chapter we shall consider selected examples of predators and parasitoids, with special reference to their life histories and behavior and some of the problems and opportunities they present. Blood-feeders and true parasites fall mainly within the realm of medical and veterinary entomology, and we shall treat them in the following chapter.

Predatory Insects

Predatory insects may exhibit one of several major strategies: They may be searchers, stalkers, or trappers. Searchers move about actively, capturing insects in the air (dragonflies, for example), on the ground (ground beetles, Fig. 12–1), or on vegetation (lady beetles, see Fig. 12–4). Stalkers lie in wait for prey, then approach it stealthily and pounce on it, in the manner of preying mantids. Trappers are well exemplified by ant lions, often called doodle bugs. These larvae build conical pits in sandy soil into which ants and other insects fall. If they do not immediately slip into the jaws of the ant lion buried at the bottom, the latter is able to toss sand at the prey and prevent it from crawling from the trap. Traps of a very different kind are made by certain caddisfly larvae, which feed on plankton and small arthropods seined from streams (Fig. 12–2). Some insects employ more than one strategy: The immature stages of dragonflies, for example, are stalkers, while the adults are aerial searchers.

The adaptations of predators are diverse, depending on their manner of prey capture. Visual predators rely on large eyes, often on mobile heads capable of scanning a wide area, as in mantids and dragonflies. Rapid flying or running mechanisms may be involved, or behavior patterns serving to "zero in" on some particular type of prey. There may be modifications of the legs for capturing or holding prey, usually the front legs, as in mantids, but even occasionally the hind legs, in the case of hangingflies (Mecoptera) (Fig. 7–18). Without exception the mouthparts are modified for chewing or piercing the exoskeleton of their prey, and in many cases digestive fluids are injected into the prey, rendering it immobile and flaccid while the body fluids are sucked out (for example, in robber flies).

One thinks of predators as consuming almost anything that comes by, but in fact the behavior of most entomophagous predators is specialized in such a way that only certain kinds of prey are taken. **Polyphagous** predators take a broad spectrum of insects occurring in their habitat. Dragonflies, for example, take a wide variety of small, flying insects; and tiger beetles, almost anything they can capture on the ground in the places where they occur. **Oligophagous** predators are more selective. The behavior of the larvae of lady beetles and lacewings (or aphid lions), for example, is such that they rarely feed on anything but aphids, scale insects, and other Homoptera. A few lady beetles even approach a condition we might describe as **monophagous;** that is, they prey primarily on a single species. These distinctions are of importance in biological control. The preying mantid may be the quintessence of predatism, but it is nevertheless polyphagous and has no predilection for pest species. The bee-wolf, a ground-nesting wasp prevalent in Europe and Africa, is oligophagous but captures great numbers of honey bees (Fig. 7–10). The ideal predator, from our point of view, is a specialist on noxious insects—and fortunately there are many of this kind.

Robber Flies as Examples of Entomophagous Predators.

Robber flies are large Diptera that act as top predators in many insect communities. They are formidable insects, with large (often green) eyes; powerful sucking beaks; long, spiny legs; and efficient mechanisms for quick, short-range flight (Fig. 12–3). Lee Rogers and Robert Lavigne, at the University of Wyoming, have made a thorough study of the robber flies inhabiting the Pawnee National Grasslands of northeastern Colorado, and the results of this study are instructive as to what must occur in many natural communities.

The Pawnee Grasslands comprise a vast area of rolling short-grass prairie, extensively grazed by cattle but still containing a diversity of wildlife, including many rodent species and those top predators of the avian world, hawks and owls. Lavigne and his co-workers have found that 21 species of robber flies inhabit the grasslands. Adults occur from April until late October and are most often seen on perches on stems or on the ground, flying off periodically in response to moving objects in their field of vision. Certain species attack and immobilize their prey in the air; others, on the ground; still others capture prey in the air and return to their perch to pierce and feed on their prey. Both polyphagous and oligophagous species are involved. One species, for example, feeds almost entirely on blister beetles, another almost entirely on grasshoppers. But the majority are more or less polyphagous. One species, the largest on the Pawnee Grasslands, was found to capture insects of seven different orders (including five species of robber flies—including members of its own species!) Yet the 21 species, in general, occur in slightly different habitats and thus take a different assortment of prey. Some prefer open rangeland, others rocky outcrops or sand strips along arroyos; one occurs primarily in the burrows of ground squirrels. They also differ in size, from small species only 4 to 6 mm long to relative giants more than 30 mm long, with considerable correlation between size of predator and size of prey. There is also some seasonal separation, some kinds occurring as adults in the spring, others during certain periods in the summer or early autumn. Thus, although there is some overlap in all of these features, together they make it possible for 21 species to live together, some of them in great abundance. The toll of insects, in this area mainly feeders on grasses and forbs, must be considerable.

Lady Beetles as Examples of Entomophagous Predators.

Since ancient times, humans have taken lady beetles from their hibernation clusters and released them in their orchards to control aphids and scale insects, for these beetles, both as larvae and as adults, are largely specialists on plant-feeding sucking insects and mites (Fig. 12–4). Yet they seem to have few qualifications for their role as the "perfect predators." The larvae are sluggish crawlers, the adults not particularly adept fliers; their eyes are small and inefficient, and they have no obvious adaptations for seizing or holding prey. What are the reasons for the success of these insects—so useful to humans that they were long ago named for "Our Lady," the Virgin Mary?

The eggs of most lady beetles are laid in clusters on vegetation, usually in places where aphids or other sucking insects are prevalent. They hatch in a few days, and the larvae seek out their prey. Having found an aggregation of relatively immobile sucking

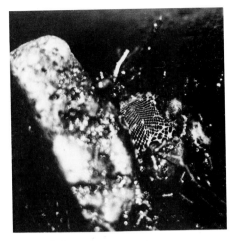

Figure 12–2
A net of the caddisfly larva *Hydropsyche* (Trichoptera) among the stones of a stream. (Photograph by Howard E. Evans.)

Figure 12–3
One of the robber flies of the Pawnee National Grasslands of Colorado, *Scleropogon picticornis*, has seized and is feeding on a grasshopper. (Courtesy Robert J. Lavigne, Wyoming Agricultural Experiment Station.)

(a)

(b)

Figure 12–4
The convergent lady beetle, *Hippodamia convergens,* so called because of the converging streaks on the pronotum: (a) an adult feeding on aphids; (b) a larva ascending a stem. (Photographs by O. Wilford Olsen, Colorado State University.)

insects, they feed on one after the other. The larva of one species, *Coccinella californica,* was reported to consume an average of 25 aphids a day, one individual eating 475 aphids before reaching full size, pupating, then continuing to eat aphids at the rate of 34 per day. Many lady beetle larvae are spangled with orange, and the bright orange or red colors of the adults are well known to everyone. These are regarded as aposematic, or "warning colors," advertising to predators the fact that lady beetles and their larvae have bitter-tasting, even poisonous blood (Chapter 5; Fig. 5–18).

How do the newly hatched larvae, which move slowly and have only a few simple eye facets on each side of their head, find their prey? C. A. Fleschner, of the Citrus Experiment Station at Riverside, California, found that larvae of a species of lady beetle that attacks citrus red mites are positively phototactic and negatively geotactic; thus they tend to move to the tops of stems, where their prey are most likely to occur. On a flat surface, such as a leaf, they move about with frequent changes of direction, occasionally stopping and moving their body in an arc. Only on actual contact is the larva able to detect its prey—even from a distance of as little as 3 mm, the larva appears unaware of its presence. After it has fed, the larva moves in a tortuous pattern, with close turns serving to place it in contact with other members of a feeding aggregation of prey (Fig. 12–5).

What if a larva happens to ascend a stem that has no potential prey on it, or only a few, as must commonly occur? Experiments by A. F. G. Dixon, then at Oxford University, showed that a slightly starved larva undergoes a temporary reversal of geotaxis, so that it is able to descend a short distance before reascending. As starvation increases, the larva descends further and further before climbing upward again. In nature, this enables a larva to explore other branches of a plant or even of neighboring plants. Under conditions of low prey density, many larvae may die of starvation.

Although larval behavior appears inefficient, adult behavior compensates for most gross mortality and permits rapid buildup of lady beetle populations when and where aphids or other sucking insects are abundant. Adults are active fliers and are attracted to outbreaks of their hosts for oviposition. Furthermore, their seasonal cycles are such that they are on the wing at times of host abundance. This synchrony is enhanced by the fact that adult lady beetles of many species are able to diapause for long periods. In hot, dry climates adults may diapause through the hot summer, the dry fall, and on through the winter, remaining in clusters on mountaintops, then migrating into the valleys in the spring when prey are abundant on growing crops (Fig. 12–6). The function of clustering behavior in diapausing lady beetles has long puzzled entomologists. Probably the major function is to bring together the two sexes for mating purposes, for mating occurs when the clusters are about to break up and the females about to disperse and lay their eggs. Adults of the next generation enter diapause when their bodies contain an abundance of fat; they are attracted to high points on the horizon, and thus the widely dispersed adults come together on mountaintops, where they join others of their species under rocks or in similar protected places. Here they may be deeply covered with snow during the winter.

Many lady beetles are restricted to a single prey species, or nearly so. In California two species have been imported from Australia, bred in the laboratory, and released in great numbers for the control of citrus pests. The best known of these is *Rodolia cardinalis,* often called the vedalia, which provided excellent control of the cottony cushion scale following its first release about 1890. A few years later a second species, which preyed on the citrus mealybug, was released, again with excellent results. There have been other examples of successful biological control with the use of lady beetles. It is now possible for homeowners to purchase lady beetles for release in their gardens. This is at best a doubtful procedure, since the beetles fly readily and may find a better source of their preferred food elsewhere.

Lady beetles are by no means the only predators of potential use in biological control. Lacewings (Neuroptera), the larvae of which are often called aphid lions, are now being reared and released in several areas. Attempts have been made to colonize paper wasps along the borders of tobacco and cotton fields, as these wasps take quantities of caterpillars to feed their larvae. In the tropics, ants serve as ubiquitous predators on plant-feeding insects and on termites, and programs to destroy ants sometimes have a considerable backlash. It seems a safe assumption that in all parts of the world insect populations are regulated to a considerable extent by predators, and in general insects play a more significant role in this regard than do birds, reptiles, and other vertebrate animals—even though their activities are much less obvious to us. The activities of parasitoids tend to be even less apparent to us than do those of predators, though in the final analysis parasitoids may play an even more important role in the natural control of insect populations.

Parasitoids

Parasitoids are often regarded as highly efficient predators, able to complete their development on a single host, which is killed only as the larva approaches completion of its development. The adult parasitoid attacks other individuals, not as food for itself, but to lay eggs and thus provide food for its offspring. In fact adults of some species do feed on the blood of the host, and a few actually kill prey and feed on it without laying eggs—demonstrating the narrow gap between predatism and parasitoidism. But in general parasitoids are **protelean,** in that adults do not feed on animal tissue.

Unlike predators, parasitoids cannot be larger than their hosts, since they develop at the expense of a single individual. They may be smaller, even much smaller, for there are many cases in which numerous parasitoid larvae develop within a single host, a phenomenon called **superparasitoidism.** Parasitoids may serve as hosts for other species, called **secondary parasitoids,** and these in turn may occasionally be attacked by **tertiary** (third-order) **parasitoids.** Some species are able to change their role, serving now as primary, now as secondary parasitoids, or now as secondary, now as tertiary (Fig. 12–7). An insect that attacks a parasitoid, at whatever level, is called a **hyperparasitoid.***

Depending on the species, adult parasitoids may lay one or more eggs either externally or internally, either in the egg of the host, in or on the larva or pupa, or less commonly in or on the adult. If oviposition is in the egg, development may be completed during the egg stage or not until the larval stage; if in the larva, development is sometimes not completed until the pupal stage. The eggs of internal parasitoids, being immersed in the body fluids of the host, often contain very little yolk and so eclose to form a very simple larva that is hardly more than a free embryo. Such insects go through **hypermetamorphosis,** eventually transforming into a more normal-appearing larva (Fig. 12–8). In some parasitoids a single egg laid by a female may divide into several embryos, each of which produces a new individual. In such cases of **polyembryony,** all individuals are identical and of the same sex. In some parasitoid Hymenop-

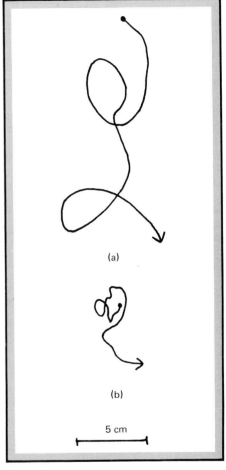

Figure 12–5
Tracks of a lady beetle larva over a one-minute period: (a) a hungry larva searching for food; (b) after feeding on an aphid. (After C. J. Banks, 1957, "The behavior of individual coccinellid larvae on plants," *British Journal of Animal Behaviour,* vol. 5, pp. 12–24.)

*In practice, many entomologists refer to parasitoids as "parasites," a point of confusion when they talk to parasitologists, who do not acknowledge them as parasites. Thus entomologists tend also to speak of "superparasitism," "hyperparasitism," and so forth. The verb *parasitize* is surely to be preferred to the awkward *parasitoidize* in most situations.

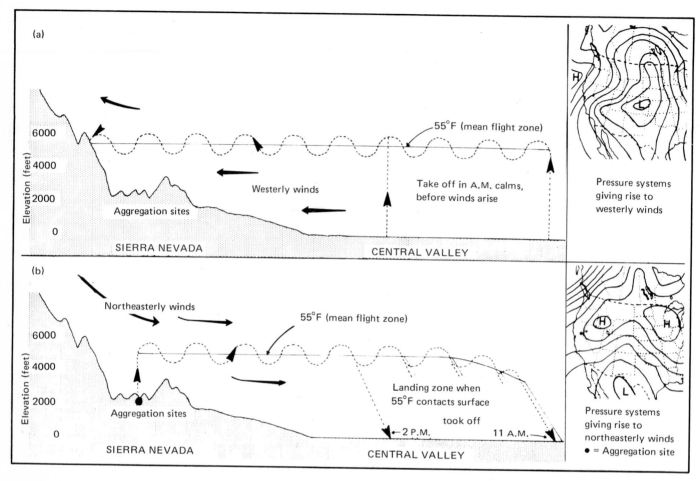

Figure 12–6
Seasonal migrations of the convergent lady beetle in California: (a) the hypothetical flight pattern toward the Sierra Nevada in May and June; (b) the flight from the Sierras to the Central Valley in February or March, presumably by the same individuals. The weather maps show the position of pressure systems at the time the flights occur. (K. S. Hagen, 1962; reproduced, with permission, from the *Annual Review of Entomology*, Volume 7. © 1962 by Annual Reviews Inc.)

tera, males are rare or sporadic in occurrence, even wholly absent, females reproducing generation after generation by parthenogenesis.

Host specificity varies greatly among parasitoids. One species of tachina fly is reported to attack approximately 100 different hosts belonging to 18 families of 3 orders. Most species attack a much more limited array of hosts and may be said to be oligophagous to varying degrees. Large ichneumons of the genus *Megarhyssa*, for example, confine their attacks to certain larvae of wood wasps, laying their eggs through the bark and wood of trees with their extremely long ovipositors (Fig. 12–9). The eggs have a very long filament at one end, and as they pass down the threadlike ovipositor, their contents are squeezed out into the filament, forming an extremely slender egg that resumes its normal shape when deposited on the host. Highly specialized parasitoids such as this would scarcely be able to attack any other host. While strictly monophagous parasitoids are uncommon (most have at least a few alternate hosts), the majority attack a limited number of hosts in one particular habitat.

When we consider the small size of many parasitoids (some egg parasitoids are less than 0.2 mm long) and the fact that the adults are short lived and the larvae are mainly internal feeders, it is easy to appreciate why these are among the most poorly known

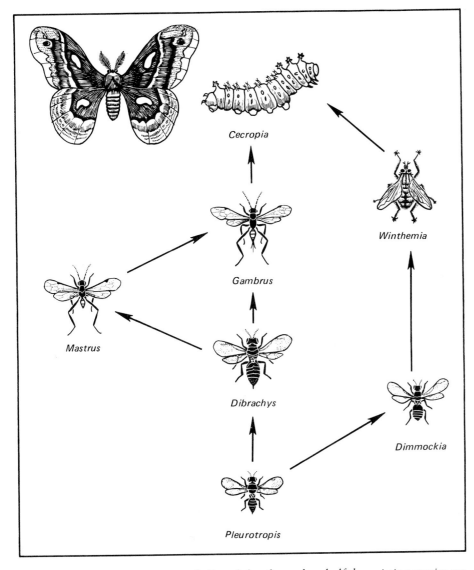

Cecropia

Winthemia

Gambrus

Mastrus

Dibrachys

Dimmockia

Pleurotropis

Figure 12–7
Some of the parasitoids and hyperparasitoids of the cecropia caterpillar. *Winthemia* is a tachina fly; the others, Hymenoptera. *Dimmockia* and *Dibrachys* are secondary parasitoids; *Pleurotropis*, tertiary; but *Dibrachys* may act as a tertiary parasitoid also, by way of *Mastrus*. (H. E. Evans, 1968, from *Life on a Little-Known Planet*, published by E. P. Dutton and Company; after L. C. Cole, 1962, in a review of Rachel Carson's *Silent Spring*. Copyright © 1962 by Scientific American, Inc. All rights reserved.)

Figure 12–8
Life stages of the chalcidoid wasp *Poropoea stollwercki*, a hypermetamorphic parasitoid of beetle larvae: (a) embryo within the egg; (b–f) first- to fifth-instar larvae. (From *Entomophagous Insects*, by C. P. Clausen. Copyright © 1940 McGraw-Hill Book Company/Weidenfeld & Nicholson Archives. Used by permission of McGraw-Hill Book Company. After F. Silvestri, 1916.)

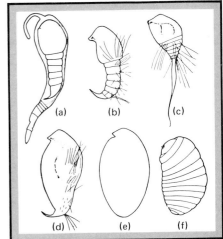

(a) (b) (c)

(d) (e) (f)

of all insects. In some groups, it is believed that fewer than half the existing species are known to science, and in many groups discrimination of species is difficult—physiological and behavioral differences are often more prominent than structural differences. We know of no group that presents more challenges to entomologists.

Life histories of parasitoids are sometimes bizarre almost beyond belief. A minute wasp called *Rielia manticida* lays its eggs inside those of preying mantids. The latter are, however, laid in a mass covered with a tough coating that protects them from most predators. Adult *Rielia* solve the problem in what seems to us an ingenious manner. They attach themselves to female mantids near their wing bases and are carried around by the mantids, sometimes for weeks (a phenomenon known as **phoresy**). From time to time they feed on the blood of the mantid through membranes at the wing bases. Fi-

Figure 12–9
The ichneumon wasp *Megarhyssa macrura* ovipositing through the bark of a tree on a wood wasp larva. (Photograph by Howard E. Evans.)

nally, when the mantid lays her eggs, they rush onto the egg mass before the frothy covering has had time to harden, lay their eggs, and try to regain their position on the mantid. Another group of small wasps, belonging to the genus *Kapala*, live as parasitoids of ant larvae. But, curiously, the adults lay their eggs in flower buds. When the buds expand, the eggs hatch to produce minute, sclerotized larvae that stand on a caudal sucker and jump into space, landing on the ground and waiting for an opportunity to attach themselves to passing worker ants, whereupon they are carried to the nest.

The majority of parasitoids belong to two orders of insects, the Diptera and Hymenoptera. Both groups contain many thousands of species, often of very diverse appearance and behavior, but we shall select only one or two examples from each order for further discussion.

Tachina Flies as Examples of Parasitoids.

A number of groups of flies contain species that are parasitoids. One large family, the Tachinidae (tachina flies), consists entirely of parasitoids that collectively attack all major orders of insects, most commonly Lepidoptera and least commonly other Diptera (Fig. 12–10). Adult tachina flies vary in size from somewhat smaller to considerably larger than a house fly; most are much more bristly than a house fly; and a few are brightly colored. Such matters as degree of host specificity and manner of oviposition vary so much within the group that few generalizations are possible, but it may be noted that nearly all are internal parasitoids of the larvae, or less commonly pupae or adults, of their hosts, and that no tachina flies are hyperparasitoids. Perhaps the majority lay their eggs externally on their hosts, and it is not uncommon to see insects bearing one or more rather large eggs of tachina flies. Larvae hatching from these eggs bore through the underside of the egg and through the host's integument and develop internally. There are, however, other methods of oviposition (or larviposition), as we shall see.

Archytas analis is a robust fly with a tawny yellow thorax and a coal-black abdomen; it is a native American species and attacks a number of native cutworms and armyworms. The adults have a long proboscis and visit a wide variety of flowers for nectar. Mating occurs soon after emergence. Fertilized eggs are retained within the female in a uterine sac, where they produce minute, active maggots, several hundred per female. These maggots are covered with small, sclerotized plates and have a cuplike membrane at their caudal end. They are deposited on plants and are able to remain in a sedentary, free-living state for several weeks. When the substrate is disturbed, the maggot assumes an upright posture on the caudal membrane and moves its head in wide circles. If no caterpillar is encountered, the original position is resumed. If, however, the maggot contacts a caterpillar, it releases its hold on the basal membrane, crawls onto the host, and eventually penetrates its cuticle. Caterpillars at various stages may be attacked, but the first-instar parasitoid maggot waits until the host pupates before molting to the second instar and beginning to feed rapidly. By the time it is ready to molt to the third instar, the host pupa is dead. The nearly mature maggot thus feeds on dead and decaying tissue until forming its own puparium inside the pupal case of the host.

Mortality among the parasitoid larvae is high. Many fail to attach themselves to a caterpillar; others attach themselves to unsuitable hosts in which they cannot develop successfully (they are, however, able to reject grossly unsuitable hosts). These and other causes of mortality are to a considerable extent offset by the high fecundity of the females. Harry W. Allen, who studied this species many years ago in Mississippi, found that populations of variegated cutworms in that state showed from 15% to 36%

Figure 12–10
A tachina fly of the genus *Trichopoda* laying eggs on the back of a green stink bug. These flies have been introduced into several parts of the world for control of this pest. (Photograph by E. Laidlaw; courtesy of the Department of Agriculture of Western Australia.)

reduction as a result of the attacks of *Archytas analis*, and in several other states this tachina fly has been credited as an important natural control agent for the cutworm. This parasitoid, like a great many others, plays a major, though subtle and largely unappreciated, role in reducing populations of native insects that compete with us for food.

Another, somewhat different, example is provided by a tachina fly called *Cyzenis albicans*, which is a native of Europe and is a parasitoid of a well-known defoliator of fruit and deciduous trees of that continent, the winter moth. The winter moth was accidentally introduced into Nova Scotia, being first discovered there in 1949. Within a few years it was causing severe defoliation over much of Nova Scotia and beginning to spread westward. The winter moth was a well-known pest in Europe and known to be attacked there by no less than 63 different natural enemies. Since none of these had arrived with their host, and no native parasitoids appeared effective against it, an effort was made to import several of these. Two became established, an ichneumon wasp and the tachina fly *Cyzenis albicans*. As a result of these introductions, the winter moth is no longer regarded as a major threat to forest and fruit trees in North America. The cost of this biological control project was estimated at $160,000. Damage to oaks in Nova Scotia had been estimated at $2 million; potential damage, several million in that province alone.

Cyzenis differs from *Archytas*, our previous example, in that females deposit small, resistant eggs on foliage, and these hatch only upon exposure to the digestive enzymes of their hosts. Females may lay up to 1300 eggs over a period of several days, and these may remain viable for several weeks. Eggs are not laid indiscriminately, however, as females are strongly attracted to foliage damaged by caterpillars. Hatching normally occurs in the midgut, and the maggots bore from the gut into the blood cavity and then move to the salivary glands, where they feed slowly until the host pupates. Then they feed more rapidly and eventually form their puparia inside the host's pupal integument.

Hymenopterous Parasitoids.   The order Hymenoptera includes several major groups of parasitoids, such as the ichneumons, the braconids, the chalcids, and several other groups, all numbering many thousands of species. Evidently members of this order have features that render them highly adapted for this mode of life. Among these may be mentioned the searching ability noted above. Using diverse cues—often odors emanating from the host or the food of the host—parasitoid wasps are able to seek these out and to lay their eggs (in most cases) directly on or in the host at a specific stage in its life cycle. The ovipositor itself is a remarkable weapon, varying from short and thick to long and slender, as required to reach the host effectively (Fig. 12–11). Frequently it is supplied with glands that secrete a fluid causing temporary paralysis of the host, rendering it quiescent while the eggs are laid. The "wasp waist" of Hymenoptera permits the ovipositors to be wielded in several directions, even directly forward in many cases. The larvae of Hymenoptera have a unique adaptation: There is no connection between the midgut and the rectum until after feeding is completed, when wastes are voided in a single mass. Thus their food is never contaminated with their own wastes.

Other adaptations include the ability of female Hymenoptera to control the sex of their offspring. As we saw in Chapter 8, males are produced from unfertilized, haploid eggs, females from fertilized eggs. As the eggs pass down the oviduct, sperm may or may not be released from the spermatheca, depending on stimuli received by the female. Thus sex ratios are readily modified to suit particular situations. True parthenogene-

Figure 12–11
The chalcidoid parasitoid *Pediobius foveolatus* ovipositing within the body of a larval Mexican bean beetle. The long spinose projections of the larva pose no difficulty for this parasitoid. (Scanning electron micrograph by M. Shepard, G. R. Carner, and J. S. Hudson, Clemson University; reprinted with permission from the *Annals of the Entomological Society of America*; copyright 1976 by the Entomological Society of America.)

Figure 12–12
A hornworm bearing the cocoons of *Apanteles* wasps. The wasps have emerged, leaving an open cap on each cocoon. The hornworm has left the tomatoes on which it was feeding and is dying on an adjacent shrub. (Photograph by O. Wilford Olsen, Colorado State University.)

sis—the development of females from unfertilized eggs, in the absence of males—has also evolved in some groups. Still other adaptive modifications in reproduction have developed, such as polyembryony, permitting the development of several to many offspring from a single act of oviposition.

The vast majority of parasitoid Hymenoptera are small to minute and are rarely observed by nonspecialists. A few, such as *Megarhyssa*, mentioned earlier, are large enough to attract much attention when they swarm around a tree infested with their hosts (Fig. 12–9). Gardeners who grow tomatoes often notice hornworms covered with small, whitish cocoons (Fig. 12–12). These contain the pupae of braconid wasps of the genus *Apanteles* that have developed inside the host. They emerge by making holes in the integument, and although the hornworm is able to live for several days, it is unable to pupate. A related species attacks the larvae of the cabbage butterfly, a pest that arrived from Europe many years ago. Ulysses Aldrovandi observed the cocoons of this insect in 1602 and was the first person to describe an insect parasitoid—though he mistook the cocoons for eggs. This same species is also important for being involved in the first international introduction of a parasitoid for control of a pest. In the 1880s cocoons were brought from England and placed in cabbage fields near Washington, D.C., and in parts of the Midwest. Within a few years the wasp was widely established, and it can still be counted on to show up in virtually every cabbage field and home garden.

Alfalfa is a major crop throughout much of North America, valued at many millions of dollars annually. In 1904 a serious pest, the alfalfa weevil (Fig. 12–13), appeared near Salt Lake City, Utah. This originally European weevil soon spread over most of the alfalfa-growing regions on the continent. The adult beetles overwinter and lay their eggs in the spring on the growing plants or on litter in the fields. The resulting larvae feed on the leaves and buds in such a way that the results of the first cutting may be greatly reduced both in quantity and in nutritive value. Later cuttings may be less affected, although some weevil larvae do often persist beyond the first cutting, and in some areas the weevils have a second generation during the summer.

Beginning in 1911, quite a number of different parasitoids were imported from Europe in the effort to control this serious pest. Several of these are now well established, including one that attacks the eggs, four that attack the larvae, one the pupae, and one the adult weevil. One of these, the larval parasitoid *Bathyplectes curculionis*, became widely established and especially effective. This ichneumon wasp overwinters in cocoons inside those of the weevil. The adults emerge in the spring at about the time the small weevil larvae are beginning to feed. They seek out the larvae and lay a single egg internally in each, the parasitoid larva feeding slowly and killing the host only after it has spun its cocoon, whereupon the ichneumon larva spins up inside the host cocoon as before. Effectiveness of *Bathyplectes* has been very uneven, depending on the season, the locality, and a number of other factors. A brief look at some of the complicating factors in this biological control program may be instructive.

Although superparasitoidism is the rule in *Apanteles*, *Bathyplectes* is larger in relation to the size of its host, and only one larva normally develops inside a host larva. However, as many as 12% of parasitized host larvae have sometimes been found to contain more than one wasp larva, thus effectively reducing the percentage of hosts that might have been destroyed. Evidently the female wasp is able to detect whether or not a weevil larva already contains an egg, but under conditions of strong competition for hosts this may break down. How the female wasp determines this fact is not known in this case, but in other cases it is known that ovipositing females leave a pheromone on the host that causes other females to reject that individual.

Another problem has been posed by the fact that a native chalcid wasp soon established itself as a hyperparasitoid and became a factor in reducing the effectiveness of *Bathyplectes*. These minute wasps overwinter as larvae inside the cocoons of their host, and the adults appear later than those of *Bathyplectes*, in fact about the time the offspring of the latter are spinning their cocoons. The adult chalcids lay their eggs in the newly formed ichneumon cocoons and destroy the pupae. These chalcid wasps attack a variety of hosts and so far have not seriously reduced the effectiveness of *Bathyplectes*.

The alfalfa weevil also has its own defense against parasitoids. In parts of the eastern United States, it was found that weevil larvae often responded to the presence of *Bathyplectes curculionis* eggs by **encapsulation;** that is, the eggs become surrounded by hemocytes that form a thick capsule. Encapsulated eggs die as a result of interference with respiration and with the chemistry of their environment (Fig. 12–14). This problem has been solved to a degree by the introduction of another species of *Bathyplectes* to the eastern states, *B. anurus*. This species is thus far immune to encapsulation by the weevil larvae. When a closely related beetle, the Egyptian alfalfa weevil, was accidentally introduced into California more recently, it was found to be partially immune to attacks of *Bathyplectes curculionis*, again as a result of encapsulation. Similar immunity to the attacks of internal parasitoids has now been described for quite a number of insects.

Still other complicating factors are introduced by climate. An insect and its natural enemies may not be equally well adapted to certain parts of the range or to certain conditions of drought or rainfall. Climatic factors may induce a lack of synchrony between host and parasitoid. At times and in places a parasitoid may find alternate hosts or may acquire competitors or hyperparasitoids. Obviously a profound knowledge of the ecology and behavior of the insects is required. Sufficient information allowing

Figure 12–13
The alfalfa weevil. (Photograph by O. Wilford Olsen, Colorado State University.)

Figure 12–14
Longitudinal section of an egg of *Bathyplectes curculionis* that has been encapsulated within the tissues of an alfalfa weevil larva. The dead embryo may be seen within the egg, which is surrounded by closely packed host tissues. (Electron micrograph by R. Berberet, Oklahoma State University; reprinted with permission from the *Annals of the Entomological Society of America*, copyright 1976 by the Entomological Society of America.)

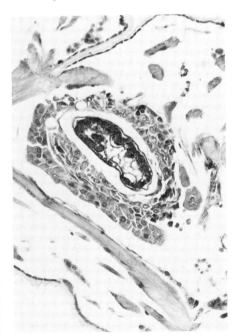

us to predict the outcome of a host–parasitoid association at any time and in any place is rarely available; and rarely are adequate funds available to probe these matters deeply.

For these reasons relatively simple systems that can be handled readily in the laboratory have received much attention. Perhaps, from these, generalizations can be made that will apply to the natural world. One such system is supplied by a minute chalcidoid wasp, *Nasonia vitripennis*, which attacks the puparia of the house fly and several species of blow flies. Because of the ease with which host and parasitoid can be reared and handled in the laboratory, *Nasonia* has been popular with ecologists, behaviorists, physiologists, and geneticists. The wasps are only about 2 mm long, and several (usually 15 to 30, but sometimes over 100) develop in a single puparium. Technically they are ectoparasitoids, since they feed externally on the pupa, though it is enclosed in a puparium. Full-grown larvae pupate inside the puparium without spinning a cocoon; the first adult emerges by cutting a hole through the puparium wall, and the others follow. Males have short wings and cannot fly. They are produced in smaller numbers than the females, the ratio being from 1:3 to 1:6. The males emerge first and mate with the emerging females. Since males do not have to search for mates and one male can inseminate several females, a sex ratio biased toward the females is highly adaptive. Courtship and copulation are surprisingly elaborate for so small an insect, involving a complex interchange of signals, chiefly with the antennae.

Unless fresh puparia are readily available, mated females must fly off to locate new hosts. Since these are generally carrion-feeders, it is perhaps not surprising that they respond to the odor of decaying flesh. When a wasp has encountered a puparium, she climbs on it, moves about, and "drums" on it with rapid strokes of her antennae. If she detects an emergence hole, she leaves and finds another. If not, she begins to tap with the tip of her abdomen, eventually finding a spot suitable for drilling with her ovipositor. Once the ovipositor has been inserted, its tip is moved about over the pupa for several minutes. If the fly pupa is dead or has detectable parasitoids feeding on it, the wasp withdraws her ovipositor and moves elsewhere. If not, she lays several eggs. After laying, the female removes all but the tip of her ovipositor, then secretes from accessory glands a fluid that hardens to form a tube. As it is hardening, the ovipositor is jabbed into the pupa several times, causing a flow of blood into the tube. The wasp then turns about and feeds on the exuding blood. All of this may take an hour or more, after which the wasp rests for a period before proceeding to another host puparium. Feeding on the blood of the host is not an uncommon behavior among chalcid wasps.

Another common phenomenon that has been demonstrated in *Nasonia* is **ovisorption.** When females are not allowed to feed, the developing eggs are resorbed, so that by the time of death (in about five days) only a few are left. When females are fed on honey, eggs continue to be produced at a low rate, but this is balanced by ovisorption, and the wasp may live for several weeks. Only when allowed access to host blood does the female produce viable eggs at a rapid rate. Ovisorption has been demonstrated in a variety of parasitoids and is highly adaptive for this mode of life, since it permits females to recycle proteins under conditions of low host densities and yet be capable of laying when hosts are available.

In 1941 Paul DeBach and Harry Smith, of the University of California at Riverside, used the *Nasonia*–house fly system to test the hypothesis that oscillations are inherent in host–parasitoid systems, as had been proposed earlier on theoretic grounds. That is, if the parasitoid (or predator) is host specific and more successful at finding hosts when these are abundant, then reduction in host numbers will inevitably result in a decrease in parasitoid numbers, permitting another increase in host numbers. In

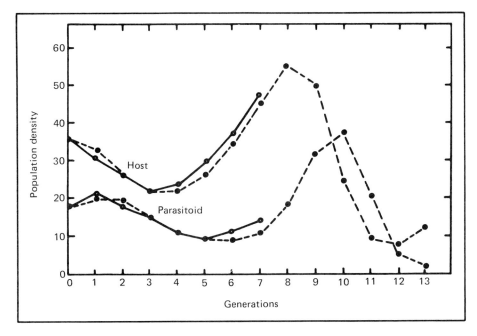

Figure 12–15
Solid lines show the population density of the parasitoid *Nasonia vitripennis* over 13 generations, in relation to that of its host, the house fly; dashed lines show predicted population densities. (based on a model proposed by A. J. Nicholson and V. A. Bailey, 1935, "The balance of animal populations," *Proceedings of the Zoological Society of London*, 1935, pp. 551–598. From P. W. Price, 1975, *Insect Ecology*; with permission of the publishers, John Wiley & Sons, Inc.; after P. DeBach and H. S. Smith, 1941, "Are population oscillations inherent in the host-parasite relation," *Ecology*, vol. 22, pp. 363–369.)

theory, these oscillations might increase in amplitude, resulting in the eventual extinction of the parasitoid, which must always be less abundant than the host. DeBach and Smith found that in their laboratory cultures there was indeed a close approximation to this model over seven generations (Fig. 12–15).

We shall have more to say about the regulation of population size in Chapter 17. Suffice it to say here that generalizations are difficult and predictions unreliable when we still have so much to learn about the biology of individual species. Seemingly simple host–parasitoid relationships are often found to involve complex adaptations that can be appreciated only after detailed study.

Summary

Predators capture and consume a succession of living individuals or prey. Predators may be searchers, stalkers, or trappers. Polyphagous predators take a broad spectrum of insects occurring in their habitat, while oligophagous predators take only prey of certain kinds. Robber flies are top predators in many insect communities. They locate prey with their large eyes; grasp it with long, spiny legs; and suck out the body fluids. In any one locality there are likely to be both polyphagous and oligophagous species; and differences in prey selection, size of the predator, and seasonal occurrence minimize competition between species.

Lady beetles feed largely on aphids and scale insects both as larvae and as adults. Larvae are nearly blind and locate prey largely by phototaxis, geotaxis, and exploratory turning movements. Adults fly considerable distances, and many species fly to mountains and cluster during the winter or through dry periods of summer, flying back to the valleys when there is green vegetation. Several species have been successfully used in biological control.

Parasitoids live as larvae in close association with a single host, which they consume slowly, causing its death only as they approach maturity. The adults are free living, the females laying their eggs on or in a succession of new hosts. Parasitoids cannot be larger than their hosts and are often much smaller, several to many developing within a single host. Many parasitoids are themselves attacked by other parasitoids, called hyperparasitoids. Many parasitoids are relatively host specific and thus of much potential value in the control of pest insects.

Tachina flies form one major group of parasitoids, most of them attacking Lepidoptera or Coleoptera. Nearly all are internal parasitoids, but methods of host finding and egg laying are diverse. One well-studied native species is important in the natural control of cutworms, and another has been introduced from Europe for control of the winter moth in Canada.

The order Hymenoptera includes several major groups of parasitoids. Some are large and conspicuous, others barely visible to the naked eye. Life histories are often complex and adaptations diverse. The role of these insects in natural control of phytophagous insects is impossible to estimate but must be very large. There are numerous examples of international transport of parasitoids for pest control, including a well-known ichneumon that attacks the alfalfa weevil. *Nasonia vitripennis*, a ubiquitous parasitoid of fly puparia, has been extensively studied with respect to host finding, oviposition, and its effect on host populations.

Selected Readings

Berberet, R. C.; K. E. Nuss; and M. L. Koch. 1976. "Capsule formation in *Hypera postica* parasitized by *Bathyplectes curculionis*." *Annals of the Entomological Society of America*, vol. 69, pp. 1029–35.

Caltagirone, L. E. 1981. "Landmark examples in classical biological control." *Annual Review of Entomology*, vol. 26, pp. 213–32.

Clausen, C. P. 1960. *Entomophagous Insects*. New York: McGraw-Hill. 688 pp.

_____. 1976. "Phoresy among entomophagous insects." *Annual Review of Entomology*, vol. 21, pp. 343–68.

DeBach, P., ed. 1964. *Biological Control of Insect Pests and Weeds*. New York: Reinhold. 844 pp.

DeBach, P. 1974. *Biological Control by Natural Enemies*. London: Cambridge University Press. 323 pp.

Hagen, K. S. 1962. "Biology and ecology of predaceous Coccinellidae." *Annual Review of Entomology*, vol. 7, pp. 289–326.

Price, P. W., ed. 1975. *Evolutionary Strategies of Parasitic Insects and Mites*. New York: Plenum Press. 224 pp.

Rogers, L. E., and R. J. Lavigne. 1972. *Asilidae of the Pawnee National Grasslands, in Northeastern Colorado*. Wyoming Agricultural Experiment Station, Science Monograph 25. 35 pp.

Swan, L. A. 1964. *Beneficial Insects*. New York: Harper & Row. 429 pp.

Vinson, S. B. 1976. "Host selection by insect parasitoids." *Annual Review of Entomology*, vol. 21, pp. 109–33.

_____, and G. F. Iwantsch. 1980. "Host regulation by insect parasitoids." *Quarterly Review of Biology*, vol. 55, pp. 143–65.

Blood-Feeders, Parasites, and Scavengers

T he majority of insects categorized as blood-feeders, parasites, or scavengers are not entomophagous but feed on warm-blooded animals or on their carcasses or droppings. Some are major pests of humans and animals because of their bites, while others actually live externally or internally in or on the body or transmit serious diseases. While people in the Western world, in times of peace, have contact with only a few of these insects (such as mosquitoes), they have been of major importance in times past and are still a threat in times of war or in places where sanitation is poor. Even scavengers play a more important role in our lives than is often appreciated. But perhaps the major reason for studying these insects is that they represent still other and often remarkable adaptations and modes of life.

Blood-Feeding Insects

In this section we shall be concerned with insects that take blood from warm-blooded animals but live apart from their hosts while not feeding. These are properly described as **hematophagous,** as opposed to true parasites, which live very intimately with their hosts. There are, as we have seen, insects that suck blood from other insects, such as robber flies and some of the chalcid wasps. But these insects normally kill their host and so are more readily classed as predators and parasitoids. There are also a few insects that attack insects larger than themselves and feed on their blood without killing them. Certain minute midges related to the punkies and no-see-ems that bite humans actually cling to the wings of butterflies and pierce their wing veins for blood. Others attack dragonflies, lacewings, and beetles; one even sucks red blood from engorged mosquitoes and might be termed a secondary hematophage. This group, the biting midges (Ceratopogonidae), also includes species that take plant juices, as well as others that confine themselves to the blood of birds and mammals, again demonstrating the difficulty of categorizing many groups of insects with respect to their feeding behavior.

Hematophagous insects are diverse in structure and behavior. Most belong to the orders Hemiptera and Diptera, but a few Lepidoptera are also blood-suckers. These are the eye-moths, a diverse assemblage of mostly tropical moths that visit the eyes of cattle and other mammals and feed on tears, pus, and sometimes blood; one Asiatic species is able to pierce the skin and suck mammalian blood (Fig. 13–1). Fleas (Siphonaptera) represent a group intermediate between blood-suckers and true parasites. The adults do take blood, but most species jump from the host after feeding, and some can survive long periods away from the host. However, some species remain imbedded in the skin of their host for long periods of time, so we shall treat them here as parasites.

All blood-feeders face certain common problems: they must find their hosts, which may occur at some distance from their breeding places; they must have equipment for piercing the skin and pumping out the blood; they must be able to feed to

Figure 13–1
The blood-sucking moth *Calpe eustrigata* of southeast Asia. The proboscis is modified for piercing the skin of mammals. (Drawing by Peter Eades, based on a photograph by Hans Bänziger.)

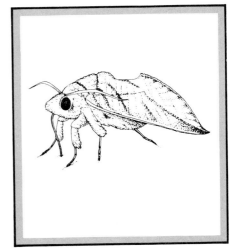

satiation without being detected (or at least destroyed) by their hosts. They must also be equipped to digest and subsist on blood, which is a rich source of protein but deficient in vitamins. The two major groups of blood-suckers have solved the latter problem in different ways. The larvae of Diptera feed in different situations from those of the adults; most of them live in water or soil and are predators or detritus feeders. Here they apparently accumulate vitamins and other essential substances in sufficient quantity to carry them through an adult, blood-feeding stage. Hemiptera that are bloodsuckers throughout their development, such as the bed bug, have in their hemocoel small bodies called **mycetomes**. These contain bacterialike organisms that live symbiotically with their hosts and are believed to produce vitamins required by the insects to supplement their diet.

Biting Flies.   Despite our progress in manipulating nature, biting flies are still very much with us and are likely to remain so. Parts of New York's Adirondack Mountains are still almost uninhabitable in early summer because of vicious hordes of black flies; backpackers in western forests are plagued by deer flies and snipe flies; and tourists in many parts of the tropics are well advised to be on guard against the vectors of malaria, yellow fever, and other diseases. The ubiquitous mosquito even invades the patios of citizens dedicated to the most gracious forms of living. Much of the problem lies in the fact that these insects breed in places remote from where they launch their attacks: swift, clear streams in the case of black flies, moist soil in the case of many deer flies and no-see-ems, and so forth. During World War II, much effort was devoted to determining the breeding places of mosquitoes in tropical and subtropical areas where troops were stationed or in combat, since each species has its own habitat preference, and each has its own characteristics as a vector or nonvector of specific diseases. Even with knowledge of breeding sites, the matter is not solved, as these may be difficult of access or may be in public watersheds or other places where chemical control is undesirable. And mosquitoes and black flies have been known to range as much as fifty or a hundred miles from their breeding places when aided by the wind.

How do biting flies find their hosts? There are literally hundreds of research papers on this subject, but they have produced no simple answers. Much depends on climatic conditions and on endogenous rhythms and responses that vary from species to species. Moderate winds often carry flies from their breeding places into other areas, although strong winds are unfavorable and induce settling. Day-fliers commonly respond to moving objects—horse flies have even been seen pursuing automobiles. Many nightfliers also have vision sufficient to detect moving objects, although they may rely more heavily on odors. Mosquitoes, black flies, and tsetse flies are sensitive to convection currents arising from the host; these currents not only produce a temperature gradient that can be followed but may also carry odors from the body. Thermotaxis may be a fairly general phenomenon but would seem less important when air temperature is close to body temperature or when the host is a turtle or other cold-blooded animal. Chemotaxis undoubtedly plays a role in host selection. Some mosquitoes, for example, attack only birds; and it is a common observation that some humans are more prone to mosquito bites than others are. It is said that Eskimos are rarely attacked by mosquitoes, a valuable adaptive feature in a habitat notorious for the abundance of these insects in the summer months.

The role of carbon dioxide as an attractant or releaser for feeding behavior is less clear. Low concentrations of CO_2 seem to activate mosquitoes rather than attract them, although no-see-ems are attracted to CO_2. Traps baited with solid CO_2 ("dry ice") sometimes collect great numbers of horse flies, though it has been argued that the flies are simply attracted to the image of the trap and are there chilled and anesthetized by the CO_2.

Feeding Behavior of Biting Flies. When a mosquito lands on the host, it first applies the tip of the labium, with its abundance of minute sensilla, to the skin. Very soon the mandibles and maxillae begin to move with alternating thrusts to pierce the skin, with them the labrum (containing the food channel) and the hypopharynx (containing the salivary channel). While the labium remains pressed against the point of entry, the stylets probe for a capillary (Fig. 13–2). At times a capillary is located immediately and the stylets pushed along it for a short distance; at other times capillaries are merely pierced and the blood sucked up as it oozes into the tissues. It is these activities that produce the pain of a mosquito bite, while the itching afterward is a reaction to the salivary fluids pumped into the wound to prevent coagulation and to initiate digestion. Blood is pumped from the host and then into the midgut of the mosquito by the pharyngeal pump. Feeding is terminated by input to the central nervous system from stretch receptors in the midgut—unless the host is alert enough to terminate it prematurely. Cutting the ventral nerve cord near the front of the abdomen results in blood meals up to four times normal size.

It should be pointed out that among most biting flies only the females are bloodsuckers. The large, feathery antennae of male mosquitoes serve as receptors for the wing sounds of the females and are thus involved in mating rather than feeding behavior. The more slender antennae of the females, as well as their palpi and tarsi, have chemo- and thermoreceptors used in finding their hosts. Also, the mouthparts of the males are not equipped to penetrate skin. This is equally true of some of the larger biting flies, such as horse flies and deer flies, in which the powerful cutting mandibles of the females are lacking in the males. It is not, however, true of species related to the house fly—for example, the stable fly, horn fly, and tsetse fly. In these flies the males and females have similar mouthparts and both sexes take blood.

Male mosquitoes and gnats commonly subsist on a dew, honeydew, and the nectar of flowers. Females frequently also use carbohydrates and amino acids from nectar and honeydew and in this way are able to prolong their life and flight range in the absence of vertebrate hosts. Some are even able to produce eggs without taking blood. These are said to be **autogenous.** Autogenous species are ordinarily able to increase their output of eggs greatly, however, following a blood meal. Autogeny is related to the feeding behavior of the larvae; when enough protein is retained through the pupal stage, the adult may have the capacity to produce at least an initial batch of eggs. This capacity is highly adaptive in areas where the supply of vertebrate hosts is unpredictable, as in the Arctic, where a number of species of mosquitoes are known to lay at least a few eggs after feeding on nectar. A few species are even able to utilize protein from their flight muscles for oogenesis and thus achieve autogeny at the expense of being able to fly. But when game is abundant, arctic biting flies take advantage of the animals in great numbers. One traveler reported hords of mosquitoes hovering around caribou and musk ox "like smoke from a row of chimney-pots."

It is a common observation that certain species of mosquitoes are most active during the day, others at dusk, others during the night (Fig. 13–3). Most of the larger biting flies, such as deer flies and stable flies, are most active during warm, sunny periods. In animals unable to defend themselves, loss of blood may be serious. Brian Hocking, of the University of Alberta, showed that in the Far North a single forearm may re-

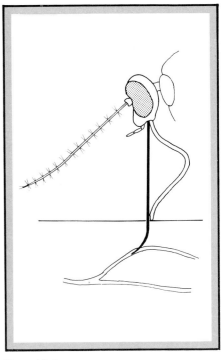

Figure 13–2
Blood feeding by a mosquito. The stylets have penetrated the skin and, after probing, have found a capillary. The labium remains in a loop outside the wound.

Figure 13–3
Feeding cycles of two species of African mosquitoes of the genus *Aedes: A,* a day feeder; *B,* a night feeder. (Adapted from A. J. Haddow, 1961, "Studies of the biting habits and medical importance of East African mosquitoes in the genus *Aedes,*" *Bulletin of Entomological Research,* vol. 50, pp. 759–777.)

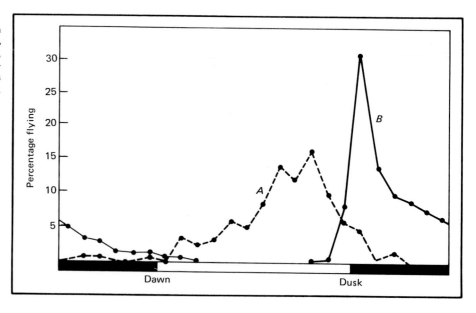

ceive as many as 289 mosquito bites per minute. From this he calculated that a totally unprotected person could receive more than 9000 bites per minute, resulting in the loss of half of the person's blood in two hours. This is hardly likely to happen, although deaths to dogs and livestock as a result of persistent mosquito attacks have occasionally been reported. Livestock much more often lose quantities of blood to horse flies. These flies are **pool-feeders** rather than **vessel-feeders;** that is, they feed from the pool of blood that oozes from the wound. Horse flies cling so firmly to the body that swishing of the tail is often insufficient to drive them off, and after they leave, the wounds often continue to bleed. Beef cattle may lose weight not only from loss of blood but from ceasing to graze while they huddle in protected places. Both humans and animals may suffer systemic illnesses resulting from allergic responses to foreign proteins in the salivary secretions of the flies during feeding. This seems to be especially true of the bites of black flies, which are also pool-feeders and have an especially vicious bite for insects so small.

Biting Flies as Vectors of Disease. The bad reputation of biting flies, however, rests primarily on their role as carriers of several major disease-producing organisms of humans and animals. The story of the discovery of the mosquito vector of yellow fever, making possible the completion of the Panama Canal, is well known. At almost the same time (about 1900) malaria was also shown to be mosquito borne. Since that time a number of other important diseases have been shown to be transmitted by biting flies; some of these are listed in Table 13–1. In several cases biting flies represent the sole method of transmission, and the parasite undergoes part of its life cycle in the fly, as in malaria (Fig. 13–4).

In modern society most of these diseases are under a considerable measure of control, either by way of vector control, control of the infection via drugs, or immunization (as in the case of yellow fever). In the absence of continual vigilance on a worldwide scale, one can easily imagine a return to earlier times, when epidemics were rampant. Even with such vigilance, there is no cause for complacency. When DDT

Table 13–1. A Partial List of Diseases of Humans and Animals Transmitted by Biting Flies

Disease	Fly vectors	Disease organism	Host	Area
Malaria (several kinds)	Mosquitoes (several species of *Anopheles*)	Protozoan (several species of *Plasmodium*)	Humans, monkeys, birds	Tropics and subtropics of world
Yellow fever	Mosquitoes (*Aedes aegypti* and other species)	Virus	Humans, monkeys	Tropics and subtropics of world
Filariasis (including elephantiasis)	Mosquitoes (several species of *Anopheles* and *Aedes*); also no-see-ems	Nematode (*Wuchereria bancrofti* and other species)	Humans, dogs, wild carnivores	Various parts of tropics and subtropics
Dengue (= break-bone fever)	Mosquitoes (species of *Culex* and *Aedes*)	Virus	Humans	Various parts of tropics and subtropics
Encephalitis	Mosquitoes (species of *Culex* and *Aedes*)	Virus	Humans	Sporadic in tropical and temperate parts of world
Equine encephalitis	Mosquitoes (species of *Aedes*)	Virus	Horses	Sporadic in tropical and temperate parts of world
Onchocerciasis (several kinds)	Black flies (species of *Simulium*)	Nematode (*Onchocerca*)	Humans, cattle	Warmer parts of world
Leishmaniasis (several kinds)	Sand flies (*Phlebotomus*)	Protozoan (several species of *Leishmania*)	Humans	Widespread in tropics and subtropics
Anthrax	Horse flies (species of *Tabanus*)	Bacterium (*Bacillus anthracis*)	Ungulates, rodents, humans	Widespread
Tularaemia	Deer flies (*Chrysops*)	Bacterium (*Francisella tularensis*)	Humans, rodents, sheep, quail	Sporadic in temperate regions
Sleeping sickness	Tsetse flies (*Glossina* species)	Protozoan (species of *Trypanosoma*)	Humans	Africa
Nagana	Tsetse flies (*Glossina* species)	Protozoan (species of *Trypanosoma*)	Domestic and wild mammals	Africa

became available, at the time of World War II, it was found that when applied to inner-wall surfaces, it killed insects for many months afterward. In 1955 the World Health Organization (WHO) initiated a program of malaria eradication throughout the world. Progress was spectacular. India, which recorded 75 million cases of malaria in 1947, reported only 125,000 cases in 1965, and in Sri Lanka (Ceylon) the yearly incidence fell from 2.8 million cases in 1946 to only 16 in 1963. But in 1975 Sri Lanka reported 500,000 cases; India, 4 million. Obviously the mosquito vectors were not to be permanently vanquished.

Within a few years mosquitoes had evolved varying degrees of resistance not only to DDT but to several other insecticides. Furthermore, the protozoan malarial parasite had, in some areas, evolved resistance to chloroquine, a drug used to treat malaria. The goal now is "containment" rather than "eradication." Control of malaria, filariasis, leishmaniasis, and other fly-borne diseases is complicated by uncontrolled urbanization in many developing countries, for sanitation is often almost nil when thousands of very poor families are crammed together in the periphery of large cities. Yellow fever remains a problem in Africa and South America in the form of "jungle

Figure 13-4
Life cycle of one of the malaria parasites, the protozoan *Plasmodium vivax*. Periods of high fever follow the bursting of red blood cells and liberation of the merozoites. Human infection is possible only through the bit of certain species of mosquitoes bearing sporozoites. (From J. W. Kimball, *Biology,* © 1978. Addison-Wesley, Reading, MA. Figure 35.8. Reprinted with permission.)

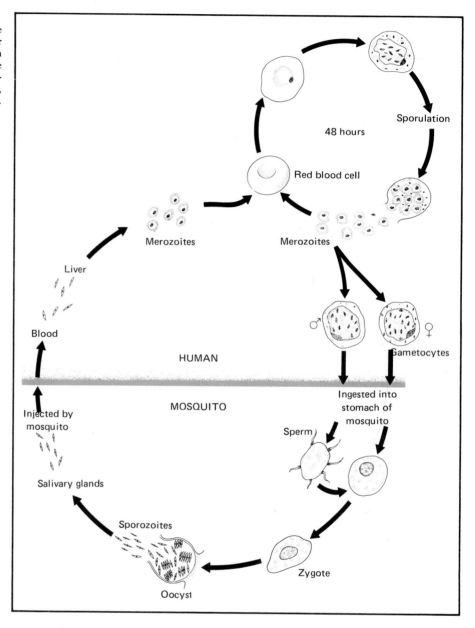

yellow fever," which is transmitted from monkey to monkey by forest-dwelling mosquitoes and is capable of infecting humans who are bitten by these mosquitoes. WHO serves as a clearinghouse for information on all communicable diseases and is prepared to funnel research and control efforts to any part of the world where they are especially needed. Anything less than an international effort would surely be inadequate. It is humbling to consider that a creature as small as a mosquito—or a no-see-em—requires such a mobilization of effort. Yet such is the case.

Sucking Bugs. All Hemiptera are sucking insects, but only a few kinds feed on the blood of vertebrates. The best known of these is the bed bug, once a common inhabitant of homes and hostelries and still a threat where sanitation is poor. Indeed, the word *bug* is said to have come from an old English word for a hobgoblin, a word that also gave us our word *bogey.* The bed bug has a painless bite, a useful adaptation for a nocturnal feeder. Many persons do, however, suffer a reaction to the anticoagulant that the bug injects into the puncture. The bed bug is not known to transmit disease, but a number of its tropical relatives, the so-called cone-nosed bugs, are vectors of Chagas's disease. This disease occurs widely in South America and is caused by a trypanosome (Protozoa); in its extreme form it may cause mental deterioration.

One of the vectors of Chagas's disease is *Rhodnius prolixus,* an insect that has become well known as the laboratory animal on which V. B. Wigglesworth developed much of his research on insect metamorphosis, as we discussed in Chapter 4. Wigglesworth also studied host finding by this insect. A hungry *Rhodnius* is aroused by air currents or heat and approaches a potential host with its antennae in movement. The antennae are covered with minute receptors that are remarkably sensitive to small temperature changes, and the bug is able to follow a temperature gradient of only a few degrees. Smell of the host may also play a role, but blinded bugs perform as well as normal ones. When the antennae touch the host, the beak is thrust forward. The insect then probes with the tip of the labium and soon inserts the stylets (mandibles and maxillae) through which saliva begins to flow. The bug then sucks up enough fluid to "taste" the host and proceeds to engorge only on receipt of favorable stimuli.

W. G. Friend and J. J. B. Smith, of the University of Toronto, have devised artificial feeding chambers for *Rhodnius* and have studied feeding and the resulting electrical resistance in the hemolymph of the bug, simultaneously recording beak movements and electrical changes (Fig. 13–5). The insect feeds on an artificial diet through a rubber membrane, and a phase contrast microscope is focused on the beak, the image being recorded on video tape. At the same time an oscilloscope records conductivity changes in the body, which are correlated with tasting and engorgement, and this is viewed on the same screen, using a split-screen arrangement. In this way the bugs can be tested for the influence of different dietary factors. Evidently adenosine triphosphate (ATP) must be present to induce engorgement; it is said to be a **phagostimulant.** The longer the bug is deprived of food, the more sensitive it becomes to low concentrations of ATP. A diet lacking ATP is sampled repeatedly, but no engorgement follows. Nucleotides similar to ATP are less effective as feeding stimulants, roughly in accordance with their molecular resemblance to ATP.

Under normal conditions an immature *Rhodnius* will take up to nine times its own body weight in blood at one feeding, while an adult will take up to three times its weight. When the gut is full, stretch receptors surrounding the gut transmit a message to the brain that causes withdrawal of the beak. In the larva, a full blood meal initiates a molt via the endocrine system, while in the adult female it initiates oogenesis and eventual oviposition. Blood is also required by the adult male, and it has been shown that fully fed females mated with unfed males lay appreciably fewer eggs.

Ectoparasites

The insect ectoparasites of mammals and birds belong primarily to two groups, the fleas (Siphonaptera) and the lice (Phthiraptera). There are, however, numerous Diptera, the so-called bat flies and louse flies (including a pest of sheep called the sheep ked, Fig. 13–6) that attack warm-blooded animals, though not humans. There is even

Figure 13–5

Friend and Smith's apparatus for studying feeding by the blood-sucking bug *Rhodnius:* b, brass mesh; dc, diet chamber; de, diet electrode; e, electrode in insect thorax; eh, electrode holder; el, electrode lead; fc, feeder chamber; ft, filling tube; r, rubber membrane. (Reprinted with permission from the *Journal of Insect Physiology,* vol. 16 J. J. B. Smith and W. G. Friend, Copyright 1970, Pergamon Press, Ltd.)

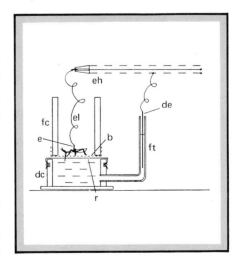

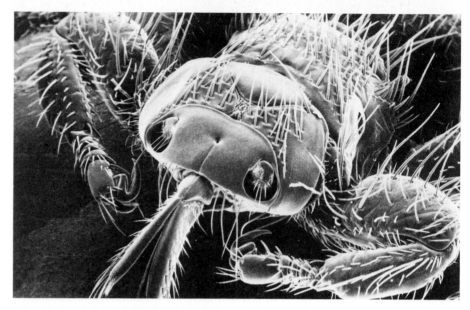

a curious beetle that is an ectoparasite of beavers, as well as a few earwigs in Africa that attack various rats and bats. All these ectoparasites have certain common features: claws or spines for clinging to the host; a body that is flattened dorsoventrally or (in the case of fleas) flattened laterally; small eyes (if any); and no wings (except in some of the louse flies). Most are blood-feeders and thus have a beak capable of piercing the host as well as other characteristics of blood-suckers. We distinguish them from typical free-living blood-feeders on the basis of their close and often permanent attachment to the host, but as already mentioned, the fleas represent a group somewhat intermediate in this respect.

Fleas. Fleas (Siphonaptera) have complete metamorphosis and are parasites only as adults, the larvae living for the most part in the nests of their hosts and feeding on detritus. Most larval fleas do require iron, however, which they obtain by feeding on the droppings of adult fleas. Mature larvae spin a silken cocoon in which the pupal stage is spent. Adult fleas are moderately host specific and are closely adapted to their hosts. Their most prominent feature, aside from the tough, spiny, compressed body, is the series of thick spines, or "combs," near the anterior end of the body. These combs help to retain the flea among the fur or feathers of the host, and the nature of the comb varies depending on the nature of the fur or feathers. The jumping mechanism is also well developed in most species. Miriam Rothschild, a British entomologist who has devoted much of her life to the study of fleas, has made a special study of the pleural arch, a depression above the insertion of the hind legs, which she regards as homologous to the wing-hinge mechanism of flying insects. The arch contains a mass of resilin, a rubberlike proteinaceous material that can store and release energy (Chapter 3, p. 74). When a flea is about to jump, the hind leg is "cocked" by compressing the resilin and engaging the coxa in a pair of sockets at its base; when these are released, the expanding resilin drives the leg downward, and the powerful leg muscles send it leaping through the air, end over end but landing head-forward.

The rat flea, which weighs between 0.2 and 0.4 mg, is said to have an average jump of about 18 cm, with the record at about 31 cm. In general, fleas infesting large mammals or birds are better jumpers, a valuable adaptation in moving from one animal to another. It is said that fleas that attack sand martins emerge from their cocoons in the nests at about the time the martins return from the south, then the fleas jump onto the birds as they hover in front of their nests. If an artificial bird with flapping wings is dangled in front of the nest, the fleas will jump onto it. On the other hand, the fleas that infest moles are poor jumpers. Stick-tight fleas, which burrow into the skin of the host and have little occasion to move about, have lost both the combs and the power to jump (Fig. 13-7).

Most fleas tend to jump from the host after engorgement, and some are able to survive away from the host for weeks or even months. It is not uncommon for persons entering a house that has been empty for some weeks to be besieged by hungry fleas if the previous occupants had an infested dog or cat. Most species have a preferred, or **primary,** host but will accept alternate hosts in the absence of the preferred host. However, certain kinds of fleas have a particularly intimate association with the host. One such species is the rabbit flea, studied in some detail by Miriam Rothschild and her colleagues.

Rabbit fleas are usually found attached to the ears of their host, where they remain for long periods with their mouthparts imbedded in the flesh. But, curiously, they are unable to breed unless the rabbit becomes pregnant or they are able to move to a pregnant female. When rabbits mate, the fleas become much more active; at this time the temperature of the rabbit's ears may rise as much as 7° C, and the fleas begin to hop about actively and may move from one rabbit to another. Following mating, the female rabbit ovulates, and the sex hormones circulating in her blood cause the fleas to attach tightly to her ears. At the same time fleas attached to bucks or nonpregnant females are more mobile and will move to pregnant females if able. About 10 days before parturition, other hormones (corticosteroids) begin to circulate in the blood of the rab-

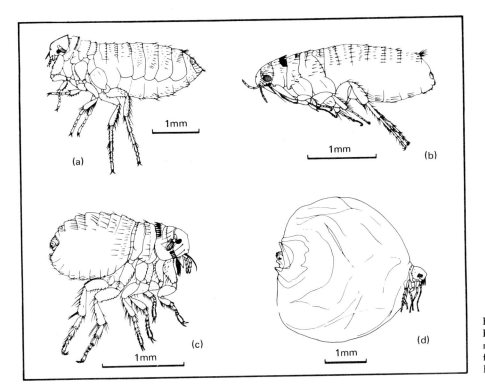

Figure 13-7
Four kinds of fleas: (a) from a swallow; (b) from a mouse; (c) the rabbit flea; (d) the stick-tight flea. (From *Parasitic Insects,* by R. R. Askew, 1971, Heinemann Educational Books Ltd.).

bit, who may now bear a heavy load of fleas. These hormones stimulate development of the eggs of the female fleas. This has been shown experimentally by injecting these hormones into bucks or even spraying them onto fleas, in both cases causing the initiation of egg development.

When the young rabbits are born, the fleas move to the face of the mother and thence to the nestlings, on which they feed voraciously, mate, and lay eggs. The hormones of the host cause other changes in the fleas' metabolism: the salivary glands and gut increase in size, and defecation occurs about five times as rapidly as before. This is also adaptive, since the flea larvae that emerge from the eggs require dried blood to supplement their diet. It is believed that growth hormones in the newborn young play a role in stimulating copulation, since fleas on adult rabbits make no attempt to copulate.

Fleas and Disease.   The rabbit flea has earned a reputation as a vector of myxomatosis, a virus infection of rabbits that has been used in the biological control of rabbits in Australia. A much more evil reputation attaches to the rat flea, which transmits murine typhus and plague. In past ages, when the rat population of cities sometimes exceeded that of the human population, outbreaks of plague (or "black death") were common and often resulted in widespread political and moral disintegration. Plague is caused by a bacterium, *Yersinia pestis*, which is pathogenic to fleas and is transmitted from rat to human or from person to person by bites of rat fleas. In the flea, the bacteria produce a blocking of the digestive tract, causing the flea to bite repeatedly but without being able to engorge, at the same time regurgitating part of its infected gut contents into the wound.

The first recorded epidemic of plague occurred in the sixth century A.D. and is said to have caused 100 million deaths. A second epidemic, in the fourteenth century, is said to have killed one-fourth of the population of Europe and over half the population of England. In 1664 to 1666 another outbreak of black death in London caused widespread death and misery and was vividly described by Daniel Defoe, Samuel Pepys, and others. After 1700, outbreaks of plague gradually declined in frequency and intensity, perhaps in part a result of better sanitation and rat control, although it was not until about 1900 that the causative organism was identified and the rat flea incriminated as the major vector.

Plague is still with us, however, and every year cases are identified in the western United States. In this instance the natural reservoir is not rats but ground squirrels, and the vectors are several species of fleas that infest those rodents. Teams in the U. S. Public Health Service monitor rodent populations carefully, especially in parks and camping areas, and when plague-infected rodents are found, the area is evacuated. Foci of plague occur in many parts of the world, and once again constant vigilance is required if plague is to remain in our history books and not in our headlines. Albert Camus's novel *The Plague* provides a frighteningly vivid scenario of this dread disease in a modern city.

Lice.   Lice (Phthiraptera) are ectoparasites par excellence (if excellence can be said to apply to such unpopular organisms). The entire life cycle is spent on the host, and most species of lice infest only one species of animal, or even only one part of the animal's body. Lice move from one host to another only when two individuals are in close bodily contact, and when the host dies, the lice die also unless they are able to transfer quickly to a new host.

There are two major groups of lice. **Biting lice** (Mallophaga) live on birds or mammals and have biting mouthparts. Their food consists primarily of bits of feathers, hairs, and epithelium. **Sucking lice** (Anoplura) have elongated, sucking stylets that can be retracted into a pouch in the head. They attack only mammals and feed on blood. Both groups have similar adaptations for clinging to the host, and the two groups are believed to have had a common ancestor. Also, many Mallophaga will take blood exuding from scratches in the skin, and some are able to pierce the skin. Teresa Clay, a British authority on lice, has discovered that some "biting lice" in fact have piercing mouthparts, indicating that mouthparts of this type have evolved at least twice independently. Thus it makes sense to place all lice in a single order, Phthiraptera, even though some books treat the two groups as separate orders.

In both biting and sucking lice the eggs are glued singly or in small custers to hairs or feathers. The larvae develop rapidly, molting three times and producing adults in only about a month's time. The more or less constant conditions on the warm-blooded host permit breeding throughout the year. In some cases females are able to breed parthenogenetically. Thus popoulations are often able to build up rapidly.

Biting lice have not been incriminated as disease vectors, but heavy infestations may produce skin irritation; loss of hair or feathers; wounds resulting from scratching by the host; sometimes lethargy and loss of weight; or, in the case of poultry lice, severe reduction in the number of eggs laid. Biting lice infest a wide variety of birds and mammals. There is even one species that lives within the pouches of pelicans, and several that infest seals in cold antarctic waters. However, whales and dolphins have no lice, and for some unknown reason there are no species attacking bats.

Sucking lice, like other obligate blood-feeders, have symbiotic bacteria that produce vitamins needed by the louse but not present in blood. These bacteria are enclosed in pouches (mycetomes) close to the midgut; but when the female is about to lay eggs, they migrate to the ovaries, and some of them enter the egg before the chorion is laid down. Thus the newly emerging larva is provided with an inoculum of these essential bacteria. It is said that these bacteria are closely attuned to the diet of the louse and that blood of an abnormal host may be lethal to them. This may account in part for the strong host specificity of sucking lice.

It has often been noted that closely related species of animals have very similar but not identical lice, suggesting that lice have evolved along with their hosts and might be used as indicators of mammalian evolution. For example, the genus *Pediculus* attacks both humans and chimpanzees; the related genus *Pedicinus*, cercopithicoid monkeys. *Phthirus*, the genus of the crab louse of humans (Fig. 13–8), also has a species on the gorilla. Curiously, the human body louse, *Pediculus humanus*, is able to live and breed on only one other animal, the pig. Several of the other parasites of pigs and humans are also interchangeable, suggesting that the drawing of conclusions regarding mammalian relationships from their parasites may sometimes be unwarranted (or if made, embarrassing).

Infestation with lice, termed **pediculosis,** can be seriously debilitating and can lower resistance to disease even in the absence of disease transmission by the lice themselves. Among domestic animals, there are sucking lice that are pests of horses, cattle, hogs, sheep, and goats. Humans are hosts for the crab louse, which is transmitted venereally, as well as the head and body louse, two species or races that tend to favor either the head or the body and are transmitted either through body contact or via infested clothing. G. H. F. Nuttall, to whom we owe a great deal of our knowledge of the lice that infest humans, once recorded 10,428 body lice on a single shirt. So prevalent were lice some generations ago that Robert Burns penned the well-known poem "To a Louse." It is said that when Theodore Hook displaced the third Lady Holland, the latter banished him from her house with the expression then current: "I

Figure 13–8
Scanning electron micrograph of the crab louse of humans. These lice are transmitted venereally, but are not known to transmit disease. (Photograph by Tyler A. Woolley, Colorado State University.)

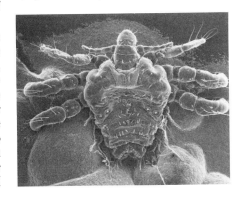

Figure 13–9
The various kind of typhus and their possible origin. Typhus rickettsiae probably originally occurred among ticks (top left) but were passed to rabbits and other small mammals; here they were picked up by fleas and transmitted among rodents and occasionally to humans (murine typhus). Human body lice transmit epidemic typhus among humans, but the relationship may be fairly recent, as infected lice (unlike fleas) die from the infection. (From *Parasitic Insects*, by R. R. Askew, 1971, Heinemann Educational Books Ltd.)

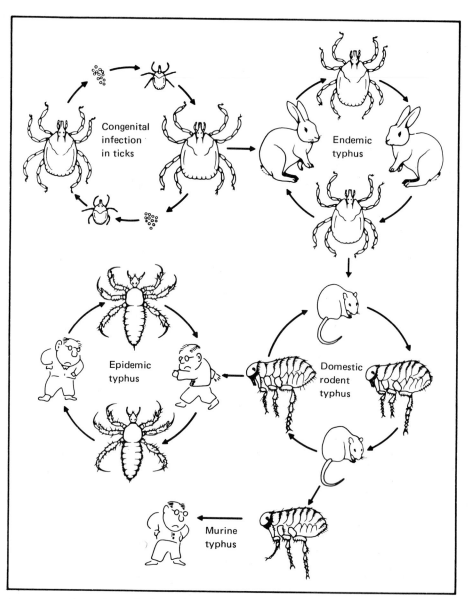

do not care three skips of a louse for you." Hook's response appeared in the papers a few days later:

> Her ladyship said when I went to her house,
> She did not regard me three skips of a louse.
> I freely forgive what the dear creature said,
> For ladies will talk of what runs in their head.

Lice and Disease.   Human head and body lice are the vectors of epidemic typhus, the causative organism of which is a rickettsia, a type of bacterium as small as some of the viruses. Typhus is a disease characterized by high fever, lethargy, and spotting of the skin; mortality varies from 10% to 80%, depending on the strain. The rickettsiae differ slightly from those causing murine typhus, which is transmitted by fleas from rodent to rodent, and sometimes to humans. However, it is usually assumed that louse-borne typhus evolved from murine typhus (Fig. 13–9). That the rickettsia is pathogenic, often fatal, to both human and louse suggests that it is poorly adapted to these hosts. The rickettsiae multiply in the midgut of the louse, and some are passed with the feces. Infection of humans occurs when louse droppings are scratched into the skin.

Typhus has long been the scourge of troops in combat or in military prisons, where there is little opportunity to keep clean or change one's clothing. Typhus is said to have been largely responsible for decimating Napoleon's army and necessitating the retreat from Moscow. During World War I, typhus was rampant on the eastern front, and over 150,000 cases were reported from Serbia, even more from Russia. Oddly, there was little typhus on the western front, even though troops were often lousy and another but less serious louse-borne disease, trench fever, was common. In World War II, an incipient epidemic in Italy was snuffed out by dusting the troops with a recently discovered chemical that was deadly to lice, DDT. Persons interested in the influence of infectious diseases on history would do well to read Hans Zinsser's classic book *Rats, Lice, and History.* In Zinsser's words:

> Swords and lances, arrows, machine guns, and even high explosives have had far less power over the fates of the nations than the typhus louse, the plague flea, and the yellow-fever mosquito. Civilizations have retreated from the plasmodium of malaria, and armies have crumbled into rabbles under the onslaught of cholera spirilla, or of dysentery and typhoid bacilli. Huge areas have been devastated by the trypanosome that travels on the wings of the tsetse fly, and generations have been harassed by the syphilis of a courtier. War and conquest and that herd existence which is an accompaniment of what we call civilization have merely set the stage for these more powerful agents of human tragedy.

Internal Parasites

Under this heading we consider various insects that undergo part of their life cycle internally in the host but do not normally cause its death. Again we shall primarily be talking about vertebrate animals as hosts; but before discussing these, it should be noted that there are a few insects that are true internal parasites of other insects. These belong to an unusual group of beetles called the Stylopoidea (or simply "stylops"). These appear to be related to other groups of beetles, such as the blister beetles, that cause the death of the host and are therefore properly considered parasitoids. Insects infested with stylops are said to be **stylopized,** and although they may have somewhat abnormal features and be unable to reproduce, they do usually reach adulthood and live a life of relatively normal length. So unusual are these insects that some persons have placed them in a separate order, called Strepsiptera, although students of Coleoptera properly claim them as beetles.

Stylops are parasites of certain Hymenoptera and Hemiptera, less commonly Orthoptera and Thysanura. Species that attack bees release newly hatched larvae on flowers, and these are sucked up by other bees when gathering nectar. When the bee regurgitates nectar from her crop into her nest cell, the parasite larvae are released and penetrate the egg and the developing embryo. After developing slowly inside the growing bee larva, they eventually become partially external when the bee pupates, protruding their heads between some of the abdominal terga. At this time the female stylops molts to a larviform adult stage, but the male becomes a pupa and must molt once more to become an adult at about the time the bee emerges as an adult. Male stylops have very short elytra and large, fan-shaped hind wings; they fly rapidly about and mate with females still residing between the terga of their host. The fertilized eggs hatch within the body of the female stylops. Later, when the bee carrying the now dead stylops is visiting flowers, the larvae burst out onto the flowers, where they are ready to be sucked up by another bee.

This is, briefly, the life history of *Stylops pacifica,* as described by E. G. Linsley and J. W. MacSwain, of the University of California at Berkeley (Fig. 13–10). Species that attack other kinds of hosts have rather different life cycles, but in each case the stylops female remains larviform, and the male is an active (though very short-lived) flying insect. The host usually lives a life of relatively normal duration, but in most cases it is sterile. In addition to **parasitic castration,** as it is called, the hosts may be **intersexes;** that is, they may have a mixture of male and female features. This is apparently a result of nutritional deficiencies produced by the parasite and allowing genes of the alternate sex to gain expression.

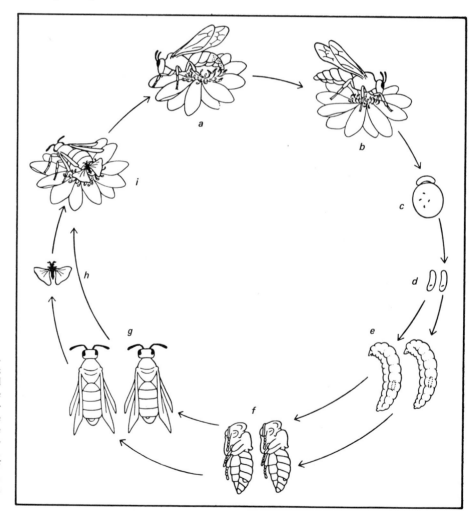

Figure 13–10
Life history of *Stylops pacifica,* a parasite of bees. At *a,* the small larvae emerge from an infected bee; at *b,* they are sucked up by another bee and carried to the nest. Here they are regurgitated into the nectar-pollen mass *(c)* and enter the egg *(d)* and develop inside the larva *(e).* Eventually they protrude partially from the pupa *(f)* and the adult *(g).* The male *(h)* emerges and flies about, inseminating a female still largely inside the host bee *(i).* (E. G. Linsley and J. W. MacSwain, 1957, University of California Publications in Entomology, vol. 11. Originally published by the University of California Press. Reprinted by permission.)

Stylops attack only a limited spectrum of hosts and do not seem especially abundant anywhere. Perhaps this is a case of overspecialization: their complex life cycles and structural adaptations permit them to succeed only under very special conditions. It is much less true of the internal parasites of warm-blooded animals, which are much more ubiquitous and broadly adapted.

Myiasis.   Nearly all the internal parasites of vertebrates belong to the order Diptera. All are **protelean**—that is, only the larvae are parasites; the adult flies, free living. Infestation with the maggots of flies is termed **myiasis,** based on the Greek word for fly, *myia.* There are many fly larvae that live in carrion or decaying organic matter but are able to live for some time in the gut or nasal passages of living animals if they gain access. This is spoken of as **facultative myiasis.** True internal parasites are able to live nowhere else and are said to produce **obligate myiasis.**

Many of the blue bottle flies or blow flies (Calliphoridae) provide good examples of facultative myiasis. These flies lay their eggs on dead animals, and the maggots consume the decaying flesh and the microorganisms it contains. The flies are also attracted to wounds in living animals if these have been neglected, and they may have a beneficial effect in cleaning out dead tissue. During the Civil War, and in fact until quite recently, blow fly maggots were used medicinally for this purpose. In addition to removing dead tissue, the maggots are believed to secrete substances that kill or inhibit bacteria as well as substances that render the wound more alkaline and stimulate phagocytosis. Unfortunately certain species of blow flies may also enter living tissue and expand the wound.

Finally, there are several kinds of blow flies that no longer live on carrion at all but lay their eggs only on wounds, where the larvae develop on living flesh, deepening the wound and preventing healing. Such an insect is the notorious screwworm, which in past years has caused the death of hundreds of thousands of cattle in the southern United States. Use of the sterile-male technique has greatly reduced the ravages of this insect (see Chapter 18).

Other blow fly maggots occur in the nests of birds, taking blood from the nestlings and in some cases burrowing into the skin. Many mammals are similarly attacked by fly maggots that infest their nests, suck their blood, and often burrow into the skin to form boils or abscesses. The tumbu fly of Africa, for example, lays its eggs in soil dampened with urine or feces, and the maggots attach themselves to the feet of passersby and burrow in to form a boil. After a week or two the fully grown maggot leaves the boil and drops to the ground to pupate. Hosts include humans, dogs, rats, and monkeys.

Infection with these or any of the many other blow flies or flesh flies can be a serious matter. Even facultative-myiasis producers can produce painful conditions of the bowels, nasal passages, and so forth, as well as psychosomatic problems, especially when the maggots leave the body to pupate. Obligate-myiasis producers such as the screwworm, the tumbu fly, and other species, remain a threat in certain areas of the tropics and subtropics. Much academic interest attaches to these insects as evolutionary precursors to the bots and warbles, which belong to other, related families of flies and have more intimate, complex relationships with their hosts.

Horse bot flies are hairy, beelike flies that fly swiftly and lay their eggs singly on hairs by "striking" the horse about the legs or flank. The eggs remain viable for some time and hatch only on being licked by the horse, whereupon the small maggots burrow into the tongue. After completing one instar in the tongue, they move to the lining of the stomach, where they attach firmly through lesions in the stomach wall (Fig. 13-11). When fully fed, they pass out with feces and pupate in the soil or feces. Cer-

Figure 13-11
Horse bots attached to the lining of the stomach of a horse. Several have been removed, showing the perforations in the stomach wall. (Photograph by Roger D. Akre, Washington State University.)

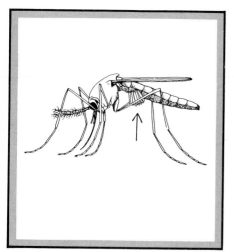

Figure 13–12
A "warble" on the back of a steer, showing the cattle grub with its spiracles opposite an opening in the skin.

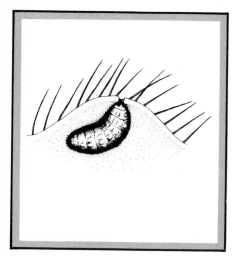

Figure 13–13
The eggs (indicated by an arrow) of the human bot fly, *Dermatobia hominis*, which have been laid on a mosquito.

tain species lay their eggs primarily around the mouth and throat of the horse and, when laying their eggs, are especially annoying to the horse. Heavily infested horses suffer from malnutrition and from ulceration of the stomach wall.

The common cattle grub, or "ox warble," belongs to a somewhat unrelated group of flies, but the adults are also hairy and beelike and lay their eggs on the hairs of their hosts. Although no pain is produced when the eggs are laid, cattle are nevertheless often terrorized by these flies and may stampede, hence the term *gadfly* that is often applied to these insects. In this instance the eggs do not have to be licked, but the larvae hatch after a few days and crawl down the hairs and into the skin. Then for several months they burrow about in the body tissues. Eventually they settle in the subcutaneous tissue of the back, making a small breathing pore to the outside. In these swellings, or "warbles," the larvae continue to grow and molt, finally leaving to pupate in the ground (Fig. 13–12).

Other species of bot flies attack sheep, deer, rabbits, and other mammals, with variations in the life cycle adaptive for that particular host. The human bot fly, not uncommon in parts of the American tropics, has one of the most unusual life cycles of all, since the adult flies capture mosquitoes or other biting flies and lay a packet of eggs on them (Fig. 13–13). When the mosquito is released and bites a person, the eggs hatch and the larvae burrow into the skin. Here they feed for several weeks, forming painful boils, before emerging and dropping to the ground to pupate. Larvae of cattle grubs occasionally infect humans and can be particularly distressing because of their tendency to wander about inside the body.

The Evolution of Insect Parasitism of Warm-Blooded Animals

As we have seen, many groups of insects feed on warm-blooded animals, and some of these constitute major pests of humans and domestic animals. There are still other groups we have not reviewed. Altogether members of seven insect orders have developed close dependence on vertebrate blood and tissues as a food source. Their adaptations are diverse, including piercing and sucking mouthparts of several kinds; large claws for clinging to the host; flattening; winglessness; and dependence on symbiotic bacteria.

Vertebrates have, in turn, developed diverse mechanisms for reducing the effectiveness of these parasites and blood-feeders. Reindeer migrate seasonally to areas where mosquitoes are less abundant; horses aggregate tightly and use their tails vigorously when biting flies are abundant; mammals and birds groom and preen, removing lice and other ectoparasites. Passive defenses include dense fur, feathers, and thick hides. It has been noted that tsetse flies do not attack zebras, though they feed readily on horses, and it has been suggested that the disruptive coloration of zebras makes it difficult for tsetse flies to orient to their hosts. On the other hand, since zebras occur in the native home of tsetse flies, they may well have evolved a chemical defense such as a surface odor repellent to the flies. It is well known that among members of any one species (humans, for example) some individuals react much more strongly than others to insect bites. It can hardly be doubted that in the course of evolution of many animals, individuals with less resistance have had reduced reproductive success, and those with greater innate resistance to bites have tended to replace them.

In turn insects have evolved ways of overcoming the defenses of vertebrates. The elephant louse, for example, has developed a remarkably long proboscis capable of piercing the thick hide of pachyderms. The elaborate "combs" of fleas are doubtless effective in preventing dislodging during grooming. Rapid and/or painless feeding has evolved in several groups as a means of ensuring satiation before the host can react. Coevolution of insects and vertebrates is as fascinating and important a subject as that of insects and plants (Chapters 9 and 10), though it has been less well studied.

How did dependence on warm-blooded animals evolve? In terms of geologic time, insects are a much older group than mammals and birds. There can be little doubt that for hundreds of millions of years insects were plant-feeders, predators, and scavengers. Some paleozoic insects had long, sucking mouthparts and are believed to have sucked plant juices (Fig. 2–3). The first associations with vertebrates may have been with ectotherms, such as amphibians and reptiles; it may have been in defense, just as certain plant-feeding Hemiptera today are capable of piercing the skin if handled. Ultimately the switch to blood must have entailed many physiological and behavioral changes, though doubtless the first blood-feeders were less efficient than contemporary mosquitoes and bed bugs.

A second evolutionary pathway may have been by way of scavenging. The nests of birds and mammals provide a source of food scraps, bits of feathers and hair, dried blood, and so forth; and fecal matter also provides a source of food for many insects. Some of these scavengers probably came to cling to their hosts, feeding on feathers and skin fragments (as biting lice still do) or riding on the host to fresh fecal droppings (as some dung beetles do). In this instance the transition to ectoparasitism first involved flattened bodies and clinging devices, with piercing and sucking mouthparts coming later (the reverse of the previous pathway); but similar physiological changes must have occurred to permit the use of blood and tissues as food. Some of the Diptera are both scavengers and flesh- or blood-feeders, the larvae living in dung or carcasses; and of course the larvae of fleas are scavengers in the nests of their hosts.

That these were the two major pathways leading to hematophagy and vertebrate parasitism is the conclusion of Jeffrey Waage, of Imperial College, England. In the first, the ancestral insects were preadapted by way of their piercing and sucking mouthparts; in the second, by way of their intimacy with vertebrates via their nests or droppings. Since switching to vertebrate tissues (doubtless at different times in the past), insects have coevolved closely with their hosts. Such an evolutionary approach to this subject may well yield clues as to how ways may be found to minimize the attacks of these ubiquitous pests.

Insects as Scavengers

The relationships between insects and living plants and animals are diverse and often complex, yet many of them are part of the experience of all of us—for who has not been bitten by a gnat or annoyed by aphids infesting carefully tended house plants? However, the roles that insects play in reducing dead organic matter to a form suitable for reutilization by other organisms are not often appreciated. If one has the patience and fortitude to study an animal carcass over a period of weeks, one cannot fail to be impressed by its gradual reduction to a few fragments of bone. Maggots are likely to be the first invaders, tearing the flesh into pieces more suitable for incursion by bacteria and by secondary invaders such as carrion beetles. As the flesh disappears, dermestid beetles appear to feed on the skin and fur. A similar succession of insects occurs in fallen trees, where, greatly abetted by wood-rotting fungi, the insects ultimately reduce the log to earth. Insects also play a role in the reduction of fallen leaves to humus (Fig. 13–14). Ants, perhaps the most abundant and ubiquitous of all terrestrial animals, play a major role in the mixing of soil, throwing up subsurface soil and carrying organic particles down into their nests.

Figure 13–14
Stages in the decomposition of a leaf: (upper left) healthy leaf; (upper right) the epidermis is broken by small soil arthropods; (lower left) fly larvae, earwigs, sowbugs, and other arthropods continue the process; (lower right) finally all but the veins is chewed away by mites, springtails, and small insect larvae, aided by bacteria.

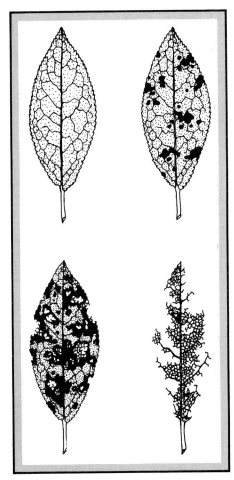

Figure 13–15
A pair of dung beetles rolling a ball of dung to a
nest they will prepare in the soil. (Photograph by
James E. Lloyd, University of Florida.)

Animals that feed on dead organic matter are said to be **saprophagous.** When the
food consists of carrion, they are said to be **necrophagous**; when it consists of fecal
material, they are said to be **coprophagous.** Study of the diet of such insects often reveals that they feed also, or even exclusively, on microorganisms or small invertebrates
in the flesh or feces. Nevertheless the result of feeding by the vast number of saprophagous insects is reduction—with an important assist from bacteria and fungi—of dead or
waste materials to part of the soil that nurtures all living things. Studies of humus on
the forest floor reveal that much of it consists of small, dark particles that in fact are
the droppings of small arthropods that have been instrumental in the initial stages of
humus formation. The role of earthworms and other invertebrates can, of course, by
no means be discounted.

Unfortunately, some of the insects that act as scavengers in nature can become
pests when we ourselves make use of dead animals. Improperly preserved meat readily
becomes infested with maggots, some of which are capable of causing intestinal
myiasis. Woolen clothing and carpets are attacked by clothes moths and carpet beetles
(dermestids), which in nature inhabit the fur of dead animals. Other household pests,
such as cockroaches, formerly scavenged in nature, where their efforts were
undoubtedly beneficial from the point of view of recycling wastes.

Some appreciation of the value of saprophages may be obtained by considering
situations where essential species are absent. Although most parts of the world are well
endowed with dung beetles (Fig. 13–15), Australia is a continent without native ungulates and without native dung beetles capable of utilizing the droppings of these animals. The droppings of kangaroos and their kin consist of small, dry pellets, and the
native dung beetles are adapted for handling these. According to Eric Matthews, who
is curator of insects at the South Australia Museum, some native dung beetles have
large, hooked claws, which they use to cling to hairs and ride on the bodies of kangaroos and wallabies. When the animals begin to defecate, the beetles move rapidly to
the emerging pellets and may even ride to the ground with them. Here they rapidly
bury the pellets and lay their eggs on them.

But cattle have become a major industry in Australia. According to D. F.
Waterhouse, former chief of the Division of Entomology of the Commonwealth Scientific and Industrial Research Organization (CSIRO), there are now 30 million cattle in
Australia, collectively producing at least 300 million dung pads per day. Since the native dung beetles cannot utilize these, they persist for months, littering the ground in
some areas so densely that good pasturing is destroyed. Also, they provide breeding
places for certain flies that spend the larval stage in dung. These are the bush fly, a
notorious pest that does not bite but congregates on the bodies of warm-blooded animals, including humans; and the buffalo fly, a biting fly that is a vicious pest of humans
and animals in northern Australia.

In contrast, Africa has many native ungulates, and there are many species of dung
beetles adapted for utilizing their droppings. Some of these are very common: more
than 7000 beetles have been counted on a single mass of fresh elephant dung. These
facts suggested to personnel of CSIRO the possibility of a unique biological control
program: the importation of coprophages from one continent to another. Beginning in
1967, several hundred thousand beetles of several species were introduced into cattle-
producing areas of northern Australia. Several have become established and are now
widespread, causing a noticeable reduction in the number of residual dung pads in several areas.

It is possible that saprophagous insects may someday prove of value in recycling
the organic wastes that are produced in increasing amounts as human populations in-

crease. Biodegradation of poultry droppings by house fly larvae results in a reduced amount of relatively odorless, granular manure, the remainder being consumed by the maggots. The pupae of the flies contain more than 50% protein and substantial amounts of fat, with an energy value higher than that of the soybean meal that is used in feeding growing chicks. Chicks fed on dried house fly pupae as a substitute for soybean meal grow normally and produce marketable birds with no unusual flavor.

Another possible source of insect proteins is secondary sewage lagoons, which are commonly inhabited by a great variety of dipterous larvae and other small arthropods. It should be possible to develop techniques for harvesting and drying this protein-rich biomass and producing a nutrient meal suitable as animal food. As of now, little attention has been paid to the possibility of utilizing insects to reduce wastes to usable by-products. Data gathered by G. F. DeFoliart, of the University of Wisconsin, indicate that on a dry-weight basis the bodies of various insects contain 50% to 70% protein, 5% to 40% fat, and minerals in considerable quantity. As he says, what is now needed is exploratory research on the feasibility of recycling wastes economically and obtaining insect protein that can be used as animal food.

Summary

Blood-feeding (hematophagous) insects are diverse in structure and behavior, but most belong to the orders Diptera and Hemiptera. Biting flies such as mosquitoes and black flies are pests in many parts of the world, and control is complicated by the fact that breeding places of the larvae are often inaccessible or in public water supplies. Host finding by biting flies is not fully understood but appears to involve thermotaxis and chemotaxis, in some cases attraction to CO_2 emanating from the host.

Male mosquitoes subsist on dew, honeydew, and nectar, and females of some species are able to lay at least a few eggs without a blood meal, although in general blood is required for oogenesis. Mosquitoes are vessel-feeders, probing capillaries with their slender stylets. Black flies and horse flies are pool-feeders, cutting a hole in the skin and lapping up the blood that exudes.

The many diseases of humans and animals transmitted by biting flies include malaria, yellow fever, filariasis, encephalitis, sleeping sickness, and others. Sucking bugs are less important as disease carriers but do transmit Chagas's disease. Feeding behavior has been well studied in certain sucking bugs.

Fleas are blood-feeders as adults, but the larvae are scavengers in the nests of their hosts. Fleas have many adaptations for parasitic life: compressed, spiny bodies; a powerful jumping mechanism; and in some species a close response to hormonal changes in the host. Fleas transmit a number of diseases, including plague, that have had a profound influence on human history in past ages.

Lice spend their entire life on the host and die when the host dies unless they can rapidly move to another host. Most are quite host specific. Biting lice feed mainly on hairs and feathers, while sucking lice have specialized piercing mouthparts and are blood-feeders. These insects have symbiotic bacteria that are transmitted to their offspring via the eggs. Epidemic typhus is a major disease transmitted by lice.

In contrast to these ectoparasites, some insects develop internally in their hosts (endoparasites). Stylops are internal parasites of other insects, which they do not kill but may render sterile. Most internal parasites attack mammals, and most are Diptera. Infection by fly maggots is termed myiasis. A number of pests of domestic animals fall in this group: the screwworm, horse bots, and cattle grubs, for example.

Myiasis producers probably evolved from carrion-feeders (necrophages) and dung-feeders (coprophages). These insects are important in their own right, playing major roles in the reduction of dead or waste organic matter to soil. Dung beetles have been used in the biological control of dung in Australia, where there are no native species capable of reducing cattle dung, the breeding place of several pest species of flies.

Selected Readings

Askew, R. R. 1971. *Parasitic Insects.* New York: American Elsevier. 316 pp.

Friend, W. G., and J. J. B. Smith. 1977. "Factors affecting feeding by blood-sucking insects." *Annual Review of Entomology,* vol. 22, pp. 309–31.

Gillett, J. D. 1972. *The Mosquito: Its Life, Activities, and Impact on Human Affairs.* New York: Doubleday. 358 pp.

Harwood, R. F., and M. T. James. 1979. *Entomology in Human and Animal Health.* Seventh edition. New York: Macmillan. 548 pp.

Hocking, B. 1971. "Blood-sucking behavior of terrestrial arthropods." *Annual Review of Entomology,* vol. 16, pp. 1–26.

Marshall, A. G. 1981. *The Ecology of Ectoparasitic Insects.* New York: Academic Press. 446 pp.

Rothschild, M. 1965. "Fleas." *Scientific American,* vol. 213, pp. 44–53.

Waage, J. K. 1979. "The evolution of insect/vertebrate associations." *Biological Journal of the Linnean Society,* vol. 12, pp. 187–224.

right, J. W.; R. F. Fritz; and J. Haworth. 1972. "Changing concepts of vector control in malaria eradication." *Annual Review of Entomology,* vol. 17, pp. 75–102.

Zinsser, H. 1934. *Rats, Lice, and History.* Boston: Little, Brown. (Paperback edition, 1960, Bantam Books, New York, 228 pp.)

Insects as Food, and Their Defenses against Being Eaten

Considering the nutritional value of insects, it is not surprising that a vast number of animals depend on them as a source of food. In modern human societies, they are rarely consumed directly as food, though we have no objection to eating trout or pheasant and thus obtaining insect protein second-hand. In ancient societies, insects were often used as major sources of food, as among the Hottentots, the Australian aborigines, and various groups of American Indians. Indeed, one Indian tribe in Colombia is said at the present time to utilize insects of at least seven different orders, the larvae of ants, bees, and wasps being especially esteemed. The bodies of termites, particularly the queens, are especially rich in fat and are prized in parts of Africa. As larger animals become increasingly scarce, humans in many areas may have to rely more and more on insects as food.

Insects as Food for Vertebrate Animals

Vertebrate animals are commonly classified as herbivores, omnivores, or carnivores. Since strict herbivores are themselves primary consumers, their main associations with insects are as competitors or as hosts (for example, cattle grubs). Many omnivores and carnivores consume insects at least some of the time, and some are mainly or wholly entomophagous (or insectivorous) (for example, bats, swallows). Many freshwater fish depend heavily on insects, not only those occurring in the water but those that fall in or fly low over the surface. This is, of course, well known to fly fishermen, who pride themselves on designing flies that resemble certain types of insects and on striking the surface film at the right time and place with the appropriate fly. Fish will commonly continue to feed for a time on a type of prey that is abundant, for example, a flight of newly emerged caddisflies. Trout tend to take food both at the surface and from the bottom, mayflies, caddisflies, and midge larvae being major items in the diet. The productivity of lakes and streams, in terms of energy made available by producers and primary consumers, is a major concern of fisheries biologists.

Mosquito fish (*Gambusia*) have been introduced into many parts of the world for control of mosquito larvae, often with much success. To turn to another group, the Amphibia, the giant toad has been widely used in warmer parts of the globe to control ground-dwelling insects. Insects make up a considerable portion of the diet of many amphibians and reptiles.

Many mammals also subsist on insects. Even large carnivores such as the coyote consume a great many grasshoppers and beetles. Two whole orders, the Chiroptera (bats) and Insectivora (moles and shrews), are largely insectivorous. Studies in central New York by W. J. Hamilton, Jr., of Cornell University, showed that short-tailed shrews consume food equivalent to 50% of their own weight each day, and about 75%

of their diet consists of insects. Surprisingly, deer mice, rodents one commonly thinks of as plant-feeders, also consume about as high a proportion of insects during the summer. In Canada, shrews and voles are major predators on a major forest pest, the larch sawfly. In some areas 37% to 98% of the cocoons are eaten by these mammals, and a portion of the surviving adults and their larvae are eaten by birds.

It is, of course, birds that have attracted the most attention as predators on insects. Several books have been written on the values of birds, often without regard for the fact that they consume many species that are beneficial to man. Being active fliers, birds do, however, have the capacity to congregate in areas of insect outbreaks, as evidenced by starlings flocking to grub-infested lawns, or warblers concentrating on trees infested with caterpillars. Perhaps the best-known instance of birds destroying noxious insects relates to gulls and the Mormon cricket. About 1850, great swarms of these crickets descended on Mormon settlements near Great Salt Lake, destroying most of their crops and threatening the settlers with starvation. Gulls appeared by the tens of thousands and devoured most of the crickets, an historical event of such importance that a statue to the gulls was later erected in Salt Lake City.

Although many birds are rather generalized feeders, most have some degree of specialization. Warblers and vireos, for example, forage for small insects among the leaves and small branches of trees, while blackbirds and thrushes forage mainly on or close to the ground. Flycatchers take flying insects they snatch in quick flights from their perches, while woodpeckers drill for the larvae of bark beetles and other wood-infesting insects. It is not always appreciated that many seed-feeders (such as finches) consume large numbers of insects during the summer season.

The numbers of insects consumed by individual birds are very large indeed. Most birds fill their crops with food about twice a day. There is one record of a flicker stomach containing over 5000 ants, another of a captive robin consuming 165 cutworms in one day, nearly twice its own weight. Nestlings commonly consume more than their own weight in a day, the food in many cases being a high-protein diet of soft-bodied insects. Studies of a nesting pair of blue tits, in England, showed that the 11 nestlings were fed from 25 to 42 times an hour, from early in the morning to late evening, for a total of 400 to 650 visits per day. The number or weight of insects required to sustain an individual bird through its lifespan would be difficult to calculate, but the figure would obviously be very high.

It should be kept in mind that, as a general rule, the feeding behavior of vertebrate animals differs considerably from that of insects. Insects tend to be programmed to respond to only a few of the innumerable potential stimuli in the world around them, for example, *Nasonia* wasps to fly puparia, horse bots to the legs and flanks of horses. Vertebrates are more broadly responsive and may sample a variety of foods before selecting those they prefer and learning where to find and how to catch those that suit them best. Many birds are known to develop a **search image**; that is, they discover a particularly edible item, remember its appearance and where to find it, and hunt in such a way that others of the same kind are quickly discovered. Experiments have shown that birds will concentrate on one type of prey to the point that its numbers have been reduced far below those of other, equally palatable species, which have been overlooked. At a certain point a new search image will, of course, be formed. Birds also become conditioned to avoid prey that sting, taste bad, or make them ill. It is not surprising that insects have developed a variety of adaptations that take advantage of these characteristics of vertebrate predators.

The Primary Defenses of Insects

It is convenient to recognize two major types of defense mechanisms. **Primary defenses** operate before a predator initiates an attack, and in fact regardless of whether or not a predator is present. They may also be thought of as **passive defenses**, in the sense that the insect is, by its appearance and actions, merely bearing a message to potential predators. **Secondary defenses** are employed at the time of an encounter with a predator; they are **active** in that the insect has to behave in some way vis-à-vis its attacker. An insect may have both primary and secondary defenses. So diverse are these mechanisms that we can do no more than mention a few well-studied examples here.

Crypsis. Crypsis is a widespread phenomenon among insects. Although often called "camouflage" or "protective coloration," in fact it implies more than that. To be cryptic (which literally means "hidden"), an insect must not only resemble its substrate, but it must also behave appropriately, for example, by resting immobile or in an appropriate posture. **Generalized crypsis** implies an overall resemblance to the background, for example, by a speckled grasshopper resting on a pebbly surface (Fig. 14–1). **Special resemblance** implies similarity to a specific object, such as a twig or a leaf (Fig. 14–2). An insect may resemble its abiotic environment, as in the case of the grasshopper cited above. Or resemblance may be to the biotic environment, usually some part of a plant, and may vary from the simple green color of a caterpillar in one's salad to bizarre body forms copying lichens, spines, or even flowers.

Some insects are able to undergo an enhancement of crypsis by assuming more than one color or by altering the environment to render themselves less easy to detect. Some mantids and grasshoppers undergo color changes when they molt, and in some cases these are induced by humidity. High humidity may cause changes to green colors, low humidity to brown. This is highly adaptive in areas where there are marked wet and dry seasons, since grasses change color similarly. The pupae of some butterflies may be either green or brown, depending on the color of the immediate environment. This is believed to be a direct effect of light reflected from the substrate. Examples of insects that modify their environment are provided by the larvae of certain leaf beetles, which cover their backs with cast skins, fecal pellets, and bits of debris, thus resembling from above nothing more than masses of detritus.

Cryptic insects must have behavioral responses that function to orient them suitably. Bark-dwelling moths with longitudinal stripes on their wings must orient up and down on striped bark, while those with transverse stripes must position themselves sideways if the stripes are to blend with patterns in the bark. Certain caterpillars that infest pine needles are green with longitudinal white stripes, rendering them difficult to observe when resting on the needles. But at the last molt they become brown, flecked with black, and from then on they rest not on the needles but on the brown twigs bearing the needles.

Crypsis does, of course, interfere with other life activities, such as feeding and mating. Thus it is especially prevalent during the diurnal resting periods of insects that are mainly active at night, especially moths. Cryptic insects do not usually aggregate, for birds are frequently able to form search images for cryptic species, and by dispersing themselves these insects increase their chances that birds will find more rewarding prey. There are varying degrees of crypsis among species, and among individuals of any one species some are more successful than others in escaping detection. It is these better-adapted individuals that, on the average, survive to produce more offspring, thus

Figure 14–1
A cryptic grasshopper resting on an appropriate substrate. Many such grasshoppers have brightly colored hind wings, such that they present an entirely different appearance when flying. (Photograph by Howard E. Evans.)

Figure 14–2
A pair of mating walkingsticks, the male on the left. (Photograph by Howard E. Evans.)

providing the basis for the evolution of more perfect crypsis. At the same time a premium is placed on the keenest hunters among the predators. Thus there is coevolution between predator and prey, resulting in improved abilities to find prey and more refined examples of crypsis.

The Effectiveness of Crypsis. It is true that examination of the stomach contents of various birds, lizards, and mammals often reveals the presence of cryptic species. However, there is now much experimental evidence that increasing degrees of crypsis result in increasing success in escaping detection by predators. For example, laboratory studies using chameleons as predators have been conducted with a species of grasshopper having two color forms, green and yellow (Fig. 14–3). Appreciably fewer green than yellow grasshoppers were eaten when placed on a green background, and vice versa.

Perhaps the best-known studies are those of H. B. D. Kettlewell, of Oxford University, concerning a common British insect, the peppered moth. These moths are whitish, with irregular black spots and stripes, and blend beautifully with lichen-covered bark. Before the industrial revolution, one occasionally found black (melanic) individuals that were conspicuous against the lichens (Fig. 14–4). But by 1895, around industrial centers in England, the melanics had become much more abundant than the "normal" form. By this time tree trunks in these areas had become blackened with soot and the lichens killed off; thus the melanics blended best with the background.

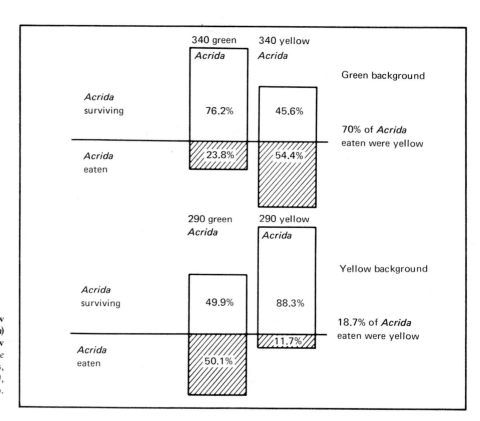

Figure 14–3
Predation by a chameleon on green and yellow grasshoppers of the same species (*Acrida turrita*) on a green background (above) and on a yellow background (below). (From M. Edmunds, *Defense in Animals*, 1974, with permission of the publishers, Longman Group Ltd.; data from S. Ergene, 1950, *Zeitschrift für vergleichende Physiologie*, vol. 32, pp. 530–51.)

(a)

(b)

Kettlewell hypothesized that birds preyed more heavily on the whitish forms in industrial areas and thus had produced rapid selection toward melanism (known to be controlled by a single dominant gene). In order to demonstrate this, Kettlewell reared large numbers of normal and melanic moths in the laboratory, marked them with colored spots, and released them in both industrial and rural areas. These were then recaptured in light traps or traps baited with pheromone. In industrial areas, many more of the melanics than normal moths were recovered, and in rural, unpolluted areas the reverse was true. He also placed moths of both types on tree trunks in selected areas and observed them from blinds. In one rural area, where tree trunks were lichen covered, 190 of the moths placed on the trunks were eaten by birds within a short time; 164 of these were dark and 26 light. Obviously moths that look conspicuous to us also appear conspicuous to birds. In recent years various pollution control programs have been instituted in England, and, as might be predicted, the melanic moths now comprise a smaller percentage of peppered moth populations in industrial areas than they formerly did. These moths have only one generation a year, so it is obvious that natural selection, in the form of predation, may provide a powerful force in the evolution of crypsis.

Figure 14-4
The peppered moth, *Biston betularia*, resting on a tree trunk, the melanic form on the left in each case, the normal form on the right: (a) on a lichen-covered trunk in the countryside, Dorset, England; (b) on a soot-covered trunk near Birmingham, England. (From the experiments of H. B. D. Kettlewell, Oxford University.)

Aposematism. **Aposematism** is a general term for signals that advertise unpleasant or dangerous attributes of an animal. The term *warning coloration* is often applied to aposematic features, but this term implies more volition than is appropriate and overlooks the fact that features other than color may be involved. Aposematic insects all have secondary defense mechanisms, such as a sting or distasteful or poisonous body fluids. Predators must either have innate avoidance responses to aposematic patterns or must learn these patterns by sampling prey. In the latter (probably much more common) instance, some individuals of each generation of an aposematic insect are likely to be sacrificed. However, some species have evolved tough bodies that cannot easily be damaged or "deflection marks" that induce the predator to bite in some nonvital part of the body.

One of the best-known aposematic patterns consists of alternating bands of black and yellow, the pattern of yellow jackets and many other stinging Hymenoptera. These bright, alternating bands are conspicuous to color-blind persons, and one assumes also to color-blind animals (such as many mammals). A second common aposematic pattern consists of orange or red bands or patches, which are especially effective signals to organisms with good color vision (which is why hunters and highway crews

(a)

(b)

Figure 14–5
A blue jay (a) feeding on a monarch butterfly and (b) becoming ill within a few moments and vomiting. (Lincoln P. Brower, 1969, "Ecological Chemistry," *Scientific American,* 220: 2, pp. 22–29. Reprinted by permission.)

Figure 14–6
The eye spots on the hind wings of a polyphemus moth, one of the giant silk moths (Saturniidae). When a resting moth is disturbed, the front wings are spread, suddenly revealing the eye spots. (Photograph by Howard E. Evans.)

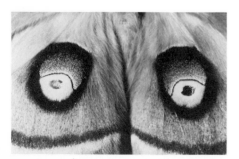

wear orange jackets). Birds, in general, have color vision similar to ours, and it is probable that these patterns evolved primarily as signals to birds. One example that quickly comes to mind is the monarch butterfly, the larvae of which feeds on milkweed. Most milkweeds contain cardiac glycosides that are retained in the bodies of caterpillars, the pupae, and on into the adult butterflies; when a bird eats a monarch it becomes ill and retches (Fig. 14–5; see further below).

Another example is provided by wingless female mutillid wasps ("velvet ants"). These insects spend much time walking over the ground in search of their hosts. Although a few species are cryptic, and colored like the sand, the majority advertise their presence with bands of brilliant orange and red. With their long, venom-laden stings and tough integument, it is not likely that many are actually sacrificed in the learning processes of predators.

It should be remembered that color vision in insects differs somewhat from that of humans and higher vertebrates, being extended on the blue end of the spectrum and into the ultraviolet and not extended appreciably into the oranges and reds. It is true that some insects use bright colors in nuptial displays (damselflies, for instance), but the colors employed are usually blues (even ultraviolet in the case of some butterflies). It is no coincidence that flowers requiring insect pollination are usually blue or yellow, or have ultraviolet nectar guides; or that flowers that rely on hummingbirds for pollination (such as scarlet gilia) are primarily red (see Chapter 11). Insects displaying orange or red markings almost without exception have secondary defense mechanisms and are advertising this fact to potential vertebrate predators—or they are mimics of such insects, as we shall discuss in the following section of this chapter.

Still a third aposematic pattern consists of paired eye spots, for example, those on the hind wings of many moths (Fig. 14–6). Since these are often associated with displays in the presence of a predator, we shall defer discussion of them to a later section (under "Secondary Defense Mechanisms"). Large eye spots presumably imply the presence of an owl, a hawk, or a large mammal, and there is experimental evidence that eye spots do in fact deter predation by small birds.

Before leaving the subject of aposematism, it should be noted that it is common for many insects in a given area to have similar aposematic patterns. In temperate regions a great many different kinds of stinging Hymenoptera, for example, have "yellow jacket" patterns. In the tropics there are often literally dozens of butterflies in the same area having similar orange wing patterns; many of these feed on poisonous or bitter-tasting plants and are themselves unpalatable. This phenomenon was first noted in Brazil by the German biologist Fritz Müller and has since been termed Müllerian mimicry, although we prefer the term **mutual aposematism**. Müller pointed out that it is advantageous for distasteful species occurring together to evolve similar patterns, since in this way predators have fewer patterns to learn and exploratory predation is spread over several species.

In contrast to cryptic species, those displaying aposematic patterns tend to occur in exposed situations and are often gregarious. Larvae of the mourning cloak butterfly provide a good example of gregarious feeders, covered as they are with long spines and streaked with red; but the chrysalids are cryptic, and prior to pupation the caterpillars crawl off in different directions, crypsis being enhanced by dispersal.

The Effectiveness of Aposematism.   There is now much evidence that aposematism "works." In a pioneering study, the British entomologist G. D. H. Carpenter offered a variety of insects to monkeys and found that 73% of the cryptic insects were eaten but only 20% of those with bright colors. Lady beetles, with their characteristic red and black patterns, are known to be avoided by many kinds of birds and mammals. The British entomologists J. F. D. Frazer and Miriam Rothschild "screened" a great number of insects for palatability, using caged animals, including diverse birds and mammals, lizards, and toads. Lady beetles fell in the highest (most unacceptable) category. Equally unacceptable was the cinnabar, a lepidopteran that has been widely used in biological control of ragwort. The caterpillars of this species are gregarious and are striped with black and yellow, while the adults have red and black wings. The blood of the cinnabar contains considerable quantities of histamine as well as certain alkaloids. In nature, cinnabar caterpillars are preyed on extensively by ground beetles and are parasitized by braconid wasps, but shrews and voles will not accept them.

One of the best-known field studies on the adaptive value of aposematism was conducted by W. W. Benson, of the University of Washington. His studies were conducted in Costa Rica on a well-known, unpalatable butterfly, *Heliconius erato*, having a bright red band on the black fore wing (Fig. 14–7). Benson captured butterflies in their gregarious, nocturnal roosts and painted out the red bands with black; in the controls he merely painted black over black, to check against any effect the paint itself might have. By recording the return of these butterflies to the roost, he found that the experimentals survived an average of 32 days, the controls an average of 52 days. Furthermore, a much higher percentage of the experimentals had wing damage resulting from predation by birds, even though they were exposed for a shorter period, on the average, than the controls (Fig. 14–8). It should be added that several other butterflies have much the same pattern as *Heliconius erato*, one of many examples of mutual aposematism in the American tropics.

Mimicry. *Mimicry* is a much-abused word. Wolfgang Wickler, in his excellent book of that title, uses the term for virtually all defense mechanisms, and many persons would agree in calling a walkingstick (for example) a "mimic." We prefer to restrict the term to examples in which a palatable species has evolved a color pattern and/or behavior similar to that of a distasteful species. This is mimicry as first enunciated by Henry Walter Bates in 1862, and thus it is often called **Batesian mimicry** to distinguish it from "Müllerian mimicry," discovered a few years later. Bates was a young Englishman traveling in the Amazon and supporting his travels, in part, by selling

Figure 14–7
Heliconius erato, a tropical American butterfly considered highly unpalatable. The front wings have a bright red band, the hind wings a yellow streak. (Drawing by Peter Eades.)

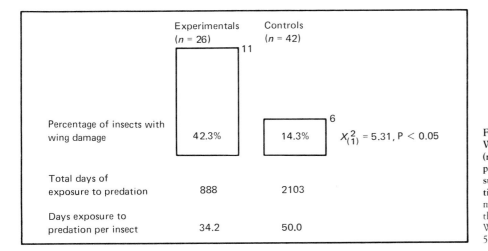

	Experimentals (n = 26)	Controls (n = 42)	
Percentage of insects with wing damage	42.3%	14.3%	$X^2_{(1)} = 5.31, P < 0.05$
Total days of exposure to predation	888	2103	
Days exposure to predation per insect	34.2	50.0	

Figure 14–8
Wing damage to *Heliconius erato* experimentals (red painted over with black) and controls (black painted over previously black areas). Controls also survived much longer, thus were exposed to predation for significantly longer periods. (From M. Edmunds, *Defense in Animals*, 1974, with permission of the publishers, Longman Group Ltd.; data from W. W. Benson, 1972, *Science,* vol. 176, pp. 530–51.)

specimens to museums and to amateur collectors. Charles Darwin was excited by Bates's discoveries, which appeared too late for inclusion in earlier editions of *The Origin of Species* but were discussed in his later publications as prime examples of natural selection at work.

The literature on mimicry has since grown by leaps and bounds—much of it tinged with skepticism until recently, when experimental evidence supporting the concept has been forthcoming in abundance. Scarcely any biological concept is able to deal with all the complexities of nature, and there is no doubt that some of the criticisms of mimicry theory have been justified. For one thing, insects cannot simply be categorized as "good to eat" or "noxious." There are all degrees of palatability, and what may be food for one predator may be poison for another. Predators vary greatly in their learning ability, their visual acuity, and so forth. If a predator is very hungry, or if there are few alternative prey available, it may be less discriminating. In short, being a mimic does not ensure a long life and abundant progeny; but under certain conditions it may ensure a life that is slightly longer, with slightly greater egg production, and that is all that natural selection requires.

The mimic best known to Americans is probably the viceroy butterfly. The viceroy is related to the white admiral but has departed radically from the color pattern of its relatives to assume the orange wing pattern of the monarch, an unrelated species belonging to a different family. Monarch caterpillars feed on milkweeds, many kinds of which contain poisons known as cardiac glycosides. These are sequestered in the body until the adult stage, and captive, hungry blue jays that feed on monarchs vomit and become ill for a short period (Fig. 14–5). Like many other plant toxins, these substances probably evolved as protective mechanisms against grazing animals. It is interesting that insects that have become adapted to feeding on milkweed—not only the monarch but the milkweed bug and the milkweed longhorn beetle—all display orange coloration. It has been found that some species of milkweeds do not contain these poisons, and the insects that feed on them are therefore more palatable. This undoubtedly accounts for the fact that there are records of birds being seen feeding on monarchs and that some persons have eaten monarchs without ill effects.

How effective a mimic is the viceroy? Jane van Zandt Brower, then at Yale University, found that inexperienced Florida jays ate viceroys readily. When a monarch was suddenly introduced, they tasted and rejected it, and after a few such experiences would not accept either viceroys or monarchs. Similar studies with toads showed that these animals quickly learn not to accept bumble bees and honey bees, and will still avoid them after three months, even without further training. Such toads also usually will not accept drone flies and robber flies that closely resemble bees. But toads trained on bees with their stings removed will readily accept the mimics.

The Evolution of Mimicry.   Many similar experiments have now been performed with diverse mimics and predators, with generally similar results. Some of the most interesting ones have confronted one of the major criticisms of mimicry theory: how can one account for the evolution of (for example) a viceroy from a black and white ancestral species without invoking a sudden saltation or total genetic reorganization? In other words, how can mimicry evolve when being "just a little bit mimetic" would seem to have no possible advantage? Lincoln Brower and his associates at Amherst College have addressed this problem, using *Heliconius erato*, mentioned above in another context. They found that birds trained to avoid this butterfly also refused to take uniformly black butterflies of similar shape as well as red-banded butterflies of a different shape. In further studies, they varied the unpalatability of the model and found that the more unpalatable it was, the more the birds tended to extend their rejection to species only rather remotely resembling the model. **Stimulus generalization** such as these experiments imply means that under certain conditions even a slight resemblance to an unacceptable model may have value in terms of natural selection. If in each generation individuals with most resemblance to the model leave even a slightly greater number of progeny than average, in time a more "perfect" mimic may evolve.

It should be noted that in many butterflies only the female sex is mimetic, the male having the "normal" color of its group. In at least some cases, colors of the male play a role in courtship and mating, and it is advantageous for males to retain patterns that release sexual behavior in the females. The tiger swallowtail of the eastern United States provides a particularly interesting case. The females of this butterfly are dimorphic, having either mainly black or yellow and black "tiger" coloration. Males always have the tiger pattern and are most successful in mating with females colored like themselves. But in the southeastern states there is a mostly black butterfly that has been shown to be highly distasteful to birds: the pipevine swallowtail. In the Southeast, tiger swallowtail females that are black mimics of this species survive longer, on the average, but yellow and black individuals have greater mating success. Thus a **balanced polymorphism** is retained. But in the North, where the pipevine swallowtail does not occur, all females are "tigers," since here there is no selective advantage in being black.

The diana fritillary, of the southeastern states, exhibits similar **sex-limited mimicry**, the females copying the pattern of the pipevine swallowtail, the males retaining the usual speckled pattern of fritillaries. In this case there is no polymorphism; but cases are known, especially among African swallowtails and relatives of the monarch, in which there are not two but several different female morphs, each having a color pattern like that of a different distasteful model. In situations involving polymorphism, the mimetic species may achieve a population size greater than that of its model, since each morph has a different model. In general, it is disadvantageous for mimics to become more plentiful than their models, since birds will encounter more mimics than models and learn much more slowly to associate distastefulness with a particular pattern.

Before leaving this subject, it should be mentioned that mimics must also behave like their models to be effective, just as a cryptic insect must "behave like a piece of bark," or whatever it resembles. Cases are known in which mimicry is primarily behavioral. For example, certain species of darkling beetles have glands at the tip of their abdomen from which they are able to spray repellent benzoquinones. When approached by a predator, these beetles assume a characteristic stance, with the abdomen high in the air, ready to discharge their secretions. Black, ground-dwelling beetles of at least two other groups also assume this stance when approached by a predator, even though they lack these glands (Fig. 14–9).

Aggressive Resemblance.
Since insects so readily evolve one or more of several cryptic, aposematic, or mimetic patterns, it is not surprising that some predatory insects have evolved coloration or behavior like that of their hosts or have evolved crypsis serving primarily to gain access to a host. Such behavior hardly qualifies as "defense"; rather it is offense. We discuss it briefly here because it has often been termed "aggressive mimicry," though it represents quite a different phenomenon from true mimicry, not only as to function but as to the nature of the signals: for these are, of course, directed toward other insects and not toward vertebrate animals.

Figure 14–9
A "circus beetle" (Tenebrionidae) in a defensive posture. In fact, this species lacks defensive glands at the tip of the abdomen and is a behavioral mimic of similar species that have such glands. (Photograph by Howard E. Evans.)

Figure 14–10
A female "femme fatale" firefly (left) that has lured the male of another species (right) and is proceeding to devour it as the pair hang upside down in vegetation. (Photograph by James E. Lloyd, University of Florida.)

Some of the best examples of aggressive resemblance occur among the many thousands of inquilines in the nests of ants and termites (Chapter 8). Many of these, beetles especially, are remarkably antlike in form, and some produce secretions that are eagerly sought by their hosts and may resemble pheromones of their hosts. Many termitophiles and myrmecophiles produce or acquire cuticular odors that copy those of their hosts. These inquilines commonly feed on foodstuffs in the nest and some actually feed on immature ants or termites.

A most remarkable instance of aggressive resemblance occurs among fireflies and has been described by James Lloyd, of the University of Florida. Male fireflies produce a pattern of flashes characteristic of their species, and it is answered at a specific interval by females of their species on the ground. Females of certain species have evolved the ability to respond to flashes of males of alien species. These males fly down to the females and are promptly consumed (Fig. 14–10).

Secondary Defense Mechanisms

Many insects have structures or behavior effective at close quarters with a predator. This applies not only to aposematic species, all of which are assumed to have undesirable qualities, but to many palatable insects. In its simplest form, it may simply involve a butterfly dashing away from a pursuing bird, or a caterpillar thrashing violently when seized by a predator. Some of the more specialized defenses are striking indeed, and we shall be more particularly concerned with these.

Flight Patterns.   For small, flying insects the best defense may be escape. Probabilities of escape may be increased by swift or evasive flight or by an abrupt color change on settling. Band-winged grasshoppers show brilliant colors of the hind wings while in flight, but when they land, these are covered by the cryptically colored front wings, causing the insect to "disappear" suddenly. Many butterflies (anglewings, in particular) have the upper wing surfaces brightly colored, but when they land only the lower surfaces are exposed, and these are commonly colored like bark or dead leaves. It is believed that **flash colors** such as these promote escape from pursuing birds.

Bats are major predators on night-flying moths. Several groups of moths have evolved receptors tuned to the ultrasonic cries that bats use in the manner of sonar in locating prey. These receptors have only two sensory neurons each, one for low-intensity sounds and one for high-intensity (closer) sounds, and since they are bilateral, the moth has means of perceiving both distance and direction (Chapter 6; Fig. 6–8b). Moths respond to bats at some distance by flying swiftly away, but at close quarters they undertake erratic flight patterns or sudden power dives (Fig. 14–11). One group of moths, the tiger moths, has gone one step further and evolved sound-producing organs that advertise to the bat their distasteful qualities. These moths respond to the approach of a bat by producing a series of clicks, and it has been demonstrated that bats veer away from clicking moths.

Death Feigning.   Since many predators are attracted to moving prey and reject dead insects, it is not surprising that many relatively defenseless insects become inert when approached. Leaf beetles and weevils are especially prone to death feigning. This is easily observed by home gardeners who control Colorado potato beetles by handpicking; for as soon as disturbed, these beetles draw their legs beneath them and drop to the ground. As soon as the disturbance is over, they climb back onto the plant and resume their feeding.

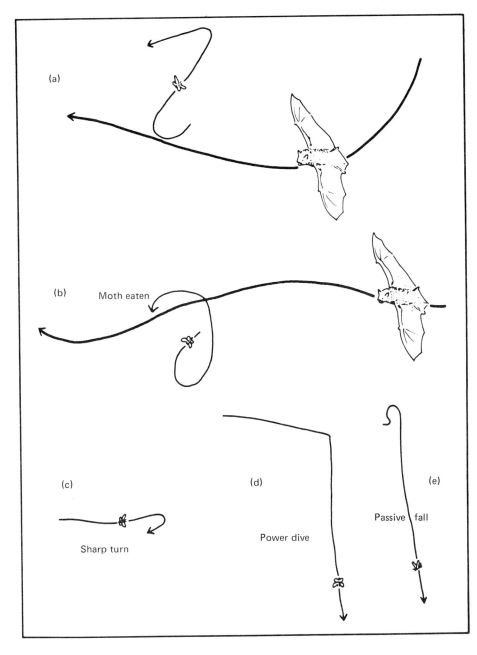

Figure 14-11
Escape tactics of moths being pursued by bats. Erratic flight paths (a and b) are sometimes, but not invariably, successful. Other escape tactics are shown in c-e. (From M. Edmunds, *Defense in Animals*, 1974, with permission of the publishers, Longman Group Ltd.; data from K. Roeder, 1965, *Scientific American*, vol. 212, no. 4, pp. 94-102.)

(a)

(b) Moth eaten

(c) Sharp turn

(d) Power dive

(e) Passive fall

Figure 14-12
Larva of the io moth (*Automeris io*, Saturniidae) feeding on willow. (Photograph by Howard E. Evans.)

Spines, Poisonous Hairs, and Stings.   Many caterpillars are hairy, and some are covered with stiff, branched spines. In some cases the tips of the hairs or spines break off easily and are capable of causing momentary irritation or a rash. This is true, for example, of the densely spiny larva of the io moth (Fig. 14-12). The sting of Hymenoptera is believed to have evolved primarily as a means of paralyzing the prey, but in bees, many ants, and social wasps it has lost this function and serves in defense (Fig. 1–14a). The nature of the venom of social species differs considerably from that of solitary forms, consisting of a mixture of pain-producing substances. In a few species (such as the honey bee) the sting is barbed and remains imbedded in the skin, causing the death of the worker and releasing a pheromone that may bring other workers to the attack (Fig. 8–13). It is interesting that the males of several groups of solitary wasps (which lack an ovipositor and therefore a sting) have the last abdominal segment prolonged into a stout spine, which is wielded much like a sting and can sometimes pierce the skin, though there are no poison glands. It is probable that such a "pseudosting" is nearly as effective as a true sting in deterring predators, especially since male wasps commonly have aposematic patterns similar to those of the females.

Detachable Body Parts. Many insects have integumentary outgrowths that readily become detached, without seriously harming the insect. These include the waxy or powdery coverings of certain aphids and whiteflies, the hairs of caddisflies, and the scales of bristletails and of moths and butterflies. Thomas Eisner and his colleagues at Cornell University have shown that moths and caddisflies are often held only momentarily by spider webs, since the scales or hairs stick to the adhesive strands and become detached from the body, permitting the insects to escape with minor damage (Fig. 14–13). In a similar way, these insects may escape from the viscid droplets on the leaves of sundews, insectivorous plants we discussed in Chapter 10. To what extent deciduous outgrowths play a role in escape from vertebrate predators is uncertain, but it is known that moths shed clouds of scales when pecked by birds. It is probable that scales first evolved as an escape mechanism and only later, in many Lepidoptera, assumed a role in conveying messages via color patterns. That is, an originally secondary defense may have become also a primary defense, reducing the frequency of predator attacks and at the same time retaining a role in escaping from such attacks.

Deflection of Attack. Many butterflies have small spots along the edge of the wing, and it is believed that these attract the attention of predators and cause them to bite at a nonessential part of the body. Butterfly collectors are aware that the wings of fritillaries and others having such deflection marks on the margin of the wing commonly have triangular beak marks at the margin. The "tails" of swallowtails may also serve a role in deflecting attack from the body, and as collectors are all too aware, these are commonly damaged unless the butterfly has recently emerged (Fig. 14–14). Hair-streaks have not only antennalike filaments at the posterior margins of the wings, but usually a color spot at the base of the filaments, simulating a head. Thus a perched hair-streak often seems to have its "head in the wrong place," causing a predator to seize only a piece of the wing.

Figure 14–13
Sticky strand of a spider's web. A moth has escaped from the web, leaving some of its scales behind. (Photograph by Thomas Eisner, Cornell University; from *Science*, vol. 146, p. 1058, copyright 1964 by the American Association for the Advancement of Science.)

Startle Displays.   Some insects, when approached closely or attacked by a predator, suddenly undergo movements, produce sounds or scents, or display colors serving to "threaten" or "bluff" a predator. The usual effect is probably to startle or to cause a momentary indecision, permitting the insect to escape. There is evidence that birds and mammals remember encounters with displaying insects and to a certain extent avoid them. Both cryptic and aposematic insects may have startle displays as a "second line of defense" in the event of attack.

Walkingsticks, which one thinks of as the ultimate in crypsis, have a variety of displays. Many tropical species have short, brightly colored hind wings that are suddenly erected, abruptly increasing their size; this sometimes may be accompanied by stridulation that is said to resemble the hissing of a snake. Michael Robinson, of the Smithsonian Tropical Research Institute, in Panama, has described a number of such cases, including one wingless species that flexes its abdomen up and down while stridulating, suggesting a scorpion or (more remotely) a snake about to strike. Some walkingsticks have sharp spines capable of pricking an aggressor, and others have caustic secretions that can actually be aimed at an aggressor (Fig. 5–20).

A number of different insects produce squeaking sounds when seized, such as long-horned beetles and mutillid wasps. Even the relatively inert, unprotected pupae of some Lepidoptera produce squeaks when handled. Some bright-colored, distasteful tiger moths produce an odorous froth from thoracic glands when disturbed. Larvae of swallowtail butterflies have eversible, hornlike glands on the thorax called **osmeteria** (Fig. 5–19). These produce a volatile substance smelling rather like rancid butter, and in fact its major component is butyric acid (Chapter 5, p. 132). Swallowtail caterpillars often have aposematic patterns, and many have eye spots that are dramatically displayed by elevating and inflating the thorax.

Eye spots deserve special mention, since they occur in a wide variety of insects and are usually associated with some type of display. We refer here not to the small, marginal deflection spots on butterfly wings, but to the large eye spots often found on the heads or thoraces of caterpillars or the hind wings of mantids, moths, and other insects. Hawk moths provide a particularly fine example. Many species are cryptically colored but have brightly colored hind wings, each containing a large eye spot. When disturbed, the moth spreads its fore wings and undergoes quivering motions, presumably serving to further startle a predator. Some moths and butterflies have spots shaded in such a way as to have remarkable resemblance to the eyes of owls or snakes.

David Blest, of University College, London, has shown that the sudden appearance of any bright color releases escape responses in certain birds. This proved to be particularly true of eye spots. Blest fed several species of birds with mealworms, using a box that would flash a light beneath, revealing a pattern on the floor of the box. The sudden flash of a pair of parallel lines or crosses caused some escape responses, but pairs of circles caused many more such responses. Eye spots shaded so as to appear three-dimensional were especially effective in "frightening" the birds (Fig. 14–15). He concluded that many small birds respond innately to the sudden appearance of eyes that might represent those of a large predator. However, Blest showed that birds do sometimes habituate to butterflies with eye spots when these prove palatable. Thus eye spots are most effective if not too widespread in nature and when displayed only under certain conditions.

Chemical Defense.   The ultimate form of defense is the use of one or more chemicals that are in some way repugnant to a predator. These may be obtained from the host plant (as in the case of the monarch butterfly) or synthesized by the insect. Defen-

Figure 14–14
A swallowtail butterfly with a large beak mark at the "tail" of the right wing. This butterfly is "mud-puddling," obtaining moisture and minerals from damp soil in the manner of many butterflies.   (Photograph by Howard E. Evans.)

Figure 14–15
Artificial eye spots used by David Blest in experiments with birds. The single circle (upper left) elicited the fewest escape responses; the shaded figures with eccentric rings (lower two), the most. (From *Mimicry in Plants and Animals*, by W. Wickler. Copyright 1968 McGraw Hill Book Co./ Weidenfeld & Nicolson Archives; based on research of A. D. Blest, 1957, *Behaviour*, vol. 11, pp. 209–56.)

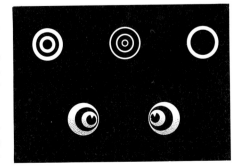

sive chemicals (called **allomones**) may be contained in the blood or may be produced by specialized exocrine glands. Many insects that exhibit chemical defense are aposematically colored, and some are mimicked by insects lacking such defense, as we have seen. Allomones have been reviewed in detail in Chapter 5 (see also Figs. 5–18 through 5–24).

Integrated Defense Systems.   Many insects have not one but several defense mechanisms. In this way they may achieve protection against different predators or against the same predator at different levels of motivation or different stages in the learning process. Commonly larvae and adults are protected quite differently; for example, the larva of the viceroy butterfly resembles a bird dropping, while the pupa resembles a dried leaf, and the adult is a mimic of a distasteful species, the monarch. In the same life stage, insects may have several "lines of defense." Walkingsticks are cryptic, but if attacked they may have startle displays or discharge irritating chemicals at the intruder. Some tiger moths have cryptic fore wings and brilliantly colored hind wings, which provide flash colors, but if actually attacked discharge noxious fluids from thoracic glands.

In a now classic study of 15 species of preying mantids on Trinidad, Jocelyn Crane, of the New York Zoological Society, found that all 15 were cryptic, some being sticklike, others leaflike, some like bark or lichens. All 15 had escape mechanisms: dropping, dodging, or jumping. As a third line of defense, several species had startle displays, differing in details but usually involving body swaying, spreading the wings, and raising the forelegs and abdomen (Fig. 14–16). As a last resort several species would actually strike the predator with the raptorial forelegs. Crane hypothesized that the elaborate displays evolved by ritualization of simultaneous tendencies to remain motionless and to escape.

Perhaps the most impressive integrated defenses occur in the colonies of social insects. Some tropical social wasps produce sounds by drumming on the carton of the nest; and if approached, they display by wagging their brightly colored abdomens or, as a last resort, stinging the intruder en masse. Yet their defense against army ants may be very different: they simply flee and found a new nest elsewhere. The nests of termites often have walls nearly as hard as stone, but if the walls are breached, a horde of specialists in chemical defense quickly arrives on the scene.

The question is often asked: if these things are true, why is it that some insects have several defense mechanisms, some have only one, and some appear to have none at all? A partial answer is to be found in the reproductive capacities. Insects such as aphids or house flies, which have no spectacular defense strategies (but like most insects may have subtle ones), compensate by having unusually high reproductive rates. They are specialists in reproduction rather than survival. Every species may be said to maintain itself in nature by achieving some sort of balance between reproductive potential and survival capacity. These are matters we shall explore further in the following chapters.

Summary

Insects are major elements in all terrestrial ecosystems. Most are primary or secondary consumers; that is, they represent links in the food chain between plants (producers) and higher levels in the food chain. A great many vertebrate animals rely heavily on insects as food, and some subsist wholly or almost wholly on insects (for example, toads, shrews, bats, flycatchers). Birds have the capacity to fly to insect aggregations,

Figure 14–16
The startle display of a preying mantid. This mantid is cryptically colored, but when disturbed it assumes an aggressive pose and threatens to strike, at the same time displaying bright colors on the hind wings.   (Photograph by Howard E. Evans.)

and an individual bird may consume many thousands of insects in the course of its lifetime. Birds often form searching images for certain prey species. Birds and other vertebrates learn readily to find certain types of prey and readily become conditioned to avoid distasteful prey.

Insects have evolved a variety of adaptations serving to reduce the success of vertebrate predators. Primary defenses operate before a predator initiates an attack, and in fact regardless of whether or not a predator is present. A common example is crypsis, often called camouflage or protective coloration. This may take the form of generalized resemblance to the background or special resemblance to an object such as a twig or leaf. Although cryptic insects are often found in the stomachs of vertebrates, there is now convincing evidence that crypsis plays a major role in reducing detection by predators.

A second type of primary defense is aposematism, often called warning coloration. Aposematic insects all have secondary defenses, such as a sting or distasteful body fluids. Common patterns include alternating bands of black and yellow, as in yellow jackets, and orange or red patterns, as in the monarch butterfly. In a given area numerous distasteful insects may assume a common color pattern, a phenomenon called mutual aposematism or Müllerian mimicry. Both laboratory and field studies confirm the value of aposematic patterns in reducing predation.

True mimicry, or Batesian mimicry, involves a palatable species bearing a color pattern similar to that of a distasteful species and thereby gaining protection. There is experimental evidence that while inexperienced birds will eat viceroys readily, for example, after a few experiences with monarchs they will also reject viceroys. Birds also experience stimulus generalization; that is, they may reject patterns only remotely like a distasteful model. This concept supplies a basis for understanding how mimicry may evolve and become perfected.

Secondary defense mechanisms are those effective at close quarters with a predator. They include specialized escape patterns, death feigning, the presence of spines or poisonous hairs or stings, detachable scales, deflection marks, and the like. Startle displays involve aggressive postures, sudden increases in apparent size, sounds, or the display of eye spots. Chemical defenses are diverse and often involve specialized exocrine glands. Some insects are able to aim a spray of irritating fluid toward a predator. Not uncommonly, insects have not one but several mechanisms acting together or separately to deter predation.

Selected Readings

Brower, L. P. 1969. "Ecological chemistry." *Scientific American*, vol. 220, no. 2, pp. 22–29.

Carpenter, G. D. H., and E. B. Ford. 1933. *Mimicry*. London: Methuen. 134 pp.

Cott, H. B. 1949. *Adaptive Coloration in Animals*. London: Methuen. 508 pp., 48 pls. Reprinted 1957; published in U.S.A. by Dover, New York.

Crane, J. 1952. "A comparative study of innate defensive behavior in Trinidad mantids." *Zoologica*, vol. 37, pp. 259–93.

Edmunds, M. 1974. *Defence in Animals*. Harlow, England: Longman. 357 pp.

Eisner, T. 1967. "Defensive use of a 'fecal shield' by a beetle larva." *Science*, vol. 158, pp. 1471–73.

———. 1965. "Insect scales are asset in defense." *Natural History*, vol. 74, no. 6, pp. 27–31.

Harvey, P. H., and P. J. Greenwood. 1978. "Anti-predator defence strategies: some evolutionary problems," in *Behavioural Ecology: An Evolutionary Approach* (J. R. Krebs and N. B. Davies, eds.), pp. 129–51. London: Blackwells.

Pasteels, J. M., and J.-C. Grégoire. 1983. "The chemical ecology of defense in arthropods." *Annual Review of Entomology*, vol. 28, pp. 263–89.

Pasteur, G. 1982. "A classificatory review of mimicry systems." *Annual Review of Ecology and Systematics*, vol. 13, pp. 169–99.

Rettenmeyer, C. W. 1970. "Insect mimicry." *Annual Review of Entomology*, vol. 15, pp. 43–74.

Wickler, W. 1968. *Mimicry in Plants and Animals*. New York: World University Library, McGraw Hill. 255 pp.

Wiklund, C., and T. Järvi. 1982. "Survival of distasteful insects after being attacked by naive birds: a reappraisal of the theory of aposematic coloration evolving through individual selection." *Evolution*, vol. 36, pp. 998–1002.

Someone has estimated that there are a billion billion insects on earth at any one time. This is not a figure we care to attack or defend! That insects dominate the land surface of the earth and provide a major challenge to humans is obvious enough. As *individuals*, insects are of interest chiefly to collectors—and there is perhaps no more rewarding hobby than collecting insects. To the entomologist, insects are best thought of in terms of **populations.** The term is a loose one; it may apply to a local grouping of individuals (the apple maggot population in an orchard or on a tree) or it may apply in a broader sense (populations of the spruce budworm in eastern Canada). Note that in the second case we used the plural, recognizing that large populations are made up of smaller ones. In every case some spatial discontinuity is implied, but

Insects and Their Environments

P A R T

also some interchange of individuals between populations. Proximity and overlap of populations depend on patchiness of their food, the insects' powers of dispersal, and the like. The term *population* is sometimes used to include several species (for example, populations of grasshoppers in the Great Plains), but this usage overlooks the fact that no two species are alike in their relationships with the environment.

Species are groupings of populations that do not interbreed with other, similar groupings; they are protected gene pools, protected in that there are mechanisms for preventing interbreeding with other species. Each species has its own biological characteristics, including its distinctive structure, physiology, behavior, and responses to temperature and day length. The accurate

S I X

identification of species and the accumulation of knowledge regarding their characteristics provide major tasks for entomologists everywhere. The role that a species occupies in nature is spoken of as its **ecological niche.** Part of the answer to the question "why are there so many insects?" is to be found in their small size and their ability to occupy such small niches. A single leaf may provide a niche for several species—perhaps a leaf miner, a gall former, a sap feeder, and a leaf roller.

A group of insects and other organisms living on a leaf—or in a tree or a woodland—constitutes a **community.** Members of a community have as a rule evolved complex interrelationships, and it is also the task of the entomologist to attempt to fathom these interrela-

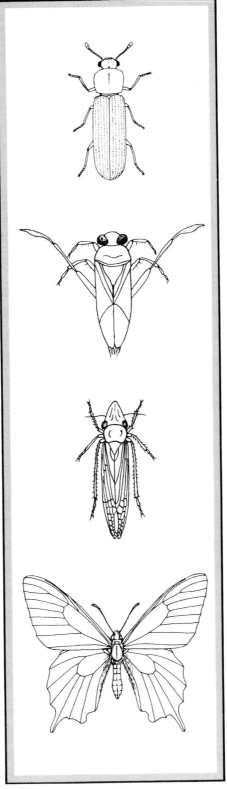

tionships. When a major community is considered in relation to its physical environment, it is termed an **ecosystem.** These five elements —individual–population–species– community–ecosystem—form a hierarchy of increasing complexity, the last two involving many organisms besides insects.

In Chapter 15 we shall focus on species and the ways they interact competitively and as members of communities and ecosystems. We shall also have something to say on the perennially interesting question of the origin of species. Problems of dispersal and migration are treated in a separate chapter, 16. In the final section of this text, focus will be shifted to populations: how they are regulated in nature and how we may hope to manage them in our forests and croplands.

CHAPTER

FIFTEEN

The Ecological Attributes of Insect Species

Note that when we speak of a *niche*, we mean not only where an insect occurs but also what it does: its place in the environment and its relation to its food, to natural enemies, to competitors, and even to physical factors such as temperature and humidity. The hypothetical leaf we mentioned above provides niches for several species; it may also be part of the niche of a lady beetle, a parasitoid wasp, or a warbler. Species occurring in the same place do not ordinarily have identical niches. No matter how similar they seem, careful study usually reveals differences in certain parameters:

"If it's true that the world ant population is 10^{15}, then it's no wonder that we never run into anyone we know."

perhaps food selection, susceptibility to natural enemies, seasonal cycle, or some other factor. If this were not the case, there would be severe interspecific competition, and the better-adapted species would eliminate the other. This **competitive exclusion principle** holds that complete competitors cannot long coexist, or, to put it another way, that a single niche cannot accommodate two species at the same time. Some qualifications are necessary regarding this concept, as we shall see.

Interspecific Competition

In a now classic series of experiments, Thomas Park, of the University of Chicago, showed that closely similar species occurring in the same habitat and eating the same food may in fact have slightly different relationships with their environments; that is, their niches are not identical. Park worked with two species of flour beetles, *Tribolium confusum* and *T. castaneum*. The two species can be bred easily in flour in the laboratory, and both species establish populations of relatively stable size so long as fresh flour is added periodically. But attempts to form a mixed culture fail. One or the other disappears in time, which one depending on subtle factors in the apparently very simple environment. Most flour contains spores of a protozoan that is a parasite of *castaneum*, with the result that populations of that species undergo a crash after a period of time. When *castaneum* is cultured alone, it is able to recover from the crash; but in a mixed culture it succumbs to competition with its more resistant relative, *confusum*.

However, when a mixed culture is formed in flour free of the parasite, *castaneum* may win the competition. Park found that at a temperature of 34° C and a relative humidity of 70%, the culture eventually becomes pure *castaneum*. If either temperature or humidity is low, say 24° and 30%, *confusum* is the winner. At an intermediate temperature (29°), *confusum* usually wins at low humidities, *castaneum* at high. In a situation in which both these variables fluctuate (as would normally be the case in nature), the two species may coexist indefinitely (Fig. 15–1). This may explain how, in nature, two species often do in fact coexist: they respond differently to environmental variables—in this sense they are not "complete competitors."

In these instances the two species have identical food requirements but respond differently to other factors. In nature there are many instances of insects using the same food but doing so at different times or under different circumstances. The case of the large ichneumons of the genus *Megarhyssa* is especially interesting. Three common species of this genus exist in the eastern United States, and all attack the larvae of the pigeon tremex, a wood wasp that bores in the trunks of trees. The largest species, *M. atrata*, has an ovipositor over 10 cm in length; another, *M. macrurus* (shown in Fig. 12–9) is somewhat smaller; and the third, *M. greenei*, is considerably smaller, with an ovipositor about 4 cm in length (Fig. 15–2). Tremex larvae tend to bore straight into the trunk and then follow the grain of the wood at various depths, so that in any tree there are likely to be larvae boring at different depths in the trunk. Female *Megarhyssa* are able to detect the location and depth of larvae and to select larvae near the maximum length of their ovipositors. Thus the three species essentially exploit three different populations of the host and are not direct competitors.

A striking example of **resource partitioning** is provided by J. Bruce Wallace, of the University of Georgia. He studied three related species of caddisflies occurring in streams in northeastern Georgia. The caddisfly larvae spin silken webs that are used to seine small organisms from the current (Fig. 12–2). The larvae live in retreats nearby and feed periodically on the diatoms, small arthropods, and other organisms that become enmeshed in the net (they are "trappers," as we used the term in Chapter 12).

Figure 15–1
Temperature and humidity as niche parameters of two species of flour beetles (*Tribolium*). As these parameters fluctuate, first one species and then the other will be favored (arrow). (From P. W. Price, *Insect Ecology*, 1975, with permission of the publishers, John Wiley & Sons, Inc.)

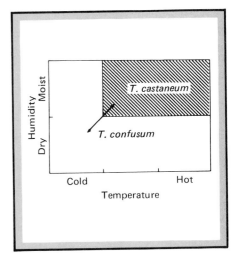

Figure 15–2
Ovipositor lengths of three species of *Megarhyssa* (Ichneumonidae), all occurring in the eastern United States. Vertical lines indicate range of variation; central horizontal lines, the means; rectangles twice the standard error. (From H. Heatwole and D. M. Davis, "Ecology of three sympatric species of parasitic insects of the genus *Megarhyssa* (Hymenoptera: Ichneumonidae)," *Ecology*, vol. 46, pp. 140–150. Copyright © 1965, the Ecological Society of America.)

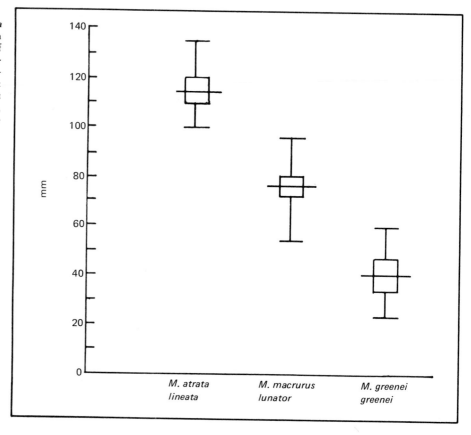

Wallace found that one species makes a coarse net, with mesh openings measuring about 134×249 μ; another a finer mesh, openings 48×79 μ; the third a still finer mesh, with openings 5×40 μ. The first species also spins its web in broader spaces among the rocks where the current is swifter, that is, in places likely to supply larger organisms. Study of gut contents confirmed that the three species do in fact feed largely on organisms of a certain size range and thus "divide up" the available food.

In Jackson Hole, Wyoming, there are five species of bee wolves (*Philanthus*, discussed in another context in Chapter 7). Although all nest in the ground and prey on bees, they effectively do not compete, since they have different seasonal cycles and/or utilize different soil types and/or use bees of different sizes (Fig. 15–3). In Chapter 12 we discussed a similar situation involving 21 species of robber flies inhabiting the Pawnee Grasslands of Colorado.

The Meaning of "Competition." We have cited several examples of resource partitioning, in which coexisting species share some but not all niche parameters. They may be said to exhibit **niche overlap,** but they do not occupy precisely the same niche. Although they are not complete competitors, they may still be competitors if a common niche parameter is in short supply.

Species having most niche parameters identical are said to be **ecological homologs.** The degree to which such homologs are able to coexist depends on the extent of

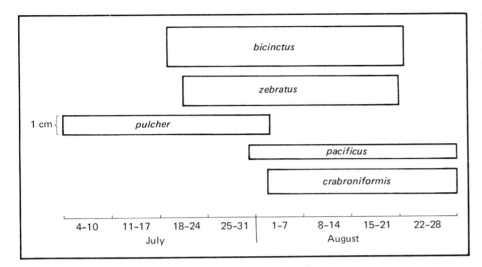

Figure 15–3

Competitive exclusion among five species of *Philanthus* (Sphecidae) occurring in Jackson Hole, Wyoming. Female size is indicated by width of the bars. *P. pulcher* females are about 1 cm long and prey on a variety of small bees and wasps. *P. zebratus* is appreciably larger and uses larger bees and wasps, while *P. bicinctus* preys mainly on bumble bees. *P. pacificus* and *crabroniformis* use similar prey to *pulcher* but nest later in the season; the females nest in very different types of soil. (H. E. Evans, 1970, "Ecological-behavioral studies of the wasps of Jackson Hole, Wyoming," *Bulletin of the Museum of Comparative Zoology, Harvard University*, vol. 140, pp. 451–511.)

the homologies. Can we conceive of any situation in which ecological homologs might occur together? If we assume that A is newly arrived in the habitat of B, the two may of course coexist for several generations. We know that virtually all resources are patchy to some extent. The more patchy the distribution, the more chance may play a role in determining what species will occupy a particular plot or even a particular bush. That is, two species with low populations and identical requirements may survive by occupying different microhabitats that they have reached by random dispersal. At times the two are likely to arrive in the same microhabitat, in which case one or the other is likely to become the sole occupant if the habitat is available for a long enough time.

There is no logical reason why ecological homologs might not evolve in widely disparate geographic areas, for example, on different islands or continents. When introducing a beneficial insect from one continent to another, it is important to ask whether it may be an ecological homolog of a species already present: will the introduced species occupy an unfilled niche, or will it simply compete with a native species? In the latter instance **competitive displacement** may occur. That is, one or the other species may reduce the effectiveness of or even eliminate the other, or conflict between the two may lower the effectiveness of both.

An example of competitive displacement is provided by minute parasitoid wasps of the genus *Aphytis* that have been used in biological control of the California red scale of citrus. Around 1900, one species of this genus (*A. chrysomphali*) was accidentally introduced and became effective in partial control of the scale. In 1948 another species (*A. lingnanensis*) was deliberately introduced and almost completely displaced the first species. In 1956 still a third species (*A. melinus*) was brought in and rapidly began to replace the second species in many areas. Actually all three species continue to survive because of the patchiness of citrus orchards and because each has slightly different niche parameters that were present when it arrived or evolved after introduction as a result of natural selection serving to reduce niche overlap (Fig. 15–4).

Obviously one must understand the life history and ecology of a species thoroughly before attempting to manipulate its populations in any way. Not uncommonly one detects an unoccupied niche—a weed introduced without herbivores, or a plant-feeder without parasitoids—in which case a careful search for an insect with precisely the right biological features may be in order. If an apparently appropriate species survives

Figure 15–4
The distribution of *Aphytis* species in California in 1961. *A. chrysomphali,* the first to be introduced, had once occupied all the areas outlined. After its introduction in 1948, *A. lignanensis* almost completely displaced it, only to be displaced itself in most areas away from the coast by *A. melinus* after the latter's introduction in 1956. By 1961 *A. chrysomphali* persisted in only a few isolated orchards. (After DeBach and Sundby, 1963, "Competitive displacement between ecological homologues," *Hilgardia,* vol. 34, pp. 105–166.)

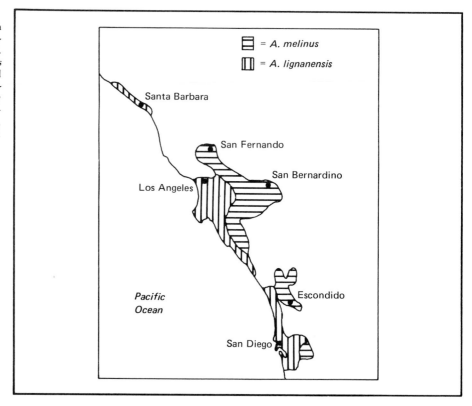

poorly under our climatic conditions, or if it attacks hosts not intended, efforts clearly have been wasted or the program may backfire. There are many problems in dealing with these matters, not the least of which is the difficulty of obtaining results in the laboratory that can be extrapolated to natural situations. A good example of successful introductions to fill unoccupied niches is provided by the control of cattle dung in Australia by African dung beetles (see Chapter 13).

Niche Breadth.   Ecologists are by no means in agreement on all matters relating to competition. Another complicating factor is **niche breadth.** In any given habitat there are likely to be **generalists** and **specialists,** the latter by definition occupying smaller niches within those of the former. A field of corn may be infested with European corn borer and corn billbugs (with slightly overlapping, restricted niches) and with grasshoppers, which are general feeders on vegetation. Coexistence is possible because corn is only a small part of the grasshoppers' food; only if the grasshoppers occur in outbreak proportions are the borers and billbugs likely to suffer seriously. If the farmer decides to grow soybeans, the grasshoppers will survive on weeds and grasses, but the more specialized feeders will have to move elsewhere or perish.

A particularly fine example of niche breadth is provided by spider wasps that prey on spiders that live in vertical burrows in sand dunes (genus *Geolycosa*). One species of wasp is a specialist on these spiders, living only in the dunes, stalking the spiders in

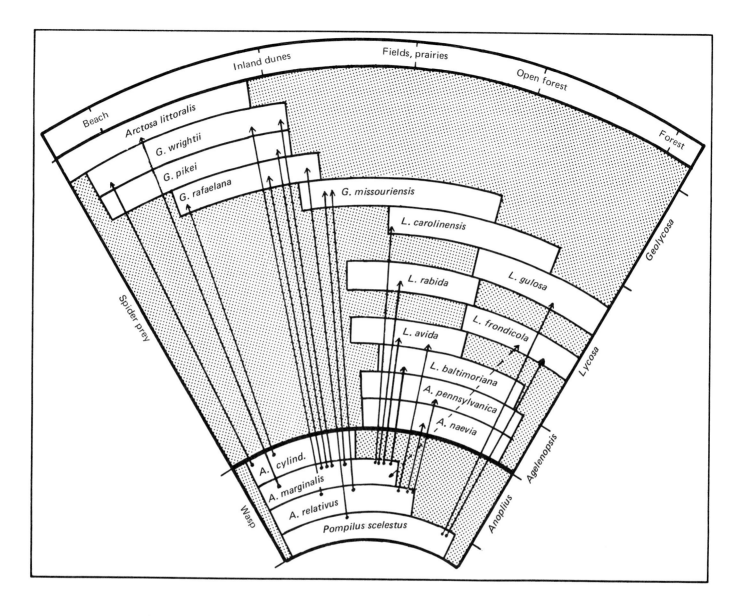

their burrows and using the burrow as their own nest. A second species is similar but larger and uses more mature spiders. Its niche is not only different in that respect, but it is broader, since other kinds of spiders in and around the dunes are sometimes attacked, and in this case the wasp must dig its own burrow in which to place the paralyzed spider before laying its egg. At least two other species sometimes use *Geolycosa* as prey, but these are wide-ranging species that employ a variety of spiders as prey and invariably dig their own burrows (Fig. 15–5). The *Geolycosa* specialist has an "inside track" on this relatively dependable prey, taking younger individuals. The other three species require larger prey and are sufficiently versatile to use a certain spectrum of other species, and two of them are by no means restricted to sand dunes. The first two are members of a community of sand dune organisms and have overlapping niches in that community; the other two are casual members of this and several other communities.

Figure 15–5
Niche breadth in four species of spider wasps. One species, *Anoplius cylindricus,* is confined to sand dunes and preys only on immature *Geolycosa* spiders that make vertical burrows in the sand. *A. marginalis* is also abundant in sand dunes but takes more mature spiders of several genera, including *Geolycosa.* The other two are only casual members of the sand dune community, both taking *Geolycosa* occasionally. (D. T. Gwynne, 1979, "Nesting biology of the spider wasps which prey on burrowing wolf spiders," *Journal of Natural History,* vol. 13, pp. 681–692, with permission of the publishers, Taylor and Francis Ltd.)

Interference Competition.   We have been discussing interspecific competition for a common resource, often food. This common phenomenon is often referred to as **exploitation competition.** Of a somewhat different nature are instances in which members of one species interact with members of a second species, not as predators but in such a way as to reduce their effectiveness as competitors. This is **interference competition.** For example, workers of Pharaoh's ant, after discovering a food source, produce an allomone that repels members of other ant species that approach the food.

A remarkable example of interference competition involving ants has recently been described by M. Möglich and G. D. Alpert, then at Harvard University. Workers of a desert ant, *Conomyrma bicolor,* pick up small stones and other objects with their mandibles and drop them into the entrances of nearby nests of honeypot ants. This results in an almost complete cessation of foraging among colonies of the honeypot ants. Curiously, when stones are dropped in the entrance experimentally, in the absence of *Conomyrma* workers, the honeypot ants forage readily. Apparently the total behavior of the *Conomyrma* workers is required to depress foraging in honeypot ants, and the stones may serve primarily as an interspecific signal.

Host Races and Sibling Species

We have spoken of generalists and specialists, and of the ability of insects to evolve rapidly and to take advantage of new opportunities. If evolution is a continuing process—as of course it is—we should expect to find many populations in the process of change. Indeed, we have already mentioned several examples of diversification within a species. In Chapter 14 we discussed the peppered moth, in which numbers of the melanic individuals increased dramatically around industrial centers in England, while the "peppered" type prevailed in rural areas, where tree trunks were lichen covered. In this instance predation by birds apparently provided the major selection pressure. We also mentioned (Chapter 12) that eastern U.S. populations of the alfalfa weevil often encapsulate and thus destroy their ichneumon parasitoids, while western populations rarely do so. Similar subtle but important differences in the evolved adaptations of populations of a single species are common, in fact more the rule than the exception.

Thus it is not surprising that many specialists that attack a number of related plants or animals tend to develop **host races,** that is, populations that are still capable of interbreeding with one another but that tend to "prefer" a particular host. Such a preference may be established over many generations through associative learning that is reinforced genetically so as to provide closer adaptation to a certain host species. Many host-specific insects mate on their host. This provides a sorting mechanism such that there is reduced gene flow between populations attacking different hosts. Bees that pollinate restricted kinds of blossoms, for example, usually mate in or near the blossoms; fruit flies commonly mate on the fruit they attack; and many parasites and parasitoids mate on the host. Some parasitoid males wait on the host in which they developed and mate with their sisters as they emerge; in some cases mating actually occurs within the host, before emergence!

An interesting example of host races is provided by the balsam fir sawfly, *Neodiprion abietis*. This insect is a pest of various conifers but consists of a number of populations that are variously regarded as closely related species, host races of a single species, or simply parts of a complex pattern of variation within one species. The different populations in many cases emerge at different seasons, show different feeding patterns and host preferences, and in some cases differ slightly in size or color. Apparently all evolved from a common ancestor, and over time populations on specific hosts acquired the ability to detoxify the defensive chemicals of that host, in somewhat the same manner that insects develop resistance to insecticides. Thus the "white fir strain" survives poorly on pine and not at all on balsam fir (Fig. 15–6, lower right). Although the various strains hybridize readily in the laboratory, they appear to do so rarely in nature.

Particularly instructive are studies of the apple maggot and its allies (genus *Rhagoletis*) by Guy L. Bush, then at the University of Texas. As we described in Chapter 7, these flies are visually attracted to red spheres. The males wait on the fruits and present visual and chemical signals to arriving females. Following mating, the female

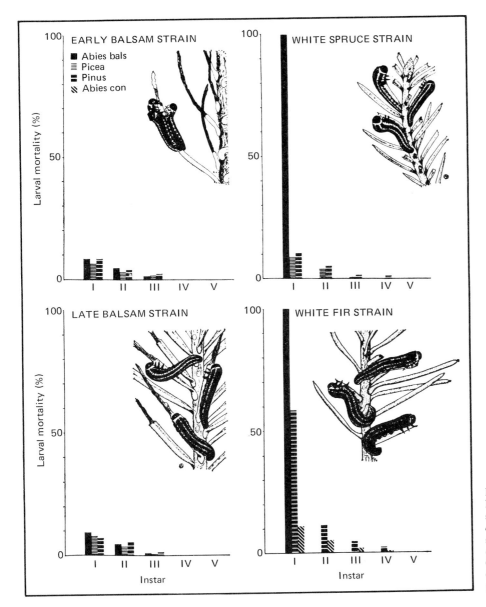

Figure 15–6
Larval mortality of four strains of the balsam fir sawfly, *Neodiprion abietis* (Diprionidae) on four different hosts. Abies bals = *Abies balsamea*, balsam fir; *Picea* = spruce; *Pinus* = pine; Abies con = *Abies concolor*, white fir. (G. Knerer and C. E. Atwood, from *Science*, vol. 179, pp. 1090–99. Copyright 1973 by the American Association for the Advancement of Science.)

searches the fruit for chemical and tactile cues relating to host suitability, and if the cues are favorable she lays her eggs.

The apple maggot is a native North American insect but the apple is not a native tree. It is believed that hawthorn (*Crataegus*) was the original host. Attacks on apples were first noted in the Hudson Valley in the 1860s, and within a few decades the species had become a pest of apples throughout the northeastern United States and eastern Canada. Since apple fruits generally mature earlier than those of hawthorn, a partial seasonal separation of the two races was soon developed. At a still later date, about 1960, a cherry race of the apple maggot appeared in Wisconsin. Since cherries mature still earlier in the year, this race also developed an earlier emergence time, so early in fact that scarcely any overlap occurs with the hawthorn race (Fig. 15–7). Members of these races are nearly identical in structure and interbreed freely in the laboratory.

In contrast, a closely related population that is restricted to blueberries is of a smaller average size and has a shorter ovipositor. Females of the blueberry form cannot oviposit successfully in apples, and although female apple maggot flies oviposit successfully in blueberries, one blueberry is usually not enough for the maggot to complete its development. Another closely related form attacks dogwood berries. Bush has shown that there is complete reproductive isolation among the blueberry, dogwood, and apple maggot flies; thus these populations have achieved the status of species. They remain, however, very similar **sibling species.** It is probable that these species diverged from an ancestral *Rhagoletis* stock many years ago via host switching and coevolution with their hosts (Fig. 15–8). Conceivably the three races of the apple maggot have an equal potential for becoming species, since they now show distinct host preferences and are partially separated by time of emergence.

Another important pest, the codling moth, is known to have coexisting host races in California. These occur on apple, plum, and walnut, females of each race showing a preference for oviposition on their specific host. These preferences appear to be based on a combination of conditioning and inherited behavior patterns. Again, the three races differ in their response to temperature, so that they emerge at a time when host fruits are available.

Knowledge of the prevalence of genetic divergence on different hosts is particularly important in entomology, since it provides the rationale for answering questions such as: What is the chance that a given pest or imported parasitoid will attack a different host or will evolve the capacity to attack such a host? How does one select the

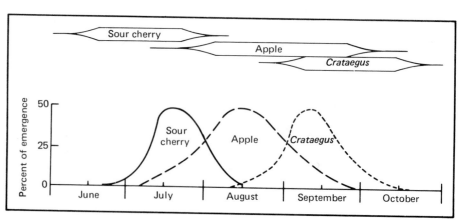

Figure 15–7
Emergence patterns of three host races of the apple maggot, *Rhagoletis pomonella* (Tephritidae) in Door County, Wisconsin. Fruiting time of the host trees is shown above. (G. L. Bush, 1975, from *Evolutionary Strategies of Parasitic Insects and Mites,* edited by P. W. Price, with permission of the publishers, Plenum Publishing Corp.)

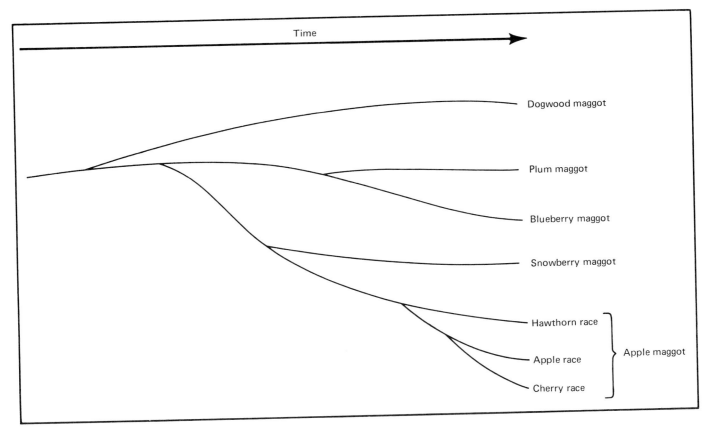

Figure 15–8
Possible paths of evolution of fruit-inhabiting *Rhagoletis* species and races. (Based on G. L. Bush, 1975, "Sympatric speciation in phytophagous parasitic insects," in *Evolutionary Strategies of Parasitic Insects and Mites*," P. W. Price, ed. pp. 187–206.)

appropriate breeding stock for biological control of weeds or pest insects? Is sufficient information available to manipulate a given species successfully? In the broad view, insect–host relationships are best seen as dynamic rather than static. Similarly, human–insect relationships are best seen as a contest between human insights and the genetic capacities of insects.

Origin of Insect Species

Models of Sympatric Speciation. Within the framework of the preceding discussion lies a further answer to the question "Why are there so many insect species?" Clearly insects are able to diversify readily, sometimes by way of host races that in the course of time may undergo reduced genetic exchange to the point that they acquire the status of full species. Many groups of insects consist of great numbers of closely related, sibling species. For example, the parasitoid genus *Apanteles* includes well over 200 species in North America, each attacking a particular complex of lepidopterous larvae (two of these were discussed in Chapter 12). On a broader scale, the family of weevils (Curculionidae) contains at least 40,000 species—more than all the species of birds, mammals, and reptiles combined; nearly all weevils are specialist feeders. Over all, perhaps half a million species of insects show a strong degree of host specificity.

There is evidence that in some cases host selection may be controlled by one or a few genes influencing chemoreception. Guy Bush and his associates have studied the genetics of two sibling species of gall-forming flies, each of which attacks a different species of composite. These flies can be hybridized in the laboratory, but the female offspring, rather than attacking either host plant indiscriminately, prefer to lay their eggs on the plant on which they were reared, presumably as a result of conditioning. But the products of backcrosses to the parent do not necessarily prefer the host on which they were reared. Approximately half the females oviposit on the host appropriate to the other of the two original species. Evidently a simple genetic mechanism, perhaps alleles at a single locus, influences the response to the odor of the host plant. Since these insects mate on the host in nature, it seems probable that this is a case of initial sorting on the host involving conditioning that is reinforced by changes in gene frequency and ultimately by selection for mutants producing a closer adaptation to the host. In time the host itself may evolve defenses against this particular insect, and there may be continued coevolution of the host and its attacker.

Speciation on different hosts may be spoken of as **allophagic** ("different food"). Often the hosts mature at different times (as in the case of the apple maggot complex), thus enhancing separating of populations. **Allochronic** speciation ("different times") may occur in other contexts, without a shift in hosts. The spring and fall field crickets of the eastern United States, for example, resemble one another closely, have similar feeding behavior, and even have similar calling songs. But one overwinters in the egg stage and matures in the fall, while the other overwinters partially grown and matures in the spring. Catherine and Maurice Tauber, of Cornell University, have shown that in green lacewings (Neuroptera), a change of a single allele produces a major change in response to photoperiod. Their studies of two closely related species of these relatively polyphagous predators suggest that the parent species was polymorphic for habitat preference, and that this was reinforced by selection for mutants, each having an emergence time more suited to one or the other habitat. Assuming a partial separation of populations on this basis, selection might further fine tune each subpopulation to its habitat and decrease gene wastage by favoring the development of reproductive isolating mechanisms.

Many species are known to exhibit local differences in gene frequencies associated with environmental gradients (for example, in temperature, rainfall, vegetation type, edaphic factors). In insects that are closely tied to their environment and do not disperse widely, selection for greater local adaptiveness may often overcome the effect of gene exchange and result in the development of mechanisms that reinforce isolation (sometimes called "area effect"). In some cases local adaptations may involve preference for a particular host plant or animal. Speciation resulting from population division via close adaptation of local populations is often spoken of as **parapatric** ("beside the place"). Clearly it may differ from allophagic speciation only in the degree of spatial separation of the initial populations. Doubtless there are many instances in nature whereby all three of these mechanisms are operative more or less simultaneously or sequentially. The three together are often described as **sympatric** models of speciation ("same area"). In contrast, most vertebrate animals (and many insects, too) are believed to have speciated **allopatrically** ("different areas"). Although students of evolution often debate the relative importance of sympatric versus allopatric speciation, clearly both concepts are needed in explaining the great diversity of insects. Nor is there any sharp distinction: even different host plants, and surely different local populations, occupy slightly different areas; and speciation in grossly different areas often involves a shift in host or in time of emergence.

Allopatric Models of Speciation.   There is no question of the effectiveness of an actual physical barrier, such as an ocean or a desert, in reducing or eliminating gene exchange between populations and causing them to evolve separately (often called **population splitting**). If the barrier is of sufficient duration and is then removed, the populations may have diverged to the point that they cannot interbreed; or if they can, the hybrids may prove less well adapted, and there will be selection for mechanisms intensifying isolation (reducing gene wastage), such as differences in courtship behavior or sex pheromones. Eventually the ranges of the two populations will stabilize, and the two may coexist without interbreeding. Presumably they will by this time not be ecological homologs, having developed different hosts or other associations with the environment during the period of isolation. If they have not, there will be geographic exclusion or evolution for some means of competitive exclusion.

Obviously there are many variables in such a process, including the mobility of the animals, the kind of barrier and its strength and duration, and the size of the populations. It must be appreciated that allopatric speciation is a process requiring long periods of time—thousands or even tens of thousands of years. Often our hypotheses must be built on knowledge of past climatic events or on knowledge of whatever fossils have been discovered in the group we are studying. We know, for example, that North America has in recent geologic time been subjected to recurrent periods of glaciation. During periods of maximum glaciation, life zones were pushed much farther south, and many plants and animals came to occupy warmer "refugia" in the southeastern states, in Mexico, or in California. In many cases isolation was sufficiently long that populations diverged to the point that they attained species status. Another effect of glaciation was to tie up much moisture in the form of ice, resulting in a lowering of the sea level. The shallow Bering Sea was evidently dry land for considerable periods, and since much of Alaska was not glaciated, there was an opportunity for interchange of animals between Eurasia and North America (Fig. 15–9). Once on a new continent, stocks were subjected to different selection pressures and influenced by differing topography, patterns of glaciation, and so forth. Many of the resemblances and differences between the faunas of Eurasia and North America are best understood in terms of such east–west and north–south movements and subsequent isolation of population fragments.

A somewhat different form of allopatric speciation occurs when an inseminated female or a small group of individuals is wafted to a point well beyond the normal range of the species. Such individuals must usually fail to survive in the new area, but when they happen to arrive in an area where there are niches that are unoccupied or in which they can compete successfully, they may thrive. Given no opportunity to breed with the parent population, members of the subpopulation may diverge to the point that they develop quite different adaptations and lose the capacity to interbreed with the parent population. This is spoken of as speciation by the **founder effect.**

Speciation by this means is believed to be by no means rare in insects. As we shall see in the next chapter, individuals may be carried great distances in air currents. Since founder populations are quite small, they represent only a small fraction of the variability present in the parent population. In the new area, population size may increase rapidly in the absence of predators and competitors. During this expansion the lessened effect of natural selection permits the establishment of new gene frequencies and mutations, often by random effects. Thus when the population eventually stabilizes, its features may be quite different from those of the parent population. It is probable that, in general, speciation occurs more rapidly by the founder effect than by subdivision of a broad initial range.

Figure 15–9
Stages in the evolution of separate species in America and in Asia from an originally Asiatic species that dispersed through the Bering Straits when this was dry land. (From H. H. Ross, *Biological Systematics,* © 1974. Addison-Wesley, Reading, MA. Figure 9.5. Reprinted with permission.)

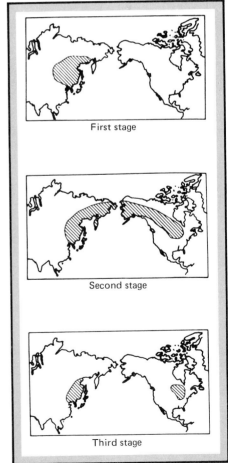

First stage

Second stage

Third stage

While the founder effect undoubtedly occurs in continental areas, its effects are most clearly seen in island archipelagos—in the Hawaiian Islands, for example. These are volcanic islands of known age, the oldest being about 5 million years old, but the youngest island, Hawaii, less than a million years old. There is no question that the entire flora and fauna arrived by immigration across the sea. It is estimated that the several thousand species of native Hawaiian insects evolved from only about 250 founder populations—probably often single inseminated females wafted in air currents, chiefly from the South Pacific. Once established, populations underwent further speciation by "island hopping" among the major islands. It is possible that the approximately 800 native species of *Drosophila* may have evolved from only two ancestral species.

Most Hawaiian *Drosophila* species are restricted to single islands, and many have come to occupy restricted and often bizarre food niches. The presence of numerous related species together in rain forest habitats has resulted in the development of elaborate reproductive isolating mechanisms. Unlike most *Drosophila*, males of many species have patterned wings and establish territories that they advertise to females by species-specific postures and patterns of wing waving. Studies of chromosome structure by Hampton Carson and his colleagues at the University of Hawaii suggest that the approximately 100 picture-winged species may have arisen by multiple "island hopping" from a small group of clear-winged species occurring on the oldest major island, Kauai. A minimum of 22 colonizations has been postulated to account for the origin of the picture-winged species (Fig. 15–10). Most of the founders have moved from northwest to southeast, which coincides with the relative time that the major islands became habitable.

Thus the evidence suggests that insect species have arisen in many different ways. Furthermore, many species consist of subpopulations among which gene exchange is

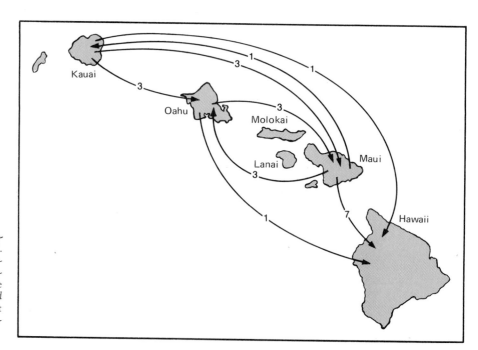

Figure 15–10
Speciation among Hawaiian picture-winged *Drosophila* by founder effect. (From *Evolution*, by T. Dobzhansky, F. J. Ayala, and G. L. Stebbins, copyright 1978, W. H. Freeman and Co.; after H. Carson et al., 1970, "The evolutionary biology of the Hawaiian Drosophilidae," in *Essays in Evolution and Genetics in Honor of Th. Dobzhansky*, M. C. Hecht and W. C. Steere, eds. New York: Appleton-Century-Crofts. pp. 437–543.)

reduced to varying degrees. Thus, while it is important to identify species correctly and to understand their biological attributes, it is also important to understand variation within species. The prevalence of host races, geographic races, and sibling species among insects presents problems and challenges above and beyond mere species abundance. Within local communities the genetics of each population is adjusted so as to enhance survival and reproductive success in the face of local variables, both biological and physical.

Distribution Patterns

No insect species occurs naturally throughout the world. The few that do now occur almost everywhere bear a close association with the human species or human artifacts, as do body lice, granary weevils, and some kinds of cockroaches. In nature all species have certain limitations to their distribution. The following four appear to be the most significant of these limiting factors.

1. *Lack of opportunity to disperse.* The fact that many pest species have been introduced to North America from other continents and now thrive there demonstrates that they evolved elsewhere and simply had no previous opportunity to cross the oceans. Now and then a species does, in fact, cross oceans and become established on a new continent under its own power. The monarch butterfly, for example, first became established in Australia about 1870. The Hawaiian Islands and other oceanic islands were populated (before the advent of humans) entirely by immigrants over the sea, as we have seen.

   The honey bee was brought from Europe to North America soon after the first colonists arrived, and it quickly escaped to form wild colonies. More recently, in 1956 the African race of the honey bee was brought to Brazil for experimental purposes and accidentally escaped to the wild. It has now spread over much of northern South America and there has been much speculation as to whether these so-called killer bees are adapted for survival in temperate North America. In fact, they interbreed with European races to produce a form called "Africanized bees," which may prove more cold resistant and less aggressive, depending on the proportion of genes from the two races.

2. *Lack of suitable habitat or host plant or animal.* An insect that is monophagous or narrowly oligophagous is necessarily confined to the range of its hosts. Many "rare" insects are so limited. The now extinct xerces blue butterfly fed on lupines that grew in sand dunes near San Francisco, a habitat now destroyed by urban sprawl. Many agricultural pests are also limited by the distribution of their hosts.

3. *Temperature.* Every species has certain temperature limits above and below which development is retarded or, at extremes, death ensues (Fig. 15–11). A few insects are able to escape lethal temperatures by migrating (see next chapter); honey bees survive the winter by producing metabolic heat in the cluster. But the ranges of most insect species are limited by the temperature regime to which they are individually adapted.

   In temperate and arctic regions, most insects spend the winter in diapause. Insects "anticipate" the cold, generally via endocrine response to shortened day length (Chapter 4). An unusually early, severe, or prolonged subfreezing temperature may produce considerable mortality among insects not adapted for these conditions. During winter diapause, physiological changes, particularly an increase in the viscosity of body fluids and the production of glycerol, prevent the destruction of tissues by ice crystals. Different populations of one species sometimes vary in

Figure 15–11
Temperature relationships of an insect (temperatures in Celsius). Much depends on the time an insect is exposed to a given temperature, the regime to which the insect is adapted, and whether or not the insect is in diapause. (Modified from B. P. Uvarov, 1931, "Insects and Climate," *Transactions of the Entomological Society of London*, vol. 79, pp. 1–247.)

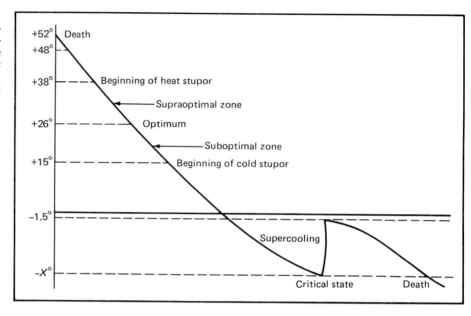

their ability to withstand cold. For example, in a race of gypsy moths in Russia, the females consistently lay their overwintering egg masses on tree trunks close to the ground, where they take advantage of snow cover to survive temperatures far below freezing for long periods.

4. *Moisture.* As with temperature, insects develop best within a certain optimum range of humidity and fail to survive under prolonged exposure to extremes. Species occurring in very dry environments survive by burrowing deeply or by utilizing metabolic water and producing very dry feces. Clothes moths are able to complete development in 20% relative humidity when the food contains only 6% water. When dry periods are seasonal, many insects undergo diapause and some are able to migrate; lady beetles do both (Chapter 12 and Fig. 12–6).

Communities and Ecosystems

Groups of species having similar ranges and similar habitat requirements form **communities,** in which the species occupy niches that are at most partially overlapping and in which diverse interrelationships develop. Some communities are sharply limited, such as those within a pond or a localized series of sand dunes, while others grade into one another, such as the communities that occur sequentially as one ascends a mountain. Species of a community vary in the size of their populations and in their niche parameters. Interactions are often expressed as a **food chain** (Figs. 15–12 and 15–13), which is also a "pyramid of numbers," since members of each level in the hierarchy cannot become more abundant than members of the level below without exhausting their food supply.

Insects are major elements in all terrestrial and freshwater communities. Most are **primary** or **secondary consumers;** that is, they represent links in the food chain between plants (producers) and consumers of higher levels. For example, a pea plant (producer) is fed on by aphids (primary consumers), which are attacked by small parasit-

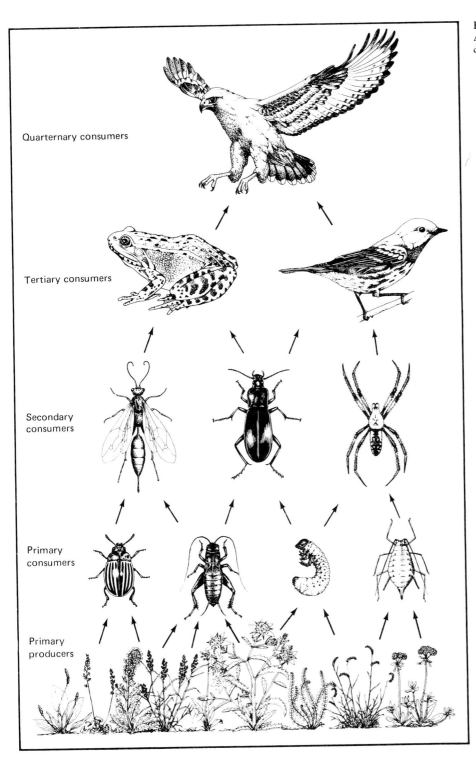

Quarternary consumers

Tertiary consumers

Secondary consumers

Primary consumers

Primary producers

Figure 15–12
A diagrammatic terrestrial food chain. Arrows indicate energy flow. (Drawings by Peter Eades.)

oid wasps (secondary consumers). The adult wasps fly about and are eaten by small birds (tertiary consumers), which in turn fall prey to hawks (fourth-level consumers). Not only are we humans top consumers in many food chains, but more important, we depend on other consumers to keep our environment from being overrun by aphids, caterpillars, mice, rabbits, and a host of other animals that, in the absence of checks, would destroy most of our foodstuffs. The low-order consumers are the most important elements in all food chains (aside from the producers!), because without them there could be no higher levels in the chain. As one writer has recently put it: "Take away the insects and terrestrial ecosystems would become empty shells within a year."

Communities change with time as a result of chance factors (forest fires, arrival of a new herbivore or predator) or of changes in physical factors (silting of a pond, accumulation of organic matter in the soil). **Ecological succession** implies the replacement of community members in time. Many of the orchards of New England, once well tended, have reverted to grasses, brambles, and small trees. If not converted to factories or suburbs, most will ultimately become mixed hardwood forests, with an undergrowth of plants able to flourish in the limited light the canopy permits to filter through. The change in the insect fauna through these successional changes is dramatic and complete.

The number of species making up a particular community depends on diverse factors. The longer the habitat has been available for colonization and the more extensive the habitat, the more species will have encountered it and will have established relationships within it. The proximity of similar communities will influence the frequency of arrival of potential occupants. Diversity of resources within the community will be of major importance, as will climatic and edaphic factors. Some of the most species-rich communities occur in tropical rain forests, vast in area, diverse in resources, and situated in areas of relatively benign, year-round climate. Relatively rich also are temperate mixed deciduous woodlands, marshes, and meadows. Coniferous forests, especially if mostly of one tree species, will be somewhat poorer in species. Such physically rigorous situations as hot springs, salt lakes, and caves may have very few species.

Human activities commonly produce rigorous conditions where only a few species can survive—activities such as bulldozing, strip mining, or lowering the water table. Many forms of agriculture involve replacement of the native flora and fauna with a monoculture that requires constant maintenance. The "community" consists of a single plant species, a limited soil fauna, and a few insectan consumers and their natural enemies—even these lead a precarious existence. Such simple communities contrast strongly with most natural environments. **Community stability** in large part depends on the development of complex relationships among its many inhabitants, including interlocking food chains, extensive food partitioning, and coevolution of the component species to form a functional whole. A stable community may be said to have achieved **homeostasis**, such that it is relatively immune to naturally occurring perturbations. Presence or absence of one species is unlikely to affect the system, and increased population size of any one species is likely to be dampened by others. Resilient though such communities are in the presence of biological disturbances, they often prove exceedingly fragile under human impact, which in a few hours may destroy myriads of complex relationships that have evolved over many years. Recent devastations in tropical rain forests provide a case in point.

The contrast between the insect fauna of a monoculture and that of a mixed environment was well demonstrated in a study by R. B. Root, of Cornell University. He grew collards in pure stand and also in isolated rows among meadow grass. Censuses of the insects in the pure stand revealed 93 species of herbivorous insects and 181 species of predators and parasitoids. The surrounding meadow was not censused, but a similar meadow was reported to contain 1584 resident insect species. More interestingly, collards grown in pure stands had a much higher herbivore load than those grown in rows among the meadow grass, mainly as a result of several persistent pests such as flea beetles and cabbageworms. Presumably the effect of these pests in the rows among meadow vegetation was lessened partly because a far greater number of generalized predators and parasitoids were there, a reflection of habitat diversity. This may not be the whole answer, however. Pure stands tend to attract insects that thrive on that crop, and these insects develop large populations rapidly and have little tendency to move elsewhere. In many cases the natural enemies of these insects require other resources (frequently nectar) and consequently are much more likely to drift elsewhere.

The relationship between species diversity and community stability is often debated by ecologists, but many would agree that simple systems are more readily subject to imbalance. The situation is comparable to that in the field of economics, where diversity of investment is an accepted manner of reducing financial risks. As energy becomes more expensive, it may be found desirable to abandon extensive, artificially maintained monocultures in favor of diversified farming and the maintenance of hedge rows, meadows, and wood plots.

Aquatic Communities. In contrast to many terrestrial situations, ponds and streams tend to be widely spaced and to differ from one another considerably in characteristics. Ponds differ in depth, in type of surrounding vegetation, and in chemical characteristics; streams differ in these features as well as in rate of flow, type of terrain traversed, and so forth. Thus diverse communities form, each consisting of groups of organisms with quite different sets of adaptations. Furthermore, aquatic communities differ from terrestrial ones in that much of the food is in the form of floating particles that require little searching but call for specialized methods of collection.

Streams tend to form sequential, overlapping communities from their source in the mountains to their union with other streams to form the sluggish, sediment-laden rivers of the plains. The nutrients of streams are derived both from stream organisms (such as algae, higher plants, and arthropods) and from input from the outside (chiefly leaves and other plant parts) (Fig. 15–13). The insects use these food sources in diverse ways. **Shredders** chew up large particles, such as leaves or bits of detritus; **scrapers** graze on algae adhering to rocks. Both groups tend to release fragments that are added to fine particles already in the stream; these are harvested by **collectors.** The most prevalent of these are the **filter feeders,** which gather particles from the water by mouth brushes (Fig. 15–14), by brushes on the legs, or by webs that are spun among the rocks (Fig. 12–2). Various Ephemeroptera, Plecoptera, Trichoptera, and Diptera are the major herbivores and detritivores. *Predators* include a diversity of beetles, Megaloptera, and some Trichoptera, Plecoptera, and Diptera, and of course fish and other vertebrates.

The study of stream ecology has become vitally important now that human population pressure has put high demands on our water resources and at the same time done so much damage to aquatic communities. Study of the disturbances to these communities produced by dams and by pollutants has become a major way of evaluating water quality and of planning future developments of water resources.

Figure 15–13
Elements in a generalized stream community.

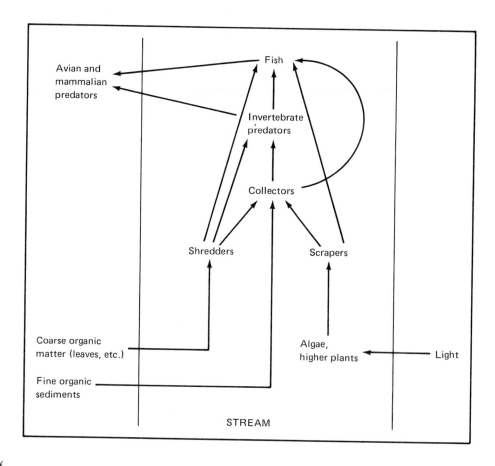

Figure 15–14
The head of the larva of a black fly, a denizen of swift streams. These filter feeders use specialized "head fans" to strain small organisms from the water. (Scanning electron micrograph by Douglas A. M. Craig, University of Alberta.)

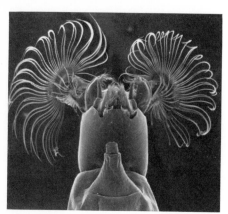

Ecosystems. No community, least of all an aquatic community, can properly be considered apart from the physical factors that influence it, including sunlight, mineral content of the water or soil, and the like. A community considered in relation to its physical environment is termed an **ecosystem.** Studies of ecosystems concern themselves with the input and outflow of energy and the circulation of materials (such as oxygen, nitrogen, or phosphorus) through the system. Emphasis is on energetics and on productivity, the latter being in terms of total biomass or of some feature under study (for example, trout in a stream or marketable beef in rangeland). Needless to say, insects are only one of many groups of organisms occurring in ecosystems, but as primary and secondary consumers they play major roles in all systems.

Ecosystems are of such complexity that they are frequently approached through **modeling**. A model is an abstraction of a system and is a means of conceptualizing a complex situation and of making predictions regarding the future behavior of the system. Models must be checked frequently against the real world, but at the same time they cannot include all attributes of ecosystems or they would lose their value as models. The person constructing a model must first have a broad knowledge of the system and be able to make preliminary assumptions as to the more important interactions. The interactions under study (for example, energy flow, population size) are treated algebraically, producing a series of equations that can be used to construct a computer model. Such a model should be constantly tested, provided with feedback from the real world, and redesigned as necessary. A valid model will frequently suggest other models or submodels and may generate new hypotheses. Further discussion of this growing area of research is outside the scope of this book; Hall and Day (1977) provide an excellent introduction to the subject.

Summary

The term *niche* implies not only where an insect occurs but also what role it plays in the environment. Species occurring at the same place and time do not ordinarily have identical niches; if they did, there would be severe interspecific competition. This is called the principle of competitive exclusion. Experiments with two species of flour beetles have shown that while both occur in the same place and use the same food, they respond differently to environmental variables and are thus not complete competitors. Coexisting species often exhibit resource partitioning; that is, their behavior is adjusted in such a way that they share the available food or other resources. Niche overlap is a term employed when some niche parameters are shared by two or more species; species with extensive niche overlap are said to be ecological homologs. Instances have been described in which such homologs replace one another competitively. Thus it is important to understand niche breadth thoroughly before undertaking the introduction of beneficial species. Generalists are defined as species having broad food niches, while specialists are those restricted to one or a few host plants or animals.

Frequently specialists tend to form host races, preferring to lay their eggs on one of two or more alternative hosts, perhaps initially as a result of olfactory conditioning. Many host-specific insects mate on the host, which results in a reduction in gene exchange between populations on different hosts. Genetic change may produce closer adaptation to the host, such that hybrids are less well adapted to either host, and this may in turn produce selection for complete reproductive isolation. The case of the apple maggot is especially well studied. The apple maggot evidently was once restricted to hawthorn. Host races have now developed on apple and cherry. In the same genus there are other forms, now regarded as different species, which attack blueberries, dogwood, and other fruits. There is evidence that speciation and race formation in this group occurred primarily as a result of host switching.

In the broad view, it is apparent that speciation among insects has occurred in several different ways, which are not mutually exclusive. These may be summarized as follows:

Allophagic (on different hosts)	
Allochronic (different times of emergence)	Sympatric ("same area")
Parapatric (area-effect genetic change)	
Population splitting (physical barriers)	Allopatric ("different areas")
Founder effect (entering a wholly new area)	

Populations occurring and interacting together constitute a community. The number of species occurring in a community may be high, but in rigorous environments or in many forms of agriculture it may be quite low. In general, species-rich communities are regarded as more stable as a result of interlocking food chains, resource partitioning, and coevolution of the component species. An ecosystem is a community considered together with its physical environment. Studies of ecosystems are largely concerned with energy flow as well as with the circulation of oxygen, minerals, and other substances through the system.

Selected Readings

Bush, G. L. 1975. "Modes of animal speciation." *Annual Review of Ecology and Systematics,* vol. 6, pp. 339–64.

Clark, L. R.; P. W. Geier; R. D. Hughes; and R. F. Morris. 1974. *The Ecology of Insect Populations.* New York: Wiley. 232 pp.

Cummins, K. W. 1973. "Trophic relations of aquatic insects." *Annual Review of Entomology,* vol. 18, pp. 183–206.

DeBach, P. 1966. "The competitive displacement and coexistence principles." *Annual Review of Entomology,* vol. 11, pp. 183–212.

Hall, C. A. S., and J. W. Day, Jr., eds. 1977. *Ecosystem Modeling in Theory and in* Practice: An Introduction with Case Histories. New York: Wiley. 684 pp.

Knerer, G., and C. E. Atwood. 1973. "Diprionid sawflies: polymorphism and speciation." *Science,* vol. 179, pp. 1090–99.

Krebs, C. 1978. *Ecology: The Experimental Analysis of Distribution and Abundance.* Second edition. New York: Harper & Row. 678 pp.

Pianka, E. R. 1978. *Evolutionary Ecology.* New York: Harper & Row. 397 pp.

Price, P. W. 1975. *Insect Ecology.* New York: Wiley. 514 pp.

Schoener, T. W. 1982. "The controversy over interspecific competition." *American Scientist,* vol. 70, pp. 586–95.

White, M. J. D. 1977. *Modes of Speciation.* San Francisco: Freeman. 455 pp.

Dispersal and Migration

E ven the most casual observations on a warm summer day reveal that the air is filled with insects. This impression is reinforced if some kind of trapping device is employed. For example, a pan painted bright yellow and filled with water will collect large numbers of aphids and other small, flying insects, many of which have been dispersing to new feeding sites and are attracted to yellow because it simulates the reflectance of green foliage. Malaise traps (unbaited, tentlike traps) collect quantities of insects flying close to the ground, while nets attached to planes collect insects high in the atmosphere and far at sea. P. A. Glick, of the U.S. Department of Agriculture, collected small arthropods as high as 4600 m—over Louisiana. Altogether he collected members of 210 families of 18 insect orders and 2 orders of Arachnida, spiders and mites. Many of these were wingless. That at least some members of this "aerial plankton" may travel great distances is shown by collections made far at sea. J. L. Gressitt and co-workers at the B. P. Bishop Museum in Honolulu have collected live insects up to 400 km from the nearest land, and small flies and even grasshoppers were taken in aerial nets more than 1500 km from land.

Many persons have also witnessed massed migrations of insects. Monarchs, painted ladies, and other butterflies often attract attention in their north–south flights through North America. Migratory locusts have caused widespread devastation ever since biblical times. C. B. Williams, a British entomologist who devoted much of his life to the study of insect migrations, reported a locust swarm in Africa that was over 1.6 km wide, over 30 m deep, and passed for about 9 hours at about 10 km/hour. He estimated that it contained about 10 billion insects.

Mass migrators and members of the aerial plankton are obviously not feeding or reproducing. Is their occurrence in the air merely accidental, or is it adaptive? If adaptive, is it part of their regular life cycle, or is it a response to abnormal conditions? How can it be adaptive for an individual when a great many must fail to find a suitable place to live and must therefore perish? What is the relation of dispersal to community structure and to the population dynamics of individual species? These are important questions, but as so often happens when dealing with living systems, there are no simple, unequivocal answers.

Many movements of insects may be considered *trivial*, such as the swarming of midges, the movements of a grasshopper from one grass clump to another, the search of a parasitoid for its host. However, such movements are often fundamental to the lifestyle of these insects. They are trivial only in the sense that they occur within the home range of the population and do not usually result in much, if any, displacement. Since such movements are commonly suppressed during periods of strong winds, it is probable that they only occasionally result in accidental dispersal.

Recent students of the subject have concluded that most dispersal is "active" and that it is an adaptation that has evolved in many species, enabling individuals to exploit new sources of food for themselves or their offspring. That is, dispersal is a behavior pattern undertaken at a particular time in an insect's life (often as a young adult), under favorable weather conditions, and in some cases in response to crowding, food stress, or other stimuli. Briefly, most dispersal may be characterized as **adaptive traveling**.

Dispersal behavior consists of three stages. First, the insect releases itself from the substrate in one of several ways in response to internal and external stimuli. A first-instar spruce budworm larva moves to the tip of a twig, then drops on a silken thread and is blown off by the wind. A young adult winged aphid spreads its wings and pushes itself into the air. Newly molted adult locusts undertake irregular flights of gradually increasing vigor and duration, then one day take off en masse to a new feeding site. The second stage consists of movements through the air, taking advantage of air currents and involving varying degrees of directional flight on the part of the insects. Finally, the insects settle in a new location, using sensory cues to locate suitable food or oviposition sites (Fig. 16–1).

Dispersal and migration should not be regarded as different phenomena. **Dispersal** is a general term for all forms of displacement to new habitats; often it is multidirectional and takes advantage of wind currents. **Migration** is usually considered a special form of dispersal. Migrating insects move in a specific direction, and movement is usually at least partially under control of the insects themselves. At a given site, insects may be arriving **immigrants** or departing **emigrants**. In practice it sometimes proves arbitrary as to whether a particular insect is said to be "dispersing" or "migrating."

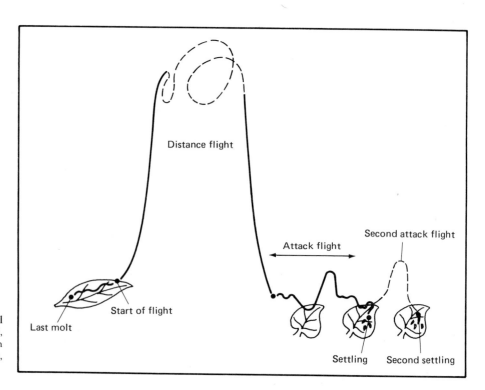

Figure 16–1
Schematic representation of the phases in dispersal of a winged aphid. (After V. Moericke, 1955, "Über die Lebensgewohnheiten der geflügellen Blattläuse." *Zeitschrift für Angewandte Entomologie,* vol 37. pp. 29–91.)

Dispersal to New Habitats

In many insects dispersal is a normal feature of the life cycle. That is, at a certain stage there is a temporary suppression of feeding and reproduction while the insect moves to a new feeding or oviposition site. For example, young adults of the mountain pine beetle fly from the tree in which they developed to other trees. Recently emerged adult mosquitoes may fly several kilometers to a source of a blood meal, and having fed may disperse further to breeding sites. Much of the evidence comes from release and recapture of marked individuals. Mosquito larvae have, for example, been reared on media containing radioactive phosphorus. Adults collected at light traps at various stations were then monitored with a Geiger counter to locate individuals that had dispersed from the marked group. Genetic mutants have also been used in mark–recapture studies. Larger insects, such as butterflies, can often be successfully marked with paints, dyes, or even tags. Swarms of locusts have been tracked by radar and even followed by planes. Single swarms have been observed for more than a week and over distances of hundreds of kilometers.

The presence of a dispersal phase in the life cycle of many insect species accounts in considerable degree for the fact that most available food sources are quickly exploited. A cattle dropping or a rabbit carcass soon acquires a fauna of maggots and beetles; a broccoli patch is quickly discovered by ovipositing cabbage butterflies. More permanent habitats, such as a young orchard or forest, gradually acquire a rich fauna of organisms, which over time develop complex interrelationships. Some insects colonize new habitats much more readily than others. Just as dandelions and a variety of other weeds quickly disperse to newly available sites, so certain insects with high reproductive rates and strong dispersal powers tend to colonize quickly any new source of food. Many of our agricultural pests fall in this category, and in many cases the dispersal powers of the pests exceeds that of their natural enemies.

Dispersal and diapause are often intimately related and under endocrine control. Some of the most detailed studies have involved the milkweed bug, *Oncopeltus fasciatus*, a subtropical insect that migrates north each summer as far as Canada. During the summer, the combination of high temperatures and long photoperiod suppresses migration; but as the season progresses, a combination of shorter day length, lower temperatures, and declining food quality causes a decrease in the titer of juvenile hormone (JH) below the point required for reproduction. This results in the onset of reproductive diapause and increased flight activity, triggering a southward migration. Further food deprivation causes a further decline in JH titer to the point that flight is suppressed, but this can be reversed by the implantation of corpora allata. In the milkweed bug, migration is evidently associated with intermediate titers of JH, but in other instances low titers of this hormone appear to be associated with flight activity. That young adults of many species are most commonly migrants may reflect the fact that the corpora allata are not yet producing JH in sufficient quantities to induce reproductive behavior.

In many aphids there is a pronounced **polymorphism**, generations of apterous (wingless), parthenogenetic, ovoviviparous females being followed by a generation of alates of very different appearance. In the alates, not only is there a gradual development of wing pads to fully developed wings in the adult, but the thorax is more clearly defined and sclerotized and there are differences in the antennae, head, and abdomen as compared to the apterous females (Fig. 16–2). The apterous females are juvenile in appearance (although reproductively mature), as a result of active corpora allata; but the corpora allata of embryos destined to become alates are suppressed such that they

Figure 16–2
Figure 16–2
Polymorphism among females of the sunflower aphid, *Aphis helianthi*. (a) Stem mother (fundatrix), which emerges from overwintered eggs and reproduces by parthenogenetic ovoviviparity to produce the first of several generations of ovoviviparous females (called viviparae) that differ from her in having longer antennae and legs (b). At certain times alates of very different appearance are produced (c). This species has alternate hosts, spending the summers on sunflowers and related annuals, the winters on dogwood (*Cornus*). In the fall, alate males are produced and mate with females that lay eggs on the winter host. (Drawings by Miriam Palmer, Colorado State University.)

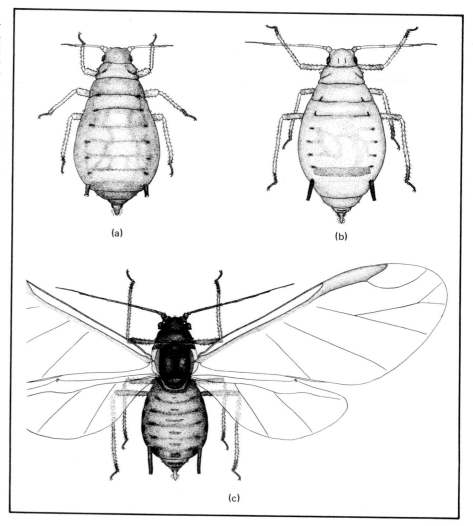

(a) (b)

(c)

develop to winged forms capable of dispersal. Experiments with the vetch aphid, *Megoura viciae*, by A. D. Lees, of Cambridge University, showed that application of juvenile hormone to larvae with developing wing pads resulted in the suppression of many of the usual alate features. Lees found that crowding was the most important factor in inducing production of alates, information on the degree of crowding being conveyed to the mother not by sight or odor but by the frequency of stimulation of tactile setae on various parts of the body.

Dispersal to Unoccupied Habitats. The ability of insects to occupy new habitats is well demonstrated by studies of small islands that have had all the insects removed. E. O. Wilson, of Harvard University, and D. S. Simberloff, now at Florida State University, undertook to defaunate several small islands in Florida Bay. These islands varied in size and in distance from the mainland and contained only one plant species, red mangrove. Each island was covered with a tent and fumigated with methyl bromide, killing all but a very few wood-boring insects and doing little damage to the trees. The islands were then censused over a period of time to determine the rate at which they were recolonized. (Each had, of course, been censused before fumigation.) The original fauna consisted of 20 to 40 species of arthropods on each island, though the species composition was by no means identical on each.

Within a remarkably short time these islands acquired faunas of diverse arthropods (Fig. 16–3). The earliest immigrants included both strong fliers (such as butterflies and wasps) and weak fliers or nonfliers (such as spiders and bark lice), the latter evidently being carried by the wind. Near islands were, predictably, colonized more quickly than far islands. Of special interest is the fact that each island acquired a fauna of approximately the same size as it had originally, though as before the species composition differed on each island. Although species number became stabilized after a few months, the species composition continued to change; that is, new colonizers sometimes replaced older tenants. Wilson and Simberloff speak of an **equilibrium number** of species, this number being determined largely by island size and distance from the mainland. As may be seen in Fig. 16–4, small, distant islands maintain a lower equilibrium number than large islands close to the mainland. This difference reflects the facts that near islands are consistently reached by larger numbers of insects, and large islands contain more living space and the possibility of developing more complex interrelationships.

Patchy terrestrial environments, such as wood plots, ponds, and meadows, are in many ways similar to islands. Fields planted to specific crops are also in many ways comparable to islands. Thus one would predict that the occupation of a field of cultivated plants by a pest would be influenced by its size and by the proximity of similar fields. While the dispersal abilities of the pest species involved must also be considered, the overwintering sites may be equally important. A species that overwinters in stubble has essentially no need to disperse if the same crop is planted in the same field

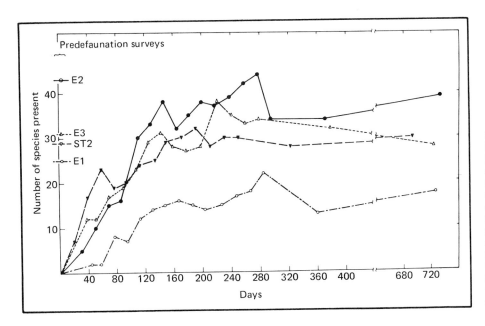

Figure 16–3
Colonization curves of four small islands in the Florida keys after defaunation with methyl bromide. The nearest island, E2 (solid line), had somewhat over 40 species prior to defaunation and within 160 days had acquired nearly the same number of (mainly different) species. The most distant island, E1, also acquired about its original number of species (20), but more slowly. The other two, of intermediate distance, also attained their approximate quota of species as shown. (D. S. Simberloff and E. O. Wilson, "Experimental zoogeography of islands. A two-year record of colonization," from *Ecology*, vol. 51, pp. 934–937. Copyright © 1970, the Ecological Society of America.)

Figure 16–4
Equilibrium model for island colonization as influenced by island size and distance. Equilibrium number is indicated at the point of crossing of two lines; for example, the lines for "small" and "far" cross at the lower left, indicating a low number of species, while the lines for "near" and "large" cross at the right, indicating a much higher equilibrium number of species. (R. H. MacArthur and E. O. Wilson, 1963, "An equilibrium theory of insular zoogeography," from *Evolution*, vol. 17, pp. 373–387, used with permission.)

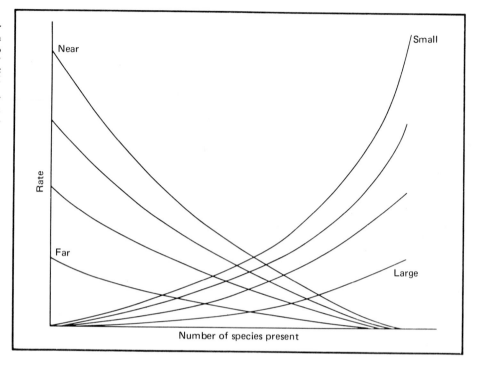

in successive years. The proximity of uncultivated areas may also be important, as these may provide overwintering sites for some pest species and living sites for various predators. The importance of distance is well illustrated by the incidence of sugar beet mild yellow virus, which declines rapidly with distance from places where beets are stored (Fig. 16–5). In this instance the aphid vectors of the virus overwinter in the storage bins and contract the virus there.

Seasonal Dispersal of Insects.   In some instances crop pests migrate from considerable distances. The fall armyworm, for example, is a resident of the tropics and subtropics, including the extreme southern United States. But in summer the moths migrate northward and lay their eggs on corn and other crops as far north as Minnesota. Although the species does not overwinter successfully at these latitudes, it remains a pest year after year, chiefly in the late summer and early fall. The corn earworm, and indeed a number of other pest and nonpest insects, extend their range northward each summer. During mild winters, many may survive to build up higher populations the following year. (Seasonal migrations of lady beetles were discussed in Chapter 12, pp. 276, and Fig. 12–6).

The beet leafhopper is a well-studied example of an insect having rather restricted spring breeding areas but capable of dispersing great distances and establishing summer breeding populations in many parts of the western United States (Fig. 16–6). Taking advantage of air currents associated with the passage of low-pressure systems, the leafhoppers may move several hundred kilometers over a period of just a few days. Sampling of populations along the migration routes reveals that many drop out along the way and that females have greater endurance than males. Endurance is evidently re-

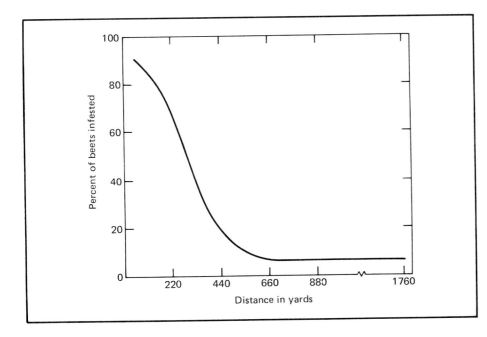

Figure 16–5
The percentage of beets infested with sugar beet mild yellow virus in relation to distance from beet storage sites, indicating the distance traveled by the aphid vectors of the disease. (R. L. Metcalf and W. H. Luckmann, 1975, *Introduction to Insect Pest Management*, with permission of the publishers, John Wiley & Sons, Inc.; data from G. D. Heathcote and A. J. Cockbain, 1966, "Aphids from mangold clamps and their importance as vectors of beet viruses," *Annals of Applied Biology*, vol. 57. pp. 321–336.)

lated to the amount of fat in their bodies, the fat providing the fuel for long-distance transport. Beet leafhoppers collected at Phoenix, Arizona, for example, had an average of 40% fat, while those collected 350 km along the dispersal path had only 25%, and those collected at 700 km only about 10%. Since the beet leafhopper is able to subsist on over 100 species of plants, including many weeds, it is able to establish populations in diverse places and to undertake short-range dispersal to beet fields where these are available.

Dispersal to Alternate Host Plants. Some of the most regular annual movements of insects involve species having alternate host plants. Many aphids, for example, spend the winter and early spring on perennial plants; the summer months, on fast-growing annuals, a more abundant source of food. In the northern United States, the green peach aphid overwinters in the egg stage on the bark of peach, plum, and cherry trees. In the spring the aphids have two or three generations of wingless individuals on the winter host, then they produce a generation of winged aphids that migrate to garden and truck crops, such as spinach, lettuce, beets, potatoes, and many others. Here numerous generations of wingless aphids are produced. In the fall winged individuals again appear, and these return to the woody plants that constitute their winter hosts. Thus far that season, all individuals have consisted of parthenogenetically reproducing, live-bearing females. In the fall, however, sexually reproducing males and females are produced, and after mating the females lay eggs that survive the winter. (For another example, see Fig. 16–2).

As mentioned earlier, aphid polymorphism is under endocrine control and involves alternate pathways of development depending on environmental cues such as population density, food quality, and photoperiod. Newly molted winged aphids, after a preflight period of variable duration, move to a leaf surface where, under suitable conditions of light and temperature, they spread their wings and move from the host

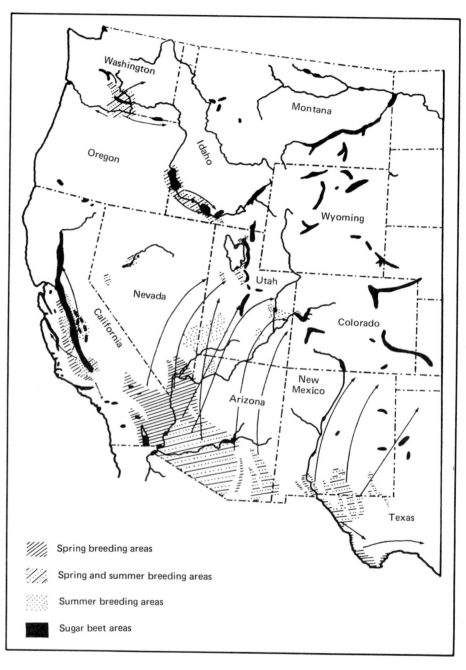

Figure 16–6
Breeding areas of the beet leafhopper, *Circulifer*
tenellus **(Cicadellidae), in relation to the major**
sugar beet growing areas. (J. R. Douglass and
W. C. Cook, "The beet leafhopper," U. S. Department of Agriculture Circular no. 942, 1954.)

plant (Fig. 16–1). At this time they are photopositive and responsive to blue wavelengths from the sky. In still air aphids may fly about 3 km per hour, but their ground speed may be greater than this when they are aided by wind currents. The green peach aphid normally flies about two hours before becoming responsive to yellow wavelengths such as those reflected from foliage. Alighting may result from response to yellow light or to decreasing light, increased wind speed, or mere exhaustion. Having landed on an unsuitable surface, such as the ground, the aphid is able to undertake short flights in search of a host plant. On a plant, the aphid probes with its beak and is able to reject plants outside the normal range of host species or otherwise unsuitable (for example, desiccated or having unacceptable physical or physiological features).

The green peach aphid can be reared on artificial media and has been found to prefer sucrose solution to pure water and to prefer sucrose–amino acid diets to sucrose alone. Although this aphid has many summer hosts, it may select different parts of specific hosts. For example, on chrysanthemum the aphids prefer to settle on young foliage, but on crucifers they prefer older leaves, perhaps because they contain less mustard oil. Dodder, a parasitic climbing plant, is an acceptable host, but not when it is growing on onions. Once the aphid has settled in an acceptable site, the flight muscles are broken down and the tissues reutilized for the production of offspring.

By no means do all aphids have alternate hosts. The greenbug, for example, overwinters on grasses as feeding immatures or, in more northern latitudes, in the egg stage. In the spring both wingless and winged females are produced, and the latter disperse to various grasses, including wheat, corn, and sorghum. Nearly all the many species of aphids are polymorphic for wing development and have a dispersing phase in their life cycle. Polymorphism for wing development also occurs in some Hymenoptera (gall wasps and certain parasitoids) also associated with dispersing and nondispersing phases in the life cycle.

There is no doubt that a great many migrants fail to find a suitable habitat and as a result perish. Most of the insects taken in aerial nets far at sea are surely doomed. The British ecologist Charles Elton reported millions of aphids on snowfields on Spitzbergen, an island far north of the Arctic Circle. In what way, then, is it advantageous for an insect to migrate, or for a female to produce offspring that will migrate? Dispersal as a normal episode in the life history is a way of forestalling future problems—overpopulation, decline in the habitat, increase in natural enemies, inbreeding—and at the same time ensuring access to new and rich sources of food. For the individual, it is a game of chance: stay, and you may do pretty well if conditions don't deteriorate; move, and you may die or you may find an abundance of fresh food and be mother of a large brood of offspring.

While a dispersing phase occurs in the normal behavior of many species, it is often hastened and enhanced in the presence of overcrowding or environmental deterioration. Much dispersal is density dependent and thus an important factor in population regulation. The evidence for this is worth examining in greater detail.

Emigration Resulting from Population Stress

As noted earlier, crowding of vetch aphids induces the production of winged individuals. Experiments with cabbage aphids by the French entomologist Lucien Bonnemaison revealed that when adult, wingless females are crowded but their offspring are removed daily, only a few of the latter develop wings. Also, when the offspring themselves are crowded, few develop wings. But when adults and offspring are crowded together, many of the young develop wings if contact is sufficiently long, demonstrating the influence of some factor from the mother.

No. of days of contact, adults–larvae	Percentage of larvae developing wings
1	0–1.5
2	0–4.0
3	8–34
4	13–74

It has been known for many years that locusts reared in crowds tend to be more active than those reared in isolation and to have lower thresholds of response to stimuli that induce movement. As immatures, they sometimes march in large groups over the ground, and as adults they fly off in large swarms. (Locusts are discussed further below, under "Long Distance, Directional Migrations.")

One effect of crowding may be to reduce the amount of food available per individual, and this in itself may influence dispersal. Poorly fed cotton stainers (tropical bugs of the order Hemiptera) tend to migrate, but well-fed females histolyze their flight muscles in favor of egg production. Incipient starvation also enhances migration by milkweed bugs. In this instance application of juvenile hormone increases the proportion of bugs making long flights. There is reason to believe that the endocrine system is involved in increasing the tendency of many insects to disperse.

Nutritional changes in plants may also influence dispersal. Toxemia caused by feeding of spotted alfalfa aphids is said to increase the production of winged individuals. Plant senescence may be an important factor in causing many aphid species to return to their winter hosts in late summer. Drought may produce wilting and lack of growth in plants, with a resulting exodus of plant-feeding insects. And of course a habitat may disappear entirely. There is an undoubted correlation between ability to migrate and the impermanence of an insect's habitat. Occupants of temporary pools, inhabitants of dung or carcasses, feeders on mushrooms or other ephemeral plants—these and many other insects must be adapted for ready dispersal.

Agricultural crops provide, for the most part, very short-lived habitats for insects. Cutting of alfalfa, for example, may cause aphids to disperse to nearby fields of soybeans. Plowing under the remains of one crop and replacing it with a different crop provides dispersal problems insects must overcome. It is, of course, implicit in the word *pest* that the species must disperse readily, build up populations rapidly, and be able to survive loss of its habitat by further dispersal.

Long-Distance, Directional Migrations

Much dispersal appears nondirectional in that the insects have little ability to control their route. Yet flight is often initiated only under certain atmospheric conditions, often at a time when winds are favorable for carrying them into more favorable areas—for example, north in summer. Sudden changes in weather conditions undoubtedly account for much of the mortality that occurs. It is principally some of the larger insects that are able to maintain a course over a considerable distance, though even these often rely to a considerable degree on the wind. It is these larger insects, also, that most often attract our attention. W. H. Hudson, in his book *The Naturalist in La Plata*, describes flights of dragonflies on the plains of Argentina as "countless millions flying like thistledown" before the wind. Swarms of locusts have plagued humankind since antiquity. And the migrations of the monarch and other butterflies often make the pages of our newspapers.

Although these insects employ persistent locomotion and maintain a more or less straight course, little is known as to how they maintain this course. Sun-compass orientation and responses to polarized light have been suggested, though not demonstrated, for distance navigation. Butterflies often fly rather close to the ground, where the effect of wind is lessened, and there is little question that they make much use of landmarks in maintaining a course.

Erik and Astrid Nielsen, working at the Archbold Biological Station in Florida, were able to shed much light on the conspicuous migrations of the great southern white (*Ascia monuste*) in that state (Fig. 16–7). At certain times these butterflies fly in streams of hundreds of individuals, only 1 to 3 m—above the ground. Flights are especially seen along the coast, and since the coast is parallelled by roads, for the most part, the Nielsens were often able to follow swarms with their jeep. They also marked butterflies with dyes of various colors, squirting them with dye solution from an oil can while the butterflies fed at flowers. In coastal Florida the larvae of the great southern white feed primarily on saltwort, which grows in salt marshes and mangrove swamps. When populations are high, young adult butterflies, after mating, fly to sources of nectar, where they form a milling cloud. It is perhaps frequency of contacts with other individuals that causes them finally to take off in a stream, following with some precision landmarks such as a shoreline, roads, or telephone lines (Fig. 16–8). When crossing a bay, the butterflies may be blown off course, but they are able to change their angle of flight so to arrive on the opposite shore at the appropriate point. The flight eventually terminates in new breeding areas, at a distance of up to 160 km from the source, though most flights are much shorter than this.

Higher-flying insects, in general, have less control over direction and make much less use of landmarks. In many cases they undoubtedly do perceive the passage of multiple images on the ground and are able to adjust their speed and flight patterns accordingly. John S. Kennedy, of Imperial College Field Station, in England, studied yellow fever mosquitoes in flight chambers having a floor of transverse movable black stripes. Kennedy found that in a light wind, with the stripes stationary, the mosquitoes flew upwind. However, their maximum speed was 150 cm per second; if the wind speed exceeded this level, they were carried backward over the stripes, and this caused them to settle. If the stripes were moving in still air, the mosquitoes flew in the direction of movement, gaining on the stripes—the equivalent of flying against the wind over stationary stripes. The perception of images on the ground thus provides information on ground speed and position.

Sequential close-up photographs of locust swarms have shown that groups of individuals within the swarm seem to be heading in different directions. When a group reaches the edge of the swarm, it tends to change direction and head back toward the center. Individual distances between locusts are apparently maintained by perception of sounds, air turbulence, and possibly sight of neighbors. Locusts on the leading edge of the swarm often settle, then take flight again as the remainder of the swarm passes over, producing a rolling motion of the swarm as a whole. Although individual locusts are able to fly at a rate of about 12 km per hour, not all their movements are in the direction of movement of the swarm, which is determined largely by the wind.

Some insects—black flies and mosquitoes, for example—often stop and refuel with nectar along the flight path. Other insects move great distances simply by burning fat in their bodies. C. B. Williams described flights of the small mottled willow moth from Morocco to England, a distance of over 3000 km, requiring four days, entirely

Figure 16–7
A migrating great southern white, *Ascia monuste* (Pieridae). (Drawing by Peter Eades.)

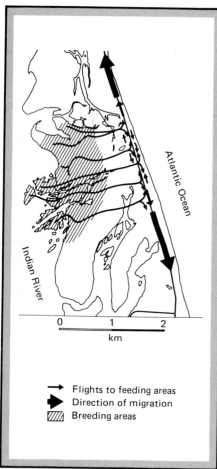

Figure 16-8
Flights of *Ascia monuste* from their breeding areas to and along offshore bars on the east coast of Florida. (After E. T. Nielsen, 1961, "On the habits of the migratory butterfly *Ascia monuste* L.," *Biologiske Meddelelser det Danske Videnskabernes Selskab*, vol. 23, no. 11, 81 pp.)

over the sea. Radar tracking of swarms has revealed that insects often migrate at night. Such insects can scarcely be using information from the ground and may well be using cues rarely demonstrated for insects; for example, geomagnetism. Clearly we still have much to learn about these matters, and what we learn about one species may not be applicable to others. A closer look at two of the better-known examples of distance migration may serve to illustrate the diversity of migratory behavior among insects.

The Monarch Butterfly.

This large and showy insect needs no introduction to North Americans. As we discussed in Chapter 14, monarch larvae feed on milkweeds, which contain toxic substances taken up by the larvae and retained by the adults. Thus the butterflies are rejected by experienced predators and advertise their noxious properties by their brilliant orange coloration. Monarchs range throughout temperate North America but cannot survive the winter north of the Gulf Coast, Mexico, and the coast of California. Each summer vast areas are repopulated by migrants from the south. After passing one to three generations in the summer home, butterflies bred in the late summer begin their migrations to certain sites where they winter year after year. The butterflies that move northward in the spring are the same individuals that moved south in the fall—a situation common in birds but fairly uncommon among insects because of their short lives. However, the individuals moving south in the fall are first-, second-, or third-generation offspring of the spring migrants. Yet, for reasons not well understood, they collect in the same overwintering areas, often on the same trees, year after year (Fig. 16–9). Indeed, certain areas along the California coast become tourist attractions because of the great masses of butterflies that cover trees winter after winter. A motel in Pacific Grove is even called "Butterfly Trees Lodge."

Butterflies in the overwintering sites do not truly hibernate. On sunny days they often spread their wings and may fly about a bit or even feed at nearby sources of nectar. Overwintering individuals have considerable fat in their bodies, and the ovaries of the females show little development until the advent of early spring. Spring migration northward begins in February or March, although it is May or June before many monarchs reach the northern states and Canada. Often they arrive before milkweeds are available and when nectar is scarce, but females evidently can live on stored fat until they are able to lay their eggs. After that they soon die. Males migrate northward too, but many die along the way, having mated in late winter or early spring further south.

Monarchs are capable of vigorous, directed flight up to about 30 km an hour; individuals have been reported to cover over 125 km/day. Marked individuals have been found to cover up to 3000 km during migration. Flight is usually close to the ground, not more than about 5m high, but the butterflies may fly over the tops of trees or buildings, and some have been taken in mountain passes up to 3500 m elevation. The nights are spent resting in vegetation. Movement northward in the spring is fairly direct, but the fall migration is more leisurely, and at this time temporary communal roosts are often established along the way. There is no doubt that monarchs take advantage of favorable winds and that on cool or rainy days, or when winds are unfavorable, they simply remain where they are. One person reported a migrating cloud of monarchs approaching Dallas, Texas, in October: "As they came closer they appeared like a gigantic, brown carpet moving rapidly along. They were riding a Texas 'norther' coming down at 15 miles an hour." Nevertheless, most persons who have studied the monarch have concluded that they are able to control their direction to a considerable extent. The navigational cues they employ are unknown, but it seems likely that they use celestial information, such as sun-compass, in maintaining such a long northward course in the spring and southward course in the fall.

Much of what we know about migrations of the monarch results from the researches of F. A. Urquhart, of the Royal Ontario Museum in Toronto. Urquhart developed a method of tagging monarchs on the fore wing in such a way that flight is not impaired. He tagged thousands of monarchs during the summer, chiefly in Ontario, and organized a group of over 200 collaborators throughout North America who did additional tagging and also recorded the presence of tagged specimens. Tags contained the address of the Royal Ontario Museum, so that other persons would return them to Urquhart. In this way the route of many fall migrants could be determined, at least in a general way. The following are three of the more striking of his records (distances are in a straight line—doubtless the monarchs followed a more tortuous path) (see also Fig. 16–10).

Released	Recovered	Distance (km)
Ontario, 3 Sept.	Texas, 18 Oct.	2110
Ontario, 12 Sept.	Mississippi, 9 Oct.	1700
Ontario, 18 Sept.	Mexico, 25 Jan.	3000

The monarch is by no means the only migratory butterfly. The red admiral and the painted lady are common seasonal migrants in both North America and Europe. C. B. Williams writes that "in the tropics butterfly migrations are on a very extensive scale, and in some areas are so frequent and conspicuous as to become sources of superstition to uneducated races." In the coastal mountains of Venezuela, William Beebe recorded "countless millions of butterflies, belonging to over 250 species, passing through from north to south almost without a break from May to September every year." Many moths also participate in distance migrations. One of the best known is the bugong moth of Australia, which flies to caves and rocky clefts in the mountains and estivates there in great numbers, often several thousand per square meter. The aborigines used to make yearly pilgrimages to these caves to feast on the moths, which are rich in fat. Migrations of Lepidoptera are usually associated with seasonal conditions (warm–cold or wet–dry) or with overpopulation and food depletion in breeding sites.

Locusts.   Locusts are grasshoppers that exhibit mass migratory behavior. The Rocky Mountain locust, which devastated parts of central North America during the nineteenth century, is now believed to be extinct (for reasons not well understood), but locusts remain prevalent in many parts of the world. About six species are now regarded as important mass migrants, although several others now and then form more limited and localized swarms. Perhaps the most notorious species is the desert locust of Africa and southwest Asia. Like other migrants, this insect is no respecter of political boundaries, and research and control have been on an international scale. For many years this international effort was centered in London and titled the Anti-Locust Research Centre. More recently the center has broadened its concern to include other insect problems and is now known as the Centre for Overseas Pest Research.

Foundation of the center in 1945 followed a series of outbreaks of the desert locust and other species in Africa. Its location in London was at least partly the influence of Boris Uvarov, a noted Russian authority on locusts who had moved to England at the time of the Russian Revolution. It was Uvarov who first clarified the point that more or less solitary "grasshoppers" can, under certain conditions, be transformed into gregarious, migratory "locusts." He termed these the **solitary** and **gregarious phases**. He

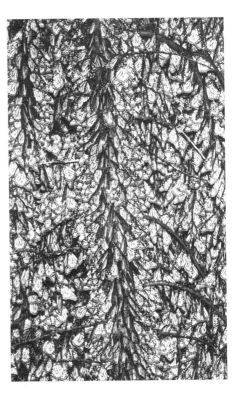

Figure 16–9
A mass of overwintering monarch butterflies on the trunk of a tree. (Photograph by George D. Lepp; courtesy of Lincoln P. Brower, University of Florida.)

Figure 16–10
Lines joining points of release and recapture of marked monarch butterflies. (Modified from F. A. Urquhart, 1960, *The Monarch Butterfly*, University of Toronto Press.)

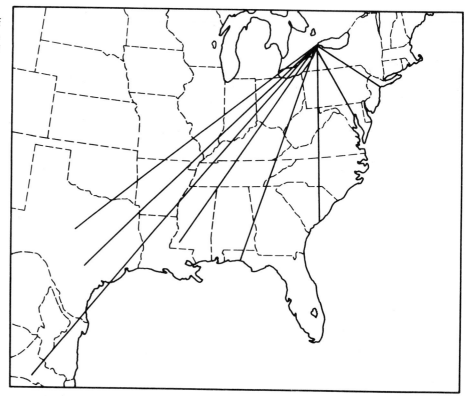

and others demonstrated that when these insects are reared under crowded conditions, the resulting adults are more active, have longer wings, are darker in color, and have minor structural differences from the "solitary" form. Furthermore, such adults produce offspring that are darker and more active than offspring of the solitary phase. Phase transformation is largely the result of increased bodily contact inducing changes in the endocrine system. Endocrine changes not only produce structural and behavioral changes but also influence the production of certain pheromones. Male desert locusts produce a pheromone that hastens development of other individuals, thus tending to synchronize the population prior to swarming. Female desert locusts produce a pheromone that attracts other females, causing the deposition of egg-pods in dense groups. This is by no means the whole story; such factors as food quality, titer of juvenile hormone, and still other stimuli may also play a role in phase transformation and the induction of swarming.

Immature hoppers often become concentrated by increase in numbers or by reduction in size of suitable habitat. Tactile contact results in increased "restlessness," with much jumping and moving about in groups. Small groups join one another and may "march" over the ground in a consistent direction, feeding continuously on grass and forbs. When they reach the adult stage, the insects begin to fly about erratically. Eventually small groups fly in response to one another and to other groups; then suddenly a great swarm develops and moves off in the direction of the wind. As mentioned earlier, individual locusts do not necessarily fly in the direction of swarm movement, which is largely controlled by the wind. Swarms of the desert locust have been reported to comprise from 10^5 to 10^{10} individuals and to vary in extent from less than 1 km^2 to about 1000 km^2. Height and shape of the swarm are determined in large part by thermal convection currents and the speed of the wind at various heights. Individual locusts fly at a speed of about 12 km/hour, but the swarm may move at a considerably greater speed than this. The sight of such a swarm is, to say the least, a frightening experience, particularly if one's source of living is threatened (Fig. 16-11).

Although swarming seems to involve several "random" elements—reduction in suitable habitat, direction of winds, for example—in fact swarming is highly adaptive for species that regularly practice it. Drying up of the habitat in the home range in certain seasons coincides with movements of winds from these areas into zones of convergence of weather systems, where widespread rains occur and produce growth and freshening of the vegetation. R. C. Rainey and his colleagues at the Centre for Overseas Pest Research mapped the distribution of swarms of the desert locust over many years and actually tracked swarms with planes. This locust is capable of invading a vast area of Africa and southwest Asia, but it breeds only in areas where seasonal rains have produced a growth of succulent vegetation. Because these rains occur in different areas at different seasons, the locust must migrate from one area to another. Spring breeding occurs along the Mediterranean Sea and east to Pakistan. In summer these areas are subject to drought, and the so-called intertropical convergence zone occurs further south. Taking advantage of prevailing winds and the resulting showers, the locusts move south of the Sahara and into southern Arabia and India. The major area of winter breeding is in the horn of Africa (Fig. 16-12).

This simplified scheme does not by any means make it possible to predict accurately the time and place that swarms will occur. Weather fronts vary in timing and intensity, and rains may be spotty or even absent for long periods. Irrigation may present the locusts with edible vegetation in an area where none might otherwise occur. There is no doubt that the system is as full of unpredictables for the locusts as it is for people living within its range. Many millions of locusts probably perish annually when they are carried to areas where they cannot develop successfully. Yet the desert locust is in many ways admirably adapted for survival in a precarious habitat, and despite efforts to understand and control it, is still as serious a threat as it was in biblical times—more so, since there are now so many more human mouths to feed.

Figure 16-11
Part of a 400-square-mile swarm of desert locusts, *Schistocerca gregaria* (Acrididae), in Ethiopia, which did extensive damage to crops in the area. (Photograph by C. Ashall; supplied by the Centre for Overseas Pest Research, London.)

Humans as a Factor in Insect Dispersal

It would be inappropriate to end this chapter without at least a brief mention of the fact that humans have themselves carried many insects to parts of the world where they did not formerly occur. All of the domestic cockroaches of temperate North America and Europe, for example, have arrived from various parts of the tropics and subtropics, in most cases long ago, by way of ships carrying cargos from their native homes. Most insects that infest grain in storage—and the parasitoids of these insects—are now worldwide in distribution as a result of commerce. Many of our major pests of agriculture, perhaps as many as half, arrived accidentally, usually in imported fruits, vegetables, or nursery stock. And of course a goodly number of parasitoids and predators have been transported from one continent to another in biological control efforts.

Reece I. Sailer, of the University of Florida, has recently compiled a list of foreign insects that have become established in the 48 contiguous United States. The list is probably far from complete, but it totals 1379 species. Of these, 287 are classed as beneficial species, many of them deliberately introduced. Of the remainder, 236 are

Figure 16–12
Breeding areas and dispersal routes of desert locusts in Africa and southwest Asia. (Zena Waloff, from *The Locust Handbook*, crown copyright 1966, published with permission of the Controller of Her Britannic Majesty's Stationery Office.)

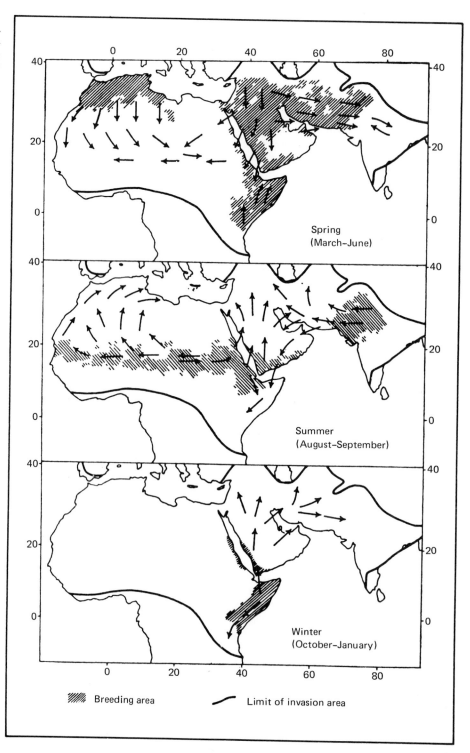

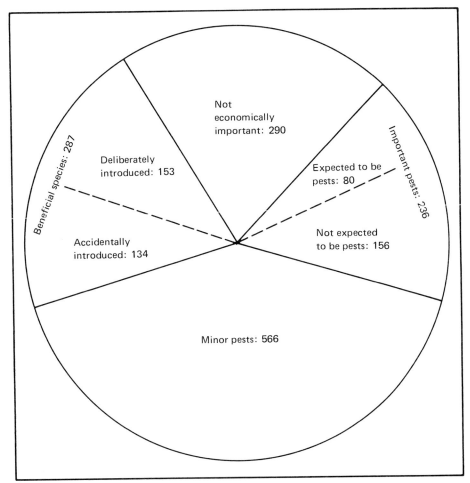

Not economically important: 290

Deliberately introduced: 153

Beneficial species: 287

Expected to be pests: 80

Important pests: 236

Accidentally introduced: 134

Not expected to be pests: 156

Minor pests: 566

Figure 16–13
Economic importance of immigrant insect species in the 48 contiguous United States. (After R. I. Sailer, 1978, "Our immigrant insect fauna." Reprinted with permission from Bulletin of the Entomological Society of America, vol. 24, pp. 3–11; copyright 1978 by the Entomological Society of America.)

classed as major pests, 566 as minor pests; only 290 are regarded as of no direct importance to humans (Fig. 16–13). Sailer points out that only 80 of the 236 major imported pests were predicted to be important; the others were of minor importance in their home country (like the Japanese beetle) and could not have been predicted to develop into serious problems. That they became serious was, of course, usually a consequence of their arrival without their natural enemies.

Since it is not usually possible to predict accurately the results of an importation, the U.S. Department of Agriculture maintains plant inspection and quarantine stations at all ports of entry, and all insects intercepted are regarded as suspect. The frequency of air travel has greatly increased the possibilities of bringing in new pests, and the problem has been made more serious by the fact that many inland cities in North America now have direct air connections to other continents. Despite the continued efforts of quarantine inspectors, it seems likely that more and more insects will extend their ranges drastically as a result of human activities.

Summary

On a warm summer day the air contains a great many insects, often to a considerable height. These insects are in most cases not there accidentally, but are moving from one site to another as part of their normal life cycle or as a result of overpopulation or habitat deterioration. Much dispersal occurs at a particular stage of life, often the young-adult stage, and involves temporary suppression of feeding and reproduction. The term *migration* is used when individuals move off in a definite direction in response to some environmental cue, although migration is simply a form of dispersal, and the decision to call a particular movement a migration is often arbitrary.

Dispersal is essential to many species so that they may find new food sources. Studies of reoccupation of small islands from which all the arthropods have been removed by fumigation reveals that the islands soon acquire faunas of about the same number of species they had originally, the number being largely determined by size of the island and its proximity to land. Patchy forests, meadows, and crop land are in many ways comparable to islands.

Many insects disperse seasonally, especially subtropical insects into temperate regions during the summer. Some insects have alternate host plants and must disperse twice a year to another host. Many aphids, for example, spend the winter on a perennial plant, the summer on annuals. Dispersal may also result from population stress, such as crowding, diminished food, nutritional factors, or simply loss of habitat. Insects occupying temporary habitats must disperse to survive.

Large-scale, long-distance migrations are especially noteworthy in butterflies and locusts. Monarch butterflies spend the winter massed on trees in the southern United States and along the California coast, migrating north as far as southern Canada in the spring. Their progeny migrate south in late summer and fall, but it is these same individuals that undertake the spring migration. Individual, marked monarchs have been found to migrate up to 3000 kilometers.

Several species of grasshoppers undertake mass migrations, and in their migratory phase are called locusts. Transformation of more or less solitary grasshoppers to gregarious, migratory locusts results from crowding: tactile contact produces endocrine changes that in turn influence behavioral and structural features and induce swarming. The desert locust breeds in areas in Africa and southwest Asia where succulent vegetation is available. Because rains occur in different areas in different seasons, the locusts migrate regularly, taking advantage of convergent winds leading to areas of increased rainfall.

Humans have been an important factor in insect dispersal, either through accidental introductions or through deliberate transport of species in biological control efforts. Well over 1000 species have been introduced into the 48 contiguous United States, and well over half of these are classed as major or minor pests, often because they arrived without their natural enemies.

Selected Readings

Dingle, H. 1972. "Migration strategies of insects." *Science*, vol. 175, pp. 1327–35.

———, ed. 1978. *Evolution of Insect Migration and Diapause*. New York: Springer-Verlag. 284 pp.

Johnson, C. G. 1969. *Migration and Dispersal of Insects by Flight*. London: Methuen. 763 pp.

Kennedy, J. S. 1975. "Insect dispersal," in *Insects, Science, and Society* (D. Pimentel, ed.), pp. 103–19. New York: Academic Press.

Kring, J. B. 1972. "Flight behavior of aphids." *Annual Review of Entomology*, vol. 17, pp. 461–92.

Lees, A. D. 1966. "The control of polymorphism in aphids." *Advances in Insect Physiology*, vol. 3, pp. 207–77.

Rainey, R. C. 1974. "Biometeorology and insect flight: some aspects of energy exchange." *Annual Review of Entomology*, vol. 19, pp. 407–39.

Simberloff, D. S., and E. O. Wilson. 1969. "Experimental zoogeography of islands: the colonization of empty islands." *Ecology*, vol. 50, pp. 278–96.

Stinner, R. E.; C. S. Barfield; J. L. Stimac; and L. Dohse. 1983. "Dispersal and movement of insect pests." *Annual Review of Entomology*, vol. 28, pp. 319–35.

Urquhart, F. A. 1960. *The Monarch Butterfly*. Toronto: University of Toronto Press. 361 pp.

Williams, C. B. 1930. *The Migration of Butterflies*. Edinburgh: Oliver & Boyd. 473 pp.

_____. 1958. *Insect Migration*. London: Collins. 235 pp.

It is difficult to conceive of a world without insects. Some we would sorely miss: the honey bee and a host of other pollinators; the many kinds that serve as food for birds and mammals; *Drosophila*, which has taught us so much about genetics and evolution; scavengers that aid in the decomposition of dung and carcasses; and of course the great horde of predators and parasitoids that assists in keeping populations of plant-feeders at reasonable levels most of the time. In many ecosystems the roles of insects are so subtle and poorly understood that it is difficult to judge their precise effects. But to the layperson, the word *insect* is almost synonymous with *pest*: all too evident are their bites and stings; their disease-carrying capacities; their appetites for crops, gardens, and shade and forest trees; their penchant for attacking products in storage (look ahead to Fig. 17–1). No amount of rationalization can minimize the importance of insects as major competitors of humans.

The impacts of insects on human economy are diverse and often indirect, such that it is difficult to quantify them in any meaningful way. Various published estimates of

The Natural and Artificial Regulation of Insect Populations

P A R T

S E V E N

losses to agriculture present a frightening picture. For the year 1974, it was estimated that insects produced crop losses in the United States of about 13%, valued at about $18 billion. Combined with losses to plant diseases and to weeds, about one-third of all crops were lost. Recent estimates of crop losses in the world as a whole are similar: about 35%. This figure does not include post-harvest losses, which are particularly severe in the tropics and in developing countries, a result of attacks by insects, rodents, and bacterial and fungal rots before the products are consumed. According to one estimate, almost half the world's foodstuffs are sacrificed to pests and diseases.

To the monetary value of these losses must be added the cost of measures used for control and containment. These have accelerated greatly as a result of spiraling energy costs. About a billion gallons of fossil fuel equivalents are used each year in the United States to produce and apply pesticides. On a world basis, consumption of pesticides totals about 4 billion pounds annually and continues to increase despite growing concern over the side effects of pesticide usage on resistance among pest populations, on the environment, and on human health.

In recent years there has been a strong reaction to increased dependence on insecticides, not only because of increased costs and impending energy shortages, but because of the development of resistant strains of insects and particularly because of concern over unforeseen effects on the environment. The latter were dramatized in Rachel Carson's influential book of 1962, *Silent Spring.* Coming as it did during the growth of the environmentalist movement, Carson's book was widely hailed by the public, though greeted with less enthusiasm by some entomologists, who had come to rely heavily on the use of chemicals for pest control.

In fact, the trend away from overreliance on chemicals was under way before 1962. Beginning about 1943, A. D. Pickett and his colleagues in eastern Canada developed approaches for harmonizing biological and chemical controls in apple orchards. In 1959 an important paper titled "The Integrated Control Concept" came from the pen of four University of California entomologists: V. M. Stern, R. F.

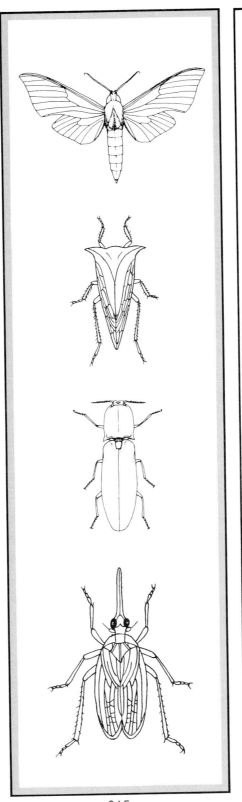

Smith, R. van den Bosch, and K. S. Hagen. In 1961, P. W. Geier and L. R. Clark, of the Commonwealth Scientific and Industrial and Research Organization, Australia, spoke to an International Congress in Warsaw on an ecological approach to insect control and introduced the term *pest management.* From these and similar beginnings evolved the modern concept of **integrated pest management** (IPM).

IPM may be defined as an approach in which all available techniques are evaluated and consolidated into a unified program to manage pest populations economically and so as to avoid undue side effects on the environment. It goes without saying that "pests" include weeds and plant pathogens as well as insects. We shall be concerned with insect pest management, which conveniently employs the same acronym, IPM. Management of pest populations is a form of applied ecology and requires a knowledge of ecological principles, particularly those concerned with population dynamics. We shall deal with some of these in the following chapter, reserving a more direct examination of the tactics of pest management for Chapter 18.

CHAPTER

SEVENTEEN

The Ecological Basis for Insect Pest Management

Figure 17–1
Insects do not always respect our efforts to control them. These volumes have been thoroughly gutted by termites. (Photograph by Bernd Heinrich, University of Vermont.)

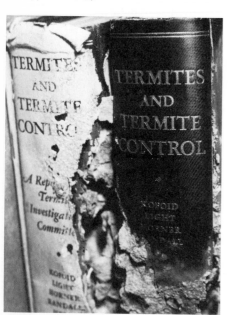

E ntomologists must have an understanding of the factors that regulate insect populations and must also be able to measure or estimate population size and its potential impact on humans. It is important to know when numbers are great enough to justify artificial controls, to be able to evaluate the effectiveness of controls, and to have some basis for anticipating outbreaks. Population size is a reflection of **fecundity,** which is subject to several sources of variation, and of **mortality,** which may occur at any stage of the life cycle and result from many different environmental or other influences. The effects of immigration and emigration must also be considered. Students of population dynamics concern themselves greatly with causes of mortality and the search for **key factors** in reducing the numbers of individuals available for reproduction. Survival, whether in a natural environment or an agroecosystem, is often relatively poor. As we sample insect populations, it is important to remember that not only the current pest population but also the future population will affect agricultural production. Large numbers of young insects can often be tolerated in a crop because natural mortality agents cause significant decreases in pest abundance before the insects become large enough to inflict serious injury. On the other hand, an initially small population may prove the focal point of a major infestation when fecundity is high and there are few dependable sources of natural mortality.

Humans often provide insects with a veritable plethora of food in the form of well-watered and fertilized crop monocultures. **Monocultures** are uniform stands of the same plant. **Polycultures** incorporate two or more types of crop plants into the same area and more closely resemble natural systems. Insect pest survival may be enhanced by the abundance of suitable food provided by agricultural monocultures. Worse yet, the natural enemies of insects may be isolated from the crop and pest insects, or the beneficial insects may be killed by insecticides. We must understand the nature of mortality factors that affect insect populations so that we can predict the eventual size of the pest population and the resultant damage.

Population Dynamics

The forces that control population size, and their effects, are spoken of as **population dynamics.** The search for key factors in determining numbers available for reproduction is abetted by preparation of a **life table,** in which the influence of each mortality factor is quantified for each life stage. For example, studies conducted in Ontario, Canada, by D. G. Harcourt and his colleagues indicated that alfalfa weevil populations in that area are principally regulated by the fungus *Zoophthora phytonomi* (Table 17–1). Although the parasitoid *Bathyplectes curculionis* had proved effective in other areas (Chapter 12, p. 283), neither it nor the parasitoids *Patasson luna* or *Tetrastichus incertus* were significant mortality factors in Ontario compared with *Zoophthora phytonomi.*

Table 17–1. Within-generation life table for the alfalfa weevil based on mean values for 15 plot years in the Quinte area of eastern Ontario, 1972–76

Stage interval	No. alive at beginning of stage*	Factor responsible for mortality	No. dying during stage*	Percent mortality	Survival rate within stage
Eggs	421	Patasson luna	1	0.2	
		Infertility	8	1.9	
		TOTAL	9	2.1	0.979
Young larvae	412	Establishment loss	106	25.7	0.743
Older larvae	306	Zoophthora phytonomi	255	83.3	
		Rainfall	6	2.0	
		TOTAL	261	85.3	0.147
Prepupae	45	Zoophthora phytonomi	7	15.6	
		Bathyplectes curculionis	2	4.4	
		Tetrastichus incertus	1	2.2	
		TOTAL	10	22.2	0.778
Pupae	35	Zoophthora phytonomi	4	11.4	0.886
Summer adults	31				
General totals			390	92.6	0.074

*Numbers per ft^2 (0.09 m^2).
Source: After Harcourt, D. G.; J. C. Guppy; and M. R. Binns, 1977, "The analysis of intrageneration changes in eastern Ontario populations of the alfalfa weevil, *Hypera postica* (Coleoptera: Curculionidae)." *Canadian Entomologist*, vol. 109, pp. 1521–1534.

Since the effectiveness of the *Zoophthora* fungus increases as host density increases, its effect is said to be **density dependent.** Such factors are of major importance in **population regulation,** that is, the maintenance of reasonably constant numbers in nature. **Density independent** factors tend to be unpredictable, catastrophic events such as unseasonal rainfall (or lack of rain), severe winter temperatures, and the like. Such factors can be important for suppression of insect populations, but they are not usually regulatory factors. Insecticides often are density-independent mortality factors, especially when used on a scheduled rather than a need basis. Thus they are effective in reducing pest population levels but are not *more* effective when density is high. When populations are monitored and insecticides are applied judiciously, insecticide use is somewhat more regulatory.

Density-dependent mortality factors help to maintain population **stability,** which is the ability of a system to absorb disturbance and return to an equilibrium state. For example, a system may be more stable in the presence of phytophagous insects when density-dependent mortality agents such as predators and parasitoids are present. Such agents prevent the herbivores from destroying their host plant, which would result in local extinction of both plant and herbivore.

When instability occurs, as a result of environmental change, an insect and its natural enemies may or may not exhibit characteristics of resilience. **Resilience** is the capacity to adapt to change or to persist in a changing environment. When agriculturalists convert natural ecosystems to monocultural systems, they provide certain herbivores with vast amounts of food. Resilient herbivore species quickly adapt to this changed environment and increase in abundance. Predators and parasitoids may also respond numerically to the increased abundance of host insects, and the plant–herbivore–predator system could stabilize at a new, higher equilibrium state. Such a shift to a new equilibrium state may equally well be brought about by other factors, such as unusual weather conditions, shifts in resistance of host plants, or the indiscriminate use of insecticides. Note that the shifts in equilibrium state, shown diagrammatically in Fig. 17–2, are not unidirectional. It is important to understand the factors responsible for such shifts so that when possible we can prevent the shift to a higher, more damaging equilibrium state or so that we can induce the shift to a lower, less damaging state. This is especially true when the new equilibrium state is above the level of damage that is economically tolerable (the EIL, economic injury level).

The Characteristics of Pest Species.   Certain species of insects have life history characteristics that make them more likely than are others to become agricultural pests. T. R. E. Southwood, of Imperial College, London, suggests that such species normally frequent habitats of low stability. Also, the insects in these species tend to be small and mobile and to reproduce rapidly. Insect species with this set of attributes are known as **r-strategists.** In contrast, **K-strategists,** which are less likely to be pests, occur in more stable habitats. *K*-strategists tend to have lower reproductive rates, are less dispersive, and are frequently larger (relative to *r*-strategists). These terms have been adapted from ecologists, who employ *r* for the intrinsic rate of increase of a population; *K*, for the carrying capacity of the environment. The assumption is that *r*-strategists make use of available resources to maximize *r*, while *K*-strategists tend to

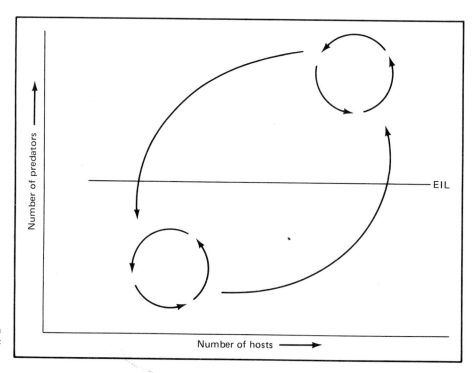

Figure 17–2
Theoretical predator and host system with multiple equilibrium positions. *EIL* = economic injury level.

Table 17-2. Some characteristics of *r*-selected and *K*-selected species		
	r-selected	*K*-selected
Type of habitat	Often unstable, impermanent	Usually more stable
Reproduction	Rapid under favorable conditions	Generally slower
	Many eggs laid	Fewer offspring produced
Development	Relatively rapid	Often slower
	Often multivoltine	Often univoltine, or extending over more than 1 year
Mortality	Often catastrophic, density independent	More constant, density dependent
Population size	Very variable in time	Fairly constant in time
Capacity to disperse	Very high	Often lower
Brood care	Absent	Sometimes present
Body size	Often very small	Often larger
Inter- and intraspecific competition	Variable, often reduced	Usually keen
Ultimate result	High productivity	Efficient use of resources

maintain more stable populations that do not closely approach carrying capacity (Table 17-2).

In general, *r*-strategists quickly utilize resources available to them, often achieving pest status and sometimes destroying their hosts before dispersing. Their resilience is seen in the rapidity with which they adapt to new resources. Natural enemies are often unable to keep up with, and to regulate, populations of *r*-strategists. Rather they are frequently suppressed by shortage of food or by unfavorable weather. Examples of *r*- strategists include aphids, house flies, and some locusts.

On the other hand, *K*-strategists rarely become so abundant as to destroy their food supply. Populations of *K*-strategists do not undergo violent fluctuations and usually are regulated by density-dependent mortality factors. In short, they tend to be stable species occurring in stable environments. Examples include the tsetse fly and the codling moth, both of which are considered to be serious pests not because they threaten to destroy their hosts, but because human values are such that we do not wish to tolerate sleeping sickness or blemished apples. Many insects that are *K*-strategists are never numerous enough to become pests.

Having said this, we must make it clear that most insects are neither extreme *r*-strategists or *K*-strategists. Rather, they may be said to occupy some point on an *r–K* continuum, although many of them tend toward one or the other extreme. *K*-strategists can often be managed through modification of their environment or disruption of reproduction, since they do not adapt quickly to change. In contrast, *r*-strategists are often controlled with insecticides, although the use of resistant hosts is likely to prove more effective in the long run because of the resilience that these pests exhibit. In all instances of pest management, it is important not to disrupt the natural enemy population, as this can lead to biotic release of the pest and attainment of higher population levels. Even species that are normally quite stable can become rather resilient, and damaging, if stable elements of the insects' habitat (such as natural enemies, plant diversity, plant resistance) are removed.

With this background, it may now be constructive to examine several examples that have been studied in some detail. Strategies of population regulation that are inherent within the population are said to be **intrinsic**; they may also be described as "self-regulating" in that they involve mechanisms that alter fecundity or rate of dispersal, depending on population density or food availability. Intrinsic factors are thus always density dependent. In contrast, **extrinsic** factors (arising outside the population) may be either density dependent (such as predators) or density independent (such as weather changes). We shall consider these three categories separately, though in some individual cases all may be operative.

Population Regulation by Intrinsic, Density-Dependent Factors. Although devastation by pest insects is sometimes all too apparent, the fact is that many species have mechanisms that lead to dispersal or to reduced population growth well before their food supply has become completely exhausted. Studies of the bean aphid by M. J. Way, of Imperial College, London, showed that wingless parthenogenetic females multiply faster in initial populations of 8 than in populations of 2, 4, 16, or 32. However, the multiplication rate decreases as populations increase (Table 17–3). He concluded that the optimal reproductive rate is realized only when the number of insects per plant is relatively low. Under crowded conditions, progeny tend to be smaller than average, to have reduced fecundity, and to include an increased number of winged individuals capable of moving to other plants or other fields (see also Chapter 16, p. 348, and Fig. 16–2). Similar results have been found in several other aphid spe-

Table 17–3. Multiplication rates of bean aphid populations developing from different numbers of initial wingless, parthenogenetic females in field cages

Initial no. of adult females	Multiplication rate		
	between days 9–17	between days 17–23	between days 23–44
2	× 29	× 20	× 17
4	× 41	× 22	× 6
8	× 46	× 15	× 3
16	× 36	× 11	× 2
32	× 27	× 7	× 2

Source: Modified from M. J. Way, 1968, "Intra-specific mechanisms with special reference to aphid populations," in *Insect Abundance*, T. R. E. Southwood, ed. Oxford: Blackwell, pp. 18–36.

cies. Mechanisms for such "self-imposed" regulation appear to reside in the genes and to be triggered by changes in the quality and flow of nutrients in the plants.

Bark beetles of the genus *Dendroctonus* are major pests of conifers, especially in the southern and western United States. Colonization of a tree involves the production of attractant pheromones and the inoculation of the tree with fungi, resulting in a mass attack on a susceptible tree (Chapter 5). Some of the factors determining susceptibility were reviewed in Chapter 10. Having reached a suitable host tree, females make galleries beneath the bark in which to lay their eggs; the larvae hatching from these eggs tunnel beneath the bark, often eventually causing the death of the tree. In some species of *Dendroctonus*, at least, females produce "territorial chirps" by means of stridulatory ridges on the posterior ends of their bodies. These signals are evidently perceived by other females, causing them either to move to another part of the tree or to dig shorter galleries in which fewer eggs are laid. Experiments by J. A. Rudinsky and R. R. Michael, of Oregon State University, showed that 81% of female western pine beetles (*Dendroctonus brevicomis*) did not approach confined, chirping females closer than 5 cm. The galleries of this species are usually spaced about 9 cm apart, suggesting that chirping plays a role in maintaining such spacing. Females also chirp more rapidly in the close presence of other females. A female alone chirped 38 times in 5 minutes, while one surrounded by other females 1.3 cm away chirped 285 times in 5 minutes (these are averages based on several replications). It has been shown that at high population densities *Dendroctonus* females also attack trees both higher and lower in the trunks than they normally do. In certain other genera of bark beetles no territorial mechanisms occur, and the large number of resulting crowded larvae scramble for the available space and food, and only some survive to maturity (scramble competition, discussed further below).

Other examples are known in which increased intensity of **intraspecific competition** results in a decrease in oviposition rate, a slowing of rate of development, or an increased tendency to emigrate. In the absence of genetic mechanisms, crowding and shortage of food may result in cannibalism or other forms of interactions that result in extensive mortality. **Cannibalism** among entomophagous species is common: larvae of solitary parasitoids, for example, often kill or suppress the development of other individuals occurring in the same host. Cannibalism has also been reported among phytophagous insects under crowded conditions—for example, in flour beetles and in codling moth larvae. **Interference competition** occurs when crowding results in incomplete or unsuccessful copulations or ovipositions or in reduced access to food. Or a few larger or more active individuals may usurp the resources, as occurs in instances of male territoriality with respect to access to females (Chapter 7). This is often called **contest competition.** Interference competition may also occur interspecifically, as we saw in Chapter 15, p. 330.

Insects occurring in limited, widely spaced habitats (such as maggots in a carcass) often exhibit **scramble competition.** The Australian entomologist A. J. Nicholson reared larvae of the sheep blow fly on homogenized cattle brains, varying the number of larvae per gram of food. Up to about 30 per gram, most larvae produced adults. At higher densities there was not enough food for all larvae to complete development, and in the resulting "scramble" only a small number reached adulthood—when over 180 per gram, most or all perished (Fig. 17–3). Although these studies were conducted in the laboratory, field studies have confirmed that many sheep carcasses receive many more blow fly eggs than they can support, resulting in much mortality. As discussed above, crowding of bean aphids and certain bark beetles may also result in diminished survival or fecundity.

Figure 17-3
Effect of competition for food among larvae of the sheep blow fly (*Lucilia cuprina*, Calliphoridae) on the number of adults produced. The food is homogenized cattle brains. (After A. J. Nicholson, 1954, "An outline of the dynamics of animal populations," *Australian Journal of Zoology*, vol. 2, pp. 1–8.)

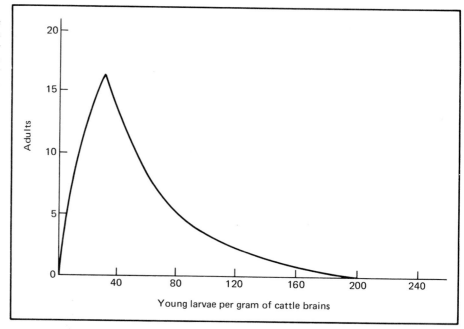

Population Regulation by Extrinsic, Density-Dependent Factors: Natural Enemies.

The importance of predators, parasitoids, and disease organisms in regulating population numbers is widely accepted (see Chapter 12). These factors are extrinsic to the populations being regulated, yet dependent on host population size for their own effectiveness. The effect of natural enemies typically is greatest following an increase in host numbers, often in the next generation (Figs. 12–15, 17–4). Hence the term *delayed density dependence,* often applied to the influence of natural enemies.

Evidence that natural enemies play an important role in maintaining population numbers well below their maxima is widespread. Virtually all life-history studies of insect species have revealed the presence of one or (much more commonly) several predators, parasitoids, or diseases that are consistently present in nature. In Chapter 12 we discussed the role of the tachina fly *Archytas* in reducing populations of the variegated cutworm, as well as the importance of the ubiquitous *Apanteles* wasps in attacking diverse caterpillars. A small chalcid wasp was found to maintain 70% to 95% parasitization of the eggs of the red-headed pine sawfly in Illinois. A complex of ichneumon wasps and tachina flies was found to produce from 7% to 97% parasitism of the black-headed budworm in Canada, resulting in population oscillations often characteristic of density-dependent factors. Examples such as these could be prolonged indefinitely.

The importance of natural enemies in regulating numbers is dramatically demonstrated in instances of **biotic release,** that is, cases in which a population is suddenly released from its natural enemies. This may occur when a predator or parasitoid is decimated by insecticides or when a species gains access to a new area where its normal enemies are absent. The Japanese beetle provides a good example of the latter. This insect reached very high populations and was incredibly destructive to lawns and to foliage following its accidental introduction in 1916 into the eastern United States,

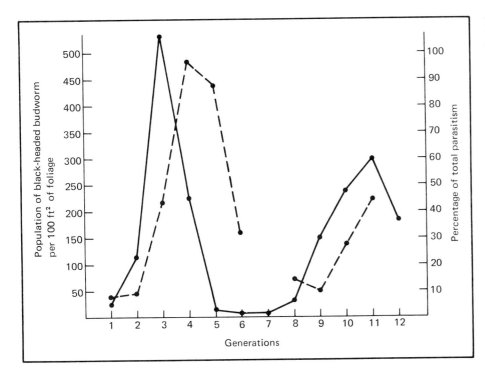

Figure 17-4
Population size of the black-headed budworm (*Acleris variana*, Noctuidae) (solid line, scale on left) in relation to that of its parasitoids (dashed line, scale on right). (After R. F. Morris, 1959, "Single-factor analysis in population dynamics," *Ecology*, vol. 40, pp. 580–588.)

where its natural enemies do not occur. The U.S.D.A. sent a team of researchers to Japan, beginning in 1920. Here the beetle is not a major pest, its numbers being kept low by several tachina flies and predatory wasps as well as by a bacterial disease of the grubs. One tachina fly species alone was found to parasitize from 20% to 90% of the beetles. Importation of several of the parasitoids—and particularly the mass production and dissemination of spores of a bacterium causing "milky disease"—gradually reduced the beetle to the status of a minor pest in most parts of the eastern United States.

Widespread use of DDT in the 1950s and 1960s in some instances resulted in the destruction of major predators and the release of certain plant-feeders from these predators. For example, DDT was found to provide excellent control of imported cabbageworms. But as the plants grow, they send out new leaves on which the butterflies lay their eggs. In the absence of further spraying, the caterpillars consume the new foliage and eat into the center of the plant, while residues in the soil continue to control ground-living predators such as carabid beetles, which frequently kill 50% to 70% of the caterpillars in unsprayed plots (Fig. 17-5).

The red-banded leafroller was considered a pest of minor importance in fruit orchards in the eastern and midwestern United States prior to the introduction of DDT, being kept under control by a variety of arthropod enemies. The sudden appearance of this native insect as a major pest in the 1950s is believed to have resulted from destruction of many of its natural enemies by DDT and related insecticides. During this same period, spider mites became major pests in orchards in many parts of the world, again apparently the result of inhibition of their natural control agents, such as lady beetles, mirid bugs, and predatory mites. Some evidence was obtained that these insecticides may also increase the nutritive value of foliage and thus increase the fecundity of spider mites. In general, however, it appeared that the influence of insecticides on

Figure 17–5
The buildup of imported cabbageworms (*Pieris rapae*, Pieridae) on plots sprayed with DDT (on 28 June) and on unsprayed plots. (After J. P. Dempster, 1975.)

Figure 17–6
The concept of population regulation "from above" versus that "from below." (Modified from A. A. Berryman, 1982, "Biological control, threshold, and pest outbreaks," from *Environmental Entomology*, vol. 11, pp. 544–549. By permission.)

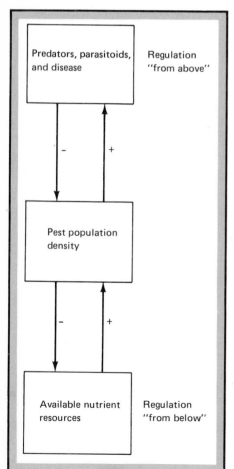

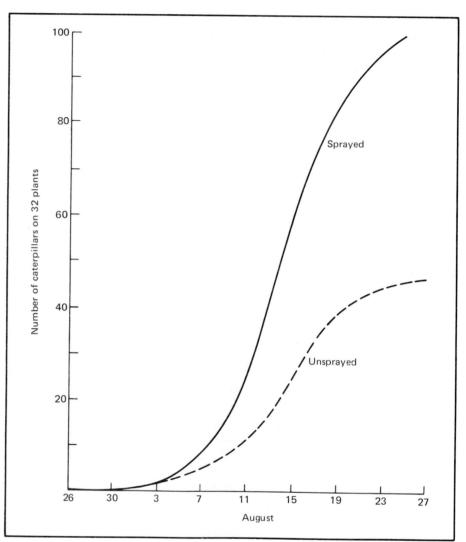

predator populations was primarily responsible for spider mite outbreaks. The British entomologist R. C. Muir demonstrated that one mirid bug eats up to 72 mites a day and can stabilize a population of about 2000 fruit tree red spider mites. But the bugs have only one generation a year, while the mites have several. Thus even a slight suppression of populations of the predator may set the stage for an outbreak of its prey. A species that does not normally attain pest status except when insecticides are employed is sometimes spoken of as a **secondary pest.**

Population Regulation by Extrinsic, Density-Dependent Factors: Relative Shortage of Food.   The effect of natural enemies may be thought of as "regulation from above." The Australian entomologist T. C. R. White has proposed that populations may alternatively be "limited from below," that is, limited by the amount of food available for normal growth and development (Fig. 17–6). Young insects and other animals require food that is rich in nitrogen, since nitrogen is required for incorporation into body protein. Herbivorous animals may appear to have plenty of food available, yet there may be a relative shortage of food for the young that is rich in available nitrogen. This is reflected in the high mortality among early instars of many insects. White conceives of most environments as being harsh in the sense that food adequate for the development of the young is in short supply. Zoophagous animals utilize food that is rich in available nitrogen, but they are also "limited from below" by the abundance of their hosts.

As we have seen in Chapter 10, water stress in plants may increase the availability of nitrogen. This may result in outbreaks of phytophagous insects (Fig. 17-7). Since the weather patterns producing water stress are independent of population density, this provides an excellent demonstration of the interplay of density-independent and density-dependent factors in nature.

Population Changes Caused by Extrinsic, Density-Independent Factors.
Physical factors in the environment may have diverse effects on populations irrespective of their density. As we have seen (Chapter 15), every species has certain ecological limits, above and below which development is retarded. At extremes of temperature, moisture, or other factors, death occurs. Sudden onset of cold may be lethal, when gradual cooling to the same temperature may be much less so; any unusual weather extreme at a nondiapausing stage may cause widespread death. Changes in soil characteristics (such as increased salinity) may influence insects directly or via their plant hosts. Similarly, reduced oxygen or increased acidity in a stream may have profound effects on the fauna. Populations suffer most when subjected to a true catastrophe, such as forest fire, flood, unseasonable frost, or sudden change in chemistry of the environment. For example, Paul Ehrlich and his associates at Stanford University reported the extinction of a local population of the butterfly *Glaucopsyche lygdamus* (one of the "blues") by a freak snowstorm in June in Colorado. At the periphery of the range, or in a suboptimal habitat, minor environmental perturbations may have major effects (Fig. 17-8).

In general, density-independent factors operate in one of four ways:

1. By direct destruction of individuals as a result of temperatures, moisture, or chemical concentrations they cannot tolerate;
2. By slowing of developmental rates such that a diapausing stage is not reached before the onset of a seasonal change;
3. By shortening the time of adults for mating and oviposition and thus reducing the number of eggs laid;
4. By inducing plant stress, resulting in a change in composition of nutrients or defensive chemicals that in turn results in increased or decreased availability of food for herbivores and ultimately for their predators and parasitoids.

In many cases population reduction will be most noticeable in subsequent generations. Thus a series of unseasonable winters or unusually dry summers may cause a severe depression of numbers. Conversely, a series of favorable years or entry into a new, highly favorable area may permit a buildup in numbers; such a population may be said to have undergone **climatic release** comparable to biotic release, discussed above. It should be remembered also that the effect of climate may be twice or three times removed from the population under study. That is, to take a hypothetical example, a predator may become rare because of scarcity of its prey, and the latter may be sparse

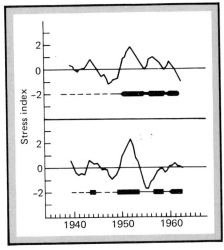

Figure 17-7
Index of water stress, calculated as varying from +2 (most stressful) to −2 (least stressful) over a 25-year period at two localities in eastern Australia. Below, records of psyllid abundance: thin solid line, psyllids present, but no outbreaks; thick black line, outbreaks of one or more species of Psyllidae; dashed line, no records for those years. Outbreaks are correlated with increased water stress. (From "An index to measure weather-induced stress of trees associated with outbreaks of psyllids in Australia," by T. C. R. White, *Ecology*, 1969, 50, 905–909. Copyright © 1969 by Duke University Press. Reprinted by permission.)

Figure 17–8
Generalized relationship between the major regulating factors on a species throughout its range. (P. W. Price, 1975, *Insect Ecology,* used with permission of the publishers, John Wiley and Sons, Inc.)

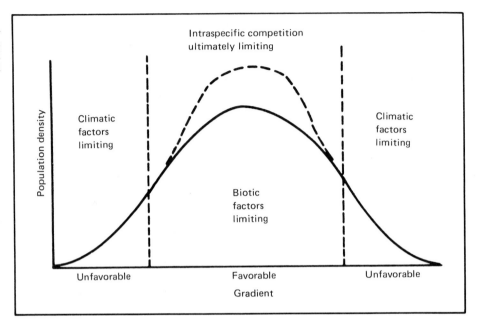

because the effect of drought has reduced the number of available host plants. Natural situations are often far more complex than this.

Some of the best examples of population regulation by climatic factors have been described by Australian entomologists. The Queensland fruit fly is a native Australian species that formerly bred in small fruits in rain forests. It is now a serious pest of peaches, pears, and other fruits grown in eastern Australia, but the severity of its attacks has varied greatly. M. A. Bateman, of the University of Sydney, studied 12 environmental parameters in experimental orchards. Of these, a single factor—summer rainfall, acting mainly by increasing survival of pupae and adults—proved of overriding influence. In favorable years larvae were produced in such numbers that intraspecific competition became severe; but in very dry summers populations approached extinction (Fig. 17–9).

One of the leading exponents of the importance of density-independent factors, H. G. Andrewartha, of the University of Adelaide, South Australia, conducted a long-term study of the abundance of the apple blossom thrips in that state. These thrips reproduce rapidly in spring and early summer and reach a population peak each year at the time of apple and rose blossoming. Their abundance varies greatly from year to year; in some years over 1000 could be found in a single blossom; in other years, a maximum of only about 200. Andrewartha and his colleagues found that several weather factors working together accounted for most of the variation in population size: winter temperatures affecting time and length of flowering of the hosts; spring temperatures influencing rate of development of the thrips; and rainfall in early spring promoting growth of the host plants and pupation and emergence of thrips in the soil. In no case did the thrips populations come close to exhausting their food supply, and influence of natural enemies appeared minimal.

Knowledge of the factors controlling population size obviously requires studies of many factors over a considerable period of time, preferably both in the field and under

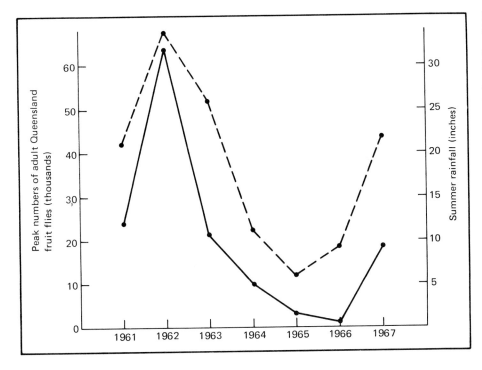

Figure 17–9
Populations of adult Queensland fruit flies (*Dacus tryoni*, Tephritidae) over seven years (solid line, scale on left) in relation to summer rainfall (dashed line, scale on right). (After M. A. Bateman, 1968, "Determinants of abundance in a population of the Queensland fruit fly," *Symposia of The Royal Entomological Society of London*, no. 4, pp. 119–131.)

controlled laboratory conditions. Since each species is unique, we have had only limited success in predicting outbreaks and in preparing to meet them in advance. Each case may require a team of highly qualified persons with long-term financial support for research unlikely to produce immediate and dramatic results. Modern society does not often see fit to direct its affluence toward such goals. Thus, in practice, we must seek short-term, pragmatic solutions to the more pressing problems.

Insect Bioeconomics

Prior to the development of modern synthetic insecticides in the mid-1900s, it was generally considered satisfactory to **suppress** insect populations sufficiently to obtain good crop yields. With the advent of modern chemical insecticides, agriculturalists found themselves with powerful new weapons with which to reduce the abundance of pest species. The notion of **eradication** became popular; species could be completely eliminated. Why was there such a change in attitude? The new insecticides, such as DDT, were so effective that it was only natural to believe that eradication was possible. However, it was not long before it became apparent that some species would not be easily eradicated. Populations of insects such as the house fly became tolerant or **resistant** to insecticides in as short a time as two years. The quantity of insecticides applied, and the frequency of application, was often increased, but the insects became more difficult to kill. Modern society's wonderful new weapons, synthetic insecticides, were quickly neutralized by natural selection for insects that could detoxify these potent poisons (Fig. 17–10).

Once agriculturalists were exposed to pest-free and damage-free crops, it was only natural to expect such levels of control. The legacy of eradication persists, in part be-

Figure 17–10
Increase in the numbers of insect pest species resistant to insecticides. (Data from U.S. Department of Agriculture.)

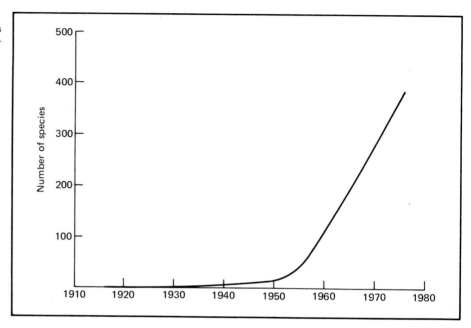

cause many species have yet to exhibit significant levels of resistance. Thus we continue to witness the desire for cosmetically perfect crops, irrespective of whether or not a few "pests" or a little "injury" really exert an economic effect.

How do insect populations affect crop yield? A generalized yield/loss relationship is shown by the curve in Fig. 17–11. The portion of the curve labeled A indicates that a number of pests can be found in a crop without a detectable reduction in crop yield. Thereafter, as the number of pests increases, there usually is a linear reduction in yield (part B). Yield stabilizes at a very low level, and additional pests cause little additive reduction in yield (part C). Some studies, however, suggest that plants may be stimulated by insect feeding and that crop yield actually increases (part A') when exposed to low levels of damage.

No single graphic model adequately describes all yield/loss relationships for all crops and insect pests. The relationship shown in Fig. 17–11 is characteristic of **indirect pests**. Indirect pests feed on foliage, stems, and plant roots, but not the portion of the crop that is marketed. When, for example, apple foliage is attacked by European red mites, or twigs are attacked by green apple aphids, or roots are attacked by woolly apple aphids, there is no immediate decrease in apple yield associated with a small amount of feeding. Some insects are called **direct pests,** because they attack the marketed produce. Apple maggots, tarnished plant bugs, and plum curculios are examples of direct pests of apple because they attack the apple fruit directly. Damage to even a single apple could be considered a loss because the apple farmer might not be able to market it. For direct pests, part A of our generalized yield/loss relationship (Fig. 17–11) may decrease immediately (A'').

A useful technique for evaluating the relative importance of pests is the **crop life table.** The use of a life table was demonstrated on Table 17–1, where various factors were shown to account for mortality of alfalfa weevils. In a crop life table, factors responsible for decreases in crop production are noted similarly. An example of a crop

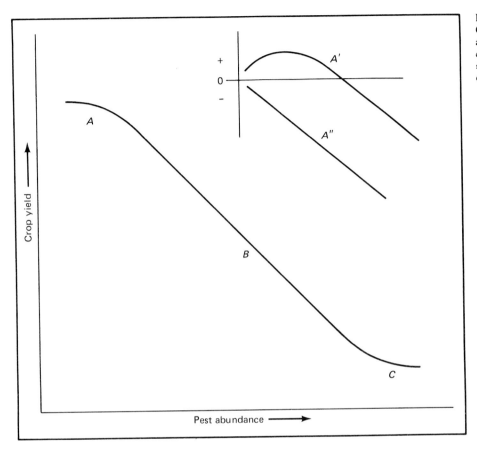

Figure 17–11
Generalized relationship of insect pest abundance
and crop yield. Portion *A–C* represents yield loss
due to indirect pests; portion *A'* represents insect
stimulation of yield; portion *A"* represents loss
due to direct pests.

life table is presented in Table 17–4. Note that in this study cutworms were responsible for a significant level of cabbage destruction while the plants were seedlings, but that other caterpillar species were more destructive later in the season. Also, the relative importance of disease, hail, or other natural forms of destruction can be documented, and a monetary cost attached to each pest. Thus the farmer can judge the importance of preventing damage by each type of pest.

Plant Compensation.   The amount of damage that occurs is usually directly related to the numerical abundance of pests. How can there sometimes be no loss of yield, or even a gain, (Fig. 17–11, A') although damage occurs? Damaged plants attempt to repair the damage caused by insect herbivores and to grow normally, a process known as **compensation.** Plant compensation is due to several factors, including the following:

1. *Removal of apical dominance.* The lateral growth of plants is inhibited by plant hormone; thus plants tend to grow upright. When clipped, the hormonal balance is disrupted, resulting in increased lateral growth and tillering. The net result may be more foliage and increased growth. Crop yield may be higher under these circumstances.

Table 17–4. Crop life table for a planting of early market cabbage, Ottawa, Canada

Growth period	Number living per plot	Mortality factor	Number dying per plot	Percent mortality	Potential revenue ($/acre)	Loss of revenue ($/acre)
Establishment	319.2	Drought	7.2	2.2	$813.96	18.36
		Cutworms	56.1	17.6		143.05
		Root maggot	1.5	0.5		3.83
		Other*	0.8	0.3		2.04
		TOTAL	65.6	20.6		$167.28
Preheading	253.6	Cutworms	8.6	2.7	$646.68	21.93
		Root maggot	9.6	3.0		24.48
		Flea beetles	0.3	0.1		0.76
		Rodents	1.3	0.4		3.32
		Clubroot	1.3	0.4		3.32
		Other*	0.9	0.3		2.29
		TOTAL	22.0	6.9		$56.10
Heading	231.6	Cabbage caterpillars	29.4	9.2	$590.58	74.97
		Root maggot	0.6	0.2		1.53
		Clubroot	3.2	1.0		8.16
		Soft rot	0.4	0.1		1.02
		TOTAL	33.6	10.5		$85.68
Harvest	198.0	Cabbage caterpillars	18.4	5.8	$504.90	46.92
		Clubroot	5.5	1.7		14.03
		Soft rot	3.0	0.9		7.65
		TOTAL	26.9	8.4		$68.60
Yield	171.1		148.1	46.5	$436.30	

Source: After D. G. Harcourt, 1970, "Crop life tables as a pest management tool," *Canadian Entomologist,* vol. 102, pp. 950–955.
*Miscellaneous factors such as frost, hail, and mechanical damage.

2. *Removal of less-productive tissue.* The photosynthetic ability of plant foliage is related to age. Intermediate-aged foliage usually is more productive than either young or old foliage. When insect herbivores selectively remove the less-productive tissue, the efficiency of the remaining tissue may increase.

3. *Increased penetration by light.* The lower leaves on a plant may be shaded by the upper leaves. Shaded leaves are less efficient photosynthetically. Partial defoliation of young, upper foliage may allow penetration of light to lower, shaded foliage and thus increase the rate of photosynthesis.

4. *Reduction in carbohydrate-induced inhibition of photosynthesis.* The rate of photosynthesis of a plant is generally less than maximum because carbohydrate accumulation produces an inhibitory feedback. When plants produce new tissue to replace that which was lost by defoliation or because apical dominance has been disrupted, carbohydrates are translocated to the new tissues. This reduces the carbohydrate-induced inhibition, and photosynthesis is stimulated.

Compensation within a crop can occur even when some plants are killed. The loss of some plants from a field is known as **reduction in stand.** Compensation can occur because plants surviving insect attack experience less competition for light, water, and nutrients; therefore, they are able to grow larger. The yield of surviving plants located near plants that have been killed may be superior to the yield of an "average" plant. If the plants killed by insect herbivores such as cutworms are spaced evenly throughout a field, a significant reduction in stand may be accompanied by no reduction in yield. However, if the mortality is aggregated, with many adjacent plants killed by insects, compensation may not be sufficient, and significant reduction in yield is more likely to occur. Unfortunately, insect distribution is more often aggregated than uniform. Grasshoppers, for example, characteristically attack the margins of fields and are infrequently found evenly distributed through the field. In arid environments reduction in stand may be damaging even where plant compensation is complete, because reduction in plant cover makes fields more susceptible to wind erosion.

Economic Factors.   Farmers seek to derive a profit from their crop, but it is not necessarily true that a pest-free crop will be more profitable. Figure 17-12 shows hypothetical crop profits in relation to effectiveness of pest control (decrease in pest abundance). For indirect pests, profits do not increase initially because moderate numbers of indirect pests are usually tolerated by crops, and control of the "pests" may not significantly improve yield. The situation is quite different with respect to direct pests, and control of direct pests often results in a rapid increase in profits, even where moderately expensive control procedures or materials are used. Profits level off as control of both indirect and direct pests increases because it becomes more and more expensive to kill the few remaining pests.

The appropriate level of pest control is determined by economic conditions. The point at which pest suppression becomes economically feasible is called the **economic injury level.** The economic injury level (EIL) can be defined as a level of damage, or abundance of insects capable of causing that damage, equal in value to the cost of suppression measures. The position of the EIL is determined principally by the damage potential of the insect and the value of the crop. As suggested earlier, potential insect pests never, or rarely, reach pest status in *some* crops. This is shown diagrammatically in Fig. 17-13(a). The average density of insects, or **equilibrium position** (EP), is well

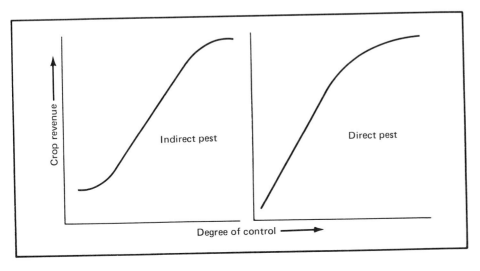

Figure 17-12
Generalized relationship of crop revenue and degree of pest control for indirect and direct pests.

Figure 17–13
Generalized trends in crop damage or pest abundance through time: (a) insect species that do not attain pest status; (b) occasional pests; (c) regular, serious pests. *EIL* = economic injury level; *ET* = economic threshold; *EP* = equilibrium position; *NEP* = new equilibrium position; * = initiation of population suppression.

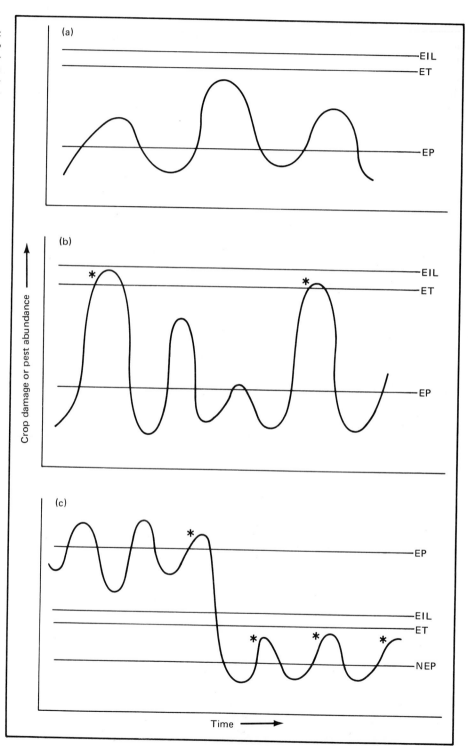

below the economic injury level. Even when insects cycle to their highest levels, they do not become damaging. Examples include alfalfa webworm in alfalfa, flea beetles in potatoes, and rose chafers on apples.

Note that the **economic threshold** (ET) is below EIL but above EP. The economic threshold, or action threshold, is a level of damage or pest abundance that serves to warn the agriculturalist of impending problems and allows the initiation of suppressive measures before the EIL is surpassed.

The relationships of the true pests to the ET and EIL are shown in Fig. 17–13(b) and (c). The pest may regularly exceed the ET and require suppressive action (Fig. 17–13b); examples include Colorado potato beetles in potato, corn earworms in corn, and boll weevils in cotton. In the extreme case (Fig. 17–13c) the EP is above the EIL, and the pest population must be suppressed and maintained at a new, lower EP. Examples of this latter relationship include codling moth in apple, green peach aphid in seed potato, and whiteflies on floricultural crops.

For many homeowners and gardeners, the value of ornamental and vegetable crops cannot be expressed by economics. Thus we can say that insects or insect damage sometimes exceed an **emotional threshold** and suppressive action is warranted, even though it may be less expensive to purchase the desired horticultural products rather than protecting them from insects.

Insect Sampling

How does the agriculturalist know that insect populations have reached the ET? To estimate the level of insect populations or damage, the agriculturalist must **sample.** Insect sampling is quite sophisticated; for a detailed discussion of sampling methods and principles the reader is referred to Southwood (1978). Methods of sampling can conveniently be divided into two types: absolute methods and relative methods.

Absolute methods are used to estimate the density of insects per unit of area. Thus we might use absolute methods to determine the number of Colorado potato beetles per acre or Mexican bean beetles per plant. Also we might determine the number of corn seedlings killed by cutworms in a field or the number of tarnished plant bug "stings" per apple.

The most frequently used type of absolute sampling method is the **unit of habitat** method. The unit of habitat to be sampled is determined by the biology of the pest insect. Therefore the abundance of codling moth larvae is determined by sampling fruit, corn earworms by sampling ears of corn, and face flies by sampling the bodies of cattle. Suction traps, brushing machines, extraction with heat or liquids, vegetation beating, visual searches, and other techniques are used to sample the unit of habitat.

Two additional absolute methods sometimes used to estimate insect populations are recapture techniques and removal trapping. **Recapture techniques** involve an initial capture of insects; they are then marked in some manner and released. The area is again sampled, and the proportion of marked individuals in the sample is used to estimate the total population in the area. **Removal trapping** requires repeated collection of individuals from an area. The rate of decline in insect abundance as the population density decreases due to sampling and removal is used to estimate the original population size.

Relative methods provide an indication of insect abundance or damage relative to other times or locations. **Visual searches** often provide relative estimates because it is difficult to assess insect numbers accurately. **Traps** are widely employed for relative estimates. Traps are usually baited with sex pheromones or food lures. Light traps disrupt the normal visual orientation of nocturnal species and can be used to capture insects that fly during the evening. Insects that are active during the day can be captured by using sticky traps painted with attractive colors. As entomologists learn more about the chemical communication of insects, pheromones become more readily available and pheromone traps will be employed more widely. Pheromone traps are useful because they are effective at detecting low densities of pests, they are quite specific, and they are economical to use. Some commonly used traps, and other methods of sampling, are shown in Figs. 17–14 and 17–15.

Plant damage is often used to estimate the relative abundance of insects. The degree of defoliation, number of clipped plants, percentage of bolls infested, and the like are usually highly correlated with the abundance of insect pests. Plant damage is the most convenient method of population estimation for some insect species because of difficulty or expense associated with sampling the insects directly. This is often the case with subterranean and nocturnal species, those that tunnel or mine foliage of plants, or those that are very numerous.

Absolute sampling methods are desirable because they are accurate and it is easy to convert from density of pests per unit area to damage potential. However, absolute methods are time consuming, they often are difficult to conduct, and they usually are quite expensive compared to relative methods. Relative methods capture an unknown, but consistent, proportion of insects present in an area. Although difficult, it is possible to convert relative estimates to absolute estimates. Relative methods are more economical in terms of time, labor, and equipment. However, to utilize relative estimates effectively, we must know the level of damage associated with a certain estimate. This requires information on yield reduction associated with relative abundance of pests, which must be collected over a period of time or from a number of locations.

Various sources of error must be borne in mind when sampling insect populations. What is actually being sampled is the proportion of those individuals whose behavior is such that they are readily captured under the prevailing conditions. The Australian entomologist P. W. Geier, for example, found that codling moths taken in flight traps were largely prereproductive, dispersing moths, while those taken in bait traps were largely mature or postreproductive females. Wind speed may greatly influence the number of individuals flying or resting high in vegetation, and light intensity may influence the response to light traps (which are much less effective when the moon is full). Obviously an intimate knowledge of the insect's life history and behavior is important in planning a sampling program, as is a working knowledge of statistics both in planning and in analyzing the results.

Why is insect sampling so important for the management of insect pests? The routine use of control procedures (usually insecticides) without regard for pest density and damage potential is economically wasteful and may result in needless destruction of beneficial insect species, contamination of the environment with toxic pesticide residues, and selection for insecticide-resistant pest species. The foundation of pest management rests on the belief that no control measure should be initiated unless a pest is present—and present in damaging, or potentially damaging, numbers. Certain exceptions to this principle exist. For example, the use of pest-resistant varieties and certain cultural practices prevent the development of damaging pest populations. Insecticide use in agriculture could be reduced significantly if agriculturalists adhered to the aforementioned principle. We are now witnessing greater interest in sampling and a subsequent reduction in insecticide use in many areas of agriculture.

(a)

(b)

(c)

(d)

Figure 17–14
Some commonly used sampling techniques: (a) sweeping sugar beets with a net; (b) suctioning individual insects from leaves; (c) a light trap; (d) a sticky trap baited with pheromone for elm bark beetles. (Part *d* courtesy Colorado State Forest Service; others John L. Capinera, Colorado State University.)

(a)

(b)

Figure 17–15
Some further examples of sampling techniques:
(a) a pheromone trap; (b) vacuuming insects from
foliage.   (John L. Capinera.)

The Strategy of Pest Management

The basic strategy behind insect pest management (IPM) is to prevent insect populations from attaining their economic injury level (EIL) while avoiding unfavorable ecological, economic, and sociological consequences. Not all pest populations approach the EIL at the same time. Also, some locations may experience pest outbreaks while others do not. When the economic threshold (ET) is reached, however, we must resort to various **tactics,** or methods of pest suppression, to keep the pest population from reaching the EIL. Sometimes tactics are utilized in advance, in anticipation of pest problems, so that the ET is never reached. Adherents of IPM principles seek to avoid preventive measures unless they are not costly or damaging to the environment; often they are necessary where damaging pest populations occur regularly or where the EIL is very low. The strategy of pest management has several components, including the following.

Identification of the Problem.   Often it is difficult to diagnose the cause of a symptom correctly. Plants and animals respond similarly to a variety of problems. For example, plant foliage will become chlorotic in response to attack by piercing-sucking insects, to infection by several plant diseases, to nutrient deficiencies, and to herbicide injury. Diseases are sometimes transmitted by insects, and nutrient-deficient plants are sometimes especially attractive to insects, so it may be difficult to ascribe the primary cause of the problem to the correct causative agent. In a crop environment we find many potential answers to the question "What is responsible for this symptom?" It takes an experienced, broadly trained individual to supply the correct answer.

Assessment of Damage.   It is easy to overreact to cosmetic injury and initiate suppressive tactics without careful assessment of damage. As indicated earlier, plants have a remarkable ability to recover from, or compensate for, insect attack. We should not be misled into initiating costly but unnecessary actions. Also, it is quite difficult to estimate accurately the level of damage. For instance, 5% or 10% foliage removal appears much more damaging than it actually is. On the other hand, mites and small insects may escape notice, and it may require thorough examination to locate nocturnal and subterranean insects. Again, there is no substitute for a knowledgeable and thorough individual when damage assessment is required.

Cost-Benefit Analysis.   Unfortunately the actual benefits derived from a particular pest suppression activity usually are not calculated. The value associated with the increased crop yield should be equal to, or exceed, the cost of the suppressive action. We often do not know the value of increased yield associated with a management action because EILs have not been calculated. This makes cost-benefit analysis difficult. Another variable that should be considered is the long-term effect of the suppressive action on the pest population. Often we can achieve season-long control by suppressing the first generation of the pest, while if we wait until later in the season the pests may be more numerous and difficult to control. Sometimes a suppressive action reduces the abundance of a pest for more than one season; thus the cost of control can be amortized over several crop harvests. On the negative side, suppressive actions may have deleterious effects on beneficial insects, which can lead to more frequent pest outbreaks, increased need for suppressive action, and increased cost.

Selection of Management Tactic.   Agriculturalists have an impressive array of pest management tactics at their disposal. The tactics available vary with crop, pest, time of year, and geographic locality. Possibly the most common tactics employed are application of insecticide and planting of pest-resistant crop varieties. These tactics are commonly used because they are effective, available, and economical and generally do not require that the user possess a great deal of entomological knowledge. Tactics such as crop rotation, modification of planting time, and release of beneficial organisms may be equally effective, but these latter tactics require more knowledge of pest life history. Pest management tactics are discussed in greater detail in Chapter 18.

Implementation of Management Tactic. A major factor influencing selection of management tactics is the ability to actually implement the chosen tactic. All too often we find a novice who is devoted to biological suppression but is unable to implement effective biological control. We witness dependence on chemical suppression tactics because these materials are readily available. However, we are beginning to see increased availability of alternative suppression methods. For example, release of parasites of dung-breeding flies into cattle feedlots is becoming more common because insectaries are making these beneficial parasites more available. Similarly, use of the bacterium *Bacillus thuringiensis* for suppression of caterpillars has become widespread because it can be purchased from most pesticide retailers.

For many suppressive tactics to be successful, we must plan ahead. Many cultural controls must be implemented before or during crop planting. For example, once seeds have been planted, it is usually impossible to apply insecticides to protect them. The decision to plant insect-resistant crop varieties obviously must be made well in advance of planting.

We should not always plan for the worst possible consequences and resort to preventive measures. Many insect populations can be monitored and damage predicted well in advance. When sufficient pest life history and sampling information is available to predict population trends and resultant damage, we should monitor the crop environment carefully and initiate suppressive actions only when needed.

Efficacy Assessment. The effectiveness of the management tactic selected can be judged only by careful monitoring of insect populations and crop yield. All too often we assume that pest suppression has been achieved because a control tactic has been initiated. We always should check to ascertain that the desired level of suppression has been attained. Again, sampling is required.

If, as part of efficacy assessment, yields from pest-infested fields are determined, valuable information on the economic justification for pest suppression can be calculated.

Follow-up Periodic Assessment. Management tactics may result in less than complete suppression of pest populations. Also, immigration of pests from unmanaged fields may occur. Crops should be monitored regularly and thoroughly to prevent unexplained losses and unpredicted pest population outbreaks.

Summary

Population size is the result of the interaction of fecundity and mortality. With respect to mortality, ecologists distinguish between factors that are density dependent (increasing their adverse effects as the population increases) and those that are density independent (without respect to density, such as weather or natural catastrophes). Stability is the ability of a system to absorb disturbance and return to an equilibrium state; resilience is the capacity to adapt to changes in the environment. Species of insects likely to become agricultural pests are those that are small, mobile, and able to reproduce rapidly. These are termed r-strategists, in contrast to K-strategists, which tend to have lower reproductive rates and to be less dispersive.

Certain factors intrinsic in populations often prevent them from coming close to exhausting their food supplies. Such self-regulating factors may be genetically controlled and triggered by changes in food quality or quantity. In certain bark beetles, behavioral mechanisms controlling density and distribution have been described. Intraspecific competition may also result in a decrease in oviposition rate, a slowing of the rate of development, or an increased tendency to migrate.

The importance of extrinsic density-dependent factors in controlling population size is well demonstrated by cases of biotic release, that is, cases in which a population is suddenly released from its natural enemies and increases greatly in size (such as the arrival of the Japanese beetle and other foreign pests in the United States). The effect of extrinsic, density-independent factors in populations is to produce fluctuations of an irregular nature. Long-term studies of the effect of climate on populations may, however, often permit predictions concerning the occurrence of population outbreaks.

Prior to the development of modern synthetic insecticides, it was considered satisfactory to suppress populations of pest species, but more recently the concept of eradication has become popular. However, this has often proved impracticable because of the development of resistance to many of these newer insecticides. The contemporary approach is in terms of pest management. This involves an assessment of loss of revenue resulting from specific agents, which may well be expressed in terms of a crop life table. It is useful to distinguish between indirect pests, which feed on foliage, stems, or roots that are not marketed, and direct pests, which attack the fruit or whatever part of the plant is marketed.

The amount of damage is usually directly related to the numerical abundance of pests. However, plants tend to compensate in various ways, so that in fact yield may not decrease and may sometimes increase. In the case of direct pests, even moderate levels of pest abundance cannot usually be tolerated, and even the use of relatively expensive controls may be justified. The appropriate level of control is determined by economic conditions. The point at which pest suppression becomes economically feasible is called the economic injury level (EIL). This differs from and is higher than the economic threshold (ET), which is the level of damage or pest abundance that serves to warn the agriculturalist of impending problems.

In order to determine whether insect populations have reached the ET, careful sampling is required. Absolute methods of sampling may involve counting the actual number of individuals per unit area or may involve mark and recapture techniques or removal trapping. Relative methods may involve visual estimates, trapping, or assessments of plant damage. The routine use of control procedures (usually insecticides) without regard to pest density is economically wasteful and may result in needless destruction of beneficial species, contamination of the environment with toxic materials, and selection of insecticide-resistant pest species.

The basic strategy behind insect pest management is to prevent insect populations from attaining their EIL. When the ET is reached, we must resort to various methods of suppression to keep the pest populations from reaching the EIL. The strategy involves several components: identifying the problem; assessment of the damage; cost-benefit analysis; selection of management tactic; implementation of this tactic; and finally an assessment of efficacy as well as periodic follow-up assessment.

Selected Readings

Dempster, J. P. 1975. *Animal Population Ecology*. New York: Academic Press. 155 pp.

Geier, P. W.; L. R. Clark; D. J. Anderson; and H. A. Nix. 1973. *Insects: Studies in Population Management*. Canberra, Australia: Ecological Society of Australia. Mem. 1. 295 pp.

Getz. W. M., and A. P. Gutierrez. 1982. "A perspective on systems analysis in crop production and insect pest management." *Annual Review of Entomology*, vol. 27, pp. 447–66.

Harcourt, D. G. 1969. "The development and use of lifetables in the study of natural insect populations." *Annual Review of Entomology*, vol. 14, pp. 175–96.

Knipling, E. F. 1979. *The Basic Principles of Insect Population Suppression and Management*. USDA Agricultural Handbook 512. 659 pp.

Metcalf, R. L. 1980. "Changing role of insecticides in crop protection." *Annual Review of Entomology*, vol. 25, pp. 219–56.

———, and W. H. Luckmann. 1982. *Introduction to Insect Pest Management*. New York: Wiley. 577 pp.

Perkins, J. H. 1982. *Insects, Experts, and the Insecticide Crisis: The Quest for New Pest Management Strategies*. New York: Plenum Press. 304 pp.

Southwood, T. R. E. 1978. *Ecological Methods*. Second edition. London: Chapman and Hall. 524 pp.

Stern, V. M. 1973. "Economic thresholds." *Annual Review of Entomology*, vol. 18, pp. 259–80.

Wellington, W. G. 1957. "The synoptic approach to studies of insects and climate." *Annual Review of Entomology*, vol. 2, pp. 143–62.

White, T. C. R. 1978. "The importance of a relative shortage of food in animal ecology." *Oecologia*, vol. 33, pp. 71–86.

The Tactics of Insect Pest Management

The insect pest manager has a number of potential management tactics available; selection should be based on economic conditions, agronomic requirements of the crop, pest situation, meteorological conditions, and many other factors. Some tactics require that the pest manager take action well in advance of the pest situation. If the manager fails to plan accordingly, then certain options are lost. Once a crop is planted, for example, it is too late to choose another variety. The frequent use of two tactics, planting of pest-resistant crop varieties and application of insecticides, reflects the desire of most agriculturalists to avoid frequent monitoring of pest populations and to eliminate the need for advance planning. Planting a resistant variety may free the farmer from a particular insect problem. If the farmer fails to utilize a resistant variety or one is not available, it is often most convenient to resort to insecticides to prevent pest outbreaks. The use of one tactic certainly does not preclude the use of others. For example, insecticides may be used if plant resistance is not completely satisfactory or if biological control agents do not provide satisfactory suppression. The integration of several tactics into an effective program requires knowledge of pest life history, crop requirements, and available suppression tactics. The following is a brief discussion of some of the tactics available to the insect manager. For a more complete discussion the reader is referred to the publications cited at the end of this and the preceding chapter.

Regulations

Government agencies sometimes establish regulations designed to prevent entry and establishment of foreign pests. Also, once pests become established, regulations may be directed at containment, suppression, or eradication of pests. This process is sometimes referred to as **regulatory control**.

Quarantines limit movement of pests, usually by preventing transport of products that could harbor pests, by requiring inspection or by requiring fumigation to kill hidden pests. The international movement of many important pests has been limited by quarantines, and inspection is required at most international borders as well as at some state boundaries. Quarantine programs are not always effective. For example, two pests that have recently arrived in North America in spite of quarantine programs are the face fly, a pest of livestock, and the cereal leaf beetle, a pest of small grains.

When quarantines are not effective and potentially damaging insects are successfully established, regulatory agencies sometimes resort to pest control activities. The principal concern regarding selection of tactics usually is effectiveness, not economy, and chemical insecticides are usually employed.

Eradication is an effort to eliminate pests from a geographic area. If the infested geographic area is relatively small, and effective suppressive tactics are available, eradication is possible. For example, the Mediterranean fruit fly has been introduced accidentally to the United States on several occasions and has been eradicated success-fully; effective population-monitoring techniques and suppressive tactics were avail-able and were applied to relatively small geographic areas. Proposed eradication pro-grams aimed at well-established pests such as boll weevil and fire ant probably are not feasible because of the extensive area infested and the disruptive effects associated with insecticide treatment of large areas.

When eradication programs are not successful or not feasible, regulatory agencies sometimes initiate containment or suppression programs. **Containment** programs attempt to limit the spread of pests through their potential habitable environment, and therefore limit economic loss. **Suppression** programs attempt to reduce pest population levels and associated damage, usually because the pest inhabits too large an area for containment. Suppression programs supported by regulatory agencies generally are limited to pest species that cause severe loss over large geographic areas; the pests are too abundant or too widespread to be managed by individual landowners. Grass-hopper control on rangeland in the western United States is an example of a regularly conducted suppression program. Federal agencies coordinate grasshopper suppression efforts, and the cost of the program is shared by ranchers and state and federal governments.

Host Resistance

Selection of a host by an insect involves four major steps: (1) finding the habitat, (2) settling, (3) sampling or tasting, and (4) ingestion (Chapter 9). Host selection is a complex process, and different stimuli are involved in long-distance and in short-dis-tance orientation. Any host-related factors that disrupt the host selection process provide the basis for **host resistance**, or unsuitability of the host for an insect.

Mechanisms of host resistance have been classified in several ways. **Physical resis-tance** refers to morphological characteristics of a host that lead to its unsuitability. Host size, shape, color, toughness, succulence, and presence of trichomes (leaf hairs) contribute to physical resistance. **Chemical resistance** refers to toxins or to nutritional and metabolic factors of a host that lead to its unsuitability. As we have seen (Chapter 10), plants frequently have physical or chemical attributes that serve important func-tions as protective agents. Plants also vary in their nutritional adequacy. Some species characteristically have low levels of essential nutrients, such as protein; others possess digestibility-reducing compounds, such as tannins, that reduce the ability of insect herbivores to use the nutrients.

An alternative approach is to attribute resistance to nonpreference, antibiosis, or tolerance. **Nonpreference** refers to insect responses to host characteristics that lead away from the use of the host for food, oviposition, or shelter. For example, the Philip-pine entomologist M. D. Pathak found that the striped borer moth, *Chilo suppressalis*, shows a strong preference for oviposition on rice varieties that are taller and have wider and smoother leaf blades. Even when borer populations are low, susceptible varieties receive large numbers of eggs while resistant varieties receive few or none. **Antibiosis** refers to deleterious effects on insect survival or life history resulting from feeding on a resistant host. One of the best-studied examples of antibiosis was initiated by S. D. Beck and his co-workers at the University of Wisconsin. Very young corn plants are resistant to survival of European corn borer because of complex toxic sub-stances called by the acronym DIMBOA. Certain inbred varieties maintain high levels of DIMBOA in their whorls at later stages of development and are relatively resistant to corn borer. **Tolerance** refers to the ability of a host to grow and reproduce normally

Figure 18–1
(a) Plants of a susceptible variety of corn severely damaged by European corn borer. (b) A resistant variety that shows little damage. Each was experimentally infested with 100 corn borer eggs per plant. (U.S. Department of Agriculture; from *Agricultural Handbook* no. 512, by E. F. Knipling, 1979.)

while supporting a pest population that would be damaging to a susceptible host. One of the best-documented examples also involves corn. B. R. Wiseman and his colleagues at the U.S. Department of Agriculture Laboratory at Tifton, Georgia, found that certain resistant varieties of corn may support as many earworms (*Heliothis zea*) as susceptible varieties do, but most remain with the silk channel (after pollination) and do not penetrate the ear.

Resistant crop varieties are used frequently for insect pest population suppression (Fig. 18–1). If a resistant variety proves to have a smaller yield than susceptible varieties, this must be more than compensated for by reduced costs of control. Plant breeders usually try to combine insect and pathogen resistance into the same cultivar, so the value of such varieties is even greater. Hessian flies and greenbug aphids are important insect pests of wheat and sorghum, respectively, and resistant cultivars are the principal tactics employed to alleviate insect damage in these crops. As is the case with any plant characteristic, however, host plant resistance may be modified by environmental conditions. Resistance of wheat varieties to Hessian fly, for example, may be reduced by high temperatures. Similarly, fertilizer application can make plants more or less susceptible to attack by insects, depending on the type of fertilizer and insect species. Also, through natural selection insects sometimes develop races that overcome the source of resistance. Host plant resistance obviously cannot be considered a panacea.

Biological Control by the Use of Predators and Parasitoids

As we saw in the previous chapter, insect populations are naturally affected by a wide variety of environmental factors, both biotic and abiotic. The action of these factors is often termed **natural control**. One aspect of natural control is **biological control**, which involves population regulation by the use of biotic agents. Biological control agents traditionally have been considered to be predators, parasitoids, and microbial pathogens, although other biotic factors such as host plant characteristics could be considered in this category. The principal distinguishing feature regarding predators, parasitoids, and pathogens is their potential to respond in a density-dependent manner.

Predators and parasitoids have long been acknowledged to be important components in natural control. In addition, there have been several outstandingly successful manipulations of predators and parasitoids to achieve better biological control. Some of the characteristics that distinguish predators from parasitoids were reviewed in Part 5, pages 272–273. Sometimes a species may be both a parasitoid and a predator. For example, adults of the chalcidoid wasp *Tetrastichus asparagi* deposit their eggs within the eggs of asparagus beetles. The wasp larvae develop within the beetle larvae, killing the beetles just prior to pupation. Thus *T. asparagi* is termed an egg-larval parasitoid. Adult *T. asparagi* also feed on the eggs of asparagus beetles. Consequently this species is also a predator.

Insect predators and parasitoids are utilized in biological control programs through introduction, conservation, and augmentation. **Introduction** of exotic predators and parasitoids entails the search for natural enemies in foreign countries and their introduction into areas where pest insects are causing damage. This is generally done where pests were accidentally introduced into new areas without their normal complement of natural enemies. Most of the spectacular successes achieved with biological control have involved introduction. Perhaps the best-known example of biological control involving a predator concerns the cottonycushion scale and vedalia beetle in California, as discussed in Chapter 12 (p. 276).

Conservation of predators and parasitoids emphasizes the importance of preserving naturally occurring beneficial species. Conservation may involve either careful use of insecticides to avoid accidental mortality among beneficial groups or manipulation of the environment to make it more suitable for them, as by planting flowering plants to provide a source of nectar for adult parasitoids.

An example of parasitoid conservation occurs in vineyards where wild blackberries are allowed to grow. The grape leafhopper, which is a serious pest of grape, is attacked by an egg parasitoid. The chalcidoid egg parasitoid, *Anagrus epos*, overwinters successfully only when an alternative host, a leafhopper that inhabits blackberry, is present. Thus the presence of blackberries and associated nonpest leafhoppers enables maintenance of the agents that attack the pestiferous grape leafhopper.

Augmentation of predators and parasitoids involves the culture and repeated release of natural enemies to suppress pest populations (Fig. 18–2). The beneficial insects may occur naturally, but at low densities, or they may be exotic species that do not persist in their new environment.

An example of parasitoid augmentation is the Mexican bean beetle suppression program developed by Allen Steinhaur and his associates at the University of Maryland. The Mexican bean beetle is a pest of several varieties of beans. Although it has not been a serious pest of soybeans, as the acreage devoted to soybean increases, Mexican bean beetle damage also increases. There is a distinct possibility that soybean growers could become "locked" into a chemical control program if insecticides are applied to soybean regularly. Experience with other crops has demonstrated that if insecticides are applied routinely, insecticide-resistant populations often develop. Also, insects that are not pests sometimes attain pest status when beneficial insects are inadvertently killed by excessive insecticide use.

To forestall these potential problems, the chalcidoid parasitoid *Pediobius foveolatus* (Fig. 12–11) is released annually to suppress Mexican bean beetle. *Pediobius* was imported from India and cannot survive Maryland winters; thus beetles and parasitoids are cultured, and parasitoids released, each spring. Since Mexican bean beetles prefer snap beans over soybeans, soybean growers are encouraged to plant a few rows of snap beans to attract emerging Mexican bean beetles away from the soybean crop. *Pediobius* then is released into the snap beans, where the Mexican bean beetles are congregated. Although the bean beetles eventually move to the soybean crop, the parasitoids become sufficiently abundant, and are sufficiently mobile, to prevent excessive damage by bean beetles.

The search for effective parasitoids is a continuing process. The navel orangeworm, an important pest of almonds in California, has plagued growers for many years. A search for natural enemies in the native home of the pest in South America resulted in the discovery of a parasitoid wasp that has been reared and liberated in California orchards by E. F. Legner, Gordon Gordh, and their associates at the University of California at Riverside (Fig. 18–3). The parasitoids are reared in gelatin capsules that are placed in the crevices of trees and opened so that the adult wasps can escape. Initial results suggest that the wasps have aided materially in depressing navel orangeworm densities well below the EIL.

In addition to suppression of pests attacking orchard and field crops, augmentation programs show particular promise for control of dung-breeding pests. Parasitoids frequently are released into poultry and cattle production facilities to alleviate dung-

Figure 18–2
Laboratory augmentation of small chalcidoid wasps of the genus *Trichogramma*. The wasps are reared on the eggs of grain moths glued to panels and later released in the field. *Trichogramma* has been used for control of insects on a variety of crops, with variable results. (U.S. Department of Agriculture; from *Agricultural Handbook* no. 512, by E. F. Knipling, 1979.)

(b)

Figure 18–3
Goniozus legneri, a wasp of the family Bethylidae that has been imported from Uruguay and released in California for control of the navel orangeworm: (a) adult wasp on the host caterpillar; (b) larvae of the parasitoid developing on the caterpillar. (Photographs by Max E. Badgley, University of California, Riverside.)

breeding fly problems because many of the fly pests are insecticide resistant. Also, it is important not to contaminate livestock with insecticides soon before they are marketed.

What are the advantages to the use of predators and parasitoids for insect pest suppression? There are three principal advantages: permanence, safety, and economy (although all may not apply to any one situation). Many predators and parasitoids, unlike *Pediobius*, are self-perpetuating. The vedalia beetle, for example, required little assistance once established. Although the initial cost of locating, culturing, and releasing predators and parasites may be quite high, experience has demonstrated that this is a very economical method of pest suppression if an appropriate predator or parasitoid can be located. Also, there is less danger of selection for resistant pest populations. Predators and parasitoids can coevolve with their hosts (in Chapter 12 we discussed the encapsulation of ichneumon larvae by alfalfa weevils). However, the development of complete immunity to parasitoids has rarely been reported. Lastly, release of predators and parasitoids has had no significant adverse environmental effects.

Is biological control through the use of natural enemies applicable to all pest situations? Predators and parasitoids certainly are not the solution to every problem, although they should be given serious consideration in most situations. Often they will be suitable as a *part* of a strategy that includes other tactics. We should look for gaps in the natural enemy complex associated with a pest and try to fill the gaps by introducing new natural enemies. Often we find that natural enemies that could be good biological control agents are already present but are inhibited by some maladjustment of the community. Sometimes we discover that insecticides have disrupted predator and parasitoid efficiency (p. 373); we should search for ways to integrate insecticides into crop protection programs with minimal disruptive effects.

The pest environment influences the probability that naturally occurring biological control will be successful. Also, the nature of the environment should be considered carefully when attempting to manage a pest situation through manipulation of predators and parasitoids. There are many types of pest situations, but the extremes are probably best represented by the forest environment and agroecosystem.

Biological Control in a Forest Environment.   In most forest environments, the long-term average insect population densities are below the economic injury level. Occasionally a pest population suddenly increases to a damaging level and then usually declines precipitously. The problem in this type of environment is to *stabilize* pest population fluctuations. It is important, therefore, to introduce density-dependent mortality factors that will exert greater impact on pest populations as they become more abundant. Density-independent factors are important because they help to maintain pest populations below injurious levels, but it is probably the absence or presence of density-dependent factors that determines population trends. The introduction of exotic predators and parasitoids may be the most appropriate tactic in the forest environment.

Successful population regulation of the winter moth in forests of eastern Canada by a tachina fly and an ichneumon wasp was discussed in Chapter 12 (p. 281). Another pest of eastern forests, the gypsy moth, has proved more intractable. Over 40 species of predators and parasitoids have been introduced and at least 12 of these are well established. The severity of periodic defoliation of trees in urban areas and in parklands has, however, led to demands for aerial applications of insecticides, and these have not favored the success of the biological control program. The severe outbreaks of 1981 have reinforced efforts to develop an integrated program for the gypsy moth, which will include biological control as an important element.

Biological Control in an Agroecosystem. In contrast, in many agroecosystems population fluctuations are not the problem; rather, the average population density exceeds the economic injury level. This can result from a number of factors: (1) the crop may have a very low EIL, such that few pests can be tolerated; (2) the crop may be a very suitable food source, promoting good pest survival and reproduction; (3) the pests may be well adapted to exploit disturbed monocultures (r-strategists); and (4) the pests may be relatively free from natural enemies. In such an environment, our objective is to reduce average pest densities rather than to reduce the severity of outbreaks. We need to introduce mortality factors that on the average destroy a portion of the pest generation that is equal to, or exceeds, the reproductive ability of the pest. Density-independent factors, in addition to density-dependent factors, are of value here. Insecticides might be more effective than predators and parasitoids in this environment, or we can utilize predators, parasitoids, or perhaps pathogens in an augmentative approach where they are applied regularly, as are insecticides.

The previous discussion considers extreme conditions, contrasting a climax forest with an annual crop. Many agroecosystems also display characteristics of a forest environment. Deciduous fruit trees and rangelands, for example, may be somewhat intermediate between the typical forest environment and agroecosystem as described above. Selection of pest suppression tactics will be influenced by the unique environmental characteristics associated with each.

The potential value of predators and parasitoids in a crop environment also is influenced by the pest–crop relationship. For many natural enemies, there must be a minimum level of hosts (pests) present in the crop for these potential control agents to be supported at effective levels. Fruit and flower crops, among others, which cannot tolerate large numbers of direct pests, may not be amenable to predator and parasitoid utilization as compared to forage crops, which often can support considerable damage.

Biological control is a tactic of wide appeal in an age of environmental awareness. Practitioners can point to important successes, but it is only fair to say that there have been many instances in which control was disappointing. Some of the problems we have alluded to in Chapters 12 and 15. Other problems arise from the difficulties of identification of these often minute organisms. When parasitoid wasps of the genus *Aphytis* were introduced for control of California red scale, results were inconsistent. Paul DeBach and his colleagues at the University of California at Riverside determined that what was thought to be one species was actually seven, each with its own biological characteristics. There remain many thousands of species of parasitoid Hymenoptera and Diptera that have yet to be discovered and studied.

Biological Control by the Use of Pathogens

Pathogens, or microbial control agents, also are important in naturally occurring biological control. The potential for pathogens in applied biological control probably exceeds that of many other pest management tactics, including predators and parasitoids. However, development of pathogen-based systems of pest management has proceeded slowly. The difficulties associated with identification, culture, and registration of microbial agents account for much of the delay in integration of pathogens into pest management programs.

In many ways, pathogens are very similar to parasitoids. They are most effective against the immature stages of insects, and adults tend to be less susceptible or immune. As is often the case with parasitoids, they develop internally, and a single host can produce many pathogens. Pathogens often have a rather specific host relationship (narrow host range), much like parasitoids. However, active host "selection" is rarely involved; the pathogenic organisms are ingested inadvertently or spread by air, water, or oviposition by parasitoids. Acute, lethal infections are common, but chronic, debilitating infections also occur, whereas the attack of parasitoids is almost always fatal.

Pathogens are very diverse and include viruses, bacteria, fungi, protozoa, and nematodes. Not all are beneficial from our point of view, for some attack useful insects such as the honey bee. The most widespread disease of adult honey bees is nosema, caused by the protozoan *Nosema apis*. American foulbrood and European foulbrood are important diseases of honey bee larvae and are caused by two species in the bacterial genus *Bacillus*. However, here we shall be concerned with microbial agents that attack noxious insects. It is easiest to discuss these in the context of their taxonomic status, since each group has its own biological characteristics.

Viruses. Particles of many insect viruses are enclosed in protein crystals called **inclusion bodies**. The principal types of **inclusion viruses** are nuclear polyhedrosis virus (NPV), granulosis virus (GV), cytoplasmic polyhedrosis virus (CPV), and entomopox virus. **Noninclusion viruses** occur in insects but are difficult to detect without an electron microscope and are not well known.

NPVs multiply in the cell nucleus. Tissues most commonly infected are the epidermis, fat body, blood cells, and tracheae in Lepidoptera. In phytophagous Hymenoptera, midgut cells are infected. Infected larvae rarely exhibit symptoms of infection until just before death, when the epidermis darkens and becomes shiny. Virus infection sometimes induces a change in larval behavior whereby the insect climbs to the highest point available prior to death. After death the integument ruptures, releasing millions of polyhedral inclusion bodies that contaminate the food plant. This enhances the probability that other insects will accidentally ingest inclusion bodies and become infected. This change in behavior is sometimes induced by other pathogens as well.

GVs multiply in both the nucleus and the cytoplasm of host cells. The fat body is the principal site of infection, but the epidermis and tracheae are sometimes infected. Symptoms of infection are similar to NPV infection. Lepidoptera are common hosts.

CPVs develop in the cytoplasm of the host cells. The midgut is infected, and Lepidoptera larvae are the usual hosts. The disease tends to be debilitating; diseased larvae exhibit loss of appetite, are smaller, and develop slowly. Insects that are infected but not killed may have a reduced reproductive capacity. CPVs are not as host specific as are the other two virus groups mentioned previously.

Entomopox viruses multiply in the cell cytoplasm of fat body and blood cells. They are found in Coleoptera, Lepidoptera, Diptera, and Orthoptera. Symptoms of infection are variable.

Viruses have considerable potential for insect pest management. They are usually quite specific, which allows entomologists to suppress pest populations without undue disruption of beneficial insects. Insect viruses can be applied with conventional insecticide application equipment. They are not harmful to mammals, and they are tolerant of many adverse environmental conditions. The main limitations to the use of viruses include our inability to produce them economically (generally they are cultured in an insect host), their susceptibility to ultraviolet radiation and certain other adverse environmental conditions, and the reluctance to apply disease-causing agents to food crops. Commercial production of insect viruses still is quite limited. "Elcar" is an example of a nuclear polyhedrosis virus that is commercially available. It is active against several species of lepidopteran pests in the genus *Heliothis*, but thus far its use is restricted to cotton.

Bacteria. Most important bacterial pathogens of insects form resistant spores and thus can survive unfavorable conditions quite well. Non-spore-forming bacteria are common in the digestive tracts of insects, but they are only pathogenic when they invade the hemocoel; this occurs under unusual conditions such as stress or injury. There has been little effort to utilize non-spore-forming bacteria as microbial agents; some are pathogenic to mammals.

The best-known insect pathogens among the spore-forming bacteria include *Bacillus thuringiensis* and *B. popilliae*. *B. thuringiensis* was first isolated from silkworm larvae by Japanese workers about 1900. There has since been extensive research on this organism, and several strains infecting a wide variety of insects are recognized; however, it is most frequently applied against lepidopterous larvae. The importance of the bacillus in microbial control was first demonstrated in North America by E. A. Steinhaus, of the University of California in Berkeley, working on control of the alfalfa caterpillar. Steinhaus, who later published important books on insect pathology, pointed out that the organism was easy to produce in large quantities and to disseminate as a "microbial insecticide." Infected caterpillars become sluggish and eventually discolored and flaccid; after death they produce odors of putrifaction. Hundreds of tons of *B. thuringiensis* spores are now being produced commercially each year, under such names as "Thuricide," "Dipel," and "Biotrol." The bacterial cells contain a toxic protein crystal as well as the spore. When dissolved in the insect, the crystal causes paralysis of the gut. However, both spore and crystal are usually required for effective action.

B. popilliae is also available commercially, under the name "Doom." It is effective against the larvae of Japanese beetles and produces "milky disease," the name resulting from the milky-white appearance of infected grubs. *B. popilliae* persists in the soil for some time, whereas *B. thuringiensis* is less persistent and must be reapplied regularly.

Fungi. Most taxonomic groups of fungi contain some insect pathogens. Entomopathogenic fungi often are not very specific. Fungal pathogens are usually transmitted from host to host by spores; unlike viruses and bacteria, fungi are capable of penetrating the insect directly through the integument. Fungi, more than any other insect pathogen, require favorable environmental conditions for development of **epizootics**, or outbreak of disease. This seems to be due to the need for high humidity that usually is required for germination of fungal spores. Fungi can often be cultured on artificial media, but infectivity is reduced. Fungi are important naturally occurring biological control agents (Fig. 18–4). Some species have been used successfully to reduce pest populations, and while availability is currently limited, we should see greater use of fungi for pest suppression in the near future. *Verticillium lecanii* is used for biocontrol of aphids and scales affecting greenhouse crops in Europe. *Beauveria bassiana* is applied

Figure 18–4
Insects that have succumbed to fungi of the genus *Entomophthora*. (a) An immature grasshopper that is infected with *E. grylli*. Infected grasshoppers typically climb to the tops of plants and die with their legs wrapped tightly around the plant. (b) A fly that has succumbed to *E. muscae*. (Photograph *a* by Howard E. Evans; photograph *b* by George C. Eickwort.)

(a)

(b)

against a variety of pests in eastern Europe, and *Metarhizium anisopliae* is available in Brazil for controlling spittlebugs on sugarcane. In the United States *Hirsutella thompsoni* is produced commercially for suppression of citrus rust mites.

Protozoa.   The phylum Protozoa is a heterogeneous assemblage of organisms. Many groups, including the flagellates, gregarines, and amoebae, may be pathogenic to insects. The order Microsporida, especially pathogens of the genus *Nosema*, has received considerable attention by entomologists. Microsporidans can be transmitted orally or via the egg. They are not very specific and often cause a chronic disease that results in lowered fecundity or sterility.

Nosema locustae affects a large number of grasshoppers and can be produced in commercial quantities by inoculating immature grasshoppers in culture. One infected grasshopper will normally produce enough spores to treat 3 to 4 acres of rangeland. *Nosema* spores usually are mixed with bran to make a bait that grasshoppers ingest readily. *Nosema* also can be applied in conjunction with a reduced amount of insecticide to provide rapid, but residual, suppression of grasshopper populations.

Nematodes.   Nematodes traditionally have been treated as microbial pathogens, but it would perhaps be more logical to consider them as microparasites and place them in a separate category. Several families of nematodes contain members that are parasitic on insects. The infective stage of the nematode penetrates the insect directly or is ingested, and reproduces inside the host. The insect is killed by the exit of the nematodes or by release of a bacterium carried by the worms.

Reesimermis nielseni nematodes appear promising for suppression of mosquito populations. Nematodes are cultured in mosquito larvae at a cost of about 10 cents per million infective juvenile nemas and are applied at a rate of about 1000 nemas per square meter of water surface.

Steinernema feltiae has a broad host range and has been used experimentally for control of many insects. The nematode penetrates the insect hemocoel and releases bacteria. The bacteria kill the host, and the nematodes feed on the bacteria and host tissue. Large numbers of infective nematodes can be produced from a single host. *S. feltiae* also can be cultured on artificial media with production costs as low as 2 cents per million nematodes. The nematodes can be applied as a foliar protectant or to the soil, with the use of standard insecticide application equipment.

Biological Control of Weeds

Weeds constitute quite another class of pests, and the possibility of controlling them with phytophagous insects has long intrigued entomologists. Programs of this nature cannot be undertaken without careful preparation, however, as there are several problems to be considered:

1. The weeds must have sufficient negative economic effect to justify the effort. Some weeds have desirable attributes, either as sources of nectar for honey bees or parasitoid wasps or as cover for game birds.
2. The insects must be monophagous or so narrowly oligophagous that they are unable to survive on desirable plants. For example, biological control of thistles is hampered by the presence of related, cultivated plants such as safflower and many garden flowers, which may be attacked.
3. The insects must be introduced without natural enemies from their native home and must survive the attacks of predators and parasitoids in their new home.

When scale insects (Coccoidea) were introduced to South Africa to control pricklypear cacti, for example, they were ineffective because of predation by lady beetles. In this instance insecticides were used to destroy these "beneficial" beetles, permitting the scale insects to destroy the cacti.

The best-known example of successful use of insects to control weeds concerns pricklypear cacti (*Opuntia* species) in Australia. Cacti are not native to Australia, but the introduction of pricklypears for ornamental purposes resulted in many parts of eastern Australia being overrun by stands of cacti so dense as to be virtually impenetrable for humans and livestock. In 1925, eggs of a small moth, appropriately called *Cactoblastis cactorum* (Pyralidae), were brought from Argentina. The larvae were reared on cacti, and over time some 3 billion eggs were released in the field. The result was a dramatic decline in the abundance of pricklypear cacti. Today, the cacti survive in small patches here and there, but the moths usually find and destroy them.

Parts of Australia were also overrun by a European weed, St. John's wort (*Hypericum perforatum*). In this instance the introduction of leaf beetles (*Chrysolina* species, Chrysomelidae) and several other insects achieved only slow and incomplete control of the weeds. St. John's wort has also become a pest in the western United States and Canada, where it is called Klamath weed. By the 1940s it had occupied about 2 million acres of rangeland. In that decade an intensive program was launched for control of Klamath weed, under the direction of Carl B. Huffaker, of the University of California in Berkeley, and James K. Holloway, of the U.S. Department of Agriculture. Two species of *Chrysolina* beetles were brought in and tested extensively to make sure they would not attack crop plants. They did not, and the beetles were released in California and later in Oregon, Washington, Idaho, Montana, and British Columbia, along with a borer and a gall midge. As a result, in most areas Klamath weed is no longer nearly as abundant as it formerly was.

Not every effort at weed control by insects has met with success, however. Attempts to control lantana in various parts of the tropics, for example, have had limited success. There are at present a number of ongoing projects, such as those for leafy spurge, alligator weed, and water hyacinth, the results of which are not yet clear.

Use of Genetics and Sterility in Insect Control

Reproduction in pest insects can be disrupted by releasing sterile or genetically altered insects into natural populations. This approach of using insects for self-destruction is known as **autocidal control**. Utilization of these techniques is dependent on our ability to produce economically large numbers of insects for release. Also, the insects released must be successful competitors with members of the natural population for mates.

Several genetic mechanisms exist that can, at least theoretically, be manipulated to reduce the reproductive capacity of pests. The use of **lethal mutations** involves introduction of genes that are lethal under certain conditions. When the specific conditions occur, a breakdown in normal physiological function occurs. For example, the inability to diapause could be introduced into a population, which would lead to mortality in winter. The use of **translocations** involves release of insects bearing chromosomes with translocations (gene exchange between nonhomologous chromosomes). Crosses between normal and translocation strains could result in death of the developing embryo because genetic information would be missing. These techniques appear promising but as yet have not been well tested under field conditions.

Autocidal control using **sterile insect release** has, however, been demonstrated to be practical and is currently in use. The genetic constitution of the natural population is not altered, but reproduction is disrupted. Any treatment that provides sterile but functional adult insects may be satisfactory; gamma irradiation is commonly used. Sterile males, females, or both may be released to achieve suppression, but usually only sterile males are used. After release sterile insects mate with normal, nonsterile adults in the population, thus reducing their reproductive potential.

The first and best-known sterile-insect program involved eradication and suppression of the screwworm, a fly attacking livestock. This program was initiated by E. F. Knipling and his associates in the U.S. Department of Agriculture. In 1954–1955, screwworm was eradicated from Curaçao, an island in the Caribbean, by releasing sterile males at the rate of about 400 flies per square mile per week for 3.5 months. A similar program in 1958–1959 resulted in the eradication of screwworm from the southeastern United States. Screwworm has not been eradicated from the southwestern United States because the pest ranges throughout Mexico and Central and South America, and reinfestation of fly-free areas occurs regularly. However, annual releases of sterile males now keep screwworm infestations in the Southwest to a very low level. Currently the U.S. Department of Agriculture and the Mexican government are engaged in an attempt to eradicate the pest in all of Mexico north of the Isthmus of Tehuantepec.

This approach may be useful against other insect pests, but there are some limitations to the tactic. The principal limiting factors for such programs are that (1) we must be able to sterilize insects, but without reducing their ability to compete effectively with nonsterile insects for mates; (2) we must be able to rear insects in large numbers, economically; and (3) the native population must reach low levels of abundance at some time in their seasonal life history (naturally or through suppression), so that we can release an advantageous ratio of sterile insects to nonsterile ones. The second of these factors might be circumvented if it were possible to develop a safe and effective chemosterilant that could be applied in the way that insecticides are applied. However, this remains for the future.

Cultural Techniques

Management of insect populations through **cultural techniques** involves modification of the environment to make it less attractive or suitable for pests through standard agronomic (cultural) practices. Cultural techniques do not require special machinery or equipment, but for cultural techniques to be successful a thorough knowledge of insect life history is essential. Cultural techniques depend on disrupting a favorable biological or physical condition; obviously we must know what is favorable for the insect pest before we can be disruptive. Common cultural techniques include tillage, sanitation, crop rotation, timing of harvest and planting, water management, and use of trap crops.

Tillage, or soil preparation, will influence thermal and moisture conditions as well as the physical structure of the soil. Also, insects may be killed directly or may be brought to the surface where they are susceptible to predation and desiccation. An example of an insect that is affected by tillage is the pale western cutworm, a pest of small grains. Females of this moth prefer bare soil for oviposition, and if tillage is delayed until after the oviposition period, cropland may escape attack. Also, caked soil is unsuitable for cutworm larvae, which burrow beneath the soil surface, so postponing tillage reduces survival of this stage. Lastly, high soil moisture, which can be induced by irrigation or by packing the soil, will drive cutworm larvae to the soil surface, where they are readily attacked by parasitoids.

Clean culture, or **sanitation**, involves removal of weeds or crop residues that might harbor insect pests. In general, crop pests are not monophagous, and weeds may serve as regular, or irregular, alternate hosts. The European corn borer, for example, attacks over 200 different plants, many of which are weeds. Weeds may harbor pests that infest crop plants as they become suitable hosts, or weeds may attract pests that then spread to crop plants as the weeds are consumed. Beet webworms, for example, are known to be much more destructive to sugar beet fields harboring lambsquarter and Russian thistle; apparently the adults prefer to oviposit on these plants.

European corn borers survive the winter as full-grown larvae in crop residue. Where corn borer populations are high, crop residue should be shredded and used for animal fodder. Destruction, or deep plowing of crop residue, will alleviate many insect problems. The trend toward reduced tillage and no tillage in American agriculture seems likely to exacerbate certain pest problems because crop residues remain to harbor insects.

Where parasitoids have potential to exert significant suppressive effect on pest populations, a thorough consideration of their biology should be made before recommending indiscriminate weed control. Many adult parasitoids are much more effective when they feed on nectar from flowering plants, including weeds. Thus certain weeds, especially those that do not harbor pests directly, may be valuable in that they supply nourishment to beneficial insects.

Alternating the type of crop grown at a given location, or **crop rotation**, often provides effective management of pest populations. Crop rotation is particularly effective where insect pests are not very mobile. Management of western corn rootworms, for example, is feasible through rotation of corn with a nonhost, such as soybeans. Rootworms overwinter as eggs in soil, and larvae that develop in spring and early summer have limited mobility. Thus eggs deposited in corn fields produce larvae that perish for lack of suitable food when soybean or another unsuitable host follows corn in a rotation sequence.

Insect pests may be abundant for a relatively brief period of the growing season or may be able to damage the crop only during a specific stage in growth. Thus it may be possible to **time planting or harvest** such that the crop is not available, or not in a susceptible stage, when the pests are active. Lack of susceptibility because of differences in timing also is called **phenological asynchrony**. A well-known example of phenological asynchrony involves the Hessian fly, a pest of wheat. It is practical to delay planting wheat in the autumn until after the brief period of adult activity, allowing the wheat crop to escape infestation.

The availability of free water or soil moisture influences the abundance of many insect pests. **Water management**, when combined with knowledge of insect response to water, can be used to regulate insect pest abundance. For example, many species of mosquitoes breed only in temporary pools of water. Modification of land contours to prevent pooling removes potential breeding sites. Alternatively, water can be permanently impounded in areas where temporary pooling occurs. Permanent impoundments are less likely to breed mosquitoes because fish and other predators that inhabit permanent water feed on the mosquito larvae, and also because some species will not oviposit in such habitats.

Sometimes insects can be lured away from the principal crop when a small area is planted to a more attractive plant, or if a small part of the crop is planted early. The attractant crop, or portion of the crop, is called a **trap crop**. When a trap crop is infested with insects, it is destroyed or treated in some manner to destroy the insects. The advantage of trap crop use is that it eliminates the need to control insects on the main crop, which reduces expense and insecticide contamination. For spruce beetle control it is possible to lure the beetles to a recently felled or girdled tree, which is highly attractive to beetles and serves as a trap. The trap crop (tree) may be injected with cacodylic acid, a poison containing arsenic, prior to felling or girdling. Thus the beetles are killed when they burrow into the treated tree, and healthy nearby trees are spared attack.

Physical and Mechanical Techniques

Some of the oldest, and in some cases most primitive, methods of insect pest suppression involve physical and mechanical techniques. Physical and mechanical suppression techniques are measures taken to destroy insect pests, to disrupt normal physiological function (other than with insecticides or similar chemicals), or to modify the environment to make it unsuitable for insect pests. Physical and mechanical suppression methods may be distinguished from cultural methods by the use of special equipment or operations in addition to the normal agronomic (cultural) procedures. Physical and mechanical methods often provide immediate and tangible results; hence they are popular. On the other hand, they may be costly or labor intensive and therefore not feasible for commercial agriculture. **Physical techniques** utilize physical properties of the environment in such a manner as to destroy insects, while **mechanical techniques** require the operation of machinery, or manual operations, to destroy pests. Obviously there are often occasions when machinery is required to modify the physical environment of a pest, so the distinction between physical and mechanical control is not always clear.

As is the case with many other suppression methods, thorough knowledge of insect life history is useful for effective pest suppression using these techniques. For example, we may need to know the physiological response of a particular insect to certain temperature or humidity conditions, or the behavioral response to certain visual stimuli.

Temperature treatments are useful for certain pests. Cold storage of fruit, especially when accompanied by modification of gaseous atmosphere, will kill some internal pests. Burning (flaming) crop stubble and field margins is useful for destroying both insect and weed pests. Reflection by aluminum mulch will repel certain flying pests, especially aphids; this is especially valuable when attempting to protect a crop from insect-borne plant disease.

Light traps are usually used for pest monitoring, but when employed in sufficient number they can effect significant reduction in pest abundance. Other attractants such as pheromone or visual sticky traps can remove pests similarly. Hand-picking is useful for home gardeners or in societies where labor is not expensive.

Chemical Modification of Behavior

Recent research has led to increased understanding of the chemical basis of insect behavior and ecology (see also Chapter 5). The manner in which many insect species locate mates, recognize food, and determine suitable oviposition sites may be regulated, at least in part, by chemical stimuli. When chemical stimuli have been identified and synthesized, it is sometimes possible to modify insect pest behavior to alleviate damage by using these chemicals. The chemicals that have received most attention, and that currently are most practical for insect management, are sex pheromones. Other chemicals that probably will prove useful in the future are other types of pheromones, kairomones, repellents, and feeding deterrents. Organic garden-ing techniques such as companionate planting purportedly have a chemical basis (see p. 404).

Sex pheromones are often used to monitor insect activity. A pheromone dispenser usually is placed in a sticky trap, and the number of insects captured is used to ascertain the need for, and timing of, suppression programs. Sex pheromones are available for many insect pests, especially Lepidoptera (Fig. 18–5). Examples include "Disparlure" for attracting gypsy moth, and "Grandlure" for boll weevil.

Sex pheromones also can be used to disrupt mating by saturating the atmosphere with the proper chemical stimulus. In a saturated atmosphere insects cannot orient properly and locate mates. Thus reproduction is inhibited and damage reduced. "Nomate PBW" is a commercially available sex pheromone formulation that disrupts reproduction in pink bollworm when the product is applied to cotton.

Aggregation pheromones are produced by some insect species to facilitate group attack of a host. Aggregation pheromones are best known from bark beetles, where attack by a large number of beetles allows the invaders to overwhelm the host's defenses (Chapter 5, pp. 123–125). Bark beetles also produce antiaggregation pheromones. Either type of pheromone could be useful for pest management. Aggrega-tion pheromones could be used to attract bark beetles to a trap, while antiaggregation pheromones could protect particular trees from attack.

Cockroaches produce an aggregation pheromone that can be used to attract them to a particular location or trap. Also, aggregation pheromone can be used to offset the repellency sometimes associated with insecticides and thereby increase insecticide efficacy.

Many insects mark their oviposition site with a pheromone; this is best known in certain parasitoid wasps and phytophagous flies. **Oviposition-deterring pheromones** keep insects from depositing too many eggs in a single host, which could result in over-crowding and insufficient food for developing larvae. Apple maggot and cherry fruit flies, for example, produce oviposition-deterring pheromones. When synthetic pheromone becomes available, it may be possible to apply it to fruit and thus protect the crop from injury.

When insects are attacked or otherwise disturbed, they may release **alarm pheromones** that warn other members of the species that danger is nearby. Alarm pheromones are known from bees, termites, aphids, and other insects. Aphid alarm pheromone also alerts attending ants to disturbance, and the ants rush to the defense of the aphids, which are valued by the ants as a source of sugary honeydew. Social insects such as ants and termites, and some presocial insects such as tent caterpillars, also produce **trail pheromones**. Alarm and trail pheromones conceivably could be used to disrupt pest behavior, but application thus far has been lacking.

Kairomones are interspecific messengers that benefit the receiver of the chemical stimulus but are deleterious to the producer. Predatory and parasitic insects apparently use kairomones as an aid in locating host insects. The chemical tricosane, for example, which is associated with corn earworm, attracts *Trichogramma* wasps to the pest insect. When tricosane is applied to a field containing corn earworm eggs, the wasps are stimulated to higher levels of activity, and a higher degree of parasitism results.

Plants also produce kairomones, which are used by herbivorous insects in locating a host plant or in allowing the insect to recognize the host (Chapter 9). In breeding for resistant crop plants, breeders should attempt to delete these chemical stimuli from the varietal phenotype.

Figure 18–5
A pink bollworm moth that has been attracted to a hollow fiber dispenser containing a synthetic sex attractant. Such fibers have been widely used in cotton fields to disrupt mating of these insects. In some cases males will attempt to copulate with the fiber. (Photograph by J. Running, copyright Albany International Corp.; from R. M. Silverstein, *Science*, vol. 213, pp. 1326–32. Copyright 1981 by the American Association for the Advancement of Science.)

It would not be necessary to kill insects if we could deter them from attacking us and the things that we value. Repellents and feeding deterrents, therefore, reduce the need for insect control. **Repellents** are chemicals that cause insects to make oriented movements away from the source of the chemical. The best-known repellent is deet, the active ingredient in mosquito repellent. **Feeding deterrents** inhibit feeding but do not necessarily repel insects. Currently there is considerable interest in azadirachtin, a chemical extracted from seeds of the neem tree, because it deters feeding by a large number of insect pests.

Certain "organic" gardening practices such as companionate planting reportedly result in insect repellence or deterrence. **Companionate planting** is intercropping of "repellent" or "confusing" plants with the crop plants. The best-known example of a companionate plant is marigold, which supposedly repels pest insects from crop plants or masks the attractant odors from plants, thereby preventing pests from locating their hosts. There is no scientific evidence to support the concept of companionate planting although increasing plant diversity through intercropping may be a successful practice. Some of the plant species reportedly useful as companionate plants, such as garlic and onions, do contain natural insecticides. Organic gardeners might achieve better levels of plant protection by applying aqueous extracts of these plants to kill pests as they appear.

Chemical Disruption of Physiology

The pest management tactic used most frequently in technologically advanced societies is application of chemical insecticides to disrupt the physiology of the pest insect. Some pests can be effectively managed without insecticides, and some crops require insecticide application for protection only occasionally. However, insecticides have been widely employed since about 1900, and although their use is likely to diminish in the future, it is not likely to be completely eliminated.

Why are insecticides widely used for pest management? When insect populations or damage approach the economic threshold rapidly, insecticides often are the only practical suppression tactic. In other words, it may take too long to augment the beneficial insect population, or it may be too late to initiate cultural or other suppression tactics. Insecticides, or pathogens that can be used like insecticides (microbial insecticides), can suppress pest populations quickly. Insecticides also offer a wide range of properties that make them useful for many different pest situations. For example, insecticide formulations are available for foliar, livestock, aquatic, soil, and stored-product pests. Relatively long-lasting or short-lived materials can be purchased. Finally, the cost of insecticides often is low relative to expected return on investment. For example, under some conditions each dollar invested in potato insect suppression results in a $29.00 increase in yield. A farmer may view insecticides as a good investment and insurance against unforeseen disaster.

Variation in insecticide properties often is related to the chemical origin of the product. Natural and synthetic inorganic insecticides are still available, but only to a limited extent; they were commonly used before the development of synthetic organics. Today, synthetic organic insecticides dominate the insecticide market, although natural organics are popular among some home gardeners. Biological insecticides, or microbial pathogens, have been discussed previously.

Natural inorganic insecticides are naturally occurring minerals. Cryolite is an example of an infrequently used natural inorganic that acts as a stomach poison for chewing insects. Sulfur is useful for suppressing mites, but it is more widely used as a fungicide. The toxicity of these two materials to mammals is very low.

Synthetic inorganic insecticides are chemical modifications of naturally occurring minerals. Examples include the arsenicals paris green, lead arsenate, and calcium arsenate. Paris green was first used against Colorado potato beetles about 1865. Lead arsenate became popular for gypsy moth control around 1890. Calcium arsenate was targeted principally against boll weevil. Arsenicals are very toxic to mammals.

Sorptive dusts are inorganics that are used for structural pest control and grain protection. The dusts usually are silicon dioxide powders that abrade and disrupt the wax layer of the insect cuticle, leading to desiccation. Sorptive dusts are not very toxic to mammals.

Natural organic insecticides are principally botanical in origin. Pyrethrum has been used as an insecticide for at least 2000 years in the form of finely ground *Chrysanthemum* flowers. The insecticidal properties are due to pyrethrins, which can be extracted from the flowers. It is less expensive to produce synthetic analogs of the active ingredients, called pyrethroids, than to extract them from plants. Pyrethroids are available commercially and have a low toxicity to mammals.

Water extracts of tobacco plants were used by American colonists for insect control prior to 1700. The alkaloid nicotine is principally responsible for the insecticidal properties. Nicotine is used in the form of nicotine sulfate and is an effective aphicide; however, it is quite toxic to mammals. Ryania and rotenone are additional examples of infrequently used plant-derived insecticides. Their mammalian toxicity is low to moderate.

Petroleum derivatives are useful insecticides and acaricides that are relatively nontoxic to mammals. Dormant oils are heavy oils applied when plants are dormant, because oil is toxic to foliage. Summer oils are lightweight oils that are safer for use on plant foliage.

Synthetic organic insecticides are widely used in commercial agriculture. The principal groups are the chlorinated hydrocarbons, organophosphates, and carbamates. Other groups, such as the insect growth regulators, probably will assume greater importance as environmental standards change, as the frequency of insecticide resistance increases, and as the need for selective insecticides becomes more acute.

Chlorinated hydrocarbons, also known as organochlorines, are some of the oldest and best-known insecticides. As the name implies, they contain chlorine, hydrogen, and carbon. Most notorious of the chlorinated hydrocarbons is DDT, one of the most useful, effective, low-cost insecticides ever produced. DDT also is quite persistent; poor biodegradability combined with overuse led to severe restrictions on its use in the United States starting in 1973. Its adverse effects were demonstrated particularly among predatory birds and mammals, which were especially susceptible to ingesting large quantities of insecticide by way of its increasing concentration in higher levels of food chains.

Other chlorinated hydrocarbons formerly used to a great extent include aldrin, dieldrin, chlordane, heptachlor, lindane, and mirex. Like DDT, most chlorinated hydrocarbons are quite residual and therefore useful where long-term control is desired. These materials were widely used as soil insecticides and as seed treatments, and some are still employed in structural pest control. In general, their use is restricted or prohibited in the United States, although they are employed in some other countries. Some chlorinated hydrocarbons such as methoxychlor and kelthane are not very persistent and are still in use.

The chemically unstable **organophosphates** are derived from phosphoric acid and have replaced many of the chlorinated hydrocarbons. They are related to military "nerve gas," and some are very toxic to mammals. Because they are chemically unstable, they are not very persistent. Common organophosphates include ronnel, disulfoton, dichlorvos, methyl parathion, diazinon, chlorpyrifos, and malathion. Their properties differ significantly. Malathion is a very safe insecticide and acaricide, while parathion is quite toxic to mammals. Dichlorvos is a fumigant that is frequently incorporated into pet collars. Ronnel is used on livestock. Chlorpyrifos is used almost exclusively for household pest control.

Many of the organophosphates are systemics. **Systemic insecticides** are absorbed and translocated through the animal or plant; ronnel and acephate are examples.

The **carbamates** are derived from carbamic acid. The first successful carbamate was carbaryl. It is widely used because it kills a broad spectrum of pests and its mammalian toxicity is fairly low. Other common carbamates include propoxur for cockroach control, aldicarb for field crop and ornamental pests, pirimicarb for aphids, and carbofuran for a variety of insects. Many carbamates also are effective nematicides. Some carbamates are quite selective and kill only certain insects. Their mammalian toxicity is extremely variable. The use of carbamates is increasing as selective suppression of pests is emphasized in pest management programs, and resistance to organophosphates becomes more common among pest species.

Insect growth regulators probably will assume an important role in pest management as they become more readily available. Insect growth regulators may act as hormones or antihormones and may have diverse effects, such as disruption of molting (Fig. 18–6; see also Chapter 10, p.243). Diflubenzuron, methoprene, and kinoprene are examples of commercially available insect growth regulators. Diflubenzuron interferes with cuticle formation, while methoprene and kinoprene act similarly to juvenile hormone.

How do insecticides kill insects? We have already mentioned the mode of action of some insect growth regulators, but surprisingly little is known about some insecticides. Organophosphates and carbamates are acetylcholinesterase inhibitors; thus they inhibit transmission of nerve impulses. Chlorinated hydrocarbons appear to disrupt nervous transmission also, but by causing a sodium–potassium imbalance in the neurons. Rotenone inactivates the enzyme glutamic acid oxidase. Nicotine affects acetylcholine receptors in some unknown manner. The pyrethroid mode of action is not understood, but nervous transmission is interrupted. Sorptive dusts cause desiccation by disrupting the cuticular wax layer. Petroleum oils have a different effect—a thin layer of oil impedes gas exchange, so insects suffocate. Thus insecticides provide a variety of ways to disrupt physiological processes in insects, ultimately leading to death.

Formulation and Application of Insecticides.   Discovery of a chemical that will disrupt the physiology of insects is only part of the process that will lead to plant and animal protection. An insecticide must be prepared in such a way that it is available for practical use; this process is called **formulation**. The components of insecticide formulations, and the common formulations available commercially, are

Figure 18–6
A pupa of the yellow mealworm (left), when treated with a juvenile hormone analog, produces an abnormal adult that is unable to reproduce (center). A normal adult mealworm is shown at the right. (U.S. Department of Agriculture; from *Agricultural Handbook* no. 512, by E. F. Knipling, 1979.)

Table 18–1. Components and Types of Insecticide Formulations

I. Components of commercial insecticide formulations

1. Active ingredient: the toxicant or poisonous substance responsible for insecticidal properties; usually given in % or lb/gal

2. Inert ingredient: the nonpoisonous diluting material; sometimes called carrier or solvent

3. Surfactant: surface active agent enhancing useful properties of the formulation; examples are wetting agents or stickers, which assist the spread of the liquid formulation over the target organism and promote retention, respectively; emulsifiers are surfactants that cause an even dispersion of one liquid throughout a second liquid, in which it is not soluble

4. Synergist: a nontoxic material that when added to a toxicant, produces a more toxic product by competing for, or inhibiting, detoxifying enzymes

II. Types of commerical insecticide formulations

1. Solution (S): a solid, liquid, or gas dissolved in a liquid to make a homogeneous mixture

2. Emulsifiable concentrate (EC): a liquid containing an emulsified active ingredient; when added to water, EC formulations normally turn milky white; usually contains a high concentration of toxicant

3. Wettable powder (WP): a dry formulation that is suspended in water; requires constant agitation to prevent the toxicant from precipitating; usually contains a high concentration of toxicant

4. Soluble powder (SP): a dry active ingredient that dissolves in water to form a solution; usually contains a high concentration of toxicant

5. Flowable (F): a solid or semisolid suspended in liquid with surfactants to provide the characteristics of a solution rather than a suspension

6. Dust (D): a dry formulation diluted with another dry material; usually contains a low concentration of toxicant

7. Granules (G): dry formulation similar to dust but consisting of larger particles; usually contains a low concentration of toxicant

8. Aerosol and fog: very fine droplets of carrier containing toxicant; produced by propellant, air blast, or heated gas; usually contains a low concentration of toxicant; provides little residual deposit

9. Smoke: very fine solid particles containing toxicant; produced by burning; little residual deposit

10. Vapor: vapors released from resin strips, or liquids that act as fumigants

11. Encapsulation: active ingredient is encapsulated (enclosed) to slow inactivation and prolong activity

12. Bait: an attractant, usually some type of food, that contains a toxicant

described in Table 18–1. Formulation accounts for many of the unique properties of insecticides and their wide applicability.

The equipment for delivering insecticide formulations to the target organisms is as diverse as the formulations available. Safety and convenience are important considerations for insecticide application in urban and suburban environments. **Aerosol applicators** utilize compressed gas as a propellant and are popular because no prepara-

Figure 18–7
A tractor-borne air sprayer designed for orchard use. (FMC Corporation, Jonesboro, Arkansas.)

Figure 18–8
Aerial spraying of insecticide on beans. (Exxon USA; photograph by Dan Guravich.)

Figure 18–9
Spraying vineyards with a suitably equipped helicopter. Downdraft is provided by the rotor blades. (Hughes Helicopters, Inc., Culver City, California.)

tion is required. However, they are very expensive to use. **Resin strips** release a vapor that usually is useful in enclosed areas only. **Compressed air sprayers** are inexpensive and easy to use but are useful for small applications only. **Hose end sprayers** are equally convenient and inexpensive and are useful for larger areas and small trees, but calibration is difficult.

Agricultural insecticide application is more concerned with efficacy and economics. Although the initial expenditure for application equipment may be quite high, large areas may be treated efficiently. Hydraulic sprayers are commonly used for many field crops. **Low-pressure hydraulic sprayers** are relatively inexpensive and versatile; they commonly are attached to tractors, trailers, and aircraft. **High-pressure hydraulic sprayers** are more expensive but provide better penetration of foliage. However, high pressure results in small droplet size, which can lead to dangerous insecticide drift. Drift also is a problem with **dusters**. Dusts provide good penetration of foliage but poor coverage. **Granule applicators** overcome the problem of drift because the insecticide is formulated in particles too large to be blown by the wind. However, granule applicators are expensive, and granules do not stick to foliage. **Air blast sprayers** inject liquid formulation into a stream of air. Large air blast sprayers are called mist blowers and are used to treat orchards, shade trees, and livestock (Fig. 18–7). Small backpack, or knapsack, air blast sprayers utilize the same principle but cannot be used for large trees.

Specialized applicators are produced for other uses, such as suppression of medically important insects and greenhouse pests. **Ultra-low-volume applicators** apply very small droplets of undiluted insecticide. Ultra-low-volume applications sometimes are used for suppression of mosquitoes, cockroaches, and grasshoppers. Application equipment is expensive, but little insecticide is required. **Thermogenerators**, or foggers, disperse aerosol droplets in a cloud of fog that is produced by vaporization from a heat source. Thermogenerators commonly are used to kill adult mosquitoes outdoors and greenhouse pests indoors.

In recent years aerial application of insecticides has become widespread, especially where there are broad acreages of single crops (Fig. 18–8). Planes have also been used extensively in forested areas, for example, in control of spruce budworm. Helicopters are especially useful in irregular terrain or where landing strips are not available (Fig. 18–9). Aerial spraying by an experienced pilot under suitable weather conditions may result in rapid coverage of wide expanses of crops with little drift to nontarget areas. However, aerial spraying in settled regions (for example, for mosquito control) may result in complaints from homeowners who object to such invasions of their privacy by noisy machines depositing unknown chemicals.

Disadvantages of Insecticides.   Insecticides provide us with useful, practical tools for insect pest suppression, but they do have disadvantages. These include selection for insecticide-resistant strains, outbreaks of secondary pests, and adverse effects on nontarget organisms. Resistance to insecticides may be the major factor to restrict insecticide use in the future (Fig. 17–10). Normal insect populations apparently contain some members that are resistant to insecticide. Insecticide use, especially when inducing high levels of mortality, selects for survival of an insecticide-resistant genotype. Under continued selection pressure the resistant strain becomes dominant and the population no longer is susceptible to insecticidal control. New insecticides with differing modes of action may be substituted, but the selection process may be repeated, sometimes with alarming speed. The magnitude of the insecticide resistance problem first became apparent with house fly populations in 1946–1947. Susceptibility to DDT was lost after only two years of selection. Since then, numerous species have exhibited partial or complete resistance to many different insecticides.

What can be done to prevent the development of resistant populations? Relaxation of selection pressure to preserve the susceptible genotype is an obvious solution. Can we allow some insect pests to exist in our crops? As we have already discussed in Chapter 17, when economic injury levels have been established we usually discover that a surprisingly large number of pests can be present with no economic loss. We should avoid broad-scale application of insecticide; untreated areas allow survival of the susceptible genotype and enhance survival of beneficial insects. Utilization of tactics other than chemical insecticides are useful to delay or prevent resistance. Alternating insecticides is sometimes recommended but may result in selection for resistance to several insecticides simultaneously.

As we have seen (Chapter 17), outbreaks of secondary pests may be induced by insecticide use. **Secondary pests** are pests that do not attain pest status in the absence of insecticide treatments but become damaging when insecticides are used. Some insects normally are checked by biological control agents. When insecticides are used, predators and parasites may be killed more effectively than secondary pest species; the pests then attain high, damaging population levels unimpeded by natural biological control agents. Red-banded leafroller is an example of an induced, or secondary, pest that now has attained key pest status in apple orchards. Mites commonly are induced pests, not only because their natural enemies have been destroyed, but because insecticides may induce physiological changes in the host plant and stimulate mite reproduction.

Pest species are not the only organisms that are affected adversely by insecticide application. We have already mentioned that beneficial predators and parasitoids may be killed by applications aimed at pest insects. Pollinators, such as honey bees, frequently are destroyed by insecticides. Wildlife, especially birds and fish, are sometimes killed by direct contact with insecticides or by eating insecticide-contaminated insects. Residues of persistent insecticides, such as DDT, have been shown to accumulate in predators. Tertiary consumers, such as raptors and humans, may ingest large quantities of insecticide. It is probably fair to assume that humans are healthier when not ingesting insecticide residues. Even plants that are presumably benefiting from insecticide application do not always escape injury. Even when applied at recommended dosages, inhibition of pollen germination and fruit drop may result.

Problems and Prospects

Agriculturalists are continually faced with the rather difficult task of producing food and fiber of sufficient quantity and quality to satisfy the world's consumers. At least in capitalist countries, this must be done at a profit. The problems of the marketplace are compounded by inflation, diversion of agricultural land and water to urban areas, and concern for environmental quality.

Entomologists contribute to agricultural production through plant and animal protection, and they contribute to human health indirectly through management of disease vectors. Entomologists have been successful at their task but cannot relax their efforts. The demand for food, fiber, and protection from disease will grow as the world's population increases. Our successes thus far have been achieved in fossil fuel–rich economies. However, agriculture, including pest management, now is faced with a new order of economic realities as fossil fuel becomes more expensive or unavailable. Economics will not allow frequent and expensive insecticide application, and environmental concerns will prohibit unnecessary residual insecticides. Thus a re-examination of pest management practices is under way.

Contemporary pest management programs are meeting the challenge of change in agricultural production. Insect pest management programs, based on the strategy outlined in Chapter 17 and the tactics discussed in this chapter, have been initiated in a number of crop agroecosystems—for example, cotton, alfalfa, and apples. The IPM program for cotton has been especially wide ranging. Cotton is attacked by several major pests, such as the boll weevil, pink bollworm, tobacco budworm, and cotton fleahopper. Heavy use of insecticides in the past resulted in the development of resistant strains of several of these pests as well as outbreaks of secondary pests such as aphids and spider mites. The current program in the United States relies on the cultivation of varieties of cotton that are both disease resistant and cold tolerant. These "short-season" varieties can be planted early and harvested early, before the major buildup of pest populations. Crop residues are destroyed or treated with insecticides to kill diapausing weevils. A variety of predators and parasitoids are being encouraged, including *Trichogramma* (Fig. 18-2). Nuclear polyhedrosis virus (NPV) and *Bacillus thuringiensis* (BT) have been widely tested, and the latter is in use. Pheromones of several pest species are available for trapping and for mating disruption, and sterilization techniques have met with some limited success with the boll weevil. Insecticide use has been reduced, and more selective chemicals are being employed. Cotton is a major crop in many parts of the world and is grown under a variety of conditions. The combination of tactics will inevitably vary, depending on climate, type of cotton grown, the complex of pests present, and the general economic situation. But IPM has brought about a fresh and hopeful approach to cotton problems.

Most IPM programs, like that of cotton, are crop oriented and involve a community of organisms, including producers, consumers of several levels, and competitors, each responding to physical factors in its own way. Preparation of crop life tables (such as Table 17-4) and computer modeling are essential steps in understanding these complex systems. When relevant data are fed into such models, including climatic and economic factors as well as biological data, intelligent decisions on the choice and timing of appropriate tactics become possible. The widespread availability of computers, coming as it does at a time when entomologists are adopting a profounder view of insect biology and population dynamics, gives us hope that we will be able to maintain a high level of agricultural production—and will do so at lower economic cost and with less disruption of our natural environment. We look forward to further expansion of the philosophy and practice of pest management into new ecosystems with continued success.

Summary

The insect pest manager has a number of tactics available, some of which must be initiated in advance of the occurrence of a pest situation. The use of one tactic does not preclude the use of others. The integration of tactics into an effective and economical program requires knowledge of life histories, crop requirements, and available suppression tactics. Several of these tactics follow:

1. *Regulations.* These may involve either quarantines, eradication, or containment programs that are enforced through state or federal legislation.
2. *Host resistance.* Resistant crop varieties are used frequently for pest population suppression, as in the case of wheat and the Hessian fly. Plant resistance may be either physical or chemical and may result in nonpreference on the part of the in-

sect or in antibiosis, that is, deleterious effects on insect survival. Tolerance refers to the ability of a host to grow normally while supporting a pest population that would be damaging to a susceptible host.

3. *Predators and parasitoids.* These agents have long been acknowledged to be important in naturally occurring biological control; there have also been many successful manipulations of these agents to achieve better control. Biological control programs may involve the introduction of natural enemies from other countries, the conservation of such enemies by manipulation of the environment, and/or the augmentation of predators and parasitoids by laboratory culture and repeated release in the field. Advantages of the use of natural enemies include permanence, safety, and economy. However, not every pest problem can be solved by biological control. In a forest environment, the introduction of exotic predators and parasitoids may be the most appropriate tactic. In agroecosystems, density-independent agents such as insecticides are sometimes more appropriate.

4. *Pathogens.* The potential for the use of microbial control agents may exceed that of many other pest management tactics, but development of this field has proceeded slowly, largely as a result of difficulties associated with identification, culture, and registration of these agents. Of particular potential value are viruses, one of which is commercially available for control of Lepidoptera on cotton; spore-forming bacteria, which have been used in control of Japanese beetle and other insects; and microsporidan protozoa, which can be used to control grasshoppers. Fungi are also important naturally occurring control agents but have been little used commercially. The use of nematodes in biological control has increased in recent years.

5. *Genetics and sterility.* Reproduction in pest insects can be disrupted by releasing sterile or genetically altered insects into the environment (autocidal control). Several genetic mechanisms exist that can be manipulated to reduce the capacity to reproduce. The best-known program involves eradication and suppression of screwworm flies.

6. *Cultural techniques.* These normal crop production practices may be modified in ways that render the environment less attractive or suitable for pests. Techniques include tillage, sanitation, crop rotation, alteration of time of planting or harvest, water management, or the planting of trap crops.

7. *Physical and mechanical techniques.* These differ from cultural practices in that they involve special equipment, such as cold storage, burning, mulching, trapping, or hand-picking.

8. *Chemical modification of behavior.* This is primarily brought about through the use of pheromones, some of which have been used commercially for monitoring insect activity or for disrupting behavior. Of potential value are pheromones involved in mating, in aggregation, in oviposition, in alarm, or in trail establishment. Kairomones, repellents, and feeding deterrents also have potential value in control.

9. *Chemical disruption of physiology.* This is the tactic most frequently employed in technologically advanced societies. It commonly involves the use of insecticides, which result in death of the insects. Insecticides often suppress insect populations quickly and may be economical in many situations. Available kinds of insecticides include natural inorganics, synthetic inorganics, natural organics, and especially synthetic organics such as the widely used chlorinated hydrocarbons, organophosphates, and carbamates. Insect growth regulators differ in that they act as hormone analogs and may have effects such as interfering with molting.

An insecticide must be prepared in such a way that it is available for practical use; this process is called formulation. Many types of equipment are available for application. Major objections to the use of insecticides include the possibility of the development of resistance, outbreaks of secondary pests, and destruction of beneficial species.

In addition to these pest management tactics, entomologists are also called on to control weeds with the use of phytophagous insects. Programs for the control of cacti in Australia with a moth and for the control of Klamath weed in northwestern North America with leaf beetles have been notably successful.

Insect pest management programs involving these tactics have been initiated in a number of agroecosystems. As a result of growing world population and increasing cost and shortages of fossil fuels, a constant reexamination of management practices is needed.

Selected Readings

Adkisson, P. L.; G. A. Niles; J. K. Walker; L. S. Bird; and H. B. Scott. 1982. "Controlling cotton's insect pests: a new system." *Science,* vol. 216, pp. 19–22.

Beck, S. D. 1965. "Resistance of plants to insects." *Annual Review of Entomology,* vol. 10, pp. 207–232.

Burges, H. D., ed. 1981. *Microbial Control of Pests and Plant Diseases, 1970–1980.* New York: Academic Press. 949 pp.

_____, and N.W. Hussey, eds. 1971. *Microbial Control of Insects and Mites.* New York: Academic Press. 861 pp.

Ferron, P. 1978. "Biological control of insect pests by entomogenous fungi." *Annual Review of Entomology,* vol. 23, pp. 409–42.

Huffaker, C. B., ed. 1980. *New Technology of Pest Control.* New York: Wiley. 500 pp.

_____, and P. S. Messenger, eds. 1976. *Theory and Practice of Biological Control.* New York: Academic Press. 788 pp.

Maxwell, F. G., and P. R. Jennings. 1980. *Breeding Plants Resistant to Insects.* College Park, Maryland: Entomological Society of America. 683 pp.

Quraishi, M. S. 1977. *Biochemical Insect Control.* New York: Wiley. 280 pp.

Smith, K. M. 1976. *Virus–Insect Relationships.* London: Longman. 291 pp.

Smith, R. H., and R. C. von Borstel. 1972. "Genetic control of insect populations." *Science,* vol. 178, pp. 1164–74.

Ware, G. W. 1978. *The Pesticide Book.* San Francisco: Freeman. 197 pp.

Waters, W. E., and R. W. Stark. 1980. "Forest pest management: concept and reality." *Annual Review of Entomology,* vol. 25, pp. 479–509.

Terrestrial Arthropods Other than Insects

nsects are only one of several groups of terrestrial arthropods, though much the largest one. In practice, entomologists are expected to be able to recognize all major groups of arthropods occurring on land and in fresh water and to be able to answer questions about them. Some of these organisms pose direct threats to humans and domestic animals through their bites, stings, or ability to transmit diseases. Others are important agricultural pests, especially the many species of plant-infesting mites. Arthropods other than insects play important roles in many terrestrial and freshwater ecosystems. Mites and springtails often occur in enormous numbers in the soil, and streams and lakes often teem with crustaceans.

The phylum *Arthropoda* is characterized by the presence of a chitinous exoskeleton and jointed appendages, the most anterior two or three pairs of which are modified for feeding. Basically the body is segmented, with some repetition of parts in each segment, but much of the segmentation may be lost, as in ticks, for example. Other basic features include a ventral, ganglionated nerve cord and dorsal, tubular heart associated with an open circulatory system. Structurally the arthropods are diverse almost beyond belief. Trilobites thrived in the ocean half a billion years ago, and when the continents became available for occupation, arthropods were among the first to take advantage of them.

There is by no means universal agreement as to the best classification of the arthropods. Indeed, it is claimed by some that the group may be polyphyletic, that is, that various included groups may not share a common ancestry (see the books of Manton and of Gupta, cited at the end of this appendix). The late S. M. Manton, of the British Museum (Natural History), suggested that three phyla should be recognized, with the trilobites forming still a fourth group of doubtful relationship to the others. We shall follow a more conservative classification and consider only two major groups (subphyla), the *Chelicerata* and the *Mandibulata*, the names being based on a fundamental difference in the mouthparts (although there are many other differences). In any classification, these groups stand in contrast to one another; they diverged long ago in geologic time and have evolved in parallel for many millions of years. Both groups have produced a diversity of forms that exploit the environment in many different ways.

Chelicerata

Chelicerates lost their antennae at an early stage of evolution and came to acquire mouthparts consisting of **chelicerae**, which are not jawlike but are in the form of pincers, fangs, or piercing organs. Just behind the chelicerae are the **pedipalps**, which are often leglike but may be modified for a variety of functions. These in turn are followed by four pairs of legs. These appendages all arise on a fused head and thorax (cephalothorax, or prosoma). In terrestrial forms, the remaining segments lack appendages (ex-

Figure A–1
A wolf spider (Arachnida, Araneae) consuming a grasshopper. (Photograph by James E. Lloyd, University of Florida.)

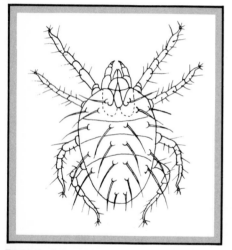

Figure A–2
A typical plant-infesting mite (Arachnida, Acari). Such mites are usually barely visible to the unaided eye.

cept sometimes terminal spinnerets) and constitute the abdomen (or metasoma). None of the terrestrial chelicerates has compound eyes, but many have two or more simple eyes at the anterior end of the body.

Terrestrial chelicerates all belong to one class, the *Arachnida*, a group that includes spiders (Fig. A–1), mites (Fig. A–2), scorpions, and other groups outlined in Table A–1. The study of arachnids constitutes the science of **arachnology**, while the study of mites in particular is called **acarology**. These fields are outside the scope of this book, but several good books are available (see references at the end of this appendix). With the exception of many mites, most arachnids are predators on invertebrates. Many mites are also predators, but others feed on plants, are scavengers, or are parasitic on insects or vertebrates. We have had occasion to mention mites as agricultural pests (pp. 373, 409), and indeed they sometimes pose serious problems, especially since chemicals applied for insect control sometimes cause biotic release of mites. The two-spotted spider mite is an especially serious pest of vegetable crops; the European red mite, of fruit trees. These and other phytophagous mites belong to a group called "spider mites" (family Tetranychidae). The mites that attack warm-blooded animals can cause various skin conditions, including mange and the well-known "chigger bites" (chiggers being immature mites of certain species). Ticks may also produce skin reactions as well as debilitation of livestock; they also transmit several diseases of humans and domestic animals, including Rocky Mountain spotted fever and Texas cattle fever. Although spiders are more often noticed than mites and ticks, as a group they have far less impact on human health and economy. The danger of the few poisonous species (in the United States, chiefly the black widow and the brown recluse) is more than offset by the role spiders play in reducing populations of flies and other small insects—and by the fascination of their colors and behavior.

Mandibulata

Mandibulates differ in many ways from chelicerates. Antennae are present, sometimes (in Crustacea) two pairs. Many have clusters of lateral eyes on the head, which may form true compound eyes, as in insects. The major mouthparts consist of mandibles that are jawlike, working in most cases from side to side. These are followed by two pairs of maxillae, which assist in manipulation of the food and often bear sensory palpi. In some groups (such as insects) the second maxillae are fused basally to form a labium, while in others (such as millipedes) the first maxillae are fused to form a labiumlike structure, and the second maxillae are absent. There is general agreement that the mandibles are not homologous to the chelicerae oɪ the maxillae to the pedipalps; embryologically, these structures are associated with different body segments.

Division of the body into tagmata is variable within the Mandibulata. In some groups (certain microscopic Crustacea) there is essentially only one tagma, while in others (such as crayfish) a cephalothorax and abdomen are present. Among terrestrial forms, when there are two tagmata they consist of head and body (as in centipedes); when there are three they consist of head, thorax, and abdomen (as in insects). Primitively, most body segments had leglike appendages, but these have undergone various modifications and reductions. The Mandibulata themselves fall into two major groups, a largely aquatic gill-breathing group, the *Crustacea*, and a mostly terrestrial group that breathes via tracheae, conveniently termed the *Tracheata*. Although crustaceans more usually fall within the province of invertebrate zoology rather than entomology, entomologists are expected to have at least a superficial grasp of the group, more especially of its terrestrial members, the sowbugs and pillbugs.

Table A–1. The Major Groups of Arachnida			
Order	General features	Feeding behavior	Importance to humans
SCORPIONES Scorpions	Abdomen with a segmented "tail" terminating in a poison gland Pedipalps large, pincerlike	Predaceous; seize prey with pincers; may also sting it	Sting may be painful, in a few species may be fatal
UROPYGI Whipscorpions Vinegaroons	Abdomen terminates in a long, bristlelike flagellum First legs very slender, whiplike Pedipalps very large, robust	Predaceous; pedipalps crush and macerate prey	Not poisonous but able to pinch and to spray irritating chemicals, chiefly acetic acid
PSEUDOSCORPIONES Pseudoscorpions	Pedipalps large, pincerlike, but abdomen short, rounded, lacking a "tail" of any kind Not over 12 mm in length	Predaceous, chiefly in litter and crevices	No major importance Sometimes phoretic on insects
SOLFUGAE Sun spiders Windscorpions	Very large, pincerlike chelicerae, directed forward Pedipalps leglike Abdomen fully segmented	Predaceous, mainly nocturnal; mainly occur in dry areas	No major importance Can pinch, but inject no toxins
OPILIONES Harvestmen Daddy longlegs	Body a single, compact unit; segmented abdomen Legs extremely long and slender; pedipalps leglike but shorter than legs	Predaceous and scavengers	No major importance May sometimes be beneficial, feeding on mites and small insects
ARANEAE Spiders	Abdomen unsegmented, separated from cephalothorax by a narrow stalk; chelicerae fanglike; abdominal spinnerets spin silk	Predaceous, either pursuing or stalking prey or trapping it in webs	Large spiders will sometimes bite, and a few are poisonous (e.g., black widow) Majority are mildly beneficial, feeding on small insects
ACARI Mites, ticks	Body a single, compact unit, unsegmented; not over 3 cm in length, usually much smaller; chelicera and pedipalps form a compact unit	Predaceous, parasitic, scavengers, or phytophagous	Ticks are ectoparasites of vertebrates and may transmit diseases Many mites are serious pests of plants, others predators, parasites, or scavengers

Crustacea.   Crustaceans are extremely diverse in structure, but generally have two pairs of antennae, maxillae that are not fused to form a labium or similar structure, and gills that function to extract oxygen from water. This type of mandibulate abounds in the ocean and in fresh waters, and vary in size from water fleas barely visible to the naked eye to giant crabs that span nearly three meters between the tips of their out-stretched claws. Water fleas, ostracods, fairy shrimps, and other minute crustaceans make up a major element in the plankton of lakes and streams and provide an impor-

Class	No. of tagmata	No. pairs walking legs	Lateral eyes	Dorsal ocelli	Musculated antennal segments
CHILOPODA Centipedes	2	15–100+	usually +	–	Many
DIPLOPODA Millipedes	2	17–180+	usually +	–	Several
PAUROPODA Pauropods	2	9	–	–	Several
SYMPHYLA Symphylids	2	12	–	–	Many
COLLEMBOLA Springtails	3	3	usually +	–	4–6
PROTURA Proturans	3	3	–	–	None
DIPLURA Diplurans	3	3	–	–	Many
INSECTA Insects	3	3	True compound eyes	Usually +	2

Table A–2. The Major Groups of Tracheata

tant source of food for larger organisms. These microscopic crustaceans are often called *Entomostraca*, although this is an unnatural grouping of diverse forms and has no taxonomic status.

Larger crustaceans, most of which have a hard carapace, belong to a group called the *Malacostraca*. The best-known members of this group are the crabs, lobsters, and crayfish. One order, the *Isopoda*, includes a number of common terrestrial forms, the sowbugs (Fig. A–3), alternately called pillbugs because of the ability of many of them to roll in a ball, or woodlice because of their propensity to inhabit damp, rotting wood. Since they breathe by gills, sowbugs are restricted to damp situations, often under

Other major features	Feeding behavior	Importance to humans
First pair of appendages behind head form poison fangs Second maxillae fused to form labium (Fig. A–4 a and b)	Predaceous	Larger species have a painful bite
Two pairs of legs on each body ring First pair of maxillae fused to form a labiumlike structure; second maxillae absent (Fig. A–4d)	Scavengers or phytophagous	No major importance Sometimes minor pests in greenhouses
Mouthparts as in millipedes Not exceeding 2 centimeters in length	Scavengers in soil and leaf litter	No major importance
Second maxillae fused to form labium Spin silk from terminal spinnerets (Fig. A–4c)	Phytophagous or scavengers	Sometimes damaging to seedling plants: ''garden centipede''
Only six abdominal segments, bearing furcula, retinaculum, and collophore (Fig. A–5); true labium, as above Entognathous; tarsi unsegmented, fused with tibiae Embryonic cleavage is total	Phytophagous or scavengers, mainly in soil	May attract attention because of great abundance, e.g., ''snow fleas'' ''Lucerne flea'' is a minor pest of plants
Mandibles styletlike; true labium Add segments as they molt Front legs thrust forward, sensory Tarsi unsegmented Entognathous	Occur deep in soil; probably predaceous	No major importance
Chewing mouthparts, mandibles two-segmented; true labium Entognathous; tarsi unsegmented May have terminal pincers or long cerci (Fig. A–6)	Predaceous or scavengers, mainly in soil	No major importance
Ectognathous; true labium; mandibles not segmented Tarsi segmented Majority have wings	Very diverse	May attack crops, domestic animals, or humans Many are beneficial or of no major importance

rocks or decaying vegetation. They have numerous legs, at least seven pairs, and only one pair of well-developed antennae, the second pair being vestigial. Although these crustaceans are often seen by gardeners and homeowners, they are scavengers and pose no threat to humans.

Tracheata. Most Tracheata have spiracles and tracheae somewhat like those of insects (Chapter 1) but these vary greatly in form and position and have undoubtedly arisen several times independently among these various terrestrial groups (outlined in Table A–2). More fundamental features include the presence of only one pair of anten-

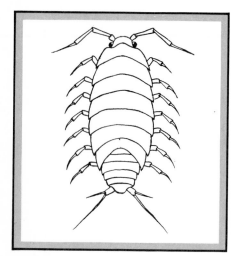

Figure A–3
A sowbug (Crustacea, Isopoda).

nae and the presence of a "lower lip," the fused first or second maxillae. Primitively the tracheates have a pair of legs on most body segments, but in two groups, the millipedes and pauropods, there has been a doubling of body segments so that each appears to have two pairs of legs. Also, several groups show a reduction or modification of legs beyond the third segment behind the head. When this occurs, and walking is concentrated in the three segments behind the head, this region is spoken of as the thorax, the remainder as the abdomen. The mode of progression of many-legged arthropods such as centipedes differs greatly from that of six-legged forms such as insects. Some larval insects have, of course, secondarily reacquired abdominal appendages, as in the prolegs of caterpillars. But in general the abdomen has become strongly differentiated from the thorax, a part of the body specialized for locomotion, while the abdomen has become modified for reproduction and for containing the major organs of digestion and excretion. Without much doubt the evolution of the thorax was major prerequisite for the development of wings.

As we have seen (Chapter 2), two orders of primitively wingless insects retain leg rudiments on the abdomen. This is also true of three other classes of six-legged tra-

Figure A–4
Some centipedes and millipedes: (a) the house centipede (Chilopoda); (b) a typical centipede (Chilopoda); (c) a millipede (Diplopoda); (d) a garden centipede (Symphyla).

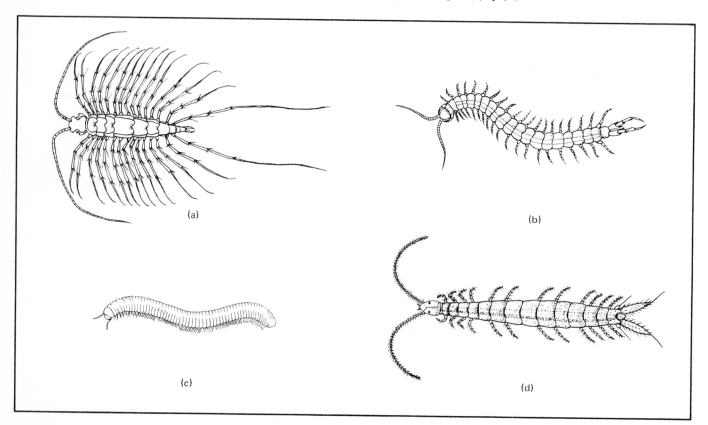

(a)

(b)

(c)

(d)

cheate arthropods, the Collembola, Protura, and Diplura. Why are not these also considered to be insects? Actually they often are, but there is reason to regard them as somewhat apart from the main line of insect evolution. All have certain features quite different from those of true insects (Table A–2), suggesting that they represent "early experiments in six-leggedness" that led to nothing else. Of the three, only the Collembola attract much attention, as they are sometimes extremely abundant in the soil, on still water, or on snow, and a few have received some notoriety as pests of plants. When these three groups are placed with the insects, as they sometimes are, they are often placed in a subclass called the Entognatha, with reference to the fact that the mouthparts are surrounded by folds of the head and thus somewhat internal. However, this obscures the fact that the springtails, proturans, and diplurans have little else in common with one another or with true insects. Recent arguments for and against the inclusion of these groups in the Insecta have been presented by Kristensen (1981) and by Boudreaux (1979).

Selected Readings

Boudreaux, H. B. 1979. *Arthropod Phylogeny with Special Reference to Insects.* New York: Wiley. 320 pp.

Cloudsley-Thompson, J. L. 1968. *Spiders, Scorpions, Centipedes, and Mites.* New York: Pergamon Press. 278 pp.

Foelix, R. F. 1982. *Biology of Spiders.* Cambridge, Mass.: Harvard University Press. 306 pp.

Gertsch, W. T. 1979. *American Spiders.* Second edition. New York: Van Nostrand Reinhold. 289 pp.

Gupta, A. P., ed. 1979. *Arthropod Phylogeny.* New York: Van Nostrand Reinhold. 762 pp.

Kaston, B. J. 1978. *How to Know the Spiders.* Third edition. Dubuque, Iowa: Brown. 272 pp.

Krantz, G. W. 1978. *A Manual of Acarology.* Second edition. Corvallis, Oregon: Oregon State University Bookstore. 509 pp.

Kristensen, N. P. 1981. "Phylogeny of insect orders." *Annual Review of Entomology,* vol. 26, pp. 135–57.

Levi, H. W., and L. R. Levi. 1968. *Spiders and their Kin.* New York: Golden Nature Guide, Golden Press. 160 pp.

Manton, S. M. 1977. *The Arthropoda: Habits, Functional Morphology, and Evolution.* Oxford, England: Clarendon Press. 527 pp.

McDaniel, B. 1979. *How to Know the Mites and Ticks.* Dubuque, Iowa: Brown. 335 pp.

Savory, T. H. 1977. *Arachnida.* Second edition. New York: Academic Press. 340 pp.

Schmitt, W. L. 1965. *Crustaceans.* Ann Arbor: University of Michigan Press. 204 pp.

Snodgrass, R. E. 1962. *A Textbook of Arthropod Anatomy.* Ithaca, N.Y.: Comstock, Cornell University Press. 363 pp.

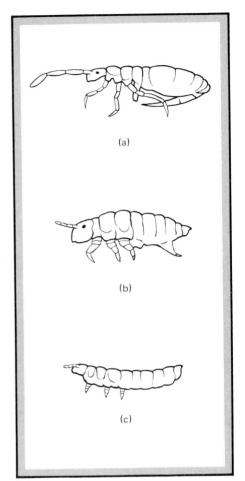

Figure A–5 (above right)
Three springtails (Collembola). (a) A species typical of those living on top of the soil; note the long legs, the terminal spring (furcula) and the tubular adhesive organ (collophore) just behind the hind legs. Species living within the soil tend to have short legs and antennae, and the spring may be short or absent (b and c). (From H. Evans, 1968, *Life on a Little-Known Planet,* copyright E. P. Dutton & Co.).

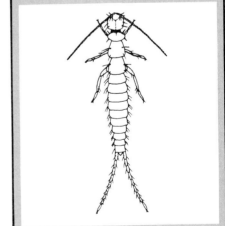

Figure A–6 (below right)
A dipluran. These delicate arthropods occur in the soil or under stones and litter in damp places. They are sometimes regarded as true insects.

Glossary

Accessory gland: A gland associated with reproductive organs of either males or females and producing substances accompanying the sperms or eggs.

Action potential: The depolarization of a nerve cell, shown as a spike on an oscilloscope.

Active space: The zone of pheromone concentration within which a response is elicited.

Aedeagus: The sclerotized median intromittent organ of a male insect.

Aeropile: The opening in the chorion (egg shell) through which air enters, often covered by a plastron.

Akipokinetic hormone: A peptide hormone released from the corpora cardiaca and influencing the fat body and the flight musculature.

Alary muscles: Muscles in the dorsal diaphragm, the contractions of which induce the flow of blood into the hemocoel surrounding the heart.

Allatectomy: An operation resulting in removal of the corpora allata.

Allometric growth: A genetically determined tendency for a certain body part to grow at a more rapid rate than other parts.

Allomone: An external chemical signal that acts between different species to benefit the producer—for example, to repel a predator.

Ametabolous: Without metamorphosis; that is, changing little in form during the course of growth and molting.

Analogy: Similarity in function filling a common need but having a different evolutionary origin.

Anamorphosis: Development of an organism in which one or more body segments are added posteriorly at each molt.

Anemotaxis: Orientation with respect to currents in air.

Anisomorphal: A defensive allomone of the walkingstick *Anisomorpha*.

Antibiosis: Any deleterious effect on insect survival resulting from feeding on a resistant host.

Aorta: A blood-containing tube that extends forward from the heart and is open anteriorly.

Apodeme: An invagination of the exoskeleton that serves as a point of muscle attachment.

Apolysis: Retraction of the epidermal cells from the inner surface of the endocuticle, the first step in molting.

Aposematism: Possession of vivid coloration that identifies an animal as having distasteful or unpleasant properties.

Appetitive behavior: Searching behavior of variable pattern, seeking an appropriate stimulus.

Apposition eye: A type of compound eye occurring in diurnal insects, in which each ommatidium is surrounded by a shield of pigment.

Apterygote: A wingless insect of a group believed never to have possessed wings in its past history.

Arolium: A padlike structure at the tip of the insect leg, between the claws.

Associative learning: Acquisition of the capacity to associate a stimulus with a reward or punishment.

Asynchronous flight muscles: Flight muscles in which contraction is not synchronized with the reception of nervous stimuli.

Autocidal control: The use of insects for self-destruction, chiefly by release of sterile individuals.

Autogenous: In blood-feeding insects, the ability to produce eggs without taking blood.

Axillary sclerite: A small sclerite at the wing base, articulating with the thorax.

Axon: A fiber of a nerve cell that carries nerve impulses away from the cell body.

Basalare: A small sclerite in the upper part of the pleuron that articulates with the axillary sclerites.

Basement membrane: A noncellular sheath separating the epidermal cells from the hemolymph.

Batesian mimicry: Resemblance of a palatable species to one that is unpalatable or has effective defenses.

Binominal nomenclature: The system of naming organisms with two names, generic and specific.

Biological control: The employment of biotic agents, such as predators, parasitoids, and disease organisms, to control populations.

Biotic release: The sudden release of a population from its natural enemies, often resulting in a population explosion.

Biotype: A population of a species that differs genetically from another population with respect to host affiliation (also called "host race").

Bivoltine: Having two generations per year.

Blastoderm: The thin cellular layer that surrounds the yolk of an egg.

Bombykol: The sex attractant pheromone of the female silkworm moth.

Brevicomin: An aggregation pheromone of the bark beetle *Dendroctonus brevicomis*.

Buccal cavity: The opening enclosed by the mouthparts, leading to the true mouth and the pharynx.

Bursa copulatrix: A pouch on the median oviduct of the female that receives the aedeagus of the male.

Bursicon: The hormone controlling tanning and expansion, produced by neurosecretory cells of the brain.

Campaniform sensillum: A sense organ consisting of a dome-shaped portion of the cuticle with associated sensory neuron; perceives stresses in the cuticle.

Cantharidin: A defense allomone of blister beetles (also known as "Spanish fly").

Cell (of the wing): A thin, membranous area surrounded by veins.

Cercus (plural, *cerci*): An antennalike sensory appendage arising from the posterior end of the abdomen.

Cervix: The largely membranous neck region of an insect, between head and thorax.

Chelicera (plural, **chelicerae**): One of the major elements in the mouthparts of spiders and related arthropods; not jawlike, but in the form of fangs, pincers, or piercing organs.

Chemoreceptor: A sense organ modified for the reception of chemical stimuli.

Chemotaxis: Orientation with respect to a chemical gradient.

Chitin: The tough, insoluble polysaccharide making up a major part of the insect procuticle.

Chordotonal organ: An elongate sense organ attached to the inner surface of the body wall and sensitive to stretching and to vibrations.

Choriogenesis: Formation of the shell (chorion) of the egg.

Circadian rhythm: An endogenous rhythm involving a response at about 24-hour intervals.

Circulatory virus: A virus that circulates within the body of an insect before being introduced into a new host.

Cleptoparasite: A "thief parasite," one that consumes the food stored by another insect in a nest.

Climatic release: Release of climatic restraints, such as a period of favorable weather or entry into a favorable region, resulting in population increase.

Clypeus: A sclerite on the front of the head, above the labrum.

Coevolution: An evolutionary change in a trait of individuals of one population in response to a trait of individuals of a second population, followed by an evolutionary response of the second population to a change in the first.

Colleterial gland: An accessory gland of the female that produces the ootheca.

Communication: The production of a signal by an individual that influences the behavior of another individual and that is mutually beneficial.

Companionate planting: The intercropping of certain repellent plants with crop plants.

Competitive exclusion principle: The concept that two species cannot long coexist if they have identical niches.

Complete metamorphosis: Striking changes between larva and adult, with an intervening pupal stage.

Conditioning: *See* Associative learning.

Contest competition: Competition involving aggressive interactions between individuals.

Coprophagous: Feeding on fecal material.

Cornicle: One of a pair of tubelike processes on the abdomen of aphids; secretes an allomone and an alarm pheromone.

Corpus allatum (plural, *corpora allata*): A small endocrine gland situated behind the brain, the source of juvenile hormone.

Corpus cardiacum (plural, *corpora cardiaca*): A small organ of nervous origin just behind the brain, associated with storage and release of PTTH and other hormones.

Corpus pedunculatum (plural, *corpora pedunculata*): *See* Mushroom body.

Coxa: The most basal segment of the insect leg, articulating with the thorax.

Crop: An expansible part of the foregut that holds food until it can be passed into the midgut.

Crypsis: Close resemblance of an animal to its physical or biotic background (also called protective coloration).

Cultural control: Modification of the environment—for example, by tillage—to make it less attractive to pests.

Cuticle: The noncellular outer portion of the integument.

Cuticulin: The tough, insoluble substance making up the outer surface of the epicuticle, containing cross-linked lipid and protein molecules.

Cytoplasmic polyhedrosis virus (CPV): A virus that develops in the cytoplasm of host cells, chiefly in the midgut.

Darwinian fitness: Differential reproduction, in terms of the number of genes an individual passes to the next generation.

Density-dependent factor: A factor that causes a level of mortality that varies with the number of individuals in the population.

Density-independent factor: A factor that causes a level of mortality that is unrelated to population density.

Deutocerebrum: The middle section of the brain, which innervates the antennae.

Diapause: A state of arrested behavior, growth, and development that occurs at one stage in the life cycle.

Direct pest: A pest insect that attacks a part of a plant that is harvested, as contrasted to an indirect pest.

Distal: Referring to the part of an appendage that is farthest from the body.

Dorsal: Referring to the upper surface (back) of an animal.

Dorsal diaphragm: A muscular sheet underlying the heart which assists in the flow of blood.

Dorsal longitudinal muscles: Muscles running longitudinally, dorsally in insect segments, in the thorax powering the downstroke of the wings of most insects.

Dorsoventral muscles: Muscles inserting on the dorsum of the thorax and originating ventrally, powering the upstroke of the wings of most insects.

Dorsum: The upper surface (back) of an animal.

Dufour's gland: An exocrine gland on the ventral, posterior part of the abdomen of female Hymenoptera, the source of pheromones serving diverse functions.

Ecdysis: Splitting and casting off of the old cuticle, the major event in molting.

Ecdysone: A molting hormone, secreted by the prothoracic glands.

Eclosion: Hatching of the egg, or emergence of the adult insect at the terminal molt.

Ecological homolog: One of two or more species having most niche parameters in common.

Economic injury level (EIL): The level of damage to a crop that is equal in value to the cost of suppressive measures.

Economic threshold (ET): The level of damage by a pest that serves to warn the agriculturalist of impending problems.

Ecosystem: A biological community considered in relation to its physical environment.

Egg-development neurosecretory hormone (EDNH): A product of the brain neurosecretory cells that stimulates vitellogenesis in the female mosquito.

Ejaculatory duct: A median duct that carries the sperm from the internal reproductive system to the exterior.

Elytron (plural, *elytra*): The hardened front wing of a beetle.

Encapsulation: The enclosure of a parasitoid larva within the blood of the host by a layer of hemocytes.

Endocrine gland: A gland that discharges its products (hormones) to the inside (as contrasted to an exocrine gland).

Endocuticle: The inner zone of the procuticle, softer and lighter in color than the exocuticle.

Endogenous activity: Nervous discharges that arise spontaneously, in the absence of stimulation.

Endopterygote: An insect that develops through the immature stages as a larva with wings retained internally as imaginal discs.

Entomophagous: Feeding on insects.

Entomopox virus: A virus that multiplies in the cell cytoplasm of fat body and blood cells in a variety of insects.

Epicuticle: The outer zone of the insect cuticle, rich in lipid and protein and lacking chitin.

Epidermis: The single outer cell layer of the body, which secretes the cuticle.

Epimorphosis: A type of development in which the insect emerges from the egg with its full complement of body segments (opposite of anamorphosis).

Equilibrium position (EP): In insect bioeconomics, the average density of a potential pest on a specific crop.

Esophagus: A tubular portion of the foregut, behind the pharynx.

Eusociality: A type of social behavior involving overlap of generations, cooperative brood care, and a caste system in which many colony members are sterile.

Exocrine gland: A gland that discharges its products to the outside (as contrasted to an endocrine gland).

Exocuticle: The outer portion of the procuticle, generally harder and darker than the inner portion (endocuticle).

Exopterygote: An insect that retains its wing pads externally through its immature stages.

Exoskeleton: A skeleton external to the remainder of the body, the muscles attaching to its inner surface.

Extrinsic: Having its origin outside the limits of an organ with which it is associated.

Farnesene: An alarm pheromone of aphids, secreted from the cornicles.

Fat body: Accumulation of large cells in the hemocoel that store metabolites and are centers of intermediary metabolism.

Femur (plural, *femora*): The third segment of the insect leg, beyond the trochanter and before the tibia.

Fibril: The contractile unit of a muscle cell (fiber).

Filter chamber: A modification of the gut of many Homoptera (such as aphids), permitting much water and some carbohydrates to bypass the midgut.

Filter feeder: An insect that seines particles from water by means of brushes or webs.

Fixed action pattern: A segment of behavior performed in a stereotyped, species-specific manner.

Flagellum: The outermost part of the antenna, beyond the scape and pedicel, usually divided into many subsegments (flagellomeres).

Follicle: A tubule of the testis in which sperm are produced.

Foulbrood: A bacterial disease of honey bee larvae and pupae.

Founder effect: Speciation resulting from the establishment of a small population in an entirely new area and the subsequent divergence of the resulting population from the parent stock.

Frontalin: A sex attractant pheromone of male bark beetles.

Galea: An apical lobe of the maxilla of an insect.

Gall: An abnormal growth on a plant, produced by stimulation of an insect or other organism and housing that organism.

Ganglion (plural, *ganglia*): A mass of nervous tissue, the basic anatomical unit of the central nervous system.

Gastric caecum (plural, *caeca*): A fingerlike, anterior extension of the midgut that serves a function in food absorption.

Genitalia: Structures associated with the release of sperm or eggs.

Germ band: A thickening of the blastoderm that produces the embryo.

Germarium: An area at the tip of the sperm follicles or ovarioles where sperm or egg formation is initiated.

Giant axon: A large-diameter axon of an interneuron that traverses several body segments and conducts messages quickly.

Glial cell: A cell surrounding the axon of a neuron.

Gonopore: The external opening of the reproductive tract.

Granulosis virus (GV): A virus that multiplies in both the nucleus and cytoplasm of host cells, usually in the fat body.

Habituation: Learning not to respond to a stimulus that provides no reward or punishment.

Hair pencils: Tufts of fine setae serving to dust pheromone-coated particles onto a member of the opposite sex.

Haltere: A modified hind wing of a fly (Diptera), acting to maintain flight stability.

Haplodiploidy: A type of parthenogenesis in which males are produced from unfertilized eggs and are therefore haploid, while the females are diploid.

Heart: A muscular tube extending dorsally and longitudinally through the insect abdomen, continuous with the aorta, serving in circulation of blood.

Hematophagous: Feeding on blood.

Hemelytron (plural, **hemelytra**): The forewing of an insect that is sclerotized basally but membranous apically (literally, half an elytron; applied chiefly to Hemiptera).

Hemimetabolous: Having incomplete metamorphosis, that is, showing gradual change from molt to molt, with externally developing wing pads.

Hemocoel: The blood-filled body cavity.

Hemocytes: Blood cells.

Hemolymph: The "blood" of insects, combining functions of the lymph and blood of vertebrates (other than respiration).

Holometabolous: Having complete metamorphosis, passing through egg, larval, pupal, and adult stages.

Homeostasis: Maintenance of a functionally steady state in the body, in the colony of social insects, or in an ecosystem.

Homology: Similarity in structure resulting from having had a common evolutionary origin.

Honeydew: The liquid excretions of sucking insects (Homoptera), consisting largely of water and sugars.

Hormone: An internal chemical signal produced by an endocrine gland and carried to the tissues by the hemolymph.

Host race: A population of a species that shows a genetically determined preference for a particular host plant or animal species.

Hydrostatic skeleton: Maintenance of body form by the pressure exerted by muscles on a fluid-filled body cavity, most important in soft-bodied larvae.

Hypermetamorphosis: A type of development in which there are two or more quite distinct larval forms sequentially.

Hyperparasitoid: An insect that is a parasitoid of a parasitoid.

Hypopharynx: A tonguelike structure in the buccal cavity, associated with the labium.

Ileum: The anterior part of the hindgut, preceding the rectum.

Imaginal disc: A group of cells set aside in the embryo and maintained through the larval stage as a center of development of adult structures.

Imago: The terminal instar, or adult.

Inclusion body: A protein crystal that encloses an insect virus.

Inclusive fitness: Net genetic representation of an individual in succeeding generations, through personal reproduction and that of individuals bearing identical genes.

Incomplete metamorphosis: Slight changes from molt to molt until wings and genitalia are fully formed in the adult.

Indirect pest: A pest insect that feeds on a part of the plant that is not marketed.

Insect growth regulator (IGR): A substance produced by a plant that mimics or antagonizes an insect hormone.

Insight learning: The ability to combine learned behavior from diverse experiences to solve a problem.

Instar: The stage of an insect's development between molts.

Instinct: Behavior performed without previous experience and without interaction with other members of the species.

Integrated pest management (IPM): An approach to the control of pests (insects, diseases, weeds) in which all available techniques are evaluated and integrated into a unified program.

Interference competition: Competition in which individuals are prevented from feeding, mating, or laying eggs as a result of the presence of other individuals.

Interneuron: A nerve cell located within the central nervous system and serving to connect other neurons.

Intersexual selection: Natural selection involving choices between the sexes, often on the basis of courtship displays.

Intrasexual selection: Natural selection involving competition among members of one sex of a species, usually for mates.

Intrinsic: Located entirely within an organ (as contrasted to extrinsic).

Ipsenol: An aggregation pheromone of bark beetles of the genus *Ips*.

Johnston's organ: An organ in the pedicel of the antenna, consisting of a cluster of chordotonal sensilla.

Juvabione: An insect growth regulator occurring in certain trees and causing abnormal development of insects feeding on the tree.

Juvenile hormone (JH): A hormone secreted by the corpora allata that maintains juvenile features in immature insects and controls certain aspects of adult physiology and behavior.

Kairomone: An interspecific chemical messenger that benefits the receiver but not the emitter.

Kin selection: Natural selection that involves inclusive fitness.

Kinesis: An undirected movement in which the speed of movement or the frequency of turning depends on the intensity of stimulation.

K-strategist: A species characterized by a low reproductive rate, increased survival mechanisms, minor tendency to disperse, and often a relatively large body size (as compared to an *r*-strategist).

Labium: The third set of mouthparts of insects (or underlip), located behind the maxillae.

Labrum: A flaplike structure anterior to the mouthparts, below the clypeus.

Latent learning: Conditioning in which the reward occurs some time following receipt of the stimulus.

Life table: A tabulation of the life stages of an insect with a cumulative record of mortality and survival.

Light compass orientation: Orientation in which a constant angle with a light source (usually the sun) is maintained.

Locustol: A primer pheromone of the desert locust that triggers development from the solitary to the gregarious form.

Malpighian tubule: An excretory tubule, opening into the gut at the junction of the midgut and hindgut.

Mandible: One of the most anterior pair of insect mouthparts, often jawlike and working from side to side.

Maxilla: One of a pair of mouthparts behind the mandibles and before the labium, bearing the maxillary palpi.

Medial (or **median**): Referring to the center, usually the midline, of an animal.

Meroistic ovary: An ovary possessing nurse cells that are connected to or accompany the oocytes.

Mesothorax: The middle segment of the insect thorax.

Metathorax: The most posterior of the three segments of the thorax.

Micropile: A pore in the chorion through which sperm enter.

Migration: A form of dispersal involving long-distance movements under at least partial control of the insects.

Mimicry: Presence of a pattern in a palatable species that closely resembles the pattern of an unpalatable species (= Batesian mimicry) (*see also* Müllerian mimicry).

Molting hormone (MH): *See* Ecdysone.

Monoculture: A uniform stand of one kind of crop plant.

Monophagous: Feeding on a single plant or animal species.

Müllerian mimicry: Presence of a similar aposematic pattern in unrelated, distasteful or poisonous species.

Multivoltine: Having several generations a year.

Mushroom body: A complex fiber tract in the anterior part of the brain, often suggesting the shape of a mushroom, associated primarily with the integration of sensory information.

Mycetome: A specialized internal organ that houses symbiotic microorganisms.

Mycoplasma: A pathogenic microorganism that passes through bacterial filters but has certain features in common with bacteria.

Myiasis: Infestation with the maggots of flies.

Myogenic flight muscles: Flight muscles that contract repeatedly as a result of mechanical stretch and do not require a nervous impulse for each contraction.

Myogenic rhythms: Rhythms produced by spontaneously active muscles.

Myrmecophile: A symbiont found in the colonies of ants, usually living at the expense of the food in the nest.

Myrmecophyte: A plant that has special cavities in which ants live.

Natural control: The maintenance of a population at nonoutbreak levels by natural environmental factors, biotic and abiotic.

Necrophagous: Feeding on dead animal matter.

Nectar guide: A streak on a flower that guides insects to nectar sources.

Neopterous: Possessing the ability to fold the wings backward over the abdomen.

Neural lamella: A fibrous, noncellular layer that surrounds and supports a ganglion.

Neurogenic flight muscles: Flight muscles that contract each time a nerve impulse is received.

Neurogenic rhythms: Rhythms maintained by spontaneously active neurons.

Neurohemal organ: An organ associated with the nervous system that stores and releases hormones.

Neuron: Nerve cell.

Neuropile: The mass of closely packed nerve cell processes comprising the central part of a ganglion.

Neurosecretory cell: A cell of the nervous system that is specialized for the production and release of hormones.

Niche: The role that a species occupies in nature; that is, its precise habitat plus its behavior in that habitat.

Notum (plural, **nota**): A dorsal sclerite of the insect thorax.

Nuclear polyhedrosis virus (NPV): A virus that multiplies in cell nuclei, chiefly in the epidermis, fat body, and blood cells.

Ocellus (plural, **ocelli**): A simple eye on the dorsal part of the head, containing a single facet.

Oligolectic: Utilizing a very limited number of plant species as sources of pollen (said chiefly of bees).

Oligophagous: Feeding on a somewhat restricted group of (often related) plant or animal species.

Ommatidium (plural, **ommatidia**): A functional unit of the compound eye, expressed externally as a facet.

Ootheca: A hardened protective structure surrounding the egg mass, composed of tanned protein and secreted by accessory glands.

Osmeterium: An eversible gland on the thorax of the larvae of swallowtail butterflies that secretes allomones.

Ostia: Segmentally arranged inlet pores in the walls of the heart.

Ovariole: One of the tubules making up the ovary, in which the eggs are formed.

Ovipositor: The egg-laying apparatus of insects, typically composed of two sets of valves or a tubular extension of the abdomen.

Ovisorption: Resorption of eggs prior to the time of oviposition.

Ovoviviparous: Producing small larvae, the eggs having hatched inside the mother.

Paedogenesis: Reproduction by larviform individuals.

Paleopterous: Lacking the ability to position the wings backward over the abdomen.

Palpus (plural, **palpi**): A paired, segmented appendage arising on the maxilla or labium and serving sensory functions associated with food ingestion.

Panoistic ovary: An ovary in which the ovarioles lack nurse cells.

Parasite: An animal that completes its development on or in another animal but does not normally kill it.

Parasitoid: An insect that lives in its immature stages in or on another insect, which it kills after completing its own feeding.

Parental investment: Behavior of a parent that increases the probability of offspring survival at the cost of the parent's ability to produce more offspring.

Parthenogenesis: Production of young from unfertilized eggs.

Pectinate: Comblike; that is, having a series of slender projections from an elongate shaft.

Pedicel: The second, usually small, segment of the antenna.

Pericardial sinus: A space around the heart, limited below by the dorsal diaphragm.

Perineurium: The layer of cells surrounding a ganglion, which secretes the neural lamella.

Peritrophic membrane: The delicate, tubular sheath that surrounds the food within the midgut.

Phagostimulant: A natural plant substance that induces feeding by an insect.

Pharate stage: A stage in which molting has occurred but the insect has not cast off the old cuticle.

Pharynx: A muscular portion of the foregut, just behind the mouth.

Phenological asynchrony: Lack of synchrony between the life cycle of a pest and the appropriate stage of its host plant.

Pheromone: An external chemical messenger that passes between individuals of the same species and controls intraspecific interactions.

Phoresy: A condition in which an individual is carried about by another individual without harming that individual.

Phylogeny: The study of the history of lines of evolution.

Physical gill: A bubble or packet of air that adheres to the body of an aquatic insect and is continuous with the tracheal air space.

Phytoecdysone: A plant product that mimics ecdysone.

Phytotoxemia: A diseaselike plant condition produced by the injection of toxic substances by insects.

Plastron: A framework of stiff, water-repellent hairs or cuticular structures on the bodies of aquatic insects, containing a film of air into which oxygen diffuses from the water.

Pleural suture: A vertical or oblique suture marking an internal ridge of the thoracic pleuron, running from the dorsal coxal articulation to the pleural wing process.

Pleural wing process: A fulcrum for the wing base, formed at the top of the internal ridge formed by the pleural suture.

Pleuron (plural, **pleura**): A lateral sclerite of the thorax.

Podite: A segment of an arthropod leg, moved by muscles inserted in its base.

Polyculture: A mixed stand of crop plants.

Polyembryony: Division of a single egg to form several identical embryos.

Polyethism: The presence of several discrete types of behavior by different groups of individuals in colonies of social insects.

Polylectic: Utilizing a variety of plant species as sources of pollen.

Polymorphism: The presence of two or more distinct, structurally different types of individuals within the same stage of one species.

Polyphagous: Feeding on a broad array of plant or animal species.

Population dynamics: The forces that control population size, and their effects.

Population regulation: The maintenance of an approximately constant population size and density, and the forces that control it.

Population resilience: The capacity of a population to adapt to change or to persist in a changing environment.

Population stability: The ability of a population to absorb disturbance and to return to an equilibrium state.

Precoccinelline: A defensive allomone produced by lady beetles during autohemorrhage.

Precocene: An insect growth regulator produced by certain plants that depresses the source of juvenile hormone.

Preimaginal conditioning: Conditioning of an immature insect that persists into the adult stage.

Prepupa: A resting stage of the last larval instar, prior to the molt to the pupal stage.

Pretarsus: The most distal segment of the insect leg, bearing the claws and arolium.

Primary defense: A defense mechanism that is continuously present, such as crypsis.

Primer pheromone: A pheromone that acts to modify the physiological state of an animal.

Proctodeum: The hindgut of insects.

Procuticle: The inner zone of the insect cuticle, containing chitin and protein, divisible into exocuticle and endocuticle.

Progressive provisioning: The supplying of food to the offspring over time, as the offspring grow.

Proleg: A fleshy, unjointed "false leg," occurring ventrally on the abdomen of caterpillars and other larval insects.

Pronotum: The dorsal, often shieldlike sclerite of the prothorax.

Proprioreceptor: A sense organ that detects the relative position of parts of an animal's own body.

Protelean parasite: An entomophagous insect that attacks its prey only when the attacking insect is immature, the adult being free living.

Prothoracic glands: Endocrine organs located in the prothorax of immature insects, secreting molting hormone.

Prothoracicotropic hormone (PTTH): A hormone secreted by neurosecretory cells of the brain and serving to activate the prothoracic glands.

Prothorax: The most anterior of the three segments of the thorax.

Protocerebrum: The largest and most anterior part of the brain, which includes the optic lobes.

Proventriculus: The portion of the foregut, just before the midgut, that controls entry of food into the midgut; often lined with sclerotized teeth that grind the food.

Proximal: Referring to the part of an appendage that is closest to the body.

Pterygote: A winged insect, or a wingless insect believed to have been derived from winged ancestors.

Pupariation: Formation of the puparium by larvae of Diptera.

Puparium: A case formed of the hardened last larval cuticle, serving as a protection for the pupa.

Qualitative defenses of plants: Toxins and small-molecular-weight compounds (such as alkaloids) that are active against the physiological systems of phytophagous insects.

Quantitative defenses of plants: Complex, digestibility-reducing substances (such as tannins) that reduce the ability of insects to feed on plants.

Queen substance: A pheromone produced by the queen honey bee and serving various functions in the hive as well as during mating and swarming flights.

Reaction chain: A continuous series of behavioral acts, each of which is dependent on completion of the preceding act.

Rectal pad: A portion of the rectum containing enlarged cells, responsible for active water and ion uptake from the contents of the rectum.

Regulatory control: The use of enforceable regulations to prevent the spread of a pest or to suppress or eradicate it.

Releaser: An environmental or communicative stimulus that triggers a fixed action pattern.

Releaser pheromone: A pheromone that acts via the central nervous system to produce a quick behavioral response.

Releasing mechanism: An innate capacity to respond in a particular way to a specific stimulus.

Resilin: A rubberlike, proteinaceous constituent of the insect procuticle.

Resting potential: The slight charge that can be measured in an unstimulated nerve cell.

Retinula cell: A monopolar sensory neuron within an ommatidium of the compound eye.

Rhabdom: The central, rodlike element in an ommatidium, consisting of several rhabdomeres, one from each retinula cell.

Round dance: A form of recruitment in the honey bee, used when a food source close to the hive is communicated to other bees.

Royal jelly: A nutritive substance produced by glands in the heads of worker honey bees and fed to the larvae. Larvae fed this diet throughout development produce queens.

r-strategist: A species characterized by having rapid development, high motility, and a high reproductive rate relative to a K-strategist.

Saprophagous: Feeding on dead organic matter.

Scape: The most basal segment of the antenna.

Sclerite: A more or less rigid cuticular plate.

Sclerotin: Cuticular protein that has been hardened and darkened through cross-linkage of the molecules.

Scolopidium: A sensillum located beneath the cuticle and modified for the reception of vibrations.

Scramble competition: Competition in which many individuals "scramble" for a limited resource, such as food.

Secondary defense: A defensive mechanism that is brought into play only in the presence of a threat—for example, an aggressive display.

Secondary pest: An insect that does not normally attain pest status except when insecticides destroy its natural enemies.

Secondary plant substance: A substance produced by a plant that plays no role in the basic metabolism of the plant.

Seminal vesicle: An expansion of the vas deferens of the male in which sperm are stored.

Sensillum (plural, sensilla): An integumental sense organ, consisting of sensory neurons and associated cuticular structures.

Sensory filtering: The process of receiving only certain specific stimuli among the many potential stimuli impinging on the body.

Serial homology: Homology within an insect due to the repetition of components of an organ system in each body segment.

Seta (plural, setae): A movable hair of the integument, typically forming a sensillum.

Sibling species: Closely related species that are difficult to distinguish by ordinary means.

Sign stimulus: A stimulus for which an animal has evolved a specific response pattern.

Social parasite: An insect that invades or lays its eggs in the nest of another insect and develops on food in the nest.

Sperm precedence: In multiple matings, the tendency for sperm from the most recent mating to fertilize the eggs.

Spermatheca: A small sac associated with the median oviduct of the female, in which sperm are stored following copulation.

Spermatocyte: A cell that divides to form the spermatozoa of male animals.

Spermatophore: A sac produced by accessory glands of male insects and transferred to the female reproductive tract, containing sperm and often proteinaceous material.

Spiracle: An external opening of the tracheal system.

Squama (plural, squamae): A membranous lobe at the extreme base of the wing of Diptera; also called calypter.

Stemmata: Simple eyes located on the sides of the head of many insect larvae.

Sternum (plural, sterna): A ventral sclerite of the insect thorax or abdomen.

Stimulus filtering: *See* Sensory filtering.

Stomodeal nervous system: A set of small ganglia and their connections, lying on the surface of the foregut.

Stomodeum: The foregut of insects.

Stylet: A thin, sclerotized lance formed of modified mouthparts, capable of piercing a plant or animal.

Stylopized: Infected by stylopoid beetles.

Stylus (plural, styli): A ventral, unsegmented appendage on the abdomen of bristletails.

Subalare: A small sclerite in the upper part of the pleuron that articulates with the axillary sclerites.

Subcuticular space: The narrow space between the endocuticle and the epidermal cells, formed during molting.

Subesophageal ganglion: A composite ganglion that innervates the mouthparts, located in the head below the digestive tract.

Subgenual organ: An organ on the tibia of many insects, consisting of a group of scolopidia and sensitive to vibrations transmitted through the legs.

Superficial cleavage: A type of embryonic development in which the cleavage nuclei migrate to the surface of the egg; cell membranes then form about each nucleus.

Supernormal stimulus: A stimulus that exceeds normal with respect to size or other properties and elicits an exaggerated response.

Superparasitoid: A parasitoid that produces several offspring per individual host.

Superposition eye: A type of compound eye occurring in nocturnal insects, in which the ommatidia are not surrounded by a shield of pigment.

Suture: A line of indentation in the cuticle, usually forming an internal strengthening ridge.

Symbiont: An organism living in intimate association with another organism.

Sympatric speciation: Division of a species into two or more descendant species within the same area.

Synapsis (plural, synapses): The point of interaction between adjacent neurons or between a neuron and a muscle or gland, involving a chemical neurotransmitter.

Synergist: A substance that enhances the effectiveness of a second substance.

Systemic insecticide: An insecticide that is absorbed by a plant or animal and transported throughout it.

Taenidia: Cuticular ridges that support the walls of tracheae.

Tagma (plural, tagmata): A cluster of associated body segments.

Tandem running: A form of recruitment behavior in which one individual follows another, maintaining antennal contact.

Tapetum: A basal layer of tracheae in an ommatidium that reflects light.

Tarsus (plural, tarsi): The segment of the insect leg distal to the tibia, usually subdivided into tarsomeres and bearing the pretarsus with its claws.

Taxis (plural, taxes): A stereotyped orientation movement directed toward or away from a source of stimulation.

Taxonomy: The practice of classifying organisms.

Tegmen (plural, tegmina): A somewhat thickened forewing, serving as a protective covering of the hindwing, as in grasshoppers and cockroaches.

Tegula (plural, tegulae): A small, sclerotized flap overlying the base of the forewing of certain insects.

Temporal memory: The ability to compensate for the passage of time during locomotory behavior such as foraging from the nest.

Tentorium: A set of apodemes that form the internal bracing of the head.

Tergum (plural, terga) (also called tergite): A dorsal sclerite of the insect body.

Termitophile: A symbiont living in the nest of termites.

Territoriality: Occupation of a site that is defended from other individuals of the same sex and species.

Testicular follicles: Tubules in which the sperm are formed.

Thecogen cell: A cell that surrounds the glial cell and the outer section of a sensory neuron.

Thorax: The central tagma of an insect's body, bearing the legs and (when present) the wings.

Tibia (plural, tibiae): The fourth segment of the insect leg, beyond the femur and before the tarsus.

Tormogen cell: An epidermal cell secreting a ring of cuticle that connects a cuticular hair to the cuticle.

Toxicogenic: Producing disease symptoms as a result of an introduced toxin.

Trachea (plural, tracheae): A cuticle-lined air-conducting tube of the insect body.

Tracheal gill: A heavily tracheated extension of the body, permitting extraction of oxygen from water.

Tracheole: A delicate tubule extending from a trachea, the site of gas exchange between the tracheal system and the tissues.

Trap crop: A crop planted because of its attractiveness to certain pests and then destroyed or treated so as to destroy the insects.

Trichogen cell: An epidermal cell that secretes a cuticular process, such as a hair.

Trichoid sensillum: A seta modified for reception of stimuli.

Trichome: A hairlike outgrowth of a plant that may serve various functions, including defense against insect attack.

Tritocerebrum: The most posterior part of the brain, which connects to the ventral nerve cord.

Trivial movements: Movements of an animal within its normal habitat, not involving dispersal.

Trochanter: The second segment of the insect leg, between the coxa and femur.

Trophogenic polymorphism: Polymorphism resulting from differences in the quantity or quality of food provided to the larvae.

Tympanic organ: The eardrumlike structure of certain insects, consisting of a thin portion of integument and associated scolopidia that perceive sound.

Univoltine: Having but a single generation a year.

Valvifer: A basal sclerite of a valve of the ovipositor, articulating with the tergum.

Vannus: A fan-shaped lobe at the posterior margin of the hindwing of certain insects.

Vector: An organism that transmits a pathogen from one host to another.

Vein (of a wing): A sclerotized rod supporting the wing membrane (collectively called the venation of a wing).

Venter: The lower or under surface of an organism.

Ventral: Referring to the lower surface of an organism.

Ventral diaphragm: A ventral muscular sheath that assists in circulating the hemolymph around the nerve cord.

Verbenone: An inhibitory pheromone produced by bark beetles of both sexes, inhibiting arrival of further individuals.

Visceral nervous system: A series of nerve fibers and ganglia closely associated with the gut and reproductive organs.

Vitellogenesis: Yolk formation in the developing egg.

Vitellogenins: Female-specific proteins synthesized by the fat body and taken up by maturing oocytes.

Viviparous: Producing living young that have been nourished internally by the mother.

Waggle dance: A form of recruitment in the honey bee in which direction and distance to a food source are indicated.

Zoophagous: Feeding on living animals.

Index